Customized Technologies for Sustainable Management of Industrial Wastewater

Scrivener Publishing
100 Cummings Center, Suite 541J
Beverly, MA 01915-6106

Publishers at Scrivener
Martin Scrivener (martin@scrivenerpublishing.com)
Phillip Carmical (pcarmical@scrivenerpublishing.com)

Customized Technologies for Sustainable Management of Industrial Wastewater

A Circular Economy Approach

Edited by

Elvis Fosso-Kankeu

Department of Metallurgy, Faculty of Engineering and the Built Environment, University of Johannesburg, Doornfontein, Johannesburg, South Africa

Vhahangwele Masindi

Department of Environmental Sciences, College of Agriculture and Environmental Sciences, University of South Africa (UNISA), Florida, South Africa

Johannes Maree

Institute for Nanotechnology and Water Sustainability (iNanoWS), College of Science Engineering and Technology (CSET), Florida Science Campus, University of South Africa, South Africa

and

Bhekie Mamba

College of Science, Engineering and Technology, University of South Africa, Pretoria, South Africa

Scrivener Publishing

WILEY

This edition first published 2025 by John Wiley & Sons, Inc., 111 River Street, Hoboken, NJ 07030, USA and Scrivener Publishing LLC, 100 Cummings Center, Suite 541J, Beverly, MA 01915, USA

For more information about Scrivener publications please visit www.scrivenerpublishing.com.

Wiley Global Headquarters
111 River Street, Hoboken, NJ 07030, USA

For details of our global editorial offices, customer services, and more information about Wiley products visit us at www.wiley.com.

The manufacturer's authorized representative according to the EU General Product Safety Regulation is Wiley-VCH GmbH, Boschstr. 12, 69469 Weinheim, Germany, e-mail: Product_Safety@wiley.com.

Library of Congress Cataloging-in-Publication Data

ISBN 978-1-394-21437-2

Cover image: Generated with AI using Adobe Firefly
Cover design by Russell Richardson

Set in size of 11pt and Minion Pro by Manila Typesetting Company, Makati, Philippines

Printed in the USA

10 9 8 7 6 5 4 3 2 1

Contents

Preface

Increased population growth and climate change are placing continuous pressure on limited freshwater resources on a global scale. The volume of wastewater discharged by industries—as well as the diversity of pollutants—has significantly increased over the years, rendering conventional treatment systems inadequate for managing effluents released into the environment. The limitations of existing treatments lie not only in the technologies' ability to abate emerging pollutants but also in the approaches used to mitigate the situation and ensure process sustainability. Embracing a circular economy based on the "Reduce-Reuse-Recycle-Restore-Recover" principle in water–wastewater treatment will minimize waste while maximizing the lifecycle value of natural resources and products.

Recently, the rising pollution of freshwater by emerging industrial pollutants has raised concerns within the research community, particularly regarding the effectiveness of current technologies in removing these contaminants. Considerable progress has been made in developing new technologies to address the issue. However, with the growing daily volume of wastewater, treatment costs have become overwhelming. This situation necessitates not only selecting the appropriate technology but also integrating multiple technologies synergistically. Such an approach can promote the reuse of recovered pollutants, thereby offsetting treatment costs and enhancing overall sustainability. The range of waste products that can be recovered depends largely on the pollutant sources and the sequential arrangement of combined technologies or unit processes. A myriad of groundbreaking works have emerged over the last few decades in this field. This book presents fourteen comprehensive chapters that explore the sustainability of industrial wastewater treatment technologies through the lens of the circular economy.

The technological imperative of this book is to highlight both the mechanisms and strategic combinations of technologies that maximize value recovery during industrial wastewater treatment, while ensuring that effluents meet the acceptable standards stipulated in various regulatory frameworks. It also covers the characteristics, purity, and potential applications of recovered products, elucidated through specific case studies in diverse contexts. This book distinguishes itself by focusing on the strategic development of technologies for the sustainable treatment of industrial wastewater on a large scale.

The editors and publisher express their gratitude to the reviewers whose constructive comments and technical insights have significantly enhanced the quality of the book. The editors also thank Martin Scrivener and the team at Scrivener Publishing for including this work in their series.

This book will be of interest to academic researchers in environmental science, chemistry, engineering, mineral processing, hydrometallurgy, and geochemistry, as well as professionals from mining plants, coal power plants, inorganic product manufacturing, environmental management, water treatment, regulatory bodies, and environmental advocacy.

Elvis Fosso-Kankeu
Vhahangwele Masindi
Johannes Maree
Bhekie Mamba
June 2025

Part I

STEPWISE TREATMENT OF INDUSTRIAL WASTEWATER USING A COMBINATION OF APPROACHES

1

A Review of the Reducing and Alkalinity-Producing System (RAPS) for Acid Mine Drainage Neutralization

Mafeto Malatji[1,2]*, Elvis Fosso-Kankeu[3] and Bhekie B. Mamba[4]

[1]Department of Mining Engineering, School of Engineering at the College of Science, Engineering and Technology, University of South Africa, Pretoria, South Africa
[2]Water and Environment Unit, Council for Geoscience, Silverton, Pretoria, South Africa
[3]Department of Metallurgy, Faculty of Engineering and Built Environment, University of Johannesburg, Doornfontein, Johannesburg, South Africa
[4]Institute for Nanotechnology and Water Sustainability (iNanoWS), College of Science Engineering and Technology (CSET), Florida Science Campus, University of South Africa, Pretoria, South Africa

Abstract
A review of the reducing and alkalinity-producing system (RAPS) for treating acid mine drainage (AMD) is presented in this paper. AMD is formed because of the oxidation of sulfide-bearing minerals in the presence of water to form an acidic solution. Characteristics of the resultant solution include a low pH (<4.5), high salinity, as well as sulfates, metals, and metalloids in elevated concentrations. AMD can create long-term environmental damage and changes that can be difficult to manage, and it is imperative that it is treated. Treatment options include active or passive treatment systems, and passive systems are cost-effective; hence, they are preferred over active systems. The RAPS that is an example of passive treatment system combines the concepts of the anaerobic wetland and anoxic limestone drain systems to treat net acidic mine water through combined biogeochemical and physical processes such as bacterial sulfate reduction and limestone dissolution. Bacterial sulfate reduction is responsible for reducing conditions in the system, and this induces metal precipitates as metal sulfides, whereas the limestone

**Corresponding author*: rmalatji@geosciece.org.za

Elvis Fosso-Kankeu, Vhahangwele Masindi, Johannes Maree and Bhekie Mamba (eds.) Customized Technologies for Sustainable Management of Industrial Wastewater: A Circular Economy Approach, (3–38) © 2025 Scrivener Publishing LLC

dissolution is responsible for the addition of alkalinity in the system, therefore raising the pH of the system. Various authors have reported the effectiveness of RAPS in treating AMD using varying approaches but maintaining the basic principles of RAPS. RAPS is advantageous in treating AMD as they encourage downward vertical flow, therefore significantly raising the interaction of water with treatment material. Other added advantages include higher available head pressures and larger cross-sectional area, and these enable them to resist Al plugging. Moreover, the reducing conditions enable the conversion of ferric ions to ferrous ions. The disadvantage of this system is that they have a limited lifespan, which is difficult to determine because of a variety of limiting factors. Moreover, RAPS is generally limited to treating AMD with low flow rates and low acidity loads. The versatility and adaptability of the RAPS present opportunities for innovation; this is because of site-specific conditions such as topography, AMD chemistry, and flow rate, which vary from region to region. The design, scale, and applicability of the RAPS can be adapted to accommodate such conditions, therefore presenting opportunities for innovation during implementation. In South Africa, there is limited information or application of the RAPS on a large scale because most of the research was conducted in a laboratory or bench scale. The successes achieved in the implementation of such systems in other parts of the world make it compelling for the piloting and upscaling of such systems in South Africa. The implementation of such systems would be advantageous to the country due to their cost-effectiveness and self-sustaining abilities. To simulate conditions or parameters that are, otherwise, difficult to achieve with field monitoring or lab experiments, geochemical modeling is used. This tool can be used to evalute geochemical processes and predict the evolution of geochemical systems over time.

***Keywords*:** Reducing and alkalinity-producing system (RAPS), acid mine drainage (AMD), bacterial sulfate reduction, limestone dissolution, geochemical modeling

1.1 Background

The mining of coal is generally environmentally sensitive as it usually involves the removal of overburden materials and the generation of coal discard dumps that are susceptible to spontaneous combustion and are responsible for the generation and release of acid mine drainage (AMD) [1–4]. Naturally, the generation of AMD is readily neutralized by buffering rocks such as clays and carbonates [5]. However, extensive coal mining activities overwhelms the buffering capacity of such rocks, and this accelerates the production and release of AMD [4, 6].

In South Africa, AMD contamination has been reported in active and abandoned mines of the Witwatersrand Gold Fields, Mpumalanga, and

KwaZulu-Natal Coal Fields along with the O'Kiep Copper District [7]. One such contamination event occurred in the town of Carolina, Mpumalanga Province; according to [8], one fateful day in 2012, the residents of Carolina experienced drinking water with an unfamiliar taste and color flowing from taps. An investigation into the incident confirmed that the water had become contaminated with sulfates and metals and the pH had dropped to 3.7, thus rendering the water unfit for human consumption [9–11]. According to the authors in [12] and [80], the investigations into the contamination incident revealed that the water quality of the Boesmanspruit Dam (Carolina's main water supply) had deteriorated, which was not equipped to treat acidic water. Further investigations revealed that the contamination was due to excessive acid mine water from surface and subsurface workings from active and abandoned mines discharging into the Boesmanspruit catchment. The cause of the contamination was flooding due to a freak storm that had occurred the previous night. This resulted in polluted water that had accumulated for decades in a nearby wetland (Boesmanspruit wetland) being released into the Boesmanspruit dam, therefore contaminating it. Water supply to the town was interrupted for 7 months, which led to civil unrest and a court case against the local authorities, as they were struggling to restore the water supply [10–12].

This review aims to document the reducing and alkalinity-producing passive treatment system by exploring its principles, processes, performance, and applicability for the remediation of polluted mine water in South Africa. The ultimate objective is to contribute toward sustainable mine water management solutions in South Africa.

1.1.1 AMD Generation

AMD is formed because of the oxidation of sulfide-bearing minerals in the presence of water to produce an acidic solution. The characteristics of the resultant solution include a low pH (<4.5), high salinity, as well as sulfates, metals, and metalloids in elevated concentrations [1, 3, 4, 13, 14]. To form AMD, a common sulfate mineral such as pyrite undergoes a process known as pyrite oxidation, *Thiobacilus* bacteria acts as a catalyst to this process. Equations 1.1 to 1.3 illustrate how pyrite oxidation unfolds [15–18].

- Firstly, an acidic solution of ferrous iron and sulfate is formed by pyrite oxidation.

$$4FeS_2(s) + 14O_2(g) + 4H_2O(l) \rightarrow 4Fe^{2+}(aq) + 8SO_4^{2-}(aq) + 8H^+(aq) \quad (1.1)$$

- Then, the oxidation of ferrous iron to ferric iron follows.

$$4Fe^{2+}(aq) + O_2(g) + 4H^+(aq) \rightarrow 4Fe^{3+}(aq) + 2H_2O\ (l) \quad (1.2)$$

- Ultimately, the formation of ferric hydroxide and acid.

$$4Fe^{3+}(aq) + 12H_2O(l) \rightarrow 4Fe(OH)_3(s) + 12H^+(aq) \quad (1.3)$$

Once started, the reactions are difficult to stop because earlier reactions give rise to later reactions, therefore resulting in a series of reactions. Typically, ferric hydroxide [$Fe(OH)_3$] gives rise to an orange color in affected water, which is referred to as "yellow-boy," therefore resulting in reduced dissolved oxygen (DO) in the affected water [1, 19].

1.1.2 Effects of AMD on the Environment

AMD can create long-term environmental damage and changes that can be difficult to manage and that can be inherited from one generation to another [19, 20]. Aquatic organisms in receiving environments can become asphyxiated because of the lack of oxygen; this is due to the precipitation of ferric hydroxide, which results in reduced dissolved oxygen in such environments [1, 19]. Furthermore, elevated concentrations of metals become bioavailable and can bioaccumulate in organisms in affected environments. Moreover, AMD leads to the pollution of surface and groundwater, poor soil quality, poor health of aquatic flora and fauna, and the distribution of potentially toxic metals and metalloids into receiving environments [4]. This is because the low pH and the elevated sulfates, metals, and metalloids exacerbate the toxicity of AMD in receiving environments, therefore enhancing their solubility, mobility, and bioavailability [14, 21–23]. The generation and quality of AMD are site-specific and are dependent on several factors including the sulfide mineral type, water and oxygen availability, pH of the medium, mine type, surface area of the exposed sulfide mineral, and microbial activity [4, 6]. The ecological and health impacts of selected metals are tabulated in Table 1.1.

Table 1.1 Ecological and health impacts of selected metals on living organisms [24].

Metal	DWA limit (mg/L)	Ecological impacts and health of acid mine drainage
Al	<0.5	Chronic dementia and Alzheimer's disease. Reduced aesthetic effects
Fe	<1	Reduced aesthetic effects and slimy coating on pipes. Higher concentration results in chronic and acute effects in minors
Mn	<0.2	Aesthetically unacceptable staining
Cu	<1	Causes nausea and vomiting; poisonous with possible fatalities
Mg	<200	Bitter taste; causes diarrhoea if excessive
Zn	<5	Bitter taste; milky appearance; causes nausea and vomiting

It is imperative that AMD generation is prevented to protect and preserve the environment and enhance sustainability. However, if prevention is not feasible and AMD occurs, then immediate remediation is required to prevent contamination in receiving environments.

1.1.3 AMD Treatment Options

Treatment options for AMD include both active and passive treatment systems [25]. The active ones continuously require regular maintenance using electrically powered equipment and chemicals to treat AMD, whereas passive systems make use of natural processes and material to treat AMD [26]. The natural processes include biological (bioreduction; biosorption), chemical (oxidation; reduction), and physical (gravity; dilution) [27, 28].

1.1.3.1 Passive Treatment Systems

Passive systems are cost-effective; hence, they are preferred over active systems. This is because they are constructed using natural materials (soil and rocks), and they use natural materials (compost and limestone) for acid neutralization, metal precipitation, and sulfate reduction. Furthermore, water flow in such systems is dependent on the natural gradient; they can be integrated into the natural landscape without looking out of place, aesthetically; and their lifespan is usually longer [29–32]. Moreover, passive systems do not require external energy input and do not require frequent monitoring and can thus be implemented in remote areas. There are various passive treatment systems options, and they include anoxic limestone drains (ALD), aerobic and anaerobic wetlands, limestone ponds, reducing and alkalinity-producing system (RAPS), and open limestone channels. Passive treatment system techniques are site-specific, and the topography, chemistry, and flow rate of the discharge guide the selection of the appropriate passive treatment system [33–37].

In this document, only the RAPS will be reviewed to propose ways in which its efficiency, performance, and longevity for AMD remediation can be improved or optimized in South Africa. Furthermore, the processes involved in the treatment, limitations, and advantages will be explored.

1.1.3.2 Selection Criteria

Several factors are considered when selecting the suitable passive treatment system; the factors include site characteristics such as the local topography, flow rate, and chemistry of the discharge [pH, acidity, alkalinity, Fe^{2+}/Fe^{3+}, Al, Mn, SO_4^{2-}, and dissolved oxygen (DO)], all of which are site-specific [33–37]. The decision tree for the design of passive treatment systems (Figure 1.1), adopted from [34], is a recommended guide for selecting the appropriate passive treatment system, along with the abovementioned factors. This is because the application of such techniques is not a "cut and paste" exercise, and there is no "cookbook" manual for these techniques according to [38]. It is recommended that the selection processes should be approached using a "phased approach" which divides the process of decision-making into three phases: the laboratory phase, bench and pilot phase, and, subsequently, the actual site implementation phase. This approach will help minimize costs (especially if most tests are done on a pilot scale rather than the final large field scale) and narrow down the best treatment options based on the water chemistry [38].

Authors in [39] recommend that passive treatment systems design should be preceded by characterizing the chemistry of the AMD influent and determining the characteristics of the site. This is done by regularly sampling AMD on the site of interest for at least a year to document the effects of seasonal variations. These samples ought to be analyzed for pH, total Fe, SO_4, Al, Mn, total acidity, and alkalinity. The flowchart provided by [40] in Figure 1.1 indicates that SAPS is the viable option for net-acidic

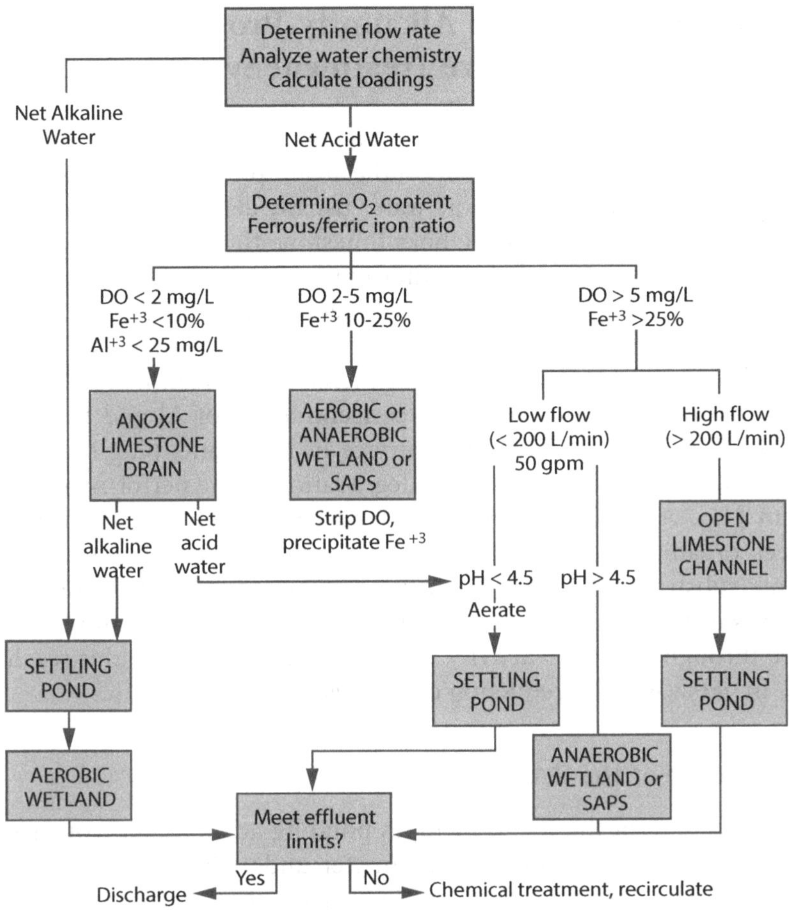

Figure 1.1 Passive treatment systems design decision tree, adopted from [34].

effluents with elevated Fe, Al, and DO concentrations, which can be preceded by a settling pond; this is to allow iron and other solids to settle before entering the system. The SAPS will provide a reducing environment as well as the addition of alkalinity and adequate residence time to reduce contaminants and precipitate metals; the SAPS will be succeeded by an aeration/settling pond into which metals will precipitate [39, 41]. The characteristics of AMD in South African coalfields usually include low to high concentrations of metals, such as Fe, Mn, Al, As, and Pb, and the elevated levels of SO4; therefore, it can be categorized as net acidic. This makes RAPS the most suitable treatment option. The mechanisms of the RAPS and its applicability will be expanded on in the sections to follow.

1.2 The Reducing and Alkalinity-Producing System (RAPS) as a Passive Treatment System

The RAPS previously referred to as SAPS (successive alkalinity-producing system) combines the mechanisms of the anaerobic wetland and ALD systems, the system treats net acidic water usually at the point of discharge (see Figure 1.2). ALDs typically bury limestone to treat anoxic water and are usually capped with soil, to exclude oxygen (O_2), retain carbon dioxide (CO_2), and encourage the hydrolysis of metals [42–44]. Because CO_2 cannot escape the system and limestone reacts with AMD, there is a build-up of bicarbonate that contributes to the generation of alkalinity [44]. In cases where the influent has appreciable concentrations of Al and Fe^{3+}, metal hydroxide precipitates armor the surface of limestone, plug the system, and retard water flow. This leads to degradation in the performance of the system [45]. A typical anaerobic wetland consists of vegetation planted in deep organic substrates (>30 cm) and a base of limestone mixed with organic substrate. The system encourages the interaction of water with organic substrates to generate reducing conditions where alkalinity is generated through microbial activity and limestone dissolution; this results in a rise in pH as well the reduction of various pollutants [5, 46].

1.2.1 Setup of the RAPS

A typical design of the RAPS includes ponding acidic water from 1 m to 3 m over 0.2 m to 0.3 m of organic matter underlain by a 0.5-m to 1-m limestone layer, which is underlain by a network of drainage pipes that are responsible for collecting treated water and transferring it to the outlet pipe [41, 47]. The RAPS employs the mechanism of a vertical flow wetland, in which contaminated water is directed downward through organic matter and limestone. As the water percolates the organic matter layer, it

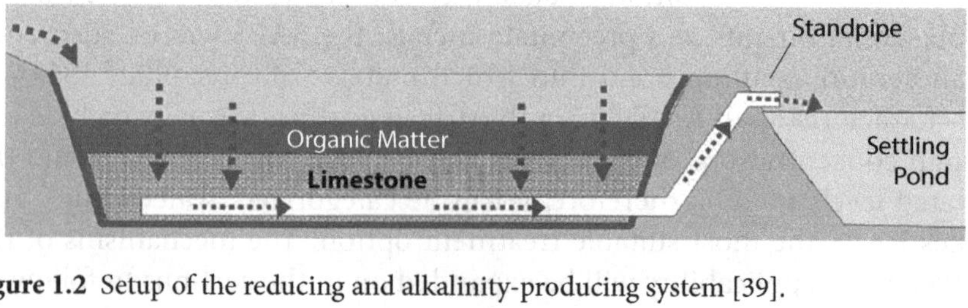

Figure 1.2 Setup of the reducing and alkalinity-producing system [39].

Table 1.2 Ideal AMD chemistry likely to be treated by RAPS [8].

Acidity level	Acidity load	Flow rate	DO	pH level	Attainable pH
<300 (mg $CaCO_3$/L)	<100 (kg $CaCO_3$/day)	<15 (L/s)	<1–3 (mg/L)	>2.5	6–8

is subjected to reducing conditions that facilitate sulfate reduction, alkalinity generation, and precipitation of metals. Furthermore, as the acidic water percolates through the limestone, dissolution occurs, resulting in the release of Ca ions into the solution, therefore generating alkalinity [39, 41]. Therefore, RAPS is suitable for water rich in ferric iron; in other systems, ferric iron armors limestone, therefore reducing their effectiveness; in a RAPS, ferric iron is reduced to ferrous iron, therefore preventing armoring [48].

The organic matter layer is composed of compost—a source of sulfate-reducing bacteria (SRB) that facilitates the system's reducing conditions. The limestone layer generates alkalinity and raises the pH in the system [17]. In summation, according to the authors in [8], RAPS (i) combines both organic and inorganic methods to attenuate AMD; (ii) it is dependent on bacterial activity and limestone dissolution to generate alkalinity; (iii) it augments reducing environments to enable sulfide precipitation and minimize armoring; (iv) it provides sites of adsorption for metal precipitates; and (v) water with circumneutral pH levels can be attained. Furthermore, RAPS ideally treats AMD with the chemistry tabulated in Table 1.2.

1.2.2 Principles of the RAPS

As with most passive treatment systems, RAPS uses combined biogeochemical and physical processes to attenuate AMD, through the processes of bacterial sulfate reduction and limestone dissolution [49, 50]. Usually, the treatment unfolds in the following manner:

- Firstly, SRB through bacterial sulfate reduction oxidizes organic matter (CH_2O in Equation 1.4) within the compost layer; this, therefore, raises the pH and generates alkalinity. Hydrogen sulfide is also produced, binding to metals and precipitating them as metal sulfides. Fe is used as an example in Equation 1.5 [51–53].

$$2CH_2O(s) + SO_4^{2-}(aq) \rightarrow 2HCO_3^-(aq) + H_2S(aq) \quad (1.4)$$

$$H_2S(aq) + Fe^{2+}(aq) \rightarrow FeS(s) + 2H^+(aq) \quad (1.5)$$

- In the limestone layer, the limestone dissolves, producing dissolved calcium and carbonic acid. The dissolved carbonic acid reacts with the limestone further to form dissolved calcium and bicarbonate alkalinity, which neutralizes the acid [51, 54, 55].

$$CaCO_3(s) + 2H^+(aq) \rightarrow Ca^{2+}(aq) + H_2CO_3(aq) \quad (1.6)$$

$$CaCO_3(s) + H_2CO_3(aq) \rightarrow Ca^{2+}(aq) + 2HCO_3^-(aq) \quad (1.7)$$

$$HCO_3^-(aq) + H^+(aq) \rightarrow H_2O(l) + CO_2(g) \quad (1.8)$$

- Finally, the metals are oxidized and hydrolyzed to form metal hydroxide precipitates; Fe is used as an example in Equations 1.9 and 1.10 [51, 56, 57].

$$Fe^{2+}(aq) + 1/4O_2(g) + H^+(aq) \rightarrow Fe^{3+}(aq) + 1/2H_2O(l) \quad (1.9)$$

$$Fe^{3+}(aq) + 3H_2O(l) \rightarrow Fe(OH)_3(s) + 3H^+(aq) \quad (1.10)$$

1.2.2.1 Sulfate-Reducing Bacteria

Authors in [57] define SRB as anaerobic and non-photosynthetic bacteria that can withstand various environmental conditions; usually, they are distinguished into the mesophilic, thermophilic, and psychrophilic groups. Moreover, these organisms are chemoorganotrophic, and, essentially, they have two core functions: firstly, they are responsible for deoxygenation in the system, therefore providing anoxic conditions and, thus, preventing iron hydroxide from precipitating in the limestone layer. Secondly, they are responsible for dissimilatory sulfate reduction, which is a process in which organic matter is oxidized by SRB to reduce sulfate to sulfide by using sulfate as a terminal electron acceptor as indicated in Equation 1.4; this results in the formation of hydrogen sulfide and bicarbonate alkalinity [57–59]. Passive treatment systems such as alkalinity-producing systems take advantage of biological sulfate reduction to treat AMD [60]. For the SRB to thrive in a passive system, an organic substrate is required, which acts as a source of carbon (energy source); compost is the most widely used substrate; other substrates include wood chips, poultry manure, sawdust, and cow manure, which are all deemed inexpensive [61–63].

The metabolic processes of SRB involve various biochemical processes which include the oxidizing of ketones, aldehydes, and alcohols; the degrading of aliphatic and aromatic hydrocarbons; the decomposing of arylnitro and arylamino compounds; and demethylation reactions and reduction of sulfate and metals; these processes are very complex and are usually catalyzed by various enzymes [28, 63]. The metabolic processes by which SRB reduces sulfate to sulfide are presented in Equations 1.11 to 1.13 [63, 64].

- Firstly, sulfate is transformed to adenosine phosphosulfate (APS); this occurs when sulfate is activated by adenosine monophosphate (AMP) and adenosine triphosphate (ATP) sulfurylase to form APS and pyrophosphates.

$$7SO_4^{2-} + 15AMP^{4-} + H^+ \rightarrow 7APS^{2-} + 4HP_2O_7^{3-} \qquad (1.11)$$

- APS is then metabolically reduced to biogenic sulfite

$$APS^{2-} + H^+ + 2e^{2-} \rightarrow HSO_3^- + AMP^{2-} \qquad (1.12)$$

- Then to biogenic-sulfide

$$HSO_3^- + 6H^+ + 6e^{2-} \rightarrow HS^- + 3H_2O \text{ } (M^{2+} \text{ - metal cation}) \qquad (1.13)$$

The conversion occurs when sulfate is actively transferred across the cytoplasmic membrane of SRB, and it is facilitated by intracellular electron mediators [63, 64]. pH plays an integral part in sulfate reduction because SRB does not thrive at low pH levels; other factors that influence the growth of SRB include temperature, substrate, metal concentration, hydraulic retention time, and concentrations of dissolved oxygen and hydrogen sulfide [52, 57].

1.2.2.2 Limestone Dissolution

When carbonate minerals, for example, calcite ($CaCO_3$) come into contact with acidic water, they dissolve and release bicarbonate and Ca ions into the solution; this process is referred to as limestone dissolution, and it requires anoxic conditions to occur (see Equation 1.6) [34, 65]. In anoxic conditions, bicarbonate concentration is higher than in oxic conditions. This is because anoxic conditions do not allow the formation of ferric hydroxide that is responsible for the armoring of carbonates in oxic conditions, therefore limiting the potential for carbonate dissolution [66, 67]. In addition, carbon dioxide plays an important part in carbonate solubility. This is because, in anoxic conditions, the partial pressure of carbon dioxide is high due to decomposing organic matter and neutralization of proton acidity [68]. This affects the maximum level of potential alkalinity; Table 1.3 shows the maximum level of potential alkalinity at various carbon dioxide partial pressures. From the table, it can be concluded that, when the pressure is high, the potential for alkalinity is high as well [34, 69, 70].

Table 1.3 Maximum level of potential alkalinity at various carbon dioxide partial pressures, adopted from [34].

PCO_2 (atm)	Alkalinity (mg/L as $CaCO_3$)
0.0003	60
0.01	220
0.05	360
0.10	475
0.20	610
0.50	850
1.00	1085

1.2.2.3 *Metal Removal Processes*

AMD is contaminated with various metals; the purpose of passive treatment systems is to provide acidic water with sufficient retention time so that biogeochemical processes can reduce metals to regulatory acceptable discharge standards. The redox potential and pH affect the biogeochemical processes that remove metals in acidic water; the processes are different for each metal. For passive treatment systems to perform efficiently, the optimal operating conditions must be sustained, and such conditions will encourage the suitable processes that will target the removal of contaminants [34]. Thus, selecting and designing of a suitable passive treatment system is site-specific and is dependent on understanding the chemistry of the AMD discharge, the flow rate, site characteristics, and the local topography [33–37].

1.2.2.3.1 Reduction

The biogeochemical processes in anaerobic conditions are different compared to aerobic conditions; this is because anaerobic conditions exclude oxygen, and, therefore, there is no oxidation of metals to form oxyhydroxide precipitates, which are highly insoluble in acidic water. Instead, in anaerobic conditions, biogeochemical processes like limestone dissolution and bacterial sulfate reduction result in the formation of bicarbonate and hydrogen sulfide; this improves the quality of the water as it reduces dissolved metal concentrations (e.g., Fe^{2+}, Mn^{2+}, Zn^{2+}, Ni^{2+}, Cu^{2+}, Cd^{2+}, and Pb^{2+}) by precipitating them as metals sulfides as shown in Equation 1.14 [34].

$$M^{2+} + H_2S + HCO_3^- \rightarrow MS + 2H_2O + 2CO_2 \tag{1.14}$$

The removal of dissolved metals as sulfide compounds is affected by various factors including the pH, the solubility product of the specific metal sulfide, and the concentrations of the reactants. The solubility products of selected metal sulfides are shown in Table 1.4. According to the authors in [34], "metal removal from mine water subjected to inflows of hydrogen sulfide occurs on an order consistent with the solubility products shown in this table." CuS forms first followed by PbS, ZnS, and CdS, and MnS forms last [34].

Table 1.4 Solubility products of various metal sulfides, adopted from [34].

Metal sulfide	Solubility product
CuS	4.0×10^{-38}
PbS	1.0×10^{-29}
ZnS	4.5×10^{-24}
CdS	1.4×10^{-23}
NiS	3.0×10^{-21}
FeS	1.0×10^{-19}
MnS	5.6×10^{-16}

1.2.2.3.2 Metal Oxidation and Hydrolysis

In acidic water, Fe^{2+}, Fe^{3+}, Mn, and Al are oxidized and hydrolyzed in aerobic conditions to decrease their concentration. These processes are dependent on various factors including oxygen availability for oxidation reactions, the pH of water, the microbial activity of catalysts and inhibitors, and the retention time. pH is the most important factor due to its influence on the solubility of metal-OH precipitates as well as oxidation and hydrolysis kinetics processes. Authors in [34] state that "The relationship between pH and metal-removal processes in passive treatment systems is complex because it differs among metals and also between abiotic and biotic processes."

The relationships are expressed as follows:

$$Fe^{3+} + 3H_2O \rightarrow Fe(OH)_3 + 3H^+ \tag{1.15}$$

$$Al^{3+} + 3H_2O \rightarrow Al(OH)_3 + 3H^+ \tag{1.16}$$

$$Fe^{2+} + 0.25O_2 + 2.5H_2O \rightarrow Fe(OH)_3 + 2H^+ \tag{1.17}$$

$$Mn^{2+} + 0.25O_2 + 1.5H_2O \rightarrow MnOOH + 2H^+ \tag{1.18}$$

Equations 1.15 and 1.16 are hydrolysis reactions only requiring water and sufficient alkalinity to neutralize the acid. Equations 1.17 and 1.18 oxidize metals before hydrolysis reactions. Passive treatment systems, ultimately, ensure that such reactions are completed, and the resultant solid precipitants are collected before they reach receiving streams.

1.2.2.4 Performance of RAPS in Treating AMD

1.2.2.4.1 Bowden Close Colliery in England

Authors in [51] implemented a RAPS at the Bowden Close Colliery site in England, at a full scale. The influent at the site was characterized by low acidity (pH < 4) and elevated concentrations of Fe and Al as depicted in Table 1.5.

The system at Bowden Close Colliery site was able to remove acidity as well as generate alkalinity; the pH was raised to near neutral levels and has never dropped below 6.4 in the first year of functioning. It was reported that alkalinity of a maximum of 320 mg/L had been added to the mine-affected waters discharging on the site. Regarding metal removal, it achieved about 85% and 93% removal of Fe and Al, respectively, and 72% removal of Zn. About 12.1 tons of sulfate load were also removed by the system. Due to the exhaustion of substrates and clogging, the performance of the system declined over time, which is estimated to be approximately 10 years [51].

1.2.2.4.2 Deerplay, Lancashire, UK

A unique design was implemented by [49] at Deerplay, Lancashire, UK, at a formerly a coal mine. The system involved the downward flow of AMD

Table 1.5 Hydrochemical characteristics for discharges at the Bowden Close Colliery site [51].

Untreated discharge no.	**Fe (mg/L)**	**Al (mg/L)**	**Zn (mg/L)**	**Alkalinity (mg/L)**	**SO_4 (mg/L)**	**pH**
1	30	10	0.3	10	300	5.5
2	8	2	0.1	50	140	6.7
3	80	50	3	0	1530	4.0

Table 1.6 AMD parameters at Deerplay, Lancashire, UK [49].

Parameters	pH	EC (μS/cm)	Alkalinity (mg/L)	Acidity (mg/L)	Cl (mg/L)	SO_4 (mg/L)	Fe (mg/L)	Al (mg/L)	Mn (mg/L)
Concentration	3.3	1588	0	165	8	863	21.5	13.6	4.25

through a layer of compost followed by an upward flow through a layer of limestone to treat the AMD. The hydrochemical parameters of the discharge are shown in Table 1.6.

The pH was raised from 3.3 to 6.7 by the system; alkalinity of approximately 100 mg/L was added by the system. A reduction of 21.5 mg/L to 15.5 mg/L (27% reduction) was observed for total Fe. Sulfates and other metals were also reduced significantly. Based on the assessment of the limestone weathering rate, the system was calculated to operate for 20 years.

1.2.2.4.3 Pennsylvania and Maryland in the United States of America

Four RAPSs were implemented in Pennsylvania and Maryland to treat polluted mine water and have been operational for 3 to 10 years as of the year 2000. Three of the systems were designed according to the typical setup of a RAPS, which includes a layer of compost atop a layer of limestone which is topped by acidic water; these systems include the Howe Bridge and the Oven Run D and E systems. The Jennings system mixes the treatment material instead of layering them; this was done to maximize the lifespan of the treatment material. This is because the lab results indicated that the acidic water from this site would deplete $CaCO_3$ in the material within 2 years if the material was layered distinctively. All the systems except the Jennings system are preceded by a settling pond/wetland before the discharge enters the main RAPS. This is to allow the settling of solids such as iron and others, which reduces the porosity of the systems. Mixing of the treatment material improves the permeability of the material; hence, a pre-treatment settling pond/wetland is not required for the Jennings site. The Howe Bridge and Jennings sites have a drainage network in the form of pipes [41].

Alkalinity generation in RAPS is primarily provided by the dissolution of limestone by releasing Ca ions into solution upon reacting with acidic water; however, alkalinity can be generated within the organic layer in the form of bicarbonate (see Equation 1.4). For 9 years, the Howe Bridge site generated as much as 219 mg/L at a rate ranging between 19.6 $gd^{-1}\,m^{-2}$ and 39.2 $gd^{-1}\,m^{-2}$; this can be attributed to both limestone dissolution and sulfate reduction. There was a substantial removal of Fe, Al, and SO_4 and minimal removal of Mn as it requires a higher pH. The Oven Run D produced alkalinity at a rate of 57.4 $gd^{-1}\,m^{-2}$ and 20.6 $gd^{-1}\,m^{-2}$, respectively, in the first and second RAPS for 5 years; overall, a combined alkalinity of as much as 69.5 mg/L was produced by the two RAPSs (see Table 1.7). There was a significant removal of Fe and Al, and the same cannot be said for Mn and SO_4 [41].

Table 1.7 Overall performance of the systems, adopted from [41].

RAPS	Net acidity, mg/L as $CaCO_3$		Ca mg/L		Fe mg/L		Mn mg/L		Al mg/L		SO_4 mg/L		Total Alk. mg/L as $CaCO_3$
	In	Out	In	Out	In	Out	In	Out	In	Out	In	Out	
Howe Bridge	323	106	193	240	193	73.7	37.3	36.0	<0.2	<0.2	1214	1102	219
Oven Run D #1 & 2	65.05	−4.33	308.5	325	19.51	1.725	28.45	26.3	1.655	0.675	1356	1340.5	69.5
Oven Run E #1 & 2	128.2	21.45	180	215.5	15.05	8.84	12.9	12.2	13.1	5.25	936.5	900.5	107
Jennings	272	−166	105	274	69.1	10.3	17.6	15.9	22.7	<0.2	772	729	438

For the Oven Run E sites, alkalinity was produced at a rate of 42.7 $gd^{-1} m^{-2}$ in the first RAPS and 15.6 $gd^{-1} m^{-2}$ in the second over 3 years of operation. Alkalinity of as much as 107 mg/L was produced by the two RAPS. Fe and Al were significantly removed, whereas Mn and SO_4 removal was minimal. The Jennings site produced alkalinity at a rate of 62.4 $gd^{-1} m^{-2}$ amounting to 438 mg/L. Fe and Al were significantly removed, whereas Mn and SO_4 removal was minimal [41].

1.2.2.4.4 Watzlaf's Simulated SAPS

Authors in [71] simulated a SAPS using a lab-scale column. The setup of the column included a limestone layer (0.30 m), a spent mushroom compost layer (0.76 m), and acidic water (0.9 m); see Figure 1.3. The setup encouraged a flow rate of 3.8 mL/min of acidic water to flow vertically downward through the treatment material for 97 weeks. The simulation managed to raise the pH, add alkalinity, and reduce various contaminants during its 97 weeks of operation until the pH in the compost layer dropped, therefore

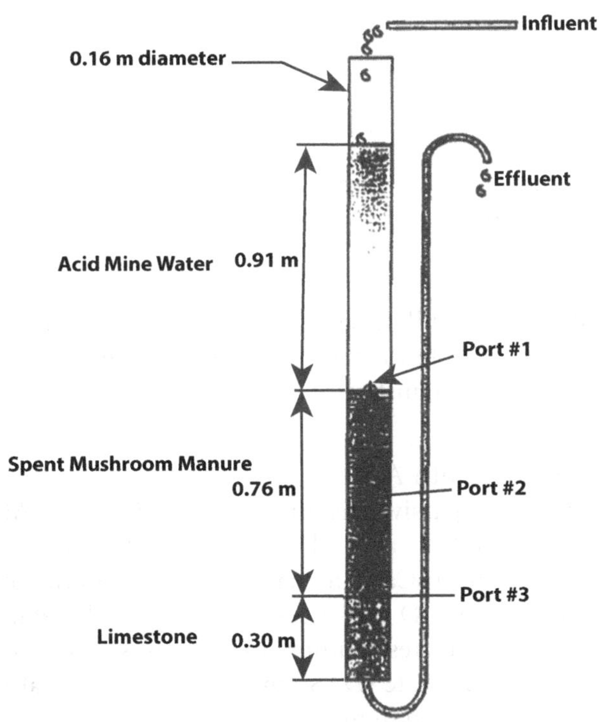

Figure 1.3 Schematic of the simulated SAPS [71].

Table 1.8 Simulated SAPS results comparing the influent and the effluent, adopted from [71].

Parameter	Simulated SAPS (97 weeks)	
	Influent	**Effluent**
pH	3.05	6.99
alkalinity	0	154
Net acidity	208	-126
Fe^{2+}	1.9	0.47
Total Fe	13.2	0.86
Al	17.6	<0.1
Mn	13.0	12.7
SO_4	605	528
Ca	91.8	192
Mg	44.9	51.0
Na	3.1	3.5
K	2.4	3.1
Anions/ cations	1.02	0.97

resulting in a decrease in sulfate reduction; this is because SRB is less active at low pH levels (<4). Table 1.8 shows the comparison of the influent and the effluent in terms of chemistry.

1.2.2.4.5 RAPS in South Africa

The implementation of passive treatment systems in South Africa is limited to only laboratory tests and pilot studies. Authors in [72] piloted a RAPS treatment system in Carolina, and, within 6 months of operation, the system neutralized AMD and successfully removed metals, achieving a 30% removal rate of sulfates. Authors in [73] investigated the efficiency of dispersed alkaline substrate (DAS) and RAPSs in attenuating AMD in a laboratory setting. Both systems achieved >95% contaminant removal (metals and metalloids). Furthermore, the systems raised the pH from a level of 3 to 8.

1.2.2.5 Advantages of RAPS

Apart from being cost-effective and self-sustaining, RAPS can be used as stand-alone treatment systems; alternatively, they can be coupled with other systems to maximize treatment efficiency according to [74]. Furthermore, they are effective in treating net acidic mine water with excessive metal content with an acid removal efficiency of about 35 $g/m^2/day$ [42]. The downward vertical flow direction significantly increases the efficacy of the system by encouraging the interaction between the water and treatment material [75]. RAPSs are advantageous as they require less land than traditional wetlands [76]. Unlike traditional ALDs that are prone to plugging by aluminum (Al), RAPS is resistant to Al plugging due to their larger cross-sectional area and high head pressures [77]. The reducing conditions facilitated by SRB enable the removal of DO in the system; this prevents the oxidation of ferrous ions and, therefore, prevents the armoring of limestone. Moreover, the reducing conditions enable the reduction of ferric ions to ferrous ions [8, 78].

1.2.2.6 Disadvantages

Passive treatment systems have a limited lifespan, which is difficult to determine, and the overall longevity of such systems is influenced by the following factors:

- (i) Passive treatment systems are prone to physical clogging; this is due to the build-up of sediments and debris as well as the accumulation of solid precipitates [51, 57].
- (ii) Organic carbon supplied by the organic substrate (compost) can become depleted, therefore threatening the survival of SRB, which are responsible for reducing pollutants [51].
- (iii) Limestone supply can also become exhausted, affecting alkalinity generation. Furthermore, hydroxide precipitates can armor limestone and reduce its effectiveness. The replacement of limestone is costly [51, 52].

In general, the systems are limited to treating AMD with low flow rates of about less than 50 L/s, low acidity of about less than 800 mg of $CaCO_3$/L, and, therefore, low acidity loads of less than 100 and 150 kg of $CaCO_3$/

day [8, 79]. Passive treatment systems produce ochre and sulfides, which can be enriched with toxic elements (As and Cd), which are detrimental if they were to be released into the environment [28]. The sizing of the reactive material is also problematic; this is because finer materials react more rapidly than coarse-grained materials, and, at the same time, they are less permeable than their coarser-grained equivalents [44]. Due to poor design and unfavorable conditions (severe winter), passive treatment systems might fail to attenuate AMD to the required discharge standards [33].

1.2.3 Novelty Opportunities of the RAPS

The versatility and adaptability of the RAPS present opportunities for innovation; this is because of site-specific conditions such as topography, AMD chemistry, and flow rate, which vary from region to region. The design, scale, and applicability of the RAPS can be adapted to accommodate such conditions, therefore presenting opportunities for innovation during implementation. Traditionally, the setup of the RAPS includes ponding acidic water atop distinguished layers of limestone and compost [41, 47]. In recent developments, the mixing of the substrates has been explored. Authors in [61] explored a mixture consisting of chicken manure, straw, and limestone to treat AMD; they observed improvements in the pH and removal of SO_4 and various metals. Authors in [80] used a mixture consisting of mushroom compost, cow manure compost, sawdust, rice straw, creek sediment, and limestone to treat AMD; their experiment showed improvements in the pH and removal of SO_4 and various metals. According to the authors in [41], the mixing of the treatment material enhances the permeability of the system. RAPS can be versatile in terms of the treatment material used; this is because the availability of treatment material is different from region to region, and substitutes can be used as the need arises; therefore, such versatility presents opportunities for innovation. Authors in [52] explored various inexpensive organic matter substrates such as compost, wood chips, poultry manure, sawdust, cow manure, and others as sources of SRB to treat contaminated mine water, all of which yielded positive results. Table 1.9 shows the performance of different substrates in bioreactors.

RAPS can be adapted to accommodate various site-specific conditions; such adaptability can present opportunities for innovation. Furthermore, the setup of the RAPS can be versatile depending on the treatment material available in specific regions; therefore, this presents opportunities for innovation.

Table 1.9 Performance of different substrates in bioreactors, adopted from [52].

		pH		SO_4^{2-}	
		Flow			
Reactor type	**SRB source**	**In**	**Out**	**In**	**Out**
Field bioreactor	Softwood, dust, hay, and cattle manure mixture	4.0–5.5	5.0–6.0	175–260	200–275
Field bioreactor	Cow manure and cut straw mixture	3.3–7.5	6.5–7.5	70–229	---
Pilot-scale bioreactor	Cattle manure, sawdust, hay, and alfalfa mixture	3.5–7.5	6.5–7.5	<60	<40
Pilot-scale bioreactor	Spent mushroom compost	3.2–6.2	6.4–7.1	1,002–2,997	831–2,387
Pilot-scale bioreactor	Methanol	2.9–3.2	6.9	1,900–2,100	832
Pilot-scale bioreactor	Rice stalks, cow manure, and limestone mixture	3.6	6.2	---	---

1.2.4 Applicability of the RAPS in South Africa

Many legacy and operational coal mines in South Africa generate AMD [72, 81–84]. As noted by the authors mentioned above, AMD in the coalfields of South Africa is usually characterized by low to high concentrations of metals, such as Fe, Mn, Al, As, and Pb and the elevated levels of SO_4. The discharge of untreated AMD into streams continues to severely degrade both habitat and water quality, producing an environment devoid of most aquatic life and unfit for desired uses. The severity and extent of damage depends upon a variety of factors including the frequency, volume, and chemistry of the drainage and the size and buffering capacity of the receiving stream.

The likely passive treatment systems for net acidic water include ALDs, open limestone drains, anaerobic wetlands, RAPSs, and sulfate-reducing bioreactors. ALDs may not be suitable to treat net acidic AMD due to potential clogging/armoring of limestone with oxides or gypsum. On the other hand, anaerobic wetlands require substantial retention time and, therefore, larger surface areas. A study by [85] indicated that the use of open limestone channels/drains can be very useful in the initial raising of the pH and removal of metals before the water enters a constructed wetland. In addition, the authors also suggested that the velocity/flow rate of the water must be high enough to prevent possible armoring or coating of the limestone. The applicability of sulfate-reducing bioreactors is also limited by a high flow rate [86].

Owing to the nature of previous legislation, the South African government has inherited environmental liabilities for most abandoned mine sites, and, hence, there is an urgent need to develop long-term sustainable remediation/management strategies [87]. In South Africa, there is limited information or application of the RAPS on a large scale because most of the research was conducted in a laboratory or bench scale [72]. The successes achieved in the implementation of such systems in other parts of the world make it compelling for the piloting and upscaling of such systems in South Africa. The implementation of such systems would be advantageous to the country due to their cost-effectiveness and self-sustaining abilities. Other added advantages include their reliance on physical (gravity; dilution), chemical (reduction; precipitation), and biological (biosorption; bioreduction) natural processes to treat AMD [27]. Furthermore, the downward vertical flow direction in these systems significantly increases the interaction of water with treatment material; this maximizes the treatment efficiency, especially in net acidic mine water with excessive metal content [42, 75]. RAPSs have a potential applicability to treat AMD in the coalfields of South Africa. They are considered to be suitable for net acidic water with elevated concentrations of metals and dissolved oxygen [86]. Moreover, South Africa is also abundant in limestone deposits that are spread out across various provinces, with approximately 1.5 to 2 billion tons; therefore, procurement of such material would be easy [88]. Other organic treatment materials such as compost, sawdust, and manure are readily available.

1.3 Geochemical Modeling for the Prediction of the Dispersion of Metals in Water Systems

Geochemical modeling can be used as a prediction tool to evaluate geochemical processes and to predict the evolution of geochemical systems over time, especially the fate of contaminants [89, 90]. According to [91], "all geochemical predictions of metal dispersion use models based upon principles of mass balance," which makes use of laws of thermodynamics, especially the first one that states that "mass cannot be created or destroyed but can be transformed from one form to another." Due to this, prediction models can simulate the behavior of soluble metals in specific environments over a specified period of time. The components of the mass balance for a particular environment include (a) transformations by transport; (b) transformations by physical or chemical processes; and (c) transformations by sources/discharges [91]. Not only does geochemical modeling simulate conditions or parameters that are, otherwise, difficult to achieve with field monitoring or lab experiments such as the long-term rates of dissolution and acid generation and speciation of elements, but it can also be used to supplement data generated from the abovementioned exercises [89, 92].

In [93], geochemical models can be used to quantify the chemical and biological processes, affecting the fate and movement of pollutants as well as the rates of mineral dissolution and precipitation. The quantity of minerals that enter or leave the flow path can be determined as well. Furthermore, future water chemistry based on simulated geochemical processes can be predicted. Moreover, author in [92] states that geochemical models can assist in (i) understanding the interrelatedness of components in receptor environments, mostly the transport and drainage of sulfidic geologic material; (ii) structuring of data and calculation of acquired parameters; (iii) speciation of dissolved and precipitated reaction products; (iv) prediction of mineral equilibrium and solubility limits; and (v) evaluation of different scenarios. Geochemical modeling is divided into equilibrium, kinetics, transport, and empirical modeling [89, 92, 94].

1.3.1 Equilibrium Models

Equilibrium models are used to predict the concentrations of various species in systems that are in a state of chemical equilibrium; they can be used to calculate mineral saturation indices, speciation as well as solubility, one example of such a model is PHREEQC [95].

1.3.1.1 PHREEQC

PHREEQC is the most widely used equilibrium geochemical model due to its wide range of applications; this is because it can perform a variety of calculations ranging from speciation, solubility, reaction path, and mass balance to reactive transport calculations [93, 96]. PHREEQC uses an ion association aqueous model to simulate chemical reactions and transport processes in natural or polluted water [95, 97, 98].

1.3.2 Kinetics Models

Kinetic models are concerned with the rates at which chemical reactions occur to reach equilibrium. Their principles can used to describe how mass is transferred between systems; examples of such models include TOUGHREACT [78, 99, 100].

1.3.2.1 TOUGHREACT

TOUGHREACT is a kinetic model used to simulate three-dimensional non-isothermal multiphase reactive transport in porous and fractured rock. Various geological, environmental, and subsurface engineering problems in such features can be simulated using this model [78, 101, 102].

1.3.3 Transport Models

Transport models are mostly used in hydrogeology to simulate the mass transport of chemical species through geological features; this is done by including possible chemical reactions in transport models; MODFLOW is an example of such a model [94, 103, 104].

1.3.3.1 MODFLOW

MODFLOW is a widely used groundwater simulation model developed by the United States Geological Survey. MODFLOW simulates three-dimensional groundwater flow and contaminant transport [104, 105].

1.3.4 Empirical Modeling

Empirical modeling relies on long-term monitoring data on a specific site for the prediction of drainage chemistry; instead of performing geochemical calculations, they describe situations. This is done by compiling historical data sets for a specific site; then, a best-fit line is drawn on a chosen parameter to relate it to aqueous concentrations and the correlating standard deviation. Assuming that there is a continuity in the processes that resulted in the correlation with a chosen parameter in different ranges, extrapolation is used to predict future trends based on current trends; basically, empirical models are statistical predictive tools [92, 106].

In summation, the occurrence of AMD in affected environments results in the release of various contaminants into receptor environments. Due to suitable conditions, contaminants can be subjected to multiple geochemical processes such as dissolution, precipitation, diffusion, adsorption, and desorption along their flow path, thus transforming the pollutants. Various geochemical models are essential tools used to assess and better understand these processes. Not only does geochemical modeling simulate conditions or parameters that are, otherwise, difficult to achieve with field monitoring or lab experiments such as the long-term rates of dissolution and acid generation and speciation of elements, but it can also be used to supplement data generated from fieldwork. Geochemical modeling can be useful in developing water management strategies in South Africa, especially the management of AMD [89, 92, 107].

1.4 Conclusion

Because RAPSs are limestone-based passive treatment systems, they are vulnerable to armoring/clogging challenges when metals precipitate. These challenges are addressed by, *inter alia*, flushing out the precipitates, constructing the systems on steeper gradients, and replenishing treatment materials. Organic compost containing *Desulfovibrio* bacteria has been recommended for microbial sulfate reduction, increased alkalinity, and metal precipitation. For effective and optimal performance of RAPS, it is advisable to use a preceding settling pond/wetland before acidic water enters the system; this is to allow the settling of solids that can reduce the permeability of the systems. Passive treatment technologies are relatively cheaper as compared to active treatment techniques, and the materials required for treatment are readily available locally in South Africa; therefore, the successful implementation of these technologies is guaranteed. Due to its advantages over the other systems, the RAPS can be able to treat the net acidic AMD generated in South African coalfields, therefore contributing to sustainable mine water solutions in the country.

References

1. CSIR, Characterising the risk of human exposure and health impacts from acid mine drainage in South Africa Deliverable for Output VII – "Final Project Report", Council for Scientific and Industrial Research, Pretoria, South Africa, 2013.

2. Lloyd, P.J., *Coal mining and the environment*, Energy Research Institute, University of Cape Town, Cape Town, 2013.
3. Munnik, V., The Social and Environmental Consequences of Coal Mining in South Africa: a case study, 2010, https://static1.squarespace.com/static/5a7859a10abd0477ecb31301/t/5c5d88f-be5ef0151ead5a97/1549633797101/2010Jan__Coal_in_SA_Social_and_Env_Impacts.pdf . Environmental Monitoring Group, Cape Town, South Africa.
4. Ojonimi, T., Asuke, F., Onimisi, M., Onuh, C., Acid Mine Drainage (AMD): an environmental concern generated by coal mining. *J. Degrad. Min. Lands Manag.*, 6, 4, 1875–1881, 2019.
5. Skousen, J., Sexstone, A., Cliff, J., Sterner, P., Calabrese, J., Ziemkiewicz, P., Acid mine drainage treatment with a combined wetland/anoxic limestone drain: greenhouse and field systems. *Proceedings of the Annual National Meeting - American Society for Surface Mining and Reclamation*, 16, 1999.
6. Feris, L. and Kotzé, L., The regulation of acid mine drainage in South Africa: law and governance perspectives. *Potchefstroom Electron. Law J.*, 17, 2104–2163, 2015.
7. Ramontja, T., Coetzee, H., Hobbs, P.J., Burgess, J.E., Thomas, A., Keet, M., Yibas, B., van Tonder, D., Netili, F., Rust, U.A., Wade, P., Maree, J., Mine Water Management in the Witwatersrand Gold Fields With Special Emphasis on Acid Mine Drainage, Pretoria, Inter-Ministerial-Committee on Acid Mine Drainage. Report of the Team of Experts for the Inter-Ministerial Committee on Acid Mine Drainage, Pretoria, South Africa, p. 146, 2011.
8. Taylor, J., Pape, S., Murphy, N., A Summary of Passive and Active Treatment Technologies for Acid and Metalliferous Drainage (AMD). *Fifth Australian workshop on acid drainage 29-31 August 2005 Fremantle, Western Australia*, Australian Centre for Minerals Extension and Research (ACMER). Earth Systems Pty Ltd, 2005.
9. Hallowes, D. and Munnik, V., The Destruction of the Highveld Part 1: Digging Coal. The groundWork Report 2016, Arrow Print, Pietermaritzburg, South Africa, 2016.
10. Shongwe, B.N., The Impact of Coal Mining on the Environment and Community Quality of Life: A Case Study Investigation of the Impacts and Conflicts Associated with Coal Mining in the Mpumalanga Province, South Africa. Master's dissertation, University of Cape Town, Cape Town, South Africa, 2018.
11. Tempelhoff, J.W.N., Ginster, M., Motloung, S., Gouws, C.M., Strauss, J.S., The 2012 acid mine drainage (AMD) crisis in Carolina's municipal water supply. *Afr. Hist. Rev.*, 46, 2, 77–107, 2015.
12. McCarthy, T.S. and Humphries, M.S., Contamination of the water supply to the town of Carolina, Mpumalanga, January 2012. *S. Afr. J. Sci.*, 109, 9/10, 1–10, 2013.

13. McCarthy, T.S., The impact of acid mine drainage in South Africa. *S. Afr. J. Sci.*, 107, 5/6, 10–7, 2011.
14. Sakala, E., Fourie, F., Gomo, M., Coetzee, H., Hydrogeological investigation of Witbank, Ermelo and Highveld Coalfields: Implications for acid mine drainage subsurface transport and attenuation, in: *Mine Water & Circular Economy*, vol. I, C. Wolkersdorfer, L. Sartz, M. Sillanpää, A. Häkkinen, (Eds.), pp. 564–572, Lappeenranta University of Technology, Lappeenranta, Finland, 2017.
15. Byrne, P., Wood, P.J., Reid, I., The impairment of river systems by metal mine contamination: A review including remediation options. *Crit. Rev. Environ. Sci. Technol.*, 42, 2017–2077, 2012.
16. Coetzee, H., van Tonder, D., Wade, P., Esterhuyse, S., van Wyk, N., Ndengu, S., Venter, Kotoane, M., *Acid mine drainage in the Witwatersrand*, p. 81, Department of Minerals and Energy, Pretoria, Council for Geoscience Report No 2007-0260, 2007.
17. Costello., C., Acid Mine Drainage: Innovative Treatment Technologies, U.S. Environmental Protection Agency Office of Solid Waste and Emergency Response Technology Innovation Office, Washington, DC, 2003.
18. URS, Passive and semi-active treatment of acid rock drainage from metal mines - state of the practice - final draft report, URS Corporation, Portland, Maine, 2003.
19. Jennings, S.R., Neuman, D.R., Blicker, P.S., *Acid Mine Drainage and Effects on Fish Health and Ecology: A Review*, Reclamation Research Group Publication, Bozeman, MT, 2008.
20. Udayahhanu, S.G. and Prasad, B., Studies on environmental impact of acid mine drainage generation and its treatment: an appraisal. *Indian J. Environ. Prot.*, 30, 11, 953–967, 2010.
21. Macías, F., Caraballo, M.A., Nieto, J.M., Rötting, T.S., Ayora, C., Natural pretreatment and passive remediation of highly polluted acid mine drainage. *J. Environ. Manage.*, 104, 93–100, 2012.
22. Miller, J.R., Gannon, J.P., Corcoran, K., Concentrations, mobility, and potential ecological risks of selected metals within compost amended, reclaimed coal mine soils, tropical South Sumatra, Indonesia. *AIMS Environ. Sci.*, 6, 4, 298–325, 2019.
23. Trumm, D. and Watts, M., Results of small-scale passive system trials to treat acid mine drainage, West Coast Region, South Island, New Zealand. *N. Z. J. Geol. Geophys.*, 53, 2-3, 227–237, 2010.
24. Masindi, V. and Muedi, K.L., *Environmental Contamination by Heavy Metals*, InTech, Rijeka, Croatia, 2018, doi: 10.5772/intechopen.76082.
25. Kefeni, K.K., Msagati, T.A.M., Mamba, B.B., *Acid mine drainage: Prevention, treatment options, and resource recovery*: A review. *J. Clean. Prod.*, 151, 475–493, 2017.
26. Johnson, D. and Hallberg, B., Acid mine drainage remediation options: a review. *Sci. Total Environ.*, 338, 1–2, 3–14, 2005.

27. Kilborn Inc, Review of passive systems for treatment of acid mine drainage phase II, MEND Report 3.14.1, Kilborn INC, Lakeshore Blvd West, Toronto, Ontario, 1996.
28. Luković, A. and Stanković, M., Passive systems for treating acid mine drainage: a general review. *Saf. Eng.*, 2, 4, 227–232, 2012.
29. Eger, P. and Wagner, J., Wetland treatment systems–how long will they really work. *Sudbury 2003 Mining and the Environment Conference*, pp. 71–76, 2003.
30. Gusek, J.J., Passive Treatment 101: An Overview of the Technologies, U.S. EPA/National Groundwater Association's Remediation of Abandoned Mine, Lands, pp. 1–13, 2008.
31. McCauley, C.A., Assessment of Passive Treatment and Biogeochemical Reactors for Ameliorating Acid Mine Drainage at Stockton Coal Mine, PhD thesis, University of Canterbury, New Zealand, 2011.
32. Ziemkiewicz, P.F., Skousen, J.G., Simmons, J., Long-term Performance of Passive Acid Mine Drainage Treatment Systems. *Mine Water Environ.*, 22, 3, 118–129, 2003.
33. Ford, K.L., Passive treatment systems for acid mine drainage, Bureau of Land Management, 2003, Technical Note 409. DigitalCommons@University of Nebraska - Lincoln.
34. Hedin, R.S., Nairn, R.W., Kleinmann, R.L.P., Information Circular 9389 : Passive treatment of coal mine drainage, Library of Congress Cataloging in Publication Data, USA, pp. 1–44, 1994.
35. Kuyucak, N., Francois, C., Martschuk, J., Successful implementation and operation of a passive treatment system in an extremely cold climate. *7th International Conference on Acid Rock Drainage (ICARD)*, R.I. Barnhisel (Ed.), St. Louis MO, March 26-30, 2006, American Society of Mining and Reclamation (ASMR), 3134 Montavesta Road, Lexington, KY 40502, pp. 980–992, 2006.
36. Neculita, C.M., Zagury, G.J., Kulnieks, V., Short-term and long-term bioreactors for acid mine drainage treatment. Association for Environmental Health and Sciences. *22nd Annual International Conference on Contaminated Soils, Sediments and Water 2006*, vol. 12, pp. 1–10, 2007.
37. Zagury, G.J. and Neculita, C., Passive Treatment of Acid Mine Drainage in Bioreactors using Sulfate-Reducing Bacteria: Critical Review and Research Needs. *J. Environ. Qual.*, 36, 1–16, 2007.
38. Gusek, J.J. and Thomas R. Wildeman, T.R.W., A new millennium of passive treatment of acid rock drainage: advances in design and construction since 1988. *J. Am. Soc. Min. Reclam.*, 2002, 1, 935–952, 2002.
39. Zipper, C.E., Skousen, J., Jage, C., Passive Treatment of Acid-Mine Drainage, in: *Acid Mine Drainage, Rock Drainage, and Acid Sulfate Soils: Causes, Assessment, Prediction, Prevention, and Remediation*, pp. 1–14, 2018.
40. Gusek, J.J., A Periodic Table of Passive Treatment for Mining Influenced Water, in: *2009 National Meeting of the American Society of Mining and*

Reclamation, R. Barnhisel (Ed.), pp. 550–562, ASMR, 3134 Montavesta Rd., Lexington, KY 40502, 2009.

41. Watzlaf, G.R., Schroeder, K.T., Kleinmann, R.L.P., Kairies, C.L., Nairn, R.W., Street, W.B., *The Passive Treatment of Coal Mine Drainage*, pp. 1–72, DOE/NETL-2004/1202, Department of Energy, United States of America, 2004.
42. Matthies, R., Jarvis, A.P., Aplin, A.C., Performance Evaluation of Two Reducing and Alkalinity Producing Systems for Coal Mine Drainage Remediation after 4 Years of Operation. *Sci. Total Environ.*, 408, 20, 4877–85, 2010.
43. Rees, B., Bowell, R., Williams, K., Passive Treatment: A Walk Away Solution? *Mine Water Environ.*, 22, 3, 118–129, 2001.
44. Younger, P.L., Passive treatment of acidic mine waters in subsurface-flow systems: exploring RAPS and permeable reactive barriers. *Land Contam. Reclam.*, 11, 2, 127–135, 2003.
45. Skousen, J., Anoxic Limestone Drains for Acid Mine Drainage Treatment. *Green Lands*, 21, 4, 30–35, 1991.
46. Skousen, J. and Ziemkiewicz, P., Performance of 116 Passive Treatment Systems for Acid Mine Drainage. *J. Am. Soc. Min. Reclam.*, 2005, 1, 1100–1133, 2016.
47. Skousen, J., Overview of passive systems for treating acid mine drainage. *Green Lands*, 26, 4, 36–45, 1996.
48. INAP, GARD Guide Chapter 7 – Mine Water Treatment, International Network for Acid Prevention, Ottawa, Canada, 2012.
49. Jarvis, A. and England, A., Operational and treatment performance of an unique Reducing and Alkalinity Producing System (RAPS) for acidic leachate remediation in Lancashire, UK, in: *Uranium in the Aquatic Environment*, B.J. Merkel, B. Planer-Friedrich, C. Wolkersdorfer (Eds.), Springer, Berlin, Heidelberg, 2002, https://doi.org/10.1007/978-3-642-55668-5_118.
50. Watzlaf, G.R., Schroeder, K.T., Kleinmann, R.L.P., Kairies, C.L., Nairn, R.W., Street, W.B., *The Passive Treatment of Coal Mine Drainage*, pp. 1–72, Department of Energy, United States of America, 2004.
51. Fabian, D., Aplin, A.C., Younger, P.L., Geochemical performance of a reducing and alkalinity-producing system (RAPS) for the passive treatment of acidic mine drainage at Bowden Close, United Kingdom, in: *Mine Water 2005 – Mine Closure, Proc. 9th International Mine Water Association Congress (IMWA)*, J. Loredo and F. Pendás (Eds.), pp. 383–387, Oviedo, Spain, 2006.
52. Kittrell, E.F., Potential acid mine drainage treatment utilizing acidophilic sulfate reducing bacteria in an upflow bioreactor. Masters Theses, p. 7245, University of Missouri, Missouri, USA, 2014, https://scholarsmine.mst.edu/masters_theses/7245.
53. Mukwevho, M.J., Maharajh, D., Chirwa, E.M.N., Evaluating the Effect of pH, Temperature, and Hydraulic Retention Time on Biological Sulphate Reduction Using Response Surface Methodology. *Water*, 12, 2662, 2020.

54. Nairn, R. and Hedin, R., Generation of alkalinity in an anoxic limestone drain. *J. Am. Soc. Min. Reclam.*, 206–219, 1992, DOI: 10.21000/JASMR92010206.
55. Franti, D.J., Acid mine drainage treatment in an open limestone channel emphasizing aeration & retention, A thesis submitted in partial fulfilment of the requirements for the degree of Master of Science in Land Rehabilitation, Montana state university, Bozeman, Bozeman, Montana, 1998.
56. Dold, B., Evolution of Acid Mine Drainage Formation in Sulphidic Mine Tailings. *Minerals*, 4, 621–641, 2014.
57. Lee, S.H., Kim, S., Jeon, B.H., Bhatnagar, A., Ji, S., Cheong, Y., Lee, G., Activity of sulfate reducing bacteria in successive alkalinity producing system: Part I - Effect of temperature. *Res. J. Chem. Environ.*, 14, 4, 67–73, 2010.
58. Joubert, H. and Pocock, G., The VitaSOFT PROCESS: A Sustainable, Long-term Treatment Option for Mining-impacted Water, WRC Report No. 2232/1/16, Water Research Commission, Gezina, Pretoria, 2016.
59. RoyChowdhury, A., Green Remediation of Acid Mine Drainage (AMD)-Impacted Soil and Water, A dissertation submitted to the faculty of Montclair State University in partial fulfilment of the requirements for the degree of Doctor of Philosophy, Montclair State University Montclair, NJ, 2016.
60. INAP, The Global Acid Rock Drainage Guide. *Mine Water Environ.*, 28, 305–310, 2009, http://www.gardguide.com.
61. Jeen, S. and Mattson, B., Evaluation of layered and mixed passive treatment systems for acid mine drainage. *Environ. Technol.*, 37, 22, 2835–2851, 2016.
62. Patel, M.D., Study of microbial contribution in alkalinity generation during treatment of acid mine drainage by laboratory successive alkalinity producing system. *Indian J. Biochem. Biophys.*, 58, 162–170, April 2021.
63. Rambabu, K., Banat, F., Pham, Q.M., Ho, S., Ren, N., Show, P., Biological remediation of acid mine drainage: Review of past trends and current outlook. *Environ. Sci. Ecotechnol.*, 2, 100024–100037, 2020.
64. Suyasa, W.B., Sudiartha, G., Pancadewi, G., Optimization of Sulphate-Reducing Bacteria for Treatment of Heavy Metals-Containing Laboratory Wastewater on Anaerobic Reactor. *Pollution*, 9, 2, 545–556, 2023.
65. Culshaw, M. and Cooper, A.H., Dissolution, in: *Encyclopedia of Engineering Geology. Encyclopedia of Earth Sciences Series*, P.T. Bobrowsky, and B. Marker, (Eds.), Springer, Cham, 2018.
66. Hammarstrom, J.M., Sibrell, P.L., Belkin, H.E., Characterization of limestone reacted with acid-mine drainage in a pulsed limestone bed treatment system at the Friendship Hill National Historical Site, Pennsylvania, USA. *Appl. Geochem.*, 18, 11, 1705–1721, 2003.
67. Santomartino, S. and Webb, J.A., Estimating the longevity of limestone drains in treating acid mine drainage containing high concentrations of iron. *Appl. Geochem.*, 22, 11, 2344–2361, 2007.

68. Cravotta, C.A., Limestone dissolution in passive systems for neutralization of acidic mine drainage, Conference: West Virginia Mine Drainage Task Force at: Morgantown WV, 2005.
69. Noiriel, C., Luquot, L., Madé, B., Raimbault, L., Gouze, P., van der Lee, J., Changes in reactive surface area during limestone dissolution: An experimental and modelling study. *Chem. Geol.*, 265, 1–2, 160–170, 2009.
70. Schweers, R.M., *In Situ* Studies of Limestone Dissolution in a Coastal Submarine Spring, A thesis submitted in partial fulfilment of the requirements for the degree of Master of Science, Department of Cell Biology, Microbiology, and Molecular Biology, College of Arts and Sciences, University of South Florida, USF Tampa Graduate Theses and Dissertations, 2015, https://digitalcommons.usf.edu/etd/6024.
71. Watzlaf, G.R., Passive treatment of acid mine drainage in down-flow limestone systems. *Conference: 14. annual national meeting of the American Society for Surface Mining and Reclamation (ASSMR)*, Austin, TX (United States), 10-15 May 1997, 1997.
72. Novhe, O., Yibas, B., Coetzee, H., Atanasova, M., Netshitungulwana, R., Modiba, M., Mashalane, T., Long-term remediation of acid mine drainage from abandoned coal mine using integrated (anaerobic and aerobic) passive treatment system in South Africa: A pilot study, *Min. Meets Water-Conflicts Solutions. Proceedings IMWA 2016*, vol. 1, Drebenstedt, Carsten, Paul, Michael (Eds.), pp. 668–675, Department of Energy, Freiberg/Germany, 2016.
73. Dube, M.G., Ramasenya, K., van Zweel, N., The use of passive treatment systems to remediate AMD from abandoned coal mines, Emalahleni, South Africa — Column Experiments, in: *Mine Water — Technological and Ecological Challenges*, E. Khayrulina, C. Wolkersdorfer, S. Polyakova, A. Bogush (Eds.), pp. 167–175, Perm State University, Perm, Russia, 2019.
74. Skousen, J., Zipper, C.E., Rose, A., Review of Passive Systems for Acid Mine Drainage Treatment. *Mine Water Environ.*, 36, 133–153, 2017.
75. RoyChowdhury, A., Sarkar, D., Datta, R., Remediation of Acid Mine Drainage-Impacted Water. *Curr. Pollut. Rep.*, 1, 131–141, 2015.
76. Florence, K., Mechanisms of the Removal of Metals from Acid and Neutral Mine Water under varying Redox Systems. A thesis submitted for the Degree of Doctor of Philosophy by Kay Florence, MSc School of Engineering, Cardiff University, United Kingdom, 2014.
77. Watzlaf, G.R. and Hyman, D.M., Limitations of Passive Systems for the Treatment of Mine Drainage. *Proceedings of the 17th Annual Association of Abandoned Mine Lands Conference*, French Lick, IN, pp. 186–199, 1995.
78. Xu, T., Sonnenthal, E., Spycher, N., Pruess, K., TOUGHREACT: A Simulation Program for Non-isothermal Multiphase Reactive Geochemical Transport in Variably Saturated Geologic Media. *Computers and Geosciences,* 32, 2, 145–165, 2004.

79. Seervi, V., Yadav, H.L., Srivastava, S.K., Jamal, A., Overview of Active and Passive systems for Treating Acid Mine Drainage. *Int. Adv. Res. J. Sci. Eng. Technol.*, 4, 5, 2017.
80. Song, H., Yimb, G., Jib, S., Neculita, C., Hwang, T., *Pilot-scale passive bioreactors for the treatment of acid mine drainage: Efficiency of mushroom compost vs. mixed substrates for metal removal. J. Environ. Manage.*, 111, 150–158, 2012.
81. Geldenhuis, S. and Bell, F.G., Acid Mine Drainage at a coal mine in the eastern Transvaal, South Africa. *Environ. Geol.*, 34, 234–242, 1998.
82. Bell, F.G., Bullock, S.E.T., Hälbich, T.F.J., Lindsay, P., Environmental impacts associated with an abandoned mine in the Witbank Coalfield, South Africa. *Int. J. Coal Geol.*, 45, 195–216, 2001.
83. McCarthy, T.S. and Pretorius, K., Coal mining on the highveld and its implications for future water quality in the Vaal river system. *Abstracts of the International Mine Water Conference Proceedings 19th – 23rd October 2009 Pretoria, South Africa Conference*, Cilla Taylor Conferences, 2009.
84. Yibas, B., Netshitungulwana, R., Novhe, O., A holistic approach towards best management practices of mine pollution impacts using a catchment area strategy, South Africa, in: *Reliable Mine Water Technology*, C. Wolkersdorfer, A. Brown, L. Figueroa (Eds.), pp. 117–122, International Mine Water Association, Golden CO, USA, 2013.
85. Ziemkiewicz, P.F., Skousen, J.G., Lovett, R., Open limestone channels for treating acid mine drainage: A new look at an old idea. *Green Lands*, 24, 36–41, 1994.
86. Skousen, J., Sexstone, A., Ziemkiewicz, P.F., Chapter 6: Acid mine drainage control and treatment, in: *Reclamation of Drastically Disturbed Lands*, 2nd Ed., American Society of Agronomy, Madison, WI, 2000.
87. Swart, E., The South African legislative framework for mine closure. *J. South African Inst. Min. Metall.*, 103, 489–492, 2003.
88. Council for Geoscience, Geological Survey of South Africa, Council for Geoscience, Pretoria, 2023.
89. Camden-Smith, B., Johnson, R.H., Camden-Smith, P., Tutu, H., Geochemical Modelling of Water Quality and Solutes Transport from Mining Environments, in: *Research and Practices in Water Quality*, pp. 167–169, IntechOpen Limited, Great Portland Street, London, W1W 5PF, United Kingdom, 2015.
90. Steyn, A.J., *Geochemical investigation of Union Colliery underground mine workings, Mpumalanga, South Africa, Submitted in fulfilment of the requirements for the degree Magister Scientiae in Geohydrology in the Faculty of Natural and Agricultural Sciences (Institute for Groundwater Studies)*, University of the Free State, Free State, South Africa, 2019.
91. Mkandawire, M., Geochemical Prediction of Metal Dispersion in Surface and Groundwater Systems, in: *Nano and Bio-Based Technologies for Wastewater Treatment: Prediction and Control Tools for the Dispersion of Pollutants in the*

Environment, E. Fosso-Kankeu (Ed.), pp. 3–40, Scrivener Publishing LLC, 100 Cummings Center, Suite 541J. Beverly, MA 01915-6106, 2019.

92. Price, W.A., *Prediction Manual for drainage chemistry from sulphuric geo;ogic materials, Mine Environment Neutral Drainage (MEND) Program Prediction Committee*, University of Western Ontario Department of Earth Sciences London, Ontario, 2009.
93. Magombedze, C., Geochemical Processes Controlling the Generation and Environmental Impacts of Acid Mine Drainage in Semi Arid Conditions: A Case Study based on Sulphide Metal Mines in Zimbabwe, Thesis for the Philosophiae Doctoral Degree (PhD). Trondheim, Norwegian University of Science and Technology, Faculty of Engineering Science and Technology, Department of Geology and Mineral Resources Engineering, 2006.
94. Bethke, C., Modelling transport in reacting geochemical systems, Geochemistry, Department of Geology, 1301 West Green Street, University of Illinois, Urbana. Illinois 61801, USA, 1997.
95. Parkhurst, D. and Appelo, C., Description of Input and Examples for PHREEQC Version 3-A Computer Program for Speciation, Batch-Reaction, One-Dimensional Transport, and Inverse Geochemical Calculations, U.S. Geological Survey Techniques and Methods, South Africa, p. 497, 2013, http://pubs.usgs.gov/tm/06/a43/.
96. van Zweel, Development of a geochemical model to predict leachate water quality associated with coal mining practices, Dissertation submitted in fulfilment of the requirements for the degree Magister Scientiae in Environmental Sciences (specialising in Hydrology and Geohydrology), Potchefstroom Campus of the North-West University, North West, South Africa, 2015.
97. Manyatshe, A., Fosso-Kankeu, E., van der Berg, D., Lemmer, N., Waanders, F., Tutu, H., Metal Speciation in the Rivers Around Potchefstroom Based on Seasonality. *Water Environ. Res.*, 90, 84–95, 20182018.
98. Netshiongolwe, K.E., Geochemical characterisation of gold tailings footprints on the Central Rand and Goldfield. Submitted in accordance with the requirements for the degree of Master of Science the subject environmental science, University of South Africa, Pretoria, South Africa, pp. 1–153, 2018.
99. Bethke, T., *Geochemical and Biogeochemical Reaction Modeling*, Cambridge University Press, University Printing House, Cambridge CB2 8BS, United Kingdom, 2022.
100. Lasaga, A.C., *Kinetic Theory in the Earth Sciences*, Princeton University Press, New York, 1998.
101. Myers, J. and Sabido, L., *Software validation plan and software validation report for toughreact version 3.0*, Center for Nuclear Waste Regulatory Analyses San Antonio, Texas, 2007.
102. Sonnenthal, E., Spycher, N., Xu, T., Zheng, L., *TOUGHREACT V4.12-OMP and TReactMech V1.0 Geochemical and Reactive-Transport User Guide*, University of California, Berkeley, CA 94720. South Africa, 2021

103. Steefel, I., DePaolo, D.J., Lichtner, P.C., Reactive transport modeling: An essential tool and a new research approach for the Earth sciences. *Earth Planet. Sci. Lett.*, 240, 3–4, 539–558, 2005.
104. Sráček, O., Černík, M., Zbyněk, *Applications of Geochemical and Reactive Transport Modeling in Hydrogeology*, Palacký University, Olomouc, 2013.
105. Langevin, D., Hughes, J.D., Banta, R., Documentation for the MODFLOW 6 Groundwater Flow Model, Sunrise Valley Drive, Reston, VA 20192, USA, 2017.
106. DWAF, *Best Practice Guideline G4: Impact Prediction*, Department of Water Affairs and Forestry, Pretoria, South Africa, 2008.
107. Perkins, E.H., Nesbitt, H.W., Gunter, W.D., St-Arnaud, L.C., Mycroft, J.R., *Critical review of geochemical processes and geochemical models adaptable for prediction of acidic drainage from waste rock*, Mine Environment Neutral Drainage (MEND) Program Prediction Committee. University of Western Ontario Department of Earth Sciences, London, Ontario, 1993.

2

Novel Hybrid Nature-Based Solutions for the Sustainable Treatment of Industrial Wastewater: Alkaline and Acid Mine Drainage

Schoeman, Y.[1,2*], Erasmus, M.[1], Naidoo, N.[3] and Oberholster, P.J.[1,2]

[1]*Faculty of Natural and Agricultural Sciences, Centre for Mineral Biogeochemistry, University of the Free State, Bloemfontein, South Africa*
[2]*Ecological Engineering Institute of Africa, University of the Free State, Bloemfontein, South Africa*
[3]*Faculty of Natural and Agricultural Sciences, Centre for Environmental Management, University of the Free State, Bloemfontein, South Africa*

Abstract
Industrial and mining wastewater treatment safeguards the environment and public health, removing hazardous pollutants like heavy metals, pathogens, and nutrients that harm aquatic life, soil quality, and human well-being. Conventional methods, though effective, are expensive, require expertise and specialized maintenance, and have high energy and carbon footprints. Nature-based solutions (NBSs) are gaining traction as sustainable alternatives, utilizing processes like wetlands, microbial bioremediation, and natural filtration. NBS offers cost-effectiveness, energy efficiency, ecosystem services, and resilience. However, decision-making challenges hinder NBS implementation, including uncertainties in system selection, configuration, and quantifying sustainability impact. Alkaline wastewater from industries like oil, gas, chemicals, and mining poses challenges due to high salt levels. Acid mine drainage (AMD) occurs when water contacts rocks with sulfur-bearing minerals, producing dissolved iron, sulfuric acid, and heavy metal runoff that pollutes water sources. Properly treating alkaline and AMD wastewater minimizes environmental impact, protects public health, and promotes planetary health. This chapter proposes a framework for hybrid

**Corresponding author*: schoeman.yolandy@gmail.com

Elvis Fosso-Kankeu, Vhahangwele Masindi, Johannes Maree and Bhekie Mamba (eds.) Customized Technologies for Sustainable Management of Industrial Wastewater: A Circular Economy Approach, (39–110) © 2025 Scrivener Publishing LLC

NBS selection, integrating a planetary health nexus approach considering water, climate, food, biodiversity, and land systems. The framework aligns with the United Nation's Sustainable Development Goals, contributing to global sustainability.

***Keywords*:** Nature-based solutions, ecological engineering, industrial waste water, alkaline waste water, acid mine drainage, microbial bioremediation, phytoremediation, phycoremediation

2.1 Introduction

2.1.1 Importance of Industrial and Mining Waste Water Treatment

Industrial and mining wastewater treatment is critical in protecting the environment and public health [1]. Discharging untreated industrial and mining wastewater into natural water bodies can introduce hazardous pollutants, including heavy metals, pathogens, and nutrients, which pose significant risks to aquatic life, soil quality, and human health [1]. These pollutants can have detrimental effects on ecosystems, causing the degradation of water quality, disruption of ecological balance, and potential contamination of drinking water sources [1, 2]. As industries continue to expand and evolve, the need for effective wastewater treatment methods becomes increasingly pressing to ensure sustainable development and the well-being of the environment and society. Conventional treatment approaches, although effective, often come with drawbacks such as high costs, specialized expertise requirements, and energy-intensive processes. Therefore, there is a growing interest in exploring alternative nature-based solution (NBS) that offer more sustainable and resource-efficient approaches to industrial and mining wastewater treatment [1, 3, 4]. This chapter focuses on treating alkaline and acid mine drainage (AMD) wastewater (with broad applicability also to industrial wastewater), addressing these mining waste water streams' unique challenges, and proposing and demonstrating novel hybrid NBS that contributes to a more sustainable and resilient future.

2.1.2 Nature of Alkaline and AMD Water Influencing Treatment Methods Toward Environmental and Public Health Protection

2.1.2.1 Alkaline Wastewater

Alkaline industrial and mine wastewater is characterized by a pH above 8 and contains acid-neutralizing bases that act as a buffer system to prevent extreme changes in water pH [5]. The sources of natural alkalinity include carbonate, bicarbonate, hydroxide, borates, phosphates, and silicates. Industrial processes such as alumina extraction, cement production, chromium ore processing, coal combustion, iron and steel manufacture, lime production, and waste incineration produce alkaline residues, which are high-concentrated industrial pollutants with large global flux [6, 7]. These residues comprise calcium, magnesium, or sodium compounds that hydrate rapidly and generate alkalinity. The dissolution of calcium or sodium hydroxides dominates the initial reaction, followed by slower reactions involving calcium silicates, magnesium hydroxides, and sodium aluminosilicates. Over time, the dissolution of carbonates becomes prevalent. The impacts of alkaline leachates on receiving waters include increased pH, oxygen depletion, high chemical oxygen demand (COD), salinity, high sulfate loading, and increased metal concentrations [6]. These impacts, coupled with rapid mineral precipitation, can resemble natural travertine deposits [8]. The risk posed by alkaline residues after environmental discharge or in after-uses depends on interactions with the local geochemical environment [6, 9–15].

2.1.2.2 Acid Mine Drainage

AMD is a condition where water runoff or infiltration has a low pH and high concentrations of dissolved natural constituents, such as metal salts, due to the weathering of specific rock-forming minerals [16]. AMD is commonly associated with mining activities [17] but can occur wherever water interacts with rocks containing sulfur-bearing minerals. The composition of AMD varies depending on factors like temperature, weather, and mineral types. It is characterized by low pH, high electrical conductivity, high concentrations of heavy metals, high levels of total dissolved solids (TDSs), and low nutrient content [17]. Visual indicators of AMD include the presence of ferruginous chromatization, also known as "yellow-boy," caused by the precipitation of iron oxyhydroxide, which can vary in color from yellow to red [18]. The primary constituents of AMD acid generation are water, sulfide minerals, and oxygen [19]. Iron sulfide minerals like pyrite and pyrrhotite are commonly associated with AMD generation, releasing sulfates, ferrous ions, and hydrogen ions upon oxidation. Additional acidity is caused by further oxidation to Fe2+ and the precipitation of hydroxy sulfates and Fe(III) oxyhydroxides.

Other sulfide minerals and mineral groups like aluminosilicates, carbonates, oxides, and sulfates can also contribute to AMD [19]. The generation of AMD is a complex process influenced by biological, chemical, hydrological, and physical factors [20]. Factors affecting the rate of acid generation include the surface area of exposed metal sulfide, water saturation, permeability, temperature, oxygen concentration, bacterial activity, chemical activity of Fe3+, and the chemical activation energy required to initiate acid generation [21]. The process of AMD formation can be summarized into four interdependent geochemical reactions, as shown in the figure below. By the fourth equation, acidophilic bacteria colonies further promote the geochemical and mineral oxidation processes resulting in acidification [22, 23].

$$2FeS_2\,(s) + 2H_2O + 7O_2\,(aq) \rightarrow 2Fe^{2+} + 4SO_4^{\;2-} + 4H^+ \tag{2.1}$$

$$4Fe^{2+} + 4H^+ + O_2 \rightarrow 4Fe^{3+} + 2H_2O \tag{2.2}$$

$$Fe^{3+} + 3H_2O \rightarrow Fe(OH)_3\,(s) + 3H^+ \tag{2.3}$$

$$FeS_2\,(s) + 14Fe^{3+} + 8H_2O \rightarrow 15Fe^{2+} + 2SO_4^{\;2-} + 16H^+ \tag{2.4}$$

2.1.3 Challenges with Conventional Treatment Methods

Conventional treatment methods for industrial and mining wastewater, such as chemical and physical treatments, have been widely employed for their effectiveness in removing pollutants [24]. However, these methods are not without their challenges. Firstly, conventional treatment approaches can be expensive, requiring significant financial investments in infrastructure, chemicals, and energy [25]. Additionally, the expertise needed to operate and maintain these treatment systems may be limited, leading to higher operational costs and potential inefficiencies [25]. Moreover, conventional methods often generate large amounts of sludge or waste by-products, which require further treatment or disposal, posing additional environmental concerns [25]. Conventional treatment processes' energy and carbon intensity also affect their ecological footprint [25]. These challenges highlight the need for alternative approaches that are cost-effective, energy-efficient, and environmentally sustainable [26]. NBS offers a promising avenue for overcoming these limitations, as they utilize natural processes and ecosystems to treat industrial wastewater. Integrating NBS into wastewater treatment strategies makes it possible to address the shortcomings of conventional methods and move toward more sustainable and resilient practices.

2.1.3.1 *Alkaline Wastewater*

Treatment methods for alkaline wastewater are similar to those of AMD. One of the main conventional methods of treatment for alkaline wastewater or leachates is neutralization or partial neutralization. The neutralization or partial neutralization of alkaline wastewater before discharge involves using acid dosing to neutralize alkaline wastewater or leachate to an approximate pH of between 8.5 and 9 [6]. Conventional treatment methods of alkaline wastewater/leachates are expensive and labor-intensive, resulting in being economically unviable to implement in post-closure plans [27]. Many conventional remediation processes significantly increase the volume of wastewater to be discharged [28]. These methods usually also require a secondary treatment to treat the wastewater to acceptable environmental standards, increasing the cost of treatment even further [27].

2.1.3.2 *Acid Mine Drainage*

Conventional AMD remediation methods typically involve adding alkaline materials to raise the pH above the threshold required by iron-oxidizing bacteria, thereby reducing the rate of acid generation [29]. Alkaline materials like lime, limestone, sodium carbonate, sodium hydroxide, magnesium oxide, and hydroxides are commonly used for neutralization. However, these conventional treatments require significant investment, including expenses for reagents, operation, and maintenance [30]. Moreover, traditional remediation processes often create new waste streams such as brines, sludge, spent media, or gaseous emissions [28]. While neutralization and precipitation methods are widely employed, they may not always be sufficient and can necessitate secondary treatments and waste disposal facilities, resulting in additional costs. Chemical neutralization methods generate large amounts of silt, which requires more landfill space for waste disposal [31]. The presence of toxic metals in the waste poses risks to soil, groundwater, and surrounding ecosystems. Therefore, efficient AMD management practices that focus on resource recovery and reuse must be implemented [31]. The choice of AMD treatment options depends on various factors, including environmental conditions, pH, AMD composition, and economic considerations [32].

2.2 Nature-Based Treatment Options for Environmental and Public Health Protection

2.2.1 Importance

NBS has gained interest as a sustainable treatment option for industrial and mining wastewater. NBS utilizes natural processes and ecosystems to remove pollutants and restore water quality, offering a cost-effective and environmentally friendly alternative to conventional methods. Effective NBS techniques include constructed wetlands, bioremediation, natural filtration systems, and phyco- and phytoremediation [33].

Constructed wetlands mimic natural wetland processes to remove pollutants by providing a habitat for microorganisms and plants that break down and absorb contaminants [34, 35]. Bioremediation uses microorganisms to degrade pollutants, making it suitable for treating various industrial pollutants [36–39]. Natural filtration systems, like sand or gravel filters, rely on physical and biological processes within porous media to remove suspended solids and pollutants [40, 41]. Phyco- and phytoremediation techniques involve using algae and plants to absorb or convert pollutants through biochemical processes [42, 43].

NBS offers advantages such as cost-effectiveness, energy efficiency, and multiple co-benefits like habitat creation and biodiversity enhancement [1, 3, 4, 33]. However, implementing NBS for industrial wastewater treatment comes with challenges, including complex decision-making and uncertainties regarding performance and long-term effectiveness [44, 45]. Overcoming these challenges requires a comprehensive understanding of ecological, social, and economic dimensions [44, 45]. AMD with applicability to alkaline wastewater treatment systems can be optimized, made more cost-effective, and require less maintenance by integrating NBS treatment alternatives [26, 46]. Appropriate NBS treatment techniques can be selected through a comprehensive framework that considers the unique characteristics and challenges associated with AMD.

2.2.2 Constructed Wetlands

Constructed wetlands are man-made wetlands designed to address water pollution [34, 35]. They offer cost-effective and low-maintenance solutions for remediation [34]. Classification of constructed wetlands involves considering hydrology, macrophytic growth type, and sub-surface flow path [47]. Free water surface flow (FWSF) wetlands have shallow basins and emergent vegetation and facilitate organic removal through settling and microbial degradation. Sub-surface flow (SSF) wetlands can be horizontal (HSSF) or vertical (VSSF) and involve anaerobic conditions. Hybrid systems combine wetland types to utilize their strengths [48].

FWSF wetlands have low flow velocity, shallow water depth, and vegetation that enhance pollutant removal [47]. They remove biochemical/biological oxygen demand (BOD), COD, total suspended solids (TSS), and pathogens with efficiencies over 70%. Nutrient removal varies but is limited. FWSF wetlands are useful for restoring polluted rivers and buffering wastewater treatment plants [34].

SSF wetlands are anaerobic and categorized as HSSF or VSSF [47]. HSSF wetlands have a horizontal flow regime, with organic matter decomposition and removal mechanisms occurring through microbial processes, sedimentation, and filtration [34, 47]. VSSF wetlands have a vertical flow regime, with water percolating through a bed of gravel, sand, and macrophytes. These systems incorporate periodic flooding and draining for nitrification [34, 47].

Hybrid constructed wetlands combine different wetland types, often HSSF and VSSF, to optimize treatment [49]. Seidel's designs from the Max Planck Institute are commonly used [47]. HSSF beds serve for elimination, whereas VSSF beds function for filtration. Hybrid systems leverage the strengths of different wetland types for efficient treatment [50].

Constructed wetland systems replicate natural wetland functions and are a cost-effective option for treating AMD [34]. Performance depends on wetland type, vegetation, microbial communities, and climate. Constructed wetlands are economical alternatives to conventional chemical treatment for AMD [34].

Case Study: Constructed Wetlands

A study conducted by Naidoo (2018) [34] investigated the use of constructed wetlands in remediating acid AMD from a disused coal mine in Dundee, KwaZulu-Natal, South Africa. The Bannockburn abandoned coal mine and constructed wetland (Figure 2.1) is located at (28°09'S; 30°10'E) and approximately 306 km north of Durban. The study aimed to investigate the efficiency of a series of constructed wetlands in remediating AMD. The constructed wetland is made up of two sets of connected wetland cells. Figure 2.2 (not to scale) is a schematic design that depicts the site's key elements. Before entering the network of constructed wetland cells, water from the mine shaft travels through a decant. Mine water is led from cells 1 and 2 to cells 3 and 4 in succession. The first segment of the constructed wetland is made up of these cells. Water is transported from cell 4 of the second set of constructed wetland cells to cells 5 and 6. This is done via a subsurface pipeline.

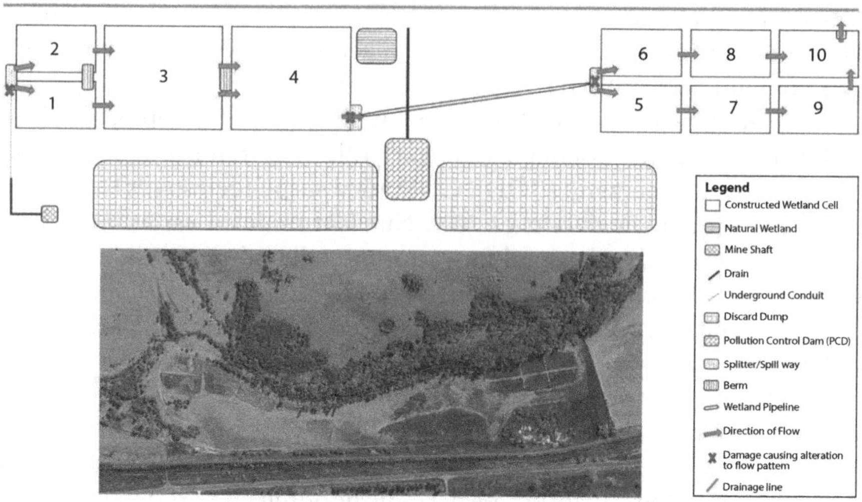

Figure 2.2 Schematic diagram: Bannockburn abandoned mine and constructed wetlands [34].

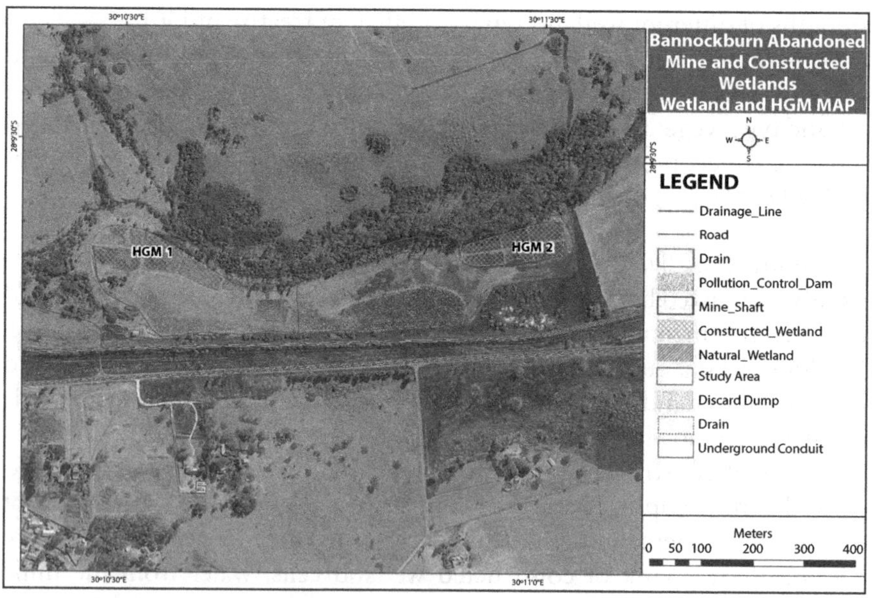

Figure 2.1 The abandoned mine, discard dumps, constructed wetlands, and natural drainage: Bannockburn abandoned mine and constructed wetlands [34].

Excess water first flows from cells 5 and 6 to cells 7 through to cell 10. Water from the latter is permitted to enter the nearby Sterkstroom River.

Given that the constructed wetland functions efficiently, it was anticipated that the flow into the river would exhibit no or acceptably low acidity. The water quality of the Bannockburn Constructed Wetland was found to be of poor quality, unfit for domestic use, and likely to negatively impact aquatic life, as per South African Water Quality Guidelines (SAWQG). The inferior water quality was not reflective of the efficacy of the wetland system in removing contaminants but instead of the contaminants emanating from the coal discard used in the base layer during the wetland construction. The elemental distribution is similar across the wetland cells, indicating that the wetland is not functioning as it should. Although the enrichment factor values were generally low, the predominance of coarse sediments suggests that contaminants are likely to be easily remobilized into the water column. The proliferation of terrestrial and alien invasive plants resulted from the inhibition of wetland conditions caused by the dysfunctional water flow regime across some wetland cells. The degradation of the system was essentially due to a lack of maintenance and monitoring for at least 5 years.

2.2.3 Bioremediation

2.2.3.1 Microbial Bioremediation

Microbial bioremediation uses living microorganisms (bacteria, algae, fungi, and yeast) to remove, concentrate, degrade, change, detoxify, or immobilize pollutants, such as chemicals and heavy metals, from contaminated soil, water, and air. The microorganisms break down pollutants into less toxic or non-toxic substances through natural metabolic processes, such as oxidation, reduction, respiration, and hydrolysis. As a result of these metabolic processes, as well as the adaptability of the microorganisms, they can be considered ideal candidates for bioremediation [36–39, 51–54]. Microbial bioremediation can be applied in various settings, including industrial sites, agricultural fields, and wastewater treatment plants and can be approached via *in situ* or *ex situ* methods [36, 53, 55, 56].

In situ microbial bioremediation involves treating the contamination at the site where it is present without excavating or removing the contaminated material. This method is typically applied when the contamination is widespread or when removing the contaminated material is impractical or expensive. *In situ* microbial bioremediation utilizes naturally occurring microorganisms or introduces specific microbial cultures to degrade or transform the contaminants into less harmful substances. It often involves the following techniques [36, 51, 57]:

- Bioattenuation (Natural Attenuation): Indigenous microbes present in the contaminated area will determine the metabolic activity and naturally degrade the contaminants.
- Bioaugmentation: Addition of specific microbial cultures to enhance the degradation process.
- Biostimulation: Stimulating the growth and activity of indigenous microorganisms by providing them with nutrients, oxygen, or other necessary factors.
- Bioventing: Supplying air or oxygen to enhance the aerobic degradation of contaminants.
- Biosparging: Injecting air or oxygen into the groundwater to promote contaminant volatilization and aerobic degradation.

Ex situ microbial bioremediation involves the removal of contaminated material from the original site and treating it in a controlled environment. The contaminated material is excavated and transported to a treatment facility, where bioremediation techniques are applied. This approach is typically used when the contamination is localized, and it may be more suitable for highly contaminated materials or specific contaminants that are difficult to treat *in situ*. *Ex situ* microbial bioremediation techniques can include the following [36, 51, 57]:

- Biopiles: Contaminated soil is piled and mixed with bulking agents and nutrients, creating an aerobic environment for microbial degradation.
- Bioreactors: Contaminated water or soil is processed in a controlled system with optimal conditions for microbial degradation.
- Composting: Organic contaminants in soil or sludge are degraded by microorganisms under controlled conditions, often combined with bulking agents and periodic turning.

Both *in situ* and *ex situ* microbial bioremediation have advantages and disadvantages, and the selection and success of the appropriate method depend on various critical factors such as the nature and extent of contamination, site characteristics, the type of microorganism(s) used, regulatory requirements, and cost considerations. More detail on some of these factors can be seen in Table 2.1 [36, 38, 51, 57].

Table 2.1 Critical factors to consider for the determination of the success of microbial bioremediation (table taken from Bala *et al.*, 2022 [38], with adaptations).

Factors	Remarks	References
Biological factors	Microorganisms competing for carbon sources can affect organic compound degradation. Contaminants and catalyst levels influence derivatization rates. Expressed enzymes can speed up or slow contaminant degradation. Enzymes must also be involved in contaminant metabolism to have an affinity for the contaminant and availability. The primary biological factors include interaction (competition, predation, and succession), population size, and composition.	[58, 59]
Oxygen	Biodegradation rates can be improved by using organisms that do not require oxygen. Anaerobic decomposition occurs as most living organisms need oxygen to survive. In most cases, hydrocarbon metabolism can be boosted by the addition of oxygen.	[60]
Nutrients	Nutrients can influence microbial growth, reproduction, biodegradation rate, and effectiveness. Optimizing the bacterial C:N:P ratio can improve biodegradation efficiency, especially when supplied with essential nutrients like N and P. Carbon, phosphorous, and nitrogen are just a few of the nutrients microorganisms need to survive. In low concentrations, biodegradation will be limited. Adding nutrients to cold environments can increase the microorganisms' metabolic activity and, thus, the biodegradation rate.	[61, 62]

(*Continued*)

Table 2.1 Critical factors to consider for the determination of the success of microbial bioremediation (table taken from Bala *et al.*, 2022 [38], with adaptations). (*Continued*)

Factors	Remarks	References
Temperature	The most important physical factor influencing microorganism survival and composition is temperature. In cold climates, natural degradation is slow, putting more pressure on microbes to clean up the contaminated area. Temperature affects the metabolic turnover of enzymes involved in degradation. Also, each compound's degradation requires a specific temperature. Temperature affects microbial physiological properties and thus speeds up or slows down bioremediation. Increased microbial activity occurs at higher temperatures.	[63, 64]
pH	A compound's acidity, alkalinity, and basicity affect microbial metabolism and the removal process. The contaminated area's pH can predict microbial growth. Even minor pH shifts have a significant impact on metabolic processes.	[65]
Site characterization and selection	Before proposing a bioremediation remedy, it is necessary to conduct adequate remedial investigation work to characterize the extent of the contamination. Site selection procedures include determining the horizontal and vertical extent of contamination, defining parameters and sampling locations, and describing sampling and analysis methods.	[66]
Metal ions	Metals are essential for bacteria and fungi, but excessive amounts inhibit cell metabolism. Metal compounds, on both a direct and indirect basis, influence degradation rates.	[67]
Microorganisms	High concentrations of some toxic compounds can harm microorganisms and slow decontamination. Toxicity varies with the toxicant, concentration, and microorganisms exposed.	[68]

Microbial bioremediation offers several advantages and disadvantages. Firstly, utilizing natural processes and organisms to degrade contaminants without harsh chemicals is environmentally acceptable. It is usually more cost-effective, particularly for large-scale contamination sites, as it requires less equipment and labor than traditional methods. Bioremediation is versatile and can be applied to a wide range of contaminants, with the ability to tailor organisms for specific pollutants. *In situ* bioremediation reduces disruption to ecosystems and minimizes risks associated with transportation. It provides a long-term solution by achieving complete degradation of contaminants and allowing organisms to adapt and continue remediation over time [36, 51, 57].

However, there are some disadvantages to consider. Bioremediation processes can be time-consuming, taking months or years to achieve complete remediation. Effectiveness is highly dependent on site-specific conditions, requiring optimization for each location. There is also uncertainty and risk associated with bioremediation outcomes, depending on the efficacy and survival of selected organisms. Bioremediation may not apply to all contaminants or extreme environmental conditions. Public perception and acceptance of bioremediation can impact decision-making due to concerns about reliability and long-term effects [36, 51, 57]. It is important to recognize that the advantages and disadvantages of bioremediation can vary depending on the specific context, site characteristics, and contaminants involved [38].

Case Study: Nitrate Bioremediation from Mining Activities

Nitrate bioremediation refers to the process of using biological agents, such as microorganisms or plants, to remove or reduce nitrates from contaminated environments. Nitrates are compounds that contain nitrogen and oxygen and are commonly found in fertilizers, agricultural runoff, explosives used in the mine industry and wastewater. Although nitrates are essential for plant growth, excessive levels can harm aquatic ecosystems and human health [69, 70].

Microbial denitrification is a complex process with multiple variables and involves the use of denitrifying bacteria to convert nitrates into nitrogen gas under anaerobic conditions. This process involves a series of biochemical reactions where denitrifying bacteria use nitrates as an electron acceptor, converting it to nitrogen gas that escapes into the atmosphere and follows the following reduction pathway: nitrate (NO_3^-) → nitrite (NO_2^-) → nitric oxide (NO) → nitrous oxide (N_2O) → nitrogen gas (N_2) [71, 72].

This case study is focused on a Diamond Mine located in the Maluti Mountains in the Southern African Kingdom of Lesotho. The mine operated two kimberlite pipes through open-cast mining, with the significant risk contaminant being nitrate (~110–120 mg/L) in their surface water due to residues of blasting with ANFO (ammonium nitrate, fuel oils) explosives. The South African National Standard (SANS) 241:2006 & 2011 for Drinking Water Class 1 (Recommended Operational levels) is 11 mg/L. Therefore, this study aimed to develop an effective microbial bioremediation strategy to reduce the nitrate to acceptable levels within the SANS 241 limits.

The treatment development consisted of various phases, starting by examining and understanding the study site and analyzing the biogeochemistry of the water in question. A microbial selection was made for the indigenous denitrifying bacteria present in the water from the study site that is capable of complete nitrate reduction, using a medium-specific for nitrate reducers and confirmed through next-generation sequencing. Laboratory columns were prepared and seeded with the enriched denitrifying bacterial population through bioaugmentation. The results obtained with the laboratory columns indicated that the proof of concept reduced the nitrate to levels within the SANS 241 allowable limits.

Results from the laboratory columns were used to design and install a site-specific pilot scale reactor. Results obtained from this pilot reactor indicated that the biological water treatment system has proven to be a viable treatment option for complete nitrate reduction (~90%–95% reduction could be maintained over time) for the water in question. It is important to note that the selection and implementation of nitrate bioremediation strategies should be carried out by experts and based on site-specific assessments to ensure effectiveness and minimize potential environmental risks.

Another study conducted by Kulshreshtha *et al.* (2010) [73] investigated the neutralization of alkaline industrial wastewater using Exiguobacterium sp. The study found that the alkaliphilic isolate Exiguobacterium sp. 12/1 could neutralize highly alkaline wastewater from beverage production when applied in a suspension of a bacterial pellet or as a lysophilized culture. The alkaliphilic bacterium was able to lower the pH of highly alkaline wastewater from a pH of 12 to 7.5 within a short span of 2 hours through the production of carboxylic acid [73]. Neutralization of highly alkaline wastewater without using an external carbon source emphasizes the potential use of Exiguobacterium sp. as an alternative to the conventional acid neutralization technique used to treat alkaline wastewater [73].

A 4-year study was conducted by Higgins *et al.* (2015) [27] as a joint venture between Aughinish Alumina and the International Aluminium Institute (IAI) to investigate the potential of a constructed wetland to treat bauxite residue leachate generated at the Aughinish Bauxite Residue Disposal Area (BRDA). The constructed wetland treated the residue leachate to a pH below 9, suitable to be discharged into the local environment. Conventional treatments for the treatment of alkaline residue leachates require frequent application and maintenance, making them labor-intensive and not economically viable for post-closure remediation [27].

2.2.3.2 Phyco- and Phytoremediation

Phycoremediation is the use of algae to treat wastes or wastewater. Algae, ranging from microscopic microalgae to larger macroalgae, can metabolize sunlight, carbon dioxide, and various nutrients. They can remove inorganic nutrients, organic compounds, pathogens, metals, and emerging contaminants. Algal cultivation can use nutrient-rich effluent from conventional treatments, reducing costs and environmental footprints. Phycoremediation is an effective treatment option for various types of waste due to algae's adaptability and tolerance. It requires minimal energy and maintenance, making it suitable for *in situ* bioremediation [42, 43, 74].

On the other hand, phytoremediation uses plants to remove, detoxify, or immobilize environmental contaminants. It involves various mechanisms such as phytoaccumulation, phytoextraction, phytostabilization, phytofiltration or rhizofiltration, rhizodegradation, phytodegradation, and phytovolatilization. Depending on the pollutants and plant species involved, these mechanisms enable plants to store, adsorb, filter, degrade, or release contaminants. Suitable plants for phytoremediation are hardy, high biomass-producing, tolerant to contaminants, and non-attractive to herbivores. Phytoremediation is cost-effective, low-maintenance, and sustainable compared to conventional treatment systems [75–81].

Both phycoremediation and phytoremediation offer environmentally friendly approaches to waste treatment, utilizing natural processes and organisms to remediate contaminants. They provide cost-effective and versatile methods for different types of pollutants and wastewater. However, the selection and optimization of specific organisms or plants, as well as consideration of site-specific conditions, are crucial for their effectiveness [43, 74–76].

2.2.4 Natural Filtration Systems

Natural filtration refers to the process where water and wastewater from industry and mining are filtered by abiotic and biotic components that exist in nature. Water and wastewater can be filtered through layers of soil, sand, rock, vegetation, and organisms found in these environments. Natural filtration relies on naturally occurring ecological and physico-chemical processes such as sedimentation, adsorption, and biodegradation to reduce or remove pollutants. These processes involve interactions between solar energy (temperature and light), wind, substrates, aquatic plants, and microorganisms [40].

Sedimentation can be defined as the partial separation of suspended solids from a liquid by settling due to gravity [41]. The sedimentation rate is dependent on the particle size, density, permeability, and the type of wastewater (industrial) [41]. Conventional wastewater treatment methods can achieve sedimentation through the use of sedimentation tanks, settling ponds/basins, and clarifiers [82]. Adsorption is defined as a mass transfer process, a phenomenon of sorption of gases or solutes by solid or liquid surfaces [44]. Physical adsorption occurs through the interaction of intermolecular forces, whereas chemical adsorption occurs through the formation and destruction of chemical bonds [44].

Biodegradation is a process where microorganisms breakdown or decay material and utilize the organic substances as a source of carbon and energy products [83]. This process can occur under aerobic conditions and anaerobic conditions. Biodegradation can be used in the environment to treat soil, water, and wastewater and preserve food and agricultural products [83].

Nature-based treatment technologies such as constructed wetlands, reed beds, and treatment ponds utilize various substrates, aquatic plants, and microorganisms to meet the remediation needs of the specific wastewater under treatment [84].

2.3 NBS for Industrial and Mining Wastewater Treatment

2.3.1 Alkaline Wastewater

NBSs provide a sustainable alternative for the remediation of alkaline wastewater. These passive treatment techniques use biological and natural processes for treatment. Some of these treatments include the use of constructed wetlands, phytoremediation, adsorption/biosorption, and extremophilic bacteria or alkaliphiles. Constructed wetlands generally consist of bedrock, soil, plants and an organic substrate. Wetland soil and vegetation aid in metal accumulation and metal precipitates remain stable in the wetland [34]. Wetland vegetation for the treatment of alkaline leachates serves many functions such as flow baffling, provision of a large surface area for inorganic precipitates for dissolution, and provision of a constant carbon source for microbial decomposers, which elevate CO_2 and thus the partial pressure of CO_2 in the wetland [85]. The effectivity of a wetland's buffering capacity depends on three factors, *viz.*, the quality of the wastewater entering the system, the flow rate of the wastewater entering the wetland, and the fluctuations in temperature and biological activity due to seasonal variation [27]. Additional CO_2 in the wetland system accelerates the precipitation of bicarbonates, thus increasing the pH [27]. Constructed wetland systems offer cost-effective and low-maintenance treatment options for the treatment of wastewater [17]. The metal adsorption capacity of biosorbents is dependent on numerous biosorbent characteristics such as surface area, pore volume, contact time, ionic strength, the temperature of the solution, pH of the solution, adsorbent dosage, and the initial metal concentration [6].

Metal adsorption is more common at acidic pH values rather than alkaline pH values due to alkali metals forming cations that are not attracted to protonated functional groups on the adsorbent surfaces [6]. Extremophilic microorganisms or alkaliphiles (bacteria growing at high pH) have noteworthy conversion potential and can degrade toxic components present in the environment under extreme conditions [73]. Their enzymatic activities enable them to gain resistance against toxic agents, assist in biodegradation, produce metabolites, and detoxify metal compounds. Biological methods using the enzymes produced by extremophilic bacteria can replace conventional treatment methods, which rely on chemical conversions [73]. The use of microorganisms or alkaliphilic bacteria has an advantage over mineral acids as neutralizing agents, in that microorganisms do not significantly increase the final volume of water to be discharged and eliminate the issues of corrosion and salt formation [73].

2.3.2 Acid Mine Drainage

NBS can provide sustainable remediation techniques for AMD. These passive treatments use natural and biological processes for treatment. Some of these include constructed wetlands, bioreactors, compost reactors, permeable reactive barriers (PRBs), and anoxic and open limestone drains and channels [86]. Constructed wetlands usually comprise bedrock, soil, plants, and an organic substrate. Wetlands can remove metals and sulfates, and metal precipitates remain stable in the wetland. These systems have high-long term performance, are a low maintenance and cost-effective treatment option, and can be used for land and habitat rehabilitation [17]. Sulfidogenic bioreactors usually comprise compost, animal manure, bark, sawdust, wood chips, spent mushroom, compost cellulose waste, or limestone. These systems can effectively remove high concentrations of metals and sulfates and allow for differential removal of metals while being cost-effective [87, 88].

Anoxic limestone drains usually comprise a sealed trench buried in the ground filled with limestone and covered with an impermeable substrate such as clay. These systems require little land area, are more cost-effective and require less maintenance compared to constructed wetlands, and are good for short-term performance [87, 88]. PRBs usually comprise wood chips, compost, bone char, compost, limestone, fly ash, dolomite, peat moss, sewage sludge, granular iron, bauxite, zerovalent iron, apatite, zeolite, ferric oxyhydroxides, or activated alumina. These systems do not require land area or waste disposal, prevent cross-contamination with surface water and excessive groundwater loss, and serve as a cost-efficient *in situ* treatment [87, 88]. Passive treatments for AMD are considered to be a low-maintenance, more cost-effective, and environmentally friendly approach compared to active treatments [17].

2.3.2.1 Sulfate-Reducing Bacteria

Sulfate-reducing bacteria (SRB) are a group of naturally occurring, chemo-organotrophic, and strict anaerobic bacteria found in many environments, including mine sites, and play a significant role in the process of AMD as they can utilize sulfate as an electron acceptor during their metabolic processes. In the case of AMD, SRB contribute to the overall process by reducing sulfate ions (SO_4^{2-}) to hydrogen sulfide (H_2S) through a series of enzymatic reactions [89–94].

The presence of SRB in AMD has both positive and negative implications. On the positive side, SRB can aid in neutralizing the acidic conditions caused by AMD. They produce hydrogen sulfide by reducing sulfate, which reacts with the acidic environment to form metal sulfides. These metal sulfides are less soluble and more stable than the original metal ions, leading to the precipitation and removal of metals from the water, thus reducing their toxicity [89, 90, 92].

However, the activities of SRB can also exacerbate certain aspects of AMD. Hydrogen sulfide produced by SRB is toxic to aquatic organisms and can contribute to the overall toxicity of the water. Additionally, SRB can further accelerate the corrosion of metal infrastructure, such as pipes and equipment, due to the production of sulfuric acid as a metabolic by-product [89].

Understanding the role of SRB in AMD is crucial for effective management strategies. Controlling the environmental conditions and implementing remediation techniques can mitigate the negative impacts of AMD [89, 91]. For example, lime or other alkaline materials can be added to neutralize the acidity and precipitate metal ions, limiting their mobility and reducing toxicity [93]. Moreover, proper waste management practices and the prevention of sulfide mineral exposure to air and water can help minimize the occurrence of AMD. In addition, organic substrates are often a limiting factor in AMD remediation when using SRB [95]. Oberholzer *et al.* (2022) [95] found that chicken feathers can be used as a substrate enhancer to boost organic carbon availability for the SRBs to sustain passive AMD treatment. This can be achieved through elevation of the pH and sulfate and metal reductions in AMD water. Even though more research is required to assess the associations between AMD water parameters, microbial diversity, and the selected substrates, Oberholzer *et al.* (2022) [95] found that chicken feathers were associated with the SRB microbial taxa, which is essential to ensure the longevity of passive AMD treatment systems.

2.3.2.2 Integrated Constructed Surface Water Wetlands and Algae Pond Systems

In a study by Oberholster *et al.* (2014) [96], the authors examine the potential of selected macroalgae species as a secondary passive AMD treatment technology in algae ponds in temperate regions during winter months when the metabolism processes in macrophytes and microorganisms in constructed wetlands decrease due to seasonal changes, e.g., shorter day length and lower water column temperature. The laboratory study exposed cultured macroalgae *K. klebsii*, *O. crassum*, and *M. tumidula* isolated from AMD field sampling sites to three different pH values. At the same time, biosorption of the metals—Al, Fe, Mn, and Zn—and glutathione-S-transferase (GST) activity was established in the different algae species at a constant water temperature of 14°C. No significant differences between the isolated macroalgae species and the three different pH ranges were observed concerning GST activity in the macroalgae. Nevertheless, differences in metal biosorption were observed at all three different pH values.

Biosorption of Al was the highest for *O. crassum*, followed by *K. klebsii* and *M. Tumidula*. No significant difference between Fe and macroalgae species was observed; however, biosorption significantly improved over the exposure time under low pH values. The study showed that the highest metal biosorption occurred in the macroalgae *O. crassum* at all three tested pH values under constant water temperature. The study showed that the macroalgae *O. crassum* can be introduced in settling ponds to act as a natural treatment mechanism. In a further study by Oberholster *et al.* (2018) [45], the authors employed selected macroalgal species under laboratory conditions to determine the bioaccumulation of sulfur (S) and other important algal growth elements such as calcium (Ca), magnesium (Mg), and phosphorus (P) from collected coal AMD water at different pH values. Following exposure of the macroalgae to AMD water for 192 hours, the macroalgae *Microspora tumidula* showed the highest bioaccumulation of S and P, which occurred at a pH value of 5. *Oedogonium crassum* showed the highest bioaccumulation of Ca at a pH of 7 and Mg at a pH of 5.

The data also showed that the accumulation efficiency of Mg by all three macroalgal species increased significantly as the pH values increased. The macroalgae species *O. crissum* showed very little increase in chl-*a* (mg/g fw) after exposure to the coal mine water after 192 hours at all three pH ranges. However, the macroalgae species *M. tumidula* show significant increases in chl-*a* (mg/g fw) at both pH values of 5 and 7. The study showed that introducing *M. tumidula* into algae ponds could play a role in passive treatment of sulfur-rich AMD. The introduction of specific macroalgae into algae ponds secondary passive AMD treatment can also play a role in the increase of alkalinity values since macroalgae utilize CO_2, sulfates, nitrates, and phosphates in the water column, to synthesize their own organic cellular material and give off free oxygen as a waste product that increases the alkalinity values within the water column.

2.3.2.3 *Algal-Bacterial Integrated Ponding System*

Rose *et al.* (1988) [97] suggested the development of an Algal Sulfate-Reducing Ponding process for Acidic and Metal Wastewater Treatment (ASPAM) utilizing the basic bioreactor design approach proposed by Oswald (1991) [98] for scale-up applications in South Africa. Wastewater ponding is a mature technology for treating large volumes, and its use as a basis for appropriate reactor design for AMD treatment includes the high rates of sulfate reduction and the precipitation of metal sulfides [97]. The integration of algal and SRB processes has been reported in the anaerobic fermentation of wastewater-grown microbial biomass produced in waste stabilization ponds as a carbon source for the treatment of tannery effluent and the linked production of sulfide [99]. This offers the opportunity of a carbon source produced during the ponding treatment of AMD, which is independent of the external carbon source supply.

2.3.2.4 Ecologically Engineered Wetlands

Although natural wetlands or wetlands with its origin from mine decant or industrial effluent are rarely optimized and used to treat AMD or alkaline effluent, they usually develop downstream from, for example, abandoned mines that decant, as in the case of the Zaalklapspruit valley bottom wetland. The Zaalklapspruit valley bottom wetland (latitude, –25.906480; longitude, 29.052827) covers an area of 135.3 ha and is heavily impacted by AMD from an abandoned underground coal mine upstream (Figure 2.3) and industrial effluent. Due to AMD upstream from the unchanneled wetland, the wetland has become a channeled valley bottom wetland over space and time. It is located in the Mpumalanga Upper Olifants River catchment (Quaternary Catchment B20G) of South Africa and forms part of a more extensive wetland system that lies along a tributary of the Zaalklapspruit stream and the Grootspruit stream. The Grootspruit stream flows into the Wilge River, approximately 35 km northwest of the Town of eMalahleni (formerly Witbank). In 2014, through ecological engineering, the Zaalklapspruit wetlands' size was increased by 9.4 ha to allow for the gravitational drainage of AMD water [100]. The enlargement of the wetland happened via the redirection of water flow using concrete structures to enlarge the surface area, resulting in improved water quality. The

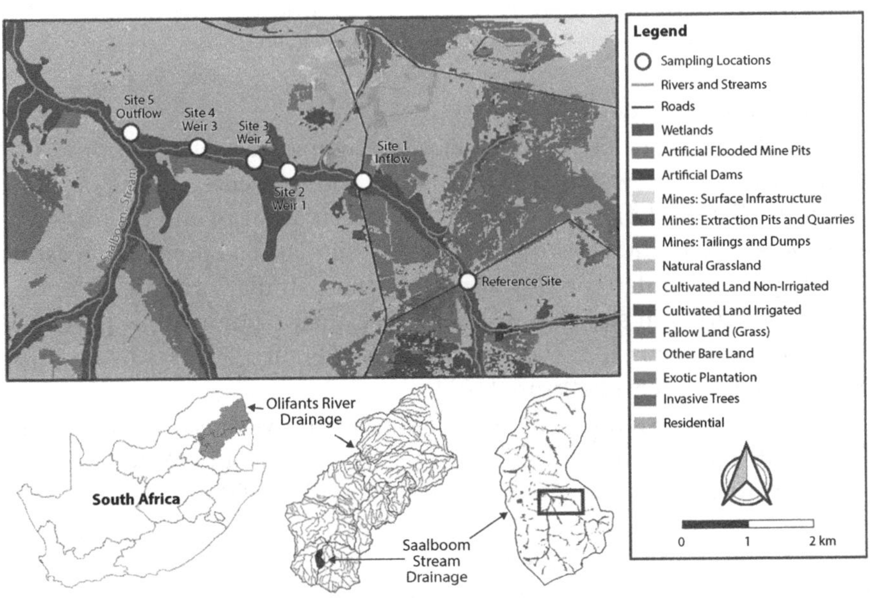

Figure 2.3 The Zaalklapspruit valley bottom wetland study area.

purpose of the concrete structures was to change the wetland's flow from a channeled to an unchanneled valley bottom wetland to increase the resident time of the surface flow and improve ecosystem services.

The wetland's main channel had four intervention points, whereas the secondary incised channel was rehabilitated at three points. These intervention points were strategically selected to deactivate the main and secondary incised channelization and reduce the water's velocity (<20 cm/s) through the wetland. At the onset of the ecologically engineered intervention, the Zaalklapspruit wetland area showed elevated metal, sulfate, and TDS concentrations with a low average pH, all indicative of an area impacted by AMD.

Over 5 years, the sites downstream from the inflow to the wetland outflow showed a significant decrease in sulfate (SO_4^{2-}), whereas alkalinity increased to a median of 128 mg L^{-1} at the outflow. The SO_4^{2-} concentration at the wetland inflow decreased from a median of 1,328 mg L^{-1} to 254 mg L^{-1} at the outflow of the wetland. Sulfate, pH, and alkalinity determined for the reference site, 2.2 km upstream of the wetland, remained relatively unchanged during the sampling period and unimpacted by mining and industrial practices or AMD. The pH, alkalinity, and silica concentrations increased, whereas the electrical conductivity (EC) decreased by 35% from a median concentration of 2,470 μS cm^{-1} at the inflow to 861 μS cm^{-1} at the outflow of the wetland. The metals primarily associated with AMD, namely, aluminium (Al) and iron (Fe), were reduced at the outflow of the wetland by respectively ~94% and ~85%.

Hydrochemical facies

Hydrochemical data of the ecologically engineered wetland from 2019 were analyzed for water type variation and presented using a Piper Diagram (Figure 2.4). This graphical representation of the chemistry of water was used to indicate the clustering of similar data points, suggesting that they have similar compositions and, thus, similar origins. Piper diagrams are helpful in showing distinct water quality populations [101] and likely drainage scenarios. The Piper diagram (Figure 2.4) was made using Aquachem version 5.1 software [102].

Figure 2.4 shows the Piper Trilinear diagram for water samples collected during 2019 at the different Zaalklapspruit wetland sites. Water types for surface water samples collected at the inflow, middle of the wetland, and the outflow of the wetland were SO_4^{2-}-type (inflow), Ca-Mg-SO_4^{2-}-type (middle wetland sites 2 and 3), Mg-SO_4^{2-}-CO_3^--HCO_3^- type (outflow), Na-CO_3^--HCO_3^--Cl^- type (NP1 piezometer), and Mg-CO_3^--HCO_3^- type (borehole adjacent to wetland site). The water types exemplified in the trilinear

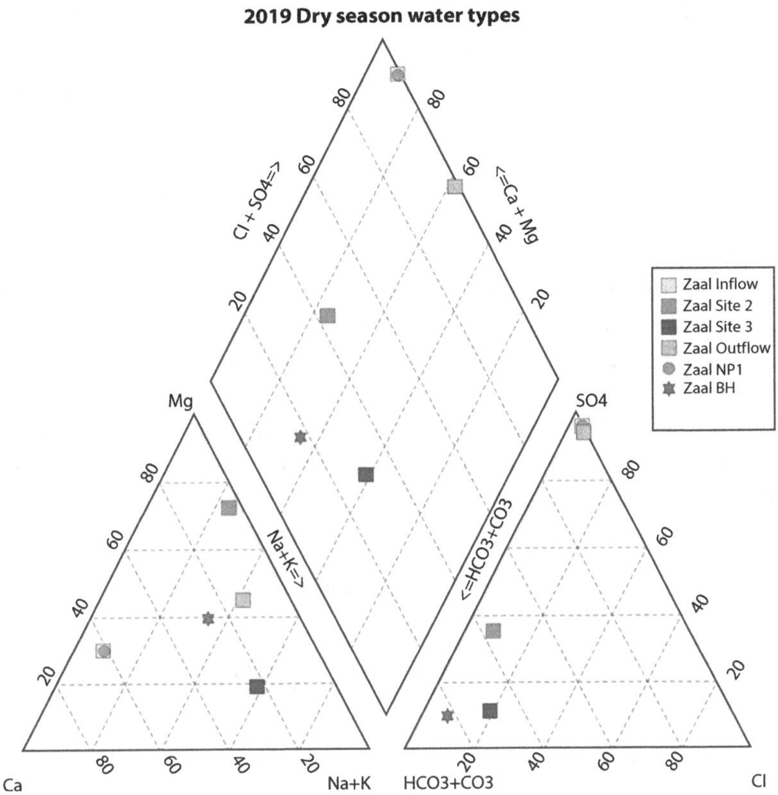

Figure 2.4 Hydrochemical of water collected from wetland surface water (square), wetland piezometer (circle), and borehole (star).

diagram (Figure 2.4) show that both the inflow site and the piezometer installed 1.5 meters below the wetland sediment at the wetland inflow were both affected by AMD. Additionally, the elevated Ca^{2+} and Mg^{2+} contents indicate that the middle section of the wetland could possibly depend on fresh groundwater discharges with increased buffering capacity (higher HCO_3^- and CO_3^-). Water rock interactions occurring in the subsurface increased the Ca^{2+} and Mg^{2+} concentrations in shallow subsurface water. Thus, its local discharge to the wetland increased the Ca^{2+} and Mg^{2+} content in the middle of the wetland site. Furthermore, the hypothesis of shallow groundwater discharging into the wetland along its flow path is given further credence by the prevalence of buffering ions HCO_3^- and CO_3^-, which arise from water that has interacted with subsurface geological material and that biological activity in the soil horizon input these two anions [103]. The similarity in groundwater type from the adjacent borehole to that of water sampled within the wetland and at its inflow suggests that locally, shallow subsurface water drains toward the wetland, providing important water resources for the wetland to withstand the desiccation during the dry period.

Stable Isotopic Variation at Zaalklapspruit Wetland During the 2019 Dry and Wet Seasons

Hydrogeological monitoring to determine the role of possible interaction occurring between groundwater and surface water was done during the 2019 dry and wet seasons. Wetland piezometers, an adjacent borehole, and wetland surface water were sampled seasonally for chemical and isotopic analysis.

During the 2019 dry season sampling trip, all water samples collected for hydrochemical analysis were used except for the limited NP1 piezometer sample. The stable isotope contents ranged between −15.2‰ to 1‰ for

Table 2.2 June 2019 dry and wet season isotope descriptive statistics.

Dry 2019	**δ^2H (‰)**	**$\delta^{18}O$ (‰)**
N	5	5
Min	−15.2	−3.3
Max	1	−0.43
Mean	−3.54	−1.226
Std Dev	6.727	1.205
Range	16.2	2.87
Wet 2019	**δ^2H (‰)**	**$\delta^{18}O$ (‰)**
N	8	8
Min	−10.1	−2.86
Max	3.8	−0.27
Mean	−5.225	−1.649
Std Dev	4.145	0.769
Range	13.9	2.59

δ^2H, and −3.3‰ to −0.43‰ for $\delta^{18}O$ (Table 2.2). The standard deviation of 6.7 for δ^2H indicates a large variation in δ^2H isotopes in the water samples relative to the arithmetic mean. These variations suggest unevenly spread δ^2H measurements. However, the slight standard deviation of the $\delta^{18}O$ isotope shows that water samples were relatively stable with respect to the oxygen isotopes.

During the 2019 wet season sampling, eight samples were collected from the wetland surface water and wetland piezometers. As shown in Table 2.2, stable isotope concentrations in all samples ranged between −10.1‰ to 3.8‰ for δ^2H and −2.86‰ to −0.27‰ for $\delta^{18}O$, respectively. Water-stable isotopes tended toward more enriched conditions, which is expected due to rapid recharge and resultant exfiltration occurring during this period due to the Piston effect [104, 105]. Figure 2.5 shows the stable isotope variation during the dry and wet seasons concerning the Global Meteoric Water Line [106]. The figure outlines the closely related water sources based on their close clustering. Based on such clustering, three groups emerge for the entire 2019 sampling period comprising i) mainly dry season wetland surface water samples with wet season inflow site; ii) wet season piezometers 1 and 2 associated with wet season wetland site 2 and dry season site 2; and iii) dry season groundwater sample with wet season piezometer 3 and wetland site 4. The clustering trend suggests a longitudinal connectivity between surface and groundwater, where samples collected at the lowest wetland sites had similarities to groundwater, whereas those collected upstream at the inflow were mainly surface water-derived samples. The preceding statement is supported by other authors, who have found a lag in groundwater discharge during the rainy season, as seen in Figure 2.5, where wet season samples approximate the dry season groundwater sampling [104].

Most of the sampled water was surface-flowing wetland water, so a more enriched signal was expected for these water sources. However, analysis values suggested minor depletion regarding ^{18}O isotope contents. As such, the depleted nature of this surface water indicates a linkage with the more depleted groundwater from the adjacent borehole. While the difference might be stark between these two water sources, it is unexpected that surface water would have any form of depletion unless it is connected with depleted water (from groundwater). In conclusion, the data show a possible interaction between groundwater and wetland surface water. Additionally, the similarity in hydrochemical and stable isotope contents implies that the downstream site 4 of the Zaalklapspruit wetland was supported by subsurface water discharges (Figure 2.5).

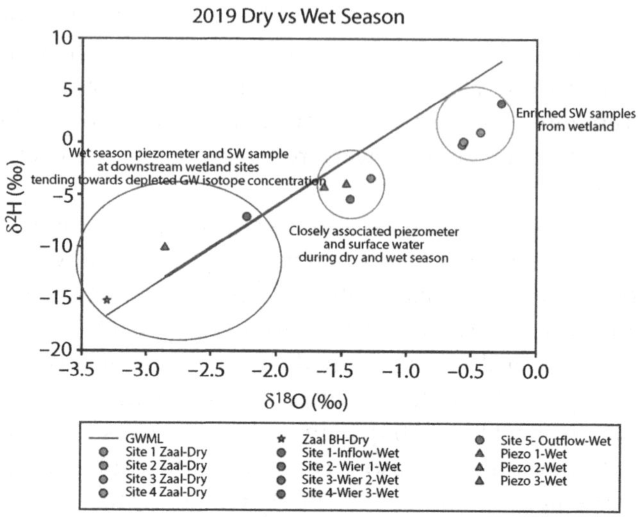

Figure 2.5 The stable isotope variation during the dry and wet seasons concerning the Global Meteoric Water Line.

2.4 Novel Hybrid NBS

2.4.1 Concept

Developing a comprehensive framework (Figure 2.6) for selecting hybrid NBS for treating alkaline and AMD wastewater is paramount. Such a framework would provide a structured and systematic approach for designing and implementing NBS, considering the unique characteristics and challenges of these types of wastewater [80, 95]. By integrating ecological engineering principles, the framework aims to select suitable NBS treatment options, integrate them as hybrid solutions, and optimize treatment systems' performance, resilience, and sustainability.

2.4.2 Framework for Selecting Hybrid NBS for Treating Alkaline and AMD Wastewater

The framework addresses key challenges of implementing NBS for alkaline and AMD wastewater treatment. It tackles decision-making challenges by promoting a multidisciplinary approach and facilitates informed decision-making based on comprehensive knowledge and expertise by considering site-specific conditions, environmental impacts, cost-effectiveness, and long-term feasibility.

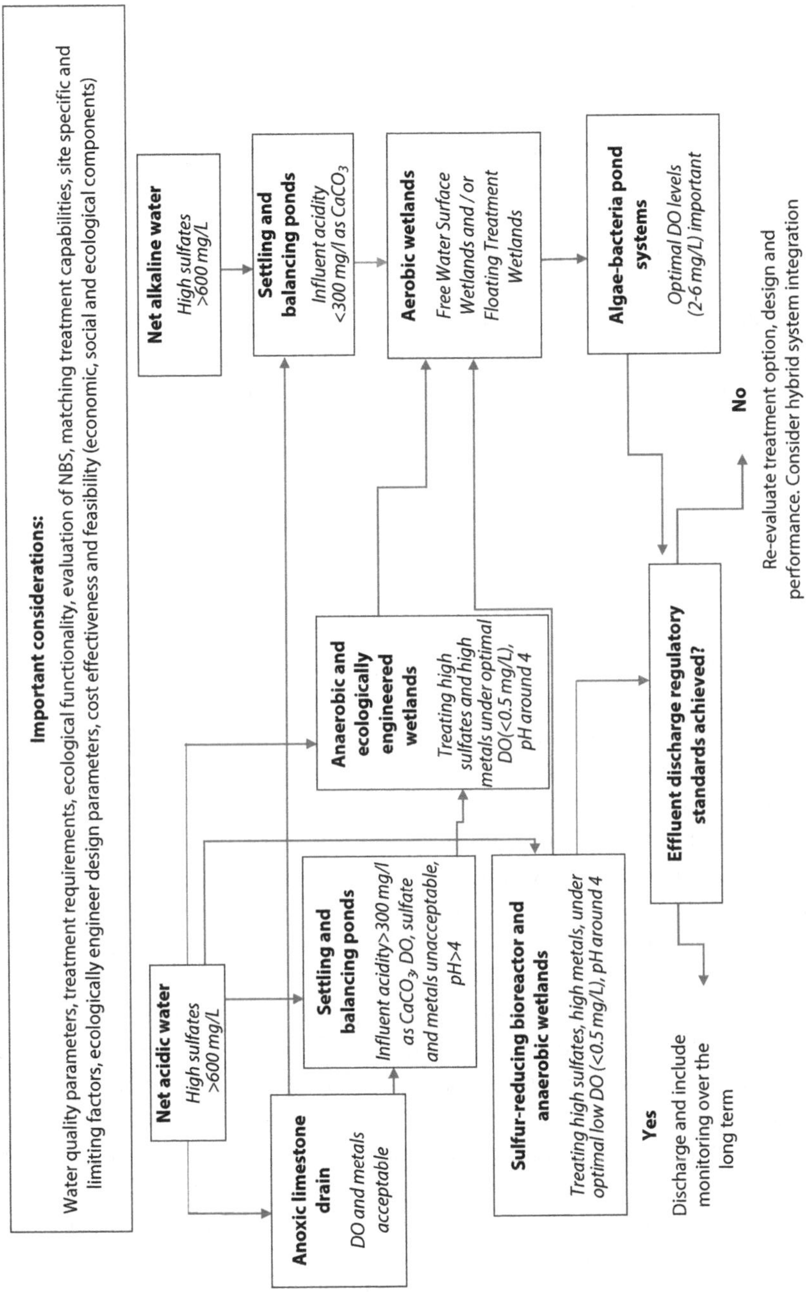

Figure 2.6 Framework for selecting hybrid NBS for treating alkaline and AMD wastewater.

When considering nature-based systems for treating alkaline and AMD, critical parameters and performance indicators must be considered to ensure effective treatment and optimization of the system.

Critical parameters include pH, metal and non-metal concentration, dissolved oxygen (DO), and EC. Also essential is pH adjustment to create suitable treatment conditions, whereas monitoring metals and non-metals helps maintain acceptable limits for effective remediation. Adequate DO levels support the growth and activity of aerobic microorganisms, and EC provides insights into water quality and treatment effectiveness. Retention time, determining the duration water stays in the system, allows for settling solids and pollutant removal. Optimizing the hydraulic loading rate ensures efficient treatment without overwhelming the system.

Performance indicators for nature-based systems in treating AMD and alkaline mine drainage include metal and non-metal removal efficiency, monitored through influent and effluent concentration measurements. Water quality parameters such as TSSs, COD, and specific conductivity provide insights into overall treatment performance.

Vegetation health indicators, including cover, diversity, and biomass, reflect system functionality and its ability to remediate mine drainage. Analyzing microbial community composition and diversity evaluates treatment performance, as certain microorganisms contribute to pollutant removal and water quality improvement. Long-term monitoring is crucial for system stability and sustainability. Continuously tracking critical parameters and performance indicators helps identify issues and ensures the effectiveness of the treatment process.

Considering these parameters and indicators facilitates the design, operation, and evaluation of nature-based systems for treating alkaline and AMD. Ongoing monitoring and optimization based on these indicators ensure efficient remediation and the long-term success of the treatment system.

Additionally, algae and bacteria offer potential biological solutions for AMD and alkaline mine water treatment. Algae, such as *Chlamydomonas reinhardtii*, *Chlorella vulgaris*, *Scenedesmus species*, *Spirulina platensis*, *Anabaena* species, and *Dunaliella salina*, demonstrate effectiveness in metal removal and adaptation to various conditions. Bacteria, including *Acidithiobacillus ferrooxidans*, *Acidithiobacillus thiooxidans*, *Leptospirillum ferrooxidans*, sulfate-reducing bacteria (SRB), and various heterotrophic bacteria, contribute to metal precipitation, acid reduction, and organic compound degradation in AMD treatment.

These algae and bacteria possess specific metabolic capabilities that enable their participation in biogeochemical processes, promoting metal removal, reducing acidity, and enhancing water quality in impacted environments.

2.4.3 Design Principles for NBS for Treating Alkaline and AMD

Ecological engineering design principles (Table 2.3), adapted from the originally developed principles [107], are crucial in developing and implementing novel hybrid NBS for sustainable industrial wastewater treatment, such as alkaline and AMD [108, 109]. These principles recognize the importance of integrating ecological processes and natural systems into engineering designs to achieve effective and sustainable solutions.

Alkaline and AMD are significant environmental challenges resulting from mining activities. Conventional treatment methods for mine drainage, such as chemical precipitation or physical separation, can be expensive, energy-intensive, and often generate secondary waste streams. In contrast, ecological engineering principles offer an innovative approach to wastewater treatment by harnessing the power of nature's self-regulating processes to reinstate ecosystem function and structure toward sustainable ecosystems. These principles emphasize using natural components and ecological interactions to achieve efficient and sustainable treatment outcomes.

By applying ecological engineering principles, various interventions can be employed in designing hybrid NBS for alkaline and AMD treatment. These interventions may include the selection of appropriate indigenous plant species, the establishment of multi-layered wetland systems, the promotion of microbial activity, and the creation of wildlife corridors. These interventions aim to enhance pollutant removal, improve ecosystem functionality, and increase overall treatment efficiency.

Key performance outcomes of these ecological engineering designs include improved water quality, increased biodiversity, enhanced ecosystem resilience, and reduced environmental impacts. Monitoring plays a critical role in assessing the effectiveness of these designs and ensuring that desired outcomes are achieved. Monitoring components may include regular assessment of water quality parameters, faunal and vegetation monitoring, and analysis of key ecological indicators. Designing hybrid NBS for alkaline and AMD treatment requires careful consideration of key design components. These components encompass factors such as the selection of appropriate plant species, the incorporation of diverse habitat structures, the provision of suitable microbial habitats, and the establishment of functional compartments within wetland systems. Integrating adaptive

Table 2.3 Design principle options for NBS treating alkaline and AMD.

Ecological engineering principle	Examples of ecological engineering options or interventions relevant to AMD and alkaline wastewater	Key performance outcomes	Key monitoring components	Key design components
Principle 1: Forcing Functions and Ecosystem Adaptation	Implement engineered wetland systems that control the inflow of mine/industrial drainage, water flow rates, and nutrient/contaminant levels to optimize treatment efficiency	Improved water quality, enhanced nutrient/contaminant removal, and efficient drainage treatment	Inflow and outflow monitoring and water quality monitoring	Control structures, engineered channels, and adjustable flow regulators
Principle 2: Energy Limitations and Sustainable Solutions	Acknowledge the energy limitations within the treatment system and implement sustainable solutions to address them. Integrate energy-efficient technologies such as solar-powered aeration systems to enhance oxygen levels within the treatment wetland, promoting biological activity for efficient nutrient cycling. This approach reduces the ecological footprint of the treatment system while achieving improved water quality, enhanced nutrient cycling, and sustainable ecosystem productivity.	Increased oxygen levels, enhanced nutrient cycling, improved ecosystem productivity, and reduced energy consumption within the treatment system	Aeration efficiency monitoring, dissolved oxygen levels, and sustainability assessments	Integration of solar-powered aeration systems, energy-efficient technologies, and sustainable solutions to overcome energy limitations while improving water treatment

(*Continued*)

Table 2.3 Design principle options for NBS treating alkaline and AMD. (*Continued*)

Ecological engineering principle	Examples of ecological engineering options or interventions relevant to AMD and alkaline wastewater	Key performance outcomes	Key monitoring components	Key design components
Principle 3: Ecosystem Connectivity and Buffer Zones	Establish buffer zones or transition zones between the wetland system and adjacent areas such as agricultural areas to filter and reduce the influx of pollutants	Reduced pollutant loading, improved water quality, and protection of wetland ecosystems	Buffer zone monitoring and water quality monitoring	Vegetation buffers, sediment traps, and constructed wetlands
Principle 4: Limiting Factors and Strategic Regeneration	Identify and strategically address the limiting factors that hinder ecosystem recovery within areas impacted by AMD and alkaline wastewater. Implement regeneration strategies tailored to these specific limiting factors, which may include addressing soil conditions, water chemistry, or habitat structure. For instance, utilize engineered wetlands that offer pH adjustment mechanisms to overcome pH-related limitations in treating AMD. This strategic regeneration approach aims to optimize treatment effectiveness, improve ecosystem health, and restore ecological balance.	Optimal treatment of limiting factors, improved metal precipitation and removal, enhanced ecosystem health, and successful ecological regeneration within the treatment area	Limiting factor assessments, pH monitoring, and ecological regeneration progress tracking	Integration of pH adjustment mechanisms, tailored strategies for addressing limiting factors, and habitat restoration techniques that enhance regeneration

(*Continued*)

Table 2.3 Design principle options for NBS treating alkaline and AMD. (*Continued*)

Ecological engineering principle	**Examples of ecological engineering options or interventions relevant to AMD and alkaline wastewater**	**Key performance outcomes**	**Key monitoring components**	**Key design components**
Principle 5: Homeostasis and Stability Enhancement	Enhance the buffer capacity of the wetland system by incorporating suitable microbial consortia, plants with high nutrient absorption capabilities, such as macrophytes or floating wetland islands	Increased nutrient uptake, improved water quality, and enhanced ecosystem stability	Plant growth monitoring, microbial functioning, and nutrient uptake monitoring	Planting of nutrient-absorbing species, microbial consortia, and floating wetland mats
Principle 6: Nutrient Cycling and Effective Recirculation	Design the treatment system to maximize nutrient cycling and efficient recirculation of water within the ecosystem. Employ natural processes and engineered components that facilitate the uptake and recycling of nutrients, such as nitrogen and phosphorus, to support the growth of native vegetation. Implement recirculation systems that minimize water loss and optimize the use of treated water within the ecosystem. This approach aims to improve water quality, enhance ecosystem stability, and reduce the environmental footprint associated with water treatment.	Enhanced nutrient cycling, improved water quality, efficient water use, and ecosystem stability through effective nutrient recirculation	Nutrient/contaminant uptake assessments, water quality monitoring, and recirculation system performance evaluations	Selection of nutrient-absorbing plant species, engineering of nutrient recirculation systems, and the integration of natural nutrient cycling processes into the treatment system

(*Continued*)

Table 2.3 Design principle options for NBS treating alkaline and AMD. (*Continued*)

Ecological engineering principle	**Examples of ecological engineering options or interventions relevant to AMD and alkaline wastewater**	**Key performance outcomes**	**Key monitoring components**	**Key design components**
Principle 7: Pulsing Systems and Timing Efficiency	Time wastewater discharge or management activities to align with natural pulses, such as seasonal variations, to maximize natural purification processes	Enhanced natural purification, improved treatment efficiency, and optimized ecosystem response	Flow monitoring, tide monitoring, and seasonal variation monitoring	Scheduling of wastewater discharge, installation of flow regulators, and utilization of tidal or seasonal cycles
Principle 8: Self-Designing Systems and Low-Maintenance Solutions	Understand the functioning of the system as a bio-intelligent system. Favor the use of indigenous plant and microbial species and natural colonization processes to allow the wetland system to self-design and establish a self-sustaining ecological community	Enhanced biodiversity, improved ecosystem resilience, and reduced maintenance requirements	Plant diversity monitoring and colonization monitoring	Selection of indigenous plant and microbial species, facilitation of natural colonization processes, and minimal intervention approach

(*Continued*)

Table 2.3 Design principle options for NBS treating alkaline and AMD. (*Continued*)

Ecological engineering principle	**Examples of ecological engineering options or interventions relevant to AMD and alkaline wastewater**	**Key performance outcomes**	**Key monitoring components**	**Key design components**
Principle 9: Effective Use of Landscape Patterns	Design and construct multi-layered wetland systems with varying hydrological conditions as guided by landscape patterns to promote emergent properties and complex ecological interactions	Increased habitat diversity, improved ecological functionality, and enhanced ecosystem resilience	Hydrological monitoring and habitat diversity monitoring	Design of multi-layered wetland systems, incorporation of varied hydrological conditions, and provision of diverse habitat structures
Principle 10: Diversity for Resilience	Establish wildlife corridors or migration pathways to promote connectivity between wetland systems and enable the movement of flora and fauna	Enhanced species dispersal, improved genetic exchange, and increased biodiversity	Faunal monitoring and vegetation monitoring	Creation of wildlife corridors, provision of migration pathways, and connectivity planning

(*Continued*)

Table 2.3 Design principle options for NBS treating alkaline and AMD. (*Continued*)

Ecological engineering principle	Examples of ecological engineering options or interventions relevant to AMD and alkaline wastewater	Key performance outcomes	Key monitoring components	Key design components
Principle 11: Smooth Transitions and Ecotones	Design nature-based treatment systems that include transitional zones or ecotones between the impacted areas and the treatment wetlands. These ecotones serve as gradual transition areas where natural processes can help alleviate the negative impacts of AMD or alkaline wastewater before they reach the main treatment system. Utilize native vegetation and natural geomorphic features to create ecotones that filter and buffer pollutants, stabilize the ecosystem, and facilitate the smooth transition of water from the impacted area to the treatment wetland.	Improved pollutant filtration and reduction, enhanced ecological stability in transitional zones, and smoother, and more effective treatment processes within the wetland	Ecotone health assessments, water quality monitoring in transition areas, and pollutant concentration monitoring	Strategic placement of transitional ecotones, selection of native vegetation, and consideration of geomorphic features to promote pollutant buffering and ecological stability
Principle 12: Synergistic Networks for Effective Solutions	Design wetland systems with distinct integrated and bio-intelligent zones or compartments, each performing specialized functions to maximize treatment efficiency and ecosystem functioning	Optimized treatment processes, enhanced pollutant removal, and improved ecosystem performance	Zonation monitoring and functional compartment monitoring	Design of distinct zones or compartments, incorporation of specialized functions, and understanding of hierarchical relationships within the wetland system

(*Continued*)

Table 2.3 Design principle options for NBS treating alkaline and AMD. (*Continued*)

Ecological engineering principle	Examples of ecological engineering options or interventions relevant to AMD and alkaline wastewater	Key performance outcomes	Key monitoring components	Key design components
Principle 13: Regional and Global Considerations	Establish partnerships and collaborations with regional and global environmental organizations, governmental agencies, and research institutions to share knowledge, data, and best practices for addressing AMD and alkaline wastewater. Engage in cross-border initiatives to address transboundary water pollution issues and promote a regional or global perspective in the design and implementation of NBS. Share findings and innovations with international networks working on similar environmental challenges	Enhanced knowledge exchange, improved regional and global cooperation, increased efficiency in addressing transboundary water pollution, and the development of standardized approaches to address AMD and alkaline wastewater on a larger scale	Collaborative research and data-sharing agreements, progress assessments, and feedback from regional and global partners	Networking and partnerships with regional and global organizations, data-sharing platforms, and transboundary pollution mitigation strategies

(*Continued*)

Table 2.3 Design principle options for NBS treating alkaline and AMD. (*Continued*)

Ecological engineering principle	**Examples of ecological engineering options or interventions relevant to AMD and alkaline wastewater**	**Key performance outcomes**	**Key monitoring components**	**Key design components**
Principle 14: Historical and Indigenous Knowledge Context	Incorporate traditional ecological knowledge from indigenous communities in the design and management of constructed wetlands. Consult with local communities to identify i.e. indigenous plant species and traditional practices that can aid in the restoration and treatment of wastewater-impacted areas	Enhanced ecological and cultural restoration, improved water quality, strengthened relationships with local communities, and increased biodiversity	Community engagement and feedback, ecological health assessments, and traditional knowledge sharing	Inclusion of indigenous plant species recommended by indigenous knowledge, cultural preservation, and collaborative ecosystem restoration strategies
Principle 15: Species Geography and Vulnerability	Conduct a thorough assessment of the local species' geographical distribution and vulnerability to AMD or alkaline wastewater. Identify key indicator species that are both resilient to the specific water quality issues and have a positive impact on ecosystem restoration. Focus on promoting the growth and protection of these indicator species, taking into account their geographical distribution and vulnerability to enhance biodiversity and ecosystem stability in the treatment area	Increased biodiversity, improved ecological functionality, and enhanced ecosystem stability through the protection and promotion of indicator species	Species population monitoring, ecological health assessments, and tracking of indicator species' response to treatment efforts	Habitat creation and protection measures tailored to indicator species, species-specific conservation strategies, and habitat restoration techniques

(*Continued*)

Table 2.3 Design principle options for NBS treating alkaline and AMD. (*Continued*)

Ecological engineering principle	Examples of ecological engineering options or interventions relevant to AMD and alkaline wastewater	Key performance outcomes	Key monitoring components	Key design components
Principle 16: Holistic Approaches and Hierarchical Organization	Implement a hierarchical organization of treatment elements within the wetland system, starting with larger-scale processes and gradually transitioning to finer-scale processes. For instance, design the wetland to include zones with different flow rates and treatment capacities, each serving a specific function in the overall treatment process. This hierarchical organization ensures that the entire ecosystem works cohesively to treat AMD and alkaline wastewater. The design should consider the holistic integration of various treatment components to achieve comprehensive and efficient results.	Optimized treatment processes, enhanced pollutant removal, improved ecosystem performance, and a cohesive and structured approach to AMD and alkaline wastewater treatment	Zonation performance assessments, water quality monitoring across hierarchical levels, and data analysis to ensure the efficiency of each treatment zone	Designing distinct zones within the wetland, incorporating specialized functions, and understanding the hierarchical relationships within the treatment system to create a holistic and organized approach

(*Continued*)

Table 2.3 Design principle options for NBS treating alkaline and AMD. (*Continued*)

Ecological engineering principle	Examples of ecological engineering options or interventions relevant to AMD and alkaline wastewater	Key performance outcomes	Key monitoring components	Key design components
Principle 17: Interactive Processes for Desired Outcomes	Implement a design that encourages interactive processes among diverse organisms within the treatment wetland. For instance, create conditions that facilitate beneficial ecological interactions, such as the symbiotic relationships between certain plant species and microbial communities. The design should foster cooperative processes, where different components of the ecosystem work together to achieve the desired outcomes, such as pollutant degradation and nutrient cycling. This approach harnesses the power of nature's interactive processes for effective AMD and alkaline wastewater treatment.	Improved pollutant degradation, enhanced nutrient cycling, and efficient treatment outcomes through the facilitation of beneficial ecological interactions	Ecological health assessments, microbial community monitoring, and the tracking of interactive processes that contribute to treatment effectiveness	Creation of conditions that promote beneficial ecological interactions, selection of plant species that support microbial communities, and the facilitation of cooperative processes within the treatment wetland

(*Continued*)

Table 2.3 Design principle options for NBS treating alkaline and AMD. (*Continued*)

Ecological engineering principle	**Examples of ecological engineering options or interventions relevant to AMD and alkaline wastewater**	**Key performance outcomes**	**Key monitoring components**	**Key design components**
Principle 18: Emergent Properties From Interactions	Design the treatment system to promote the emergence of unique and valuable properties resulting from interactions between different components of the ecosystem. For instance, create conditions that encourage the synergistic interactions between specific plant species, microorganisms, and abiotic factors within the wetland. The design should focus on fostering these interactions to achieve emergent properties, such as enhanced nutrient cycling, efficient pollutant removal, and improved ecosystem resilience. By harnessing emergent properties, the treatment system can achieve more effective results in addressing AMD and alkaline wastewater.	Enhanced nutrient cycling, improved pollutant removal, and increased ecosystem resilience through the emergence of valuable properties resulting from ecological interactions	Assessment of emergent properties, ecosystem functionality monitoring, and tracking of interactions that contribute to the system's effectiveness	Design elements that encourage synergistic interactions, selection of plant species that facilitate emergent properties, and the creation of conditions for emergent outcomes within the treatment wetland

(*Continued*)

Table 2.3 Design principle options for NBS treating alkaline and AMD. (*Continued*)

Ecological engineering principle	Examples of ecological engineering options or interventions relevant to AMD and alkaline wastewater	Key performance outcomes	Key monitoring components	Key design components
Principle 19: Information Storage of Systems	Leverage the inherent capacity of ecological systems to store information about past environmental conditions and treatment system performance. Design the treatment ecosystem to incorporate natural indicators, such as specific plant species or microbial communities, that retain ecological memory of past water quality, pollutant levels, and ecosystem health. These ecological information storage mechanisms provide a historical record within the ecosystem, enabling adaptive management and decision-making based on the stored ecological data.	Enhanced ecological memory, data-informed adaptive management, and improved responsiveness to changing environmental conditions within the treatment system	Ecological memory assessments, observation of ecological indicators, and tracking of changes in stored ecological data	Selection of indicator species with information storage capabilities, promotion of ecological memory zones, and the integration of ecological data storage within the treatment ecosystem
Principle 20: Regenerative Design and Open Systems	Implement a regenerative landscape design that promotes the natural recovery and enhancement of ecosystem functions in areas affected by AMD and alkaline wastewater. Create open and interconnected systems that allow for the flow of water and ecological interactions, facilitating the restoration of affected ecosystems. This approach focuses on the rejuvenation of damaged areas and aims to restore ecosystem health, improve water quality, and support sustainable land reclamation.	Enhanced ecosystem recovery, improved water quality, and sustainable land reclamation through regenerative design and open systems	Ecosystem health indicators, water quality monitoring, and tracking of land reclamation progress	Regenerative landscape design elements, strategic planting of native species, and ecological restoration techniques that promote open and interconnected systems

(*Continued*)

Table 2.3 Design principle options for NBS treating alkaline and AMD. (*Continued*)

Ecological engineering principle	**Examples of ecological engineering options or interventions relevant to AMD and alkaline wastewater**	**Key performance outcomes**	**Key monitoring components**	**Key design components**
Principle 21: Planetary Health Nexus	Develop a comprehensive ecological engineering strategy that recognizes the interconnectedness of water, climate, food, rewilding, and land within the context of treating AMD and alkaline wastewater. Design the treatment system to not only address water quality but also consider its broader impacts on climate, food production, rewilding efforts, and land use. Ensure that the treatment system contributes to a positive planetary health nexus by restoring ecosystems, promoting climate resilience, supporting sustainable food production, and rewilding efforts while reclaiming and improving the quality of affected land	Improved planetary health, enhanced ecosystem restoration, climate resilience, sustainable food production, rewilding success, and land reclamation within the treatment system	Comprehensive ecosystem health assessments, climate impact evaluations, food production assessments, rewilding progress tracking, and land reclamation monitoring	Holistic ecological engineering that integrates water treatment with broader ecological and planetary health considerations, sustainable food production systems, and rewilding initiatives

(*Continued*)

Table 2.3 Design principle options for NBS treating alkaline and AMD. (*Continued*)

Ecological engineering principle	**Examples of ecological engineering options or interventions relevant to AMD and alkaline wastewater**	**Key performance outcomes**	**Key monitoring components**	**Key design components**
Principle 22: Technological Synergy With Natural Systems	Foster a harmonious synergy between advanced technologies (including artificial intelligence) and natural systems in the treatment of AMD and alkaline wastewater. Utilize innovative sensor networks, real-time data analytics, and automation technologies to monitor and optimize the performance of the treatment system. These technologies should complement the natural processes, ensuring efficient pollutant removal, water quality improvement, and ecosystem restoration. The integration of technology enhances the effectiveness of ecological engineering, promoting sustainable solutions in addressing environmental challenges.	Improved treatment performance, data-informed decision-making, and enhanced ecological and environmental outcomes through the harmonious integration of technology and natural systems	Advanced sensor networks, real-time data analysis, and technology-nature interaction assessments	Integration of technology for data collection and analysis, automation systems for adaptive management, and technology-driven optimization of the treatment system
Principle 23: Anticipatory Design for Future Systems	Design treatment systems with adaptability and future challenges in mind, considering potential shifts in water quality and environmental conditions	Enhanced system resilience, adaptive management, and future-proofing	Scenario modeling and adaptive management frameworks	Flexible system designs, predictive modeling, and adaptive management strategies

management strategies and continuous monitoring also contribute to the success of these designs, allowing for timely responses and improvements based on changing conditions.

The ecological engineering design principles offer a promising approach for the sustainable treatment of alkaline and AMD. By integrating ecological processes and natural systems, these principles enable the development of hybrid NBS that are cost-effective, energy-efficient, and environmentally acceptable. They emphasize the importance of working with nature to achieve desired treatment outcomes while promoting the regeneration and conservation of ecosystems.

2.4.4 Importance of Addressing Sustainable Development Goals (SDGs) and Contributing to the Global Sustainability Agenda

Industrial wastewater, particularly alkaline and AMD, poses significant environmental and public health challenges. Addressing these challenges requires innovative and sustainable solutions that treat wastewater effectively and align with the broader goals of global sustainability. This is where the novel hybrid NBS and the novel planetary health nexus integrating water-climate-food-rewilding-land components come into play.

The proposed NBS framework (Table 4.4) integrates the planetary health nexus, offering a comprehensive approach to determine, quantify, optimally design through ecological engineering principles, and demonstrate the significance and impact of treatment options for AMD and alkaline water systems as part of greater functioning landscapes. By considering the interconnectedness of these components, the framework ensures that the treatment solution simultaneously tackles multiple dimensions of sustainability on a broader landscape level. This comprehensive approach allows decision-makers, researchers, and practitioners to design, develop and implement solutions that effectively address the challenges posed by industrial and AMD wastewater while aligning with the broader global sustainability goals. The framework ensures a holistic evaluation of the treatment process by encompassing multiple dimensions of sustainability, including water resources, energy efficiency, food production, biodiversity conservation, and land management. The significance of this framework lies in its ability to cover all of the United Nations' SDGs. The SDGs provide a global blueprint for achieving sustainability in various aspects of human life, and addressing industrial wastewater treatment aligns with several SDGs. By showcasing the application of the framework to optimize a natural wetland

Table 2.4 Novel hybrid NBS framework as an integrated water-climate-food-rewilding-land nexus (as the planetary health nexus) and the SDG framework.

SDG	SDG number	Water nexus	Climate nexus	Food nexus	Rewilding nexus	Land nexus	Actions	Accounting framework component
No Poverty	SDG 1	Water availability	Energy access	Agricultural practices	Regeneration of natural habitats	Sustainable land management	Assess the potential socio-economic benefits of the AMD or alkaline treatment project, considering the water, climate, food, biodiversity, and land nexus	Cost-Benefit Analysis and Social Impact Assessment
Zero Hunger	SDG 2	Irrigation	Renewable energy	Sustainable farming	Biodiversity conservation	Land productivity	Explore the potential for utilizing treated water from the wetland system in sustainable agriculture practices, considering the interconnectedness of water, climate, food, biodiversity, and land	Water Footprint Assessment, Agricultural Regenerative Assessment, and Life Cycle Assessment

(Continued)

Table 2.4 Novel hybrid NBS framework as an integrated water-climate-food-rewilding-land nexus (as the planetary health nexus) and the SDG framework. (*Continued*)

SDG	SDG number	Water nexus	Climate nexus	Food nexus	Rewilding nexus	Land nexus	Actions	Accounting framework component
Good Health and Well-being	SDG 3	Access to clean water	Health and well-being benefits	Nutritious food	Ecosystem regeneration	Land use planning	Assess the potential health benefits of improved water quality resulting from AMD or alkaline treatment for the well-being of local communities, considering the implications for water, climate, food, biodiversity, and land	Health Impact Assessment and Water Quality Monitoring
Quality Education	SDG 4	Water for sanitation	Renewable energy	Local school feeding programs	Ecological connectivity	Land conservation	Engage in educational programs and awareness campaigns to promote understanding of the interconnections between water, climate, food, biodiversity, and land in the context of AMD or alkaline treatment and wetland regeneration	Education Outcome Measurement and Awareness Campaign Evaluation

(*Continued*)

Table 2.4 Novel hybrid NBS framework as an integrated water-climate-food-rewilding-land nexus (as the planetary health nexus) and the SDG framework. (*Continued*)

SDG	SDG number	Water nexus	Climate nexus	Food nexus	Rewilding nexus	Land nexus	Actions	Accounting framework component
Gender Equality	SDG 5	Water for hygiene	Gender-inclusive energy	Women in agriculture	Ecosystem preservation	Land ownership	Ensure equal participation and representation of women in the AMD or alkaline treatment project and related activities, considering the water, climate, food, biodiversity, and land nexus	Gender Impact Assessment and Women Empowerment Indicators
Clean Water and Sanitation	SDG 6	Water management	Water treatment	Water for agriculture	Wetland conservation	Watershed management	Treat and remediate AMD or alkaline water to improve water quality, ensure clean and safe water resources for human and ecological systems, and consider the implications for climate, food, biodiversity, and land	Water Quality Monitoring and Ecosystem Services Valuation

(*Continued*)

Table 2.4 Novel hybrid NBS framework as an integrated water-climate-food-rewilding-land nexus (as the planetary health nexus) and the SDG framework. (*Continued*)

SDG	SDG number	Water nexus	Climate nexus	Food nexus	Rewilding nexus	Land nexus	Actions	Accounting framework component
Affordable and Clean Energy	SDG 7	Water for energy	Renewable energy	Energy-efficient farming	Ecological regeneration	Sustainable energy sources	Explore opportunities for renewable energy generation within the wetland system, considering the water, climate, food, biodiversity, and land nexus	Energy Efficiency Assessment and Renewable Energy Potential Analysis
Decent Work and Economic Growth	SDG 8	Water for industries	Green energy sources	Sustainable livelihoods	Reintroduction of species	Sustainable land use	Assess the economic opportunities and benefits associated with AMD or alkaline treatment, wetland regeneration, and sustainable land-use practices, considering the interconnectedness of water, climate, food, biodiversity, and land	Economic Impact Assessment and Job Creation Monitoring

(*Continued*)

Table 2.4 Novel hybrid NBS framework as an integrated water-climate-food-rewilding-land nexus (as the planetary health nexus) and the SDG framework. (*Continued*)

SDG	SDG number	Water nexus	Climate nexus	Food nexus	Rewilding nexus	Land nexus	Actions	Accounting framework component
Industry, Innovation, and Infrastructure	SDG 9	Water for manufacturing and reuse	Clean energy sources	Efficient food systems	Ecosystem resilience	Land development	Foster innovation and explore sustainable technologies for effective AMD or alkaline treatment, wetland regeneration, and land management, considering the water, climate, food, biodiversity, and land nexus	Technological Innovation Assessment and Infrastructure Impact Analysis
Reduced Inequalities	SDG 10	Equitable water access	Energy equity	Fair distribution of food	Conservation of biodiversity	Land tenure	Ensure equitable distribution of the benefits of the AMD or alkaline treatment project and wetland regeneration, considering the water, climate, food, biodiversity, and land nexus	Equity Analysis and Social Inclusion Metrics

(*Continued*)

Table 2.4 Novel hybrid NBS framework as an integrated water-climate-food-rewilding-land nexus (as the planetary health nexus) and the SDG framework. (*Continued*)

SDG	SDG number	Water nexus	Climate nexus	Food nexus	Rewilding nexus	Land nexus	Actions	Accounting framework component
Sustainable Cities and Communities	SDG 11	Urban or rural water management	Sustainable energy	Local food production	Urban or rural greening	Land-use efficiency	Integrate wetland regeneration and sustainable land-use practices into urban planning and development strategies, considering the water, climate, food, biodiversity, and land nexus	Urban Resilience Assessment and Sustainable Urban Development Indicators
Responsible Consumption and Production	SDG 12	Water footprint	Sustainable production	Circular food systems	Habitat connectivity	Sustainable land practices	Implement sustainable practices in AMD or alkaline treatment, wetland regeneration, and land management, considering the interconnectedness of water, climate, food, biodiversity, and land	Environmental Impact Assessment and Resource Efficiency Metrics

(*Continued*)

Table 2.4 Novel hybrid NBS framework as an integrated water-climate-food-rewilding-land nexus (as the planetary health nexus) and the SDG framework. (*Continued*)

SDG	SDG number	Water nexus	Climate nexus	Food nexus	Rewilding nexus	Land nexus	Actions	Accounting framework component
Climate Action	SDG 13	Water-related disasters	Renewable energy use	Climate-smart agriculture	Carbon sequestration	Land adaptation	Assess the wetland's role in climate change mitigation through carbon sequestration, greenhouse gas emissions reduction, and adaptation measures, considering the water, climate, food, biodiversity, and land nexus	Climate Change Impact Assessment and Carbon Accounting
Life Below Water	SDG 14	Aquatic ecosystem health	Aquatic energy sources	Sustainable fisheries	Aquatic biodiversity	Riparian and Terrestrial Interchange management	Restore and protect aquatic ecosystems within the wetland system, promoting biodiversity conservation and sustainable water management, considering the water, climate, food, biodiversity, and land nexus	Ecological Risk Assessment and Marine Biodiversity Indicators

(*Continued*)

Table 2.4 Novel hybrid NBS framework as an integrated water-climate-food-rewilding-land nexus (as the planetary health nexus) and the SDG framework. (*Continued*)

SDG	SDG number	Water nexus	Climate nexus	Food nexus	Rewilding nexus	Land nexus	Actions	Accounting framework component
Life on Land	SDG 15	Terrestrial ecosystem health	Renewable land-based energy	Sustainable land use	Wildlife conservation	Land regeneration	Restore and protect terrestrial ecosystems within the wetland system, promoting biodiversity conservation, sustainable land management, and the interrelationships between water, climate, food, biodiversity, and land	Biodiversity Impact Assessment and Habitat Regeneration Monitoring
Peace, Justice, and Strong Institutions	SDG 16	Water governance	Clean energy governance	Access to food	Conservation of natural resources	Land rights	Ensure transparent and inclusive governance structures for the AMD or alkaline treatment project and wetland regeneration, promoting peace, justice, and strong institutions and considering the water, climate, food, biodiversity, and land nexus	Governance Performance Metrics and Social Justice Indicators

(*Continued*)

Table 2.4 Novel hybrid NBS framework as an integrated water-climate-food-rewilding-land nexus (as the planetary health nexus) and the SDG framework. (*Continued*)

SDG	SDG number	Water nexus	Climate nexus	Food nexus	Rewilding nexus	Land nexus	Actions	Accounting framework component
Partnerships for the Goals	SDG 17	Collaborative water management	Joint energy initiatives	Food security partnerships	Ecosystem regeneration partnerships	Land management partnerships	Establish partnerships and collaborations between stakeholders from various sectors, including government, academia, civil society, and the private sector, to jointly address the water, climate, food, biodiversity, and land nexus in the context of AMD or alkaline treatment and wetland regeneration	Partnership Assessment and Multi-stakeholder Collaboration Evaluation

system through ecological engineering principles for AMD and alkaline water treatment, it serves as a pioneering example of how novel NBS can contribute to the global sustainability agenda.

One of the key benefits of the framework is its ability to measure and evaluate the performance of the treatment system. By incorporating accounting frameworks within the matrix, the impact of actions taken can be quantified, providing valuable insights into the effectiveness of the treatment process and can be used to calculate the return on investment and even the value of regenerated natural capital. This enables decision-makers to make informed choices and optimize the treatment system based on measurable outcomes.

Moreover, the framework demonstrates its significance by aligning with all of the UN's SDGs. By covering the SDGs, the framework showcases how treating AMD and alkaline water can address various societal and environmental challenges, contributing to the broader sustainability agenda. This highlights the potential of the framework to generate positive impacts on social well-being, economic growth, and environmental conservation. Another advantage of the framework is its ability to facilitate holistic evaluation. By accounting for social, economic, and environmental factors, decision-makers can assess the broader benefits and potential trade-offs associated with different treatment options. This enables them to make decisions considering the overall well-being of communities and ecosystems, promoting sustainable outcomes. Additionally, the framework offers scalability and replicability. It can be adapted to different geographical contexts and applied to various industrial wastewater treatment scenarios. This allows for sharing knowledge and successful practices, enabling the widespread application of sustainable solutions for AMD and alkaline water treatment.

The integration of the novel planetary health nexus within the proposed framework provides decision-makers, researchers, and practitioners with a powerful tool to optimize natural wetland systems for treating AMD and alkaline water. By considering multiple sustainability dimensions, measuring performance, aligning with the SDGs, enabling holistic evaluation, and offering scalability and replicability, the framework contributes to developing and implementing sustainable solutions. It effectively addresses the challenges posed by industrial and mining wastewater while supporting the global sustainability agenda.

The proposed index presented in Table 2.5 supports the development and application of the SDG framework (Table 2.5) by providing a structured and comprehensive approach to addressing the challenges of AMD and alkaline water treatment. It covers a wide range of key indicators, including

Table 2.5 Index supporting the development of the novel hybrid NBS as an integrated water-climate-food-rewilding-land nexus (as the planetary health nexus) and the SDG framework.

Objective/indicator	Criteria	Options/interventions
Water quality	pH (6.5–8.5)	AMD and alkaline wastewater treatment through optimized and hybrid wetland system
Dissolved oxygen (>5 mg/L)	AMD and alkaline treatment through optimized hybrid NBS systems	
Metal concentrations (comply with regulatory standards)		
Climate resilience	Stable water availability	Vegetation management and rewilding efforts in wetland system
Minimal temperature fluctuations	Vegetation management and rewilding efforts in wetland system	
Food production	Crop yield (20% increase)	Sustainable agricultural practices in the surrounding areas

(Continued)

Table 2.5 Index supporting the development of the novel hybrid NBS as an integrated water-climate-food-rewilding-land nexus (as the planetary health nexus) and the SDG framework. (*Continued*)

Objective/indicator	Criteria	Options/interventions
Nutrient content (30% improvement)	Sustainable agricultural practices in the surrounding areas	
Irrigation efficiency (25% increase)	Sustainable agricultural practices in the surrounding areas	
Biodiversity	Species richness (20% increase)	Vegetation management, rewilding efforts, and reinstating ecological functioning and structure in the wetland system
Habitat quality improvement	Vegetation management and rewilding efforts in wetland system	
Ecological connectivity enhancement		
Land Management	Sustainable land use practices adoption (50% increase)	Land use planning and regulation measures
Soil erosion reduction (30% decrease)	Land use planning, regulation measures	
Minimized land degradation		

(*Continued*)

Table 2.5 Index supporting the development of the novel hybrid NBS as an integrated water-climate-food-rewilding-land nexus (as the planetary health nexus) and the SDG framework. (*Continued*)

Objective/indicator	**Criteria**	**Options/interventions**
Energy	Renewable energy share	Renewable energy installations for wetland system and associated infrastructure
Energy efficiency improvement	Renewable energy installations for wetland system and associated infrastructure	
Economic and infrastructure	Job creation	Job creation and training programs for wetland ecosystem management
Sustainable infrastructure development	Sustainable infrastructure development for wetland treatment systems and visitor facilities	

(*Continued*)

Table 2.5 Index supporting the development of the novel hybrid NBS as an integrated water-climate-food-rewilding-land nexus (as the planetary health nexus) and the SDG framework. (*Continued*)

Objective/indicator	Criteria	Options/interventions
Sustainable consumption and production	Waste management practices	Waste management initiatives within wetland ecosystem and surrounding communities
Sustainable product usage	Waste management initiatives within wetland ecosystem and surrounding communities	
Community and Inclusivity	Community engagement	Community engagement and capacity-building for inclusive and sustainable wetland management
Social and traditional knowledge inclusivity	Community engagement and capacity-building for inclusive and sustainable wetland management	

(*Continued*)

Table 2.5 Index supporting the development of the novel hybrid NBS as an integrated water-climate-food-rewilding-land nexus (as the planetary health nexus) and the SDG framework. (*Continued*)

Objective/indicator	**Criteria**	**Options/interventions**
Data availability	Water quality	Regular sampling and analysis from monitoring stations
Climate data	Access to historical climate data and ongoing monitoring	
Agricultural practices	Farm surveys, agricultural extension data, and government records	
Biodiversity assessments	Field surveys, ecological studies, and biodiversity databases	
Land use mapping	Land use mapping records, satellite imagery, and geographic information systems	
Energy consumption	Monitoring energy consumption and renewable energy generation data	

(*Continued*)

Table 2.5 Index supporting the development of the novel hybrid NBS as an integrated water-climate-food-rewilding-land nexus (as the planetary health nexus) and the SDG framework. (*Continued*)

Objective/indicator	Criteria	Options/interventions
Employment and infrastructure	Employment records, infrastructure development plans, and project monitoring	
Community engagement	Stakeholder consultations, community surveys, and participatory processes	
Establish baselines	Water quality	Baseline assessment using historical data and field surveys
Climate conditions	Baseline assessment using historical data and field surveys	
Agricultural productivity		
Biodiversity levels		
Land use practices		
Energy consumption		
Employment		
Infrastructure		
Community dynamics		

(*Continued*)

Table 2.5 Index supporting the development of the novel hybrid NBS as an integrated water-climate-food-rewilding-land nexus (as the planetary health nexus) and the SDG framework. (*Continued*)

Objective/indicator	**Criteria**	**Options/interventions**
Stakeholder engagement	Stakeholder engagement and vulnerability index	Engage stakeholders from local communities, government agencies, environmental organizations, academia, industry representatives, wetland ecology experts, water management experts, agriculture experts, energy experts, infrastructure experts, land-use planning experts, and community development experts
Decision-making analysis and selection	Decision criteria	Analyze integrated assessment, stakeholder inputs, and decision-making criteria to select optimized wetland system for AMD treatment and alkaline water

water quality, climate resilience, food production, biodiversity, land management, energy, economic and infrastructure development, sustainable consumption and production, and community inclusivity, ensuring that multiple aspects of sustainability are considered. The specified decision criteria provide clear benchmarks to assess the effectiveness and feasibility of different options and interventions, guiding the decision-making process and aligning strategies with desired outcomes.

The identified options and interventions offer specific measures to address objectives and criteria, such as implementing optimized wetland systems for AMD treatment, sustainable agricultural practices, and land use planning. The inclusion of data availability recognizes the importance of reliable information for decision-making, enabling evidence-based assessments and progress monitoring. Baseline assessments based on historical data and field surveys provide a starting point to measure progress and evaluate the impact of interventions, while stakeholder engagement fosters collaboration and participatory decision-making. The decision analysis process enhances transparency and accountability by evaluating assessments, stakeholder inputs, and decision criteria to identify the most appropriate options. The framework emphasizes regular implementation and effectiveness monitoring, allowing for adaptive management based on emerging challenges, changing SDG priorities, and evolving conditions.

By integrating these elements, the index supports the framework's overall development, support and application by providing a structured approach to address the complex challenges of AMD and alkaline water treatment. It ensures a holistic consideration of multiple sustainability dimensions and enables evidence-based decision-making and continuous improvement toward achieving the SDGs.

2.5 Conclusion

In conclusion, the challenges associated with implementing NBS are highly relevant to treating AMD and alkaline water. However, a framework for selecting hybrid NBS can effectively address these challenges. By integrating the novel planetary health nexus approach, which considers the interconnectedness of water, climate, food, rewilding, and land, a holistic understanding of the complex interactions involved in treating AMD and alkaline water can be achieved.

The framework tackles decision-making challenges by adopting a multidisciplinary approach and involving experts from various fields. It provides a structured process for considering site-specific conditions, environmental impacts, cost-effectiveness, and long-term feasibility. Collaboration among stakeholders facilitates informed decision-making based on comprehensive knowledge and expertise.

The framework also addresses uncertainties related to treatment system selection and configuration. By considering the specific characteristics of AMD and alkaline water sites, as well as pollutant types and treatment objectives, the most suitable NBS approaches can be selected. Adaptive management, continuous monitoring, and evaluation enable adjustments based on the performance and effectiveness of the treatment systems.

The framework also contributes to the quantification and demonstration of sustainability impacts. Considering the water-climate-food-rewilding-land nexus recognizes the interconnected nature of various elements and allows for a comprehensive assessment of ecological, economic, and social impacts. Robust assessment methodologies and metrics can be developed to quantify and demonstrate the significance of the treatment, garnering support and investment.

Furthermore, the proposed framework aligns with the UNs' SDGs as it covers multiple dimensions of sustainability. By integrating various SDGs, such as clean water and sanitation, climate action, zero hunger, life on land, and sustainable cities and communities, the framework contributes to the global sustainability agenda.

Future research directions should focus on enhancing our understanding of the performance and effectiveness of hybrid NBS for treating AMD and alkaline water. This includes optimizing the selection and configuration of treatment systems, refining monitoring and evaluation methodologies, and assessing these approaches' long-term sustainability and cost-effectiveness. Exploring innovative technologies within the framework, such as nanomaterials or bioelectrochemical systems, holds promise for further advancements.

In conclusion, the framework for selecting hybrid NBS for treating AMD and alkaline water effectively addresses the challenges associated with NBS implementation. Integrating the planetary health nexus approach, quantifying and demonstrating the significance of the treatment, aligning with the SDGs, and applying ecological engineering principles provides a comprehensive and sustainable solution for AMD, alkaline water, and other water-related issues. Further research will contribute to refining and advancing this framework, enhancing its effectiveness and applicability in treating these environmental challenges.

References

1. Belle, G., Schoeman, Y., Oberholster, P., Potential Toxic-Element Pollution in Surface Water and Its Implications for Aquatic and Human Health: Source–Pathway–Receptor Model. *Water, 15*, 17, 3100, **2023**.
2. Xu, D.-M., Zhan, C.-L., Liu, H.-X., Lin, H.-Z., A critical review on environmental implications, recycling strategies, and ecological remediation for mine tailings. *Environ. Sci. Pollut. Res., 26*, 35657–35669, **2019**.
3. Chausson, A., Turner, B., Seddon, D., Chabaneix, N., Girardin, C.A., Kapos, V., Key, I., Roe, D., Smith, A., Woroniecki, S., Mapping the effectiveness of nature-based solutions for climate change adaptation. *Global Change Biol., 26*, 11, 6134–6155, **2020**.
4. Howard, P., Stevens, T., Jenner, N., Rubio, J., Maddox, T., Smale, R., Newland, J., *Forest-Smart Mining: Guidance to Applying Nature-Based Solutions in the Large-Scale Mining Sector*, World Bank Group, Washington, D.C., **2022**.
5. Baccella, S., Cerichelli, G., Chiarini, M., Ercole, C., Fantauzzi, E., Lepidi, A., Toro, L., Veglio, F., Biological treatment of alkaline industrial waste waters. *Process Biochem., 35*, 6, 595–602, **2000**.
6. Gomes, H.I., Mayes, W.M., Rogerson, M., Stewart, D.I., Burke, I.T., Alkaline residues and the environment: a review of impacts, management practices and opportunities. *J. Clean. Prod., 112*, 3571–3582, **2016**.
7. Remeteiová, D., Ružičková, S.; Mičková, V.; Heželová, M.; Pikna, Ľ., Treatment of strong alkaline wastewater from neutral leaching of EAF dust by precipitation and ion exchange. *Appl. Water Sci., 13*, 6, 1–9, **2023**.
8. Gao, W., Zhang, J., Zhang, W., Sun, D., Guo, J., Zhao, S., Zeng, Y., Liu, X., Hydrochemical characteristics and driving factors of travertine deposition in Huanglong, Sichuan, SW, China. *Water Sci. Technol., 87*, 5, 1232–1249, **2023**.
9. Bogush, A., Stegemann, J.A., Wood, I., Roy, A., Element composition and mineralogical characterisation of air pollution control residue from UK energy-from-waste facilities. *Waste Manage. (Oxford), 36*, 119–129, **2015**.
10. Bozorg-Haddad, O., *Economical, Political, and Social Issues in Water Resources*, Elsevier, 2021.
11. Freese, K., Miller, R., J Cutright, T., Senko, J., Review of chromite ore processing residue (COPR): past practices, environmental impact and potential remediation methods. *Curr. Environ. Eng., 1*, 2, 82–90, **2014**.
12. Liang, X., Lu, Y., Li, Z., Yang, C., Niu, C., Su, X., Bentonite/carbon composite as highly recyclable adsorbents for alkaline wastewater treatment and organic dye removal. *Microporous Mesoporous Mater., 241*, 107–114, **2017**.
13. Piatak, N.M., Parsons, M.B., Seal II, R.R., Characteristics and environmental aspects of slag: A review. *Appl. Geochem., 57*, 236–266, **2015**.
14. Tank, S.K. and Chippa, R., Analysis of water quality of Halena block in Bharatpur area. *Int. J. Sci. Res. Publ., 3*, 3, 57–62, **2013**.
15. Yao, Z.T., Ji, X., Sarker, P., Tang, J., Ge, L., Xia, M., Xi, Y., A comprehensive review on the applications of coal fly ash. *Earth Sci. Rev., 141*, 105–121, **2015**.

16. Jacobs, J.A., Lehr, J.H., Testa, S.M., *Acid mine drainage, rock drainage, and acid sulfate soils: causes, assessment, prediction, prevention, and remediation*, John Wiley & Sons, 2014,
17. Naidoo, N., *The remediation of acid mine drainage in the East Rand Mining Basin of the Witwatersrand Goldfields, Johannesburg, Gauteng, South Africa: Characterisation of pre- and post-treatment surface and subsurface impacts*, Durban; University of KwaZulu-Natal, South Africa, 2022.
18. Favas, P., Sarkar, S., Rakshit, D., Venkatachalam, P., Prasad, M., Acid Mine Drainages From Abandoned Mines: Hydrochemistry, Environmental Impact, Resource Recovery, and Prevention of Pollution., in: *Environmental materials and waste*, pp. 413–462, Academic Press, Cambridge, MA, USA, 2016.
19. Nordstrom, D.K., Blowes, D.W., Ptacek, C.J., Hydrogeochemistry and microbiology of mine drainage: An update. *Appl. Geochem.*, *57*, 3–16, **2015**.
20. Magowo, W., The utilisation of cellulosic biomass in the treatment of acid mine drainage and the subsequent production of fermentable sugars for bioprocessing, Master's thesis, University of the Witwatersrand, Johannesburg, South Africa, 2015.
21. Pereira, L.B., Vicentini, R., Ottoboni, L.M., Changes in the bacterial community of soil from a neutral mine drainage channel. *PLoS One*, *9*, 5, e96605, **2014**.
22. Bwapwa, J., Jaiyeola, A., Chetty, R., Bioremediation of acid mine drainage using algae strains: A review. *S. Afr. J. Chem. Eng.*, *24*, 62–70, **2017**.
23. Thisani, S.K., Kallon, D.V.V., Byrne, P., Geochemical classification of global mine water drainage. *Sustainability*, *12*, 24, 10244, **2020**.
24. Singh, B.J., Chakraborty, A., Sehgal, R., A systematic review of industrial wastewater management: Evaluating challenges and enablers. *J. Environ. Manage.*, *348*, 119230, **2023**.
25. Masindi, V., Foteinis, S., Renforth, P., Ndiritu, J., Maree, J., Tekere, M., Chatzisymeon, E., Challenges and avenues for acid mine drainage treatment, beneficiation, and valorisation in circular economy: A review. *Ecol. Eng.*, *183*, 106740, **2022**.
26. Oberholster, P.J., Schoeman, Y., Botha, A.-M., Assessing the treatment capacity of an ecological engineered wetland (Zaalklapspruit) receiving AMD over a period of 9 years. Closed Cycles and the Circular Society 2023. *The power of ecological engineering. International Conference of the International Ecological Engineering Society., Chania, Greece, 1 to 5 October, 2023*, Chania, Greece, 2023.
27. Higgins, D., Courtney, R., Curtin, T., Clune, L., Use of constructed wetlands for treating BRDA leachate and run-off, in: *Bauxite Residue Valorisation and Best Practices, Leuven, Belgium*, pp. 53–60, **2015**.
28. Simate, G.S. and Ndlovu, S., Acid mine drainage: Challenges and opportunities. *J. Environ. Chem. Eng.*, *2*, 3, 1785–1803, **2014**.

29. Akcil, A. and Koldas, S., Acid Mine Drainage (AMD): causes, treatment and case studies. *J. Clean. Prod.*, *14*, 12-13, 1139–1145, **2006**.
30. Rezaie, B. and Anderson, A., Sustainable resolutions for environmental threat of the acid mine drainage. *Sci. Total Environ.*, *717*, 137211, **2020**.
31. Chen, Z., Cuervo, D.P., Müller, J.A., Wiessner, A., Köser, H., Vymazal, J., Kästner, M., Kuschk, P., Hydroponic root mats for wastewater treatment—a review. *Environ. Sci. Pollut. Res.*, *23*, 15911–15928, **2016**.
32. Kefeni, K.K., Msagati, T.A., Mamba, B.B., Acid mine drainage: Prevention, treatment options, and resource recovery: A review. *J. Cleaner Prod.*, *151*, 475–493, **2017**.
33. Souliotis, I. and Voulvoulis, N., Operationalising nature-based solutions for the design of water management interventions. *Nat.-Based Solutions*, *2*, 100015, **2022**.
34. Naidoo, N., *The ecological and functional assessment of wetlands: case study of a constructed wetland*, MSc thesis, Durban: KwaZulu-Natal, South Africa, 2018.
35. Scholz, M. and Lee, B., Constructed wetlands: a review. *Int. J. Environ. Stud.*, *62*, 4, 421–447, **2005**.
36. Abatenh, E., Gizaw, B., Tsegaye, Z., Wassie, M., The role of microorganisms in bioremediation-A review. *Open J. Environ. Biol.*, *2*, 1, 038–046, **2017**.
37. Abo-Alkasem, M.I., Hassan, N.H., Abo Elsoud, M.M., Microbial bioremediation as a tool for the removal of heavy metals. *Bull. Natl. Res. Cent.*, *47*, 1, 1–15, **2023**.
38. Bala, S., Garg, D., Thirumalesh, B.V., Sharma, M., Sridhar, K., Inbaraj, B.S., Tripathi, M., Recent strategies for bioremediation of emerging pollutants: a review for a green and sustainable environment. *Toxics*, *10*, 8, 484, **2022**.
39. Ojha, N., Karn, R., Abbas, S., Bhugra, S., Bioremediation of industrial wastewater: A review, in: *IOP Conference Series: Earth and Environmental Science*, IOP Publishing, p. 012012, 2021.
40. Zacharia, A., Outwater, A.H., Van Deun, R., Natural Wastewater Treatment Systems for Prevention and Control of Soil-Transmitted Helminths, in: *Water Quality-Science, Assessments and Policy*, IntechOpen, London, United Kingdom, 2020.
41. Zueva, S.B., Current legislation and methods of treatment of wastewater coming from waste electrical and electronic equipment processing, in: *Waste Electrical and Electronic Equipment Recycling*, pp. 213–240, Woodhead Publishing, Cambridge, United Kingdom, 2018.
42. Balzano, S., Sardo, A., Blasio, M., Chahine, T.B., Dell'Anno, F., Sansone, C., Brunet, C., Microalgal metallothioneins and phytochelatins and their potential use in bioremediation. *Front. Microbiol.*, *11*, 517, **2020**.
43. Gupta, S.K., Sriwastav, A., Ansari, F.A., Nasr, M., Nema, A.K., Phycoremediation: an eco-friendly algal technology for bioremediation and bioenergy production, in: *Phytoremediation potential of bioenergy plants*, pp. 431–456, **2017**.

44. Hu, H. and Xu, K., Physicochemical technologies for HRPs and risk control, in: *High-risk pollutants in wastewater*, pp. 169–207, Elsevier, Amsterdam, Netherlands, 2020.
45. Oberholster, P.J., Cheng, P.-H., Botha, A.-M., Hobbs, P., Hill, L., Assessment of selected macroalgae for use in a biological hybrid system for treating sulphur in acid mine drainage (AMD). *J. Appl. Phycol.*, *30*, 1361–1370, **2018**.
46. Oberholster, P.J., Schoeman, Y., Truter, J.C., Botha, A.-M., Using periphyton assemblage and water quality variables to assess the ecological recovery of an ecologically engineered wetland affected by acid mine drainage after a dry spell. *Processes*, *10*, 5, 877, **2022**.
47. Vymazal, J., Constructed wetlands for treatment of industrial wastewaters: A review. *Ecol. Eng.*, *73*, 724–751, **2014**.
48. Vymazal, J., Horizontal sub-surface flow and hybrid constructed wetlands systems for wastewater treatment. *Ecol. Eng.*, *25*, 5, 478–490, **2005**.
49. Vymazal, J., The use of hybrid constructed wetlands for wastewater treatment with special attention to nitrogen removal: a review of a recent development. *Water Res.*, *47*, 14, 4795–4811, **2013**.
50. Maucieri, C., Barbera, A.C., Vymazal, J., Borin, M., A review on the main affecting factors of greenhouse gases emission in constructed wetlands. *Agric. For. Meteorol.*, *236*, 175–193, **2017**.
51. Adams, G.O., Fufeyin, P.T., Okoro, S.E., Ehinomen, I., Bioremediation, biostimulation and bioaugmention: a review. *Int. J. Environ. Biorem. Biodegrad.*, *3*, 1, 28–39, **2015**.
52. Kaur, S. and Sarao, L., Microbial Bioremediation, in: *New Vistas In Microbial Sciences-1 Publisher: Integrated Publications*, pp. 47–66, **2022**.
53. Kensa, V.M., Bioremediation-an overview. *I Control Pollut.*, *27*, 2, 161–168, **2011**.
54. Ojuederie, O.B. and Babalola, O.O., Microbial and plant-assisted bioremediation of heavy metal polluted environments: a review. *Int. J. Environ. Res. Public Health*, *14*, 12, 1504, **2017**.
55. Chandran, H., Meena, M., Sharma, K., Microbial biodiversity and bioremediation assessment through omics approaches. *Front. Environ. Chem.*, *1*, 570326, **2020**.
56. Das, S. and Dash, H.R., Microbial Bioremediation: A Potential Tool for Restoration of Contaminated Areas, in: *Microbial biodegradation and bioremediation*, pp. 1–21, Elsevier, Amsterdam, Netherlands, 2014.
57. Praveen, R. and Nagalakshmi, R., Review on bioremediation and phytoremediation techniques of heavy metals in contaminated soil from dump site. *Mater. Today Proc.*, 68, Part 5, 1562–1567, **2022**.
58. Alegbeleye, O.O., Opeolu, B.O., Jackson, V.A., Polycyclic aromatic hydrocarbons: a critical review of environmental occurrence and bioremediation. *Environ. Manage.*, *60*, 758–783, **2017**.

59. Sodhi, K.K., Kumar, M., Singh, D.K., Insight into the amoxicillin resistance, ecotoxicity, and remediation strategies. *J. Water Process Eng.*, *39*, 101858, **2021**.
60. Mazumder, M.A.R., Jubayer, M.F., Ranganathan, T.V., Biodegradation of plastics by microorganisms, in: *Biotechnology for Zero Waste: Emerging Waste Management Techniques*, pp. 123–141, **2022**.
61. Kebede, G., Tafese, T., Abda, E.M., Kamaraj, M., Assefa, F., Factors influencing the bacterial bioremediation of hydrocarbon contaminants in the soil: mechanisms and impacts. *J. Chem.*, *2021*, 1–17, **2021**.
62. Mupambwa, H.A. and Mnkeni, P.N.S., Optimizing the vermicomposting of organic wastes amended with inorganic materials for production of nutrient-rich organic fertilizers: a review. *Environ. Sci. Pollut. Res.*, *25*, 10577–10595, **2018**.
63. Ren, X., Zeng, G., Tang, L., Wang, J., Wan, J., Wang, J., Deng, Y., Liu, Y., Peng, B., The potential impact on the biodegradation of organic pollutants from composting technology for soil remediation. *Waste Manage. (Oxford)*, *72*, 138–149, **2018**.
64. Sharma, P., Singh, S.P., Parakh, S.K., Tong, Y.W., Health hazards of hexavalent chromium (Cr (VI)) and its microbial reduction. *Bioengineered*, *13*, 3, 4923–4938, **2022**.
65. Rajkumar, R. and Kurinjimalar, C., Microbes and plant mineral nutrition, in: *Microbiological Activity for Soil and Plant Health Management*, pp. 111–132, **2021**.
66. Sangwan, S. and Dukare, A., Microbe-mediated bioremediation: an eco-friendly sustainable approach for environmental clean-up, in: *Advances in Soil Microbiology: Recent Trends and Future Prospects: Volume 1: Soil-Microbe Interaction*, pp. 145–163, **2018**.
67. Padhan, D., Rout, P.P., Kundu, R., Adhikary, S., Padhi, P.P., Bioremediation of heavy metals and other toxic substances by microorganisms, in: *Soil bioremediation: an approach towards sustainable technology*, pp. 285–329, **2021**.
68. Boopathy, R., Factors limiting bioremediation technologies. *Bioresour. Technol.*, *74*, 1, 63–67, **2000**.
69. Al-Rajhi, S.H., Al-Gelawi, M.H., Aziz, G.M., Purification and characterization of nitrate reductase (NAR) from Pseudomonas sp. SH7 isolate. *J. Biotechnol. Res. Center*, *4*, 2, 23–36, **2010**.
70. Ghiglione, J.-F., Gourbiere, F., Potier, P., Philippot, L., Lensi, R., Role of respiratory nitrate reductase in ability of Pseudomonas fluorescens YT101 to colonize the rhizosphere of maize. *Appl. Environ. Microbiol.*, *66*, 9, 4012–4016, **2000**.
71. Alvarez, L., Bricio, C., Blesa, A., Hidalgo, A., Berenguer, J., Transferable denitrification capability of Thermus thermophilus. *Appl. Environ. Microbiol.*, *80*, 1, 19–28, **2014**.
72. Hoshino, T., Terahara, T., Tsuneda, S., Hirata, A., Inamori, Y., Molecular analysis of microbial population transition associated with the start of

denitrification in a wastewater treatment process. *J. Appl. Microbiol.*, *99*, 5, 1165–1175, **2005**.
73. Kulshreshtha, N.M., Kumar, A., Dhall, P., Gupta, S., Bisht, G., Pasha, S., Singh, V., Kumar, R., Neutralization of alkaline industrial wastewaters using Exiguobacterium sp. *Int. Biodeterior. Biodegrad.*, *64*, 3, 191–196, **2010**.
74. Phang, S.-M., Chu, W.-L., Rabiei, R., Phycoremediation, in: *The algae world*, pp. 357–389, **2015**.
75. Ansari, A.A., Naeem, M., Gill, S.S., AlZuaibr, F.M., Phytoremediation of contaminated waters: An eco-friendly technology based on aquatic macrophytes application. *Egypt. J. Aquat. Res.*, *46*, 4, 371–376, **2020**.
76. Farraji, H., Robinson, B., Mohajeri, P., Abedi, T., Phytoremediation: Green technology for improving aquatic and terrestrial environments. *Nippon. J. Environ. Sci.*, *1*, 1–30, **2020**.
77. Jesus, J.M., Danko, A.S., Fiúza, A., Borges, M.-T., Phytoremediation of salt-affected soils: a review of processes, applicability, and the impact of climate change. *Environ. Sci. Pollut. Res.*, *22*, 6511–6525, **2015**.
78. Kafle, A., Timilsina, A., Gautam, A., Adhikari, K., Bhattarai, A., Aryal, N., Phytoremediation: Mechanisms, plant selection and enhancement by natural and synthetic agents. *Environ. Adv.*, 12, 8, 100203, **2022**.
79. Khan, A.U., Khan, A.N., Waris, A., Ilyas, M., Zamel, D., Phytoremediation of pollutants from wastewater: A concise review. *Open Life Sci.*, *17*, 1, 488–496, **2022**.
80. Nedjimi, B., Phytoremediation: a sustainable environmental technology for heavy metals decontamination. *SN Appl. Sci.*, *3*, 3, 286, **2021**.
81. Yan, A., Wang, Y., Tan, S.N., Mohd Yusof, M.L., Ghosh, S., Chen, Z., Phytoremediation: a promising approach for revegetation of heavy metal-polluted land. *Front. Plant Sci.*, *11*, 359, **2020**.
82. Costa, J.A.V. and de Morais, M.G., Chapter 1. An Open Pond System for Microalgal Cultivation, in: *Biofuels from algae*, pp. 1–22, Elsevier, 2014.
83. Poznyak, T., Oria, J.I.C., Poznyak, A.S., *Ozonation and biodegradation in environmental engineering: Dynamic neural network approach*, Elsevier, Amsterdam: Netherlands, 2018.
84. Rozkošný, M., Kriška, M., Šálek, J., Bodík, I., Istenič, D., *Natural technologies of wastewater treatment*, Europe: Global Water Partnership Central and Eastern Europe, Bratislava, Slovakia, **2014**.
85. Higgins, D., Curtin, T., Pawlett, M., Courtney, R., The potential for constructed wetlands to treat alkaline bauxite-residue leachate: Phragmites australis growth. *Environ. Sci. Pollut. Res.*, *23*, 24305–24315, **2016**.
86. Naidu, G., Ryu, S., Thiruvenkatachari, R., Choi, Y., Jeong, S., Vigneswaran, S., A critical review on remediation, reuse, and resource recovery from acid mine drainage. *Environ. Pollut.*, *247*, 1110–1124, **2019**.
87. Dama-Fakir, P., Sithole, Z., Van Niekerk, A., Dateling, J., Maree, J., Rukuni, T., Mtombeni, T., Ruto, S., Zikalala, N., Hughes, C., *Mine water treatment*

technology selection tool: Users' Guide, Water Research Commission, Pretoria, **2017**.

88. Moodley, I., Sheridan, C., Kappelmeyer, U., Akcil, A., Environmentally sustainable acid mine drainage remediation: Research developments with a focus on waste/by-products. *Miner. Eng.*, *126*, 207–220, **2018**.
89. Ayangbenro, A.S., Olanrewaju, O.S., Babalola, O.O., Sulfate-reducing bacteria as an effective tool for sustainable acid mine bioremediation. *Front. Microbiol.*, 9, 1986, **2018**.
90. Bao, Y., Jin, X., Guo, C., Lu, G., Dang, Z., Sulfate-reducing bacterial community shifts in response to acid mine drainage in the sediment of the Hengshi watershed, South China. *Environ. Sci. Pollut. Res.*, *28*, 2822–2834, **2021**.
91. Costa, M.C., Martins, M., Jesus, C., Duarte, J., Treatment of acid mine drainage by sulphate-reducing bacteria using low cost matrices. *Water air Soil Pollut.*, *189*, 149–162, **2008**.
92. Dong, Y., Gao, Z., Di, J., Wang, D., Yang, Z., Wang, Y., Xie, Z., Study on the Effectiveness of Sulfate Reducing Bacteria to Remove Heavy Metals (Fe, Mn, Cu, Cr) in Acid Mine Drainage. *Sustainability*, *15*, 6, 5486, **2023**.
93. RoyChowdhury, A., Sarkar, D., Datta, R., Remediation of acid mine drainage-impacted water. *Curr. Pollut. Rep.*, *1*, 131–141, **2015**.
94. Schaider, L.A. and Hauri, J.F., Remediation of acid mine drainage with sulfate reducing bacteria. *J. Chem. Educ.*, *86*, 2, 216, **2009**.
95. Oberholzer, M.M., Oberholster, P.J., Ndlela, L.L., Botha, A.-M., Truter, J.C., Assessing alternative supporting organic materials for the enhancement of water reuse in subsurface constructed wetlands receiving acid mine drainage. *Recycling*, *7*, 3, 41, **2022**.
96. Oberholster, P.J., Cheng, P.-H., Botha, A.-M., Genthe, B., The potential of selected macroalgal species for treatment of AMD at different pH ranges in temperate regions. *Water Res.*, *60*, 82–92, **2014**.
97. Rose, P., Boshoff, G., Van Hille, .P., Wallace, L.C.M., Dunn, K.M., Duncan, J.R., An Integrated Algal Sulphate Reducing High Rate Ponding Process for the Treatment of Acid Mine Drainage Wastewaters. *Biodegradation*, *9*, 3-4, 247–257, **1998**.
98. Oswald, W., Waste treatment by pond systems, Engineering aspects, in: *Proc. IAWPRC Conference on Appropriate Waste Management Technologies*, Perth, Australia, 1991.
99. Boshoff, G., Duncan, J., Rose, P., An algal-bacterial integrated ponding system for the treatment of mine drainage waters. *J. Appl. Phycol.*, *8*, 442, **1996**.
100. Oberholster, P.J., De Klerk, A., Chamier, J., Cho, M., Crafford, J., De Klerk, L., Dini, J., Harris, K., Holness, S., Le Roux, W., *Assessment of the Ecological Integrity of the Zaalklapspruit Wetland in Mpumalanga (South Africa) Before and After Rehabilitation: The Grootspruit Case Study: Report to the Water Research Commission*, Water Research Commission, Pretoria, South Africa, 2016.

101. Appelo, C.A.J. and Postma, D., *Geochemistry, groundwater and pollution*, CRC press, 2004.
102. Services, S.W., *Aquachem version 5.1 software*, 65, 2008.
103. Chenini, F., Implications of water quality on irrigation practices under water scarcity, in: *Water Scarcity in the Mediterranean: Perspectives Under Global Change*, pp. 161–172, **2010**.
104. Bell, F., Bullock, S., Hälbich, T., Lindsay, P., Environmental impacts associated with an abandoned mine in the Witbank Coalfield, South Africa. *Int. J. Coal Geol.*, *45*, 2-3, 195–216, **2001**.
105. Vasconcelos, V.V., What maintains the waters flowing in our rivers? Rethinking hydrogeology to improve public policy. *Appl. Water Sci.*, *7*, 4, 1579–1593, **2017**.
106. Craig, H., Isotopic variations in meteoric waters. *Science*, *133*, 3465, 1702–1703, **1961**.
107. Mitsch, W.J. and Jørgensen, S.E., *Ecological engineering and ecosystem restoration*, John Wiley & Sons, Hoboken, NJ, USA, 2003.
108. Schoeman, Y., Oberholster, P.J., Erasmus, M., Towards Sustainable Water Management: A Novel Classification System for Ecological Engineering Solutions as new-order Anthromes in Human-Modified Landscapes, in: *WATERNET International Conference, Zanzibar, 25 to 27 October, 2023*, Zanzibar, 2023.
109. Schoeman, Y.O., P.J, The Ecological Engineering Nexus Accounting Framework: a tool for impact valuation of ecological engineering projects. Closed Cycles and the Circular Society 2023, in: *The power of ecological engineering. International Conference of the International Ecological Engineering Society, Chania, Greece, 1 to 5 October, 2023*, Chania, Greece, 2023.

3

Use of Chemical and Physical Techniques in Stepwise Treatment of Industrial Wastewater

Zvinowanda, Caliphs[1]* and Ncube Pauline[2]

[1]*Department of Chemical Sciences, Faculty of Science, University of Johannesburg, Doornfontein Campus, Doornfontein, Johannesburg, South Africa*
[2]*Department of Chemical Engineering, Faculty of Engineering & The Built Environment, Doornfontein Campus, Doornfontein, Johannesburg, South Africa*

Abstract
Industrial wastewater is the aqueous discharge released after using clean water in various stages of an industrial operation. The aqueous discharge contains more pollutants than the initial raw water or feed water used. The release of industrial wastewater is regulated in most countries and may not be released into the environment before it is treated and analyzed to meet specified water quality parameter limits. Wastewater generated from industrial operations tends to have a complex matrix and a wide range of contaminants that may not respond to a one-step regime of treatment conditions. More than one step of different treatment conditions may be required to successfully treat Industrial wastewater to expected pollutant concentration limits.

The broad objective of this book chapter is to discuss various modern industrial wastewater techniques being used to treat wastewater with complex matrices. The chapter discusses the physio-biochemical principles of pollutant removal from the wastewater, the sequences in which the techniques are applied, the technical and economic benefits of the processes, and the advantages and disadvantages of each technique regarding the quality of the overall industrial wastewater.

The types of industrial wastewater treatment to be reviewed in this chapter include wastewater from pharmaceutical, abattoir, power generation, food processing, and agricultural processing industries. Currently, wastewater from these

**Corresponding author*: czvinowanda@uj.ac.za

Elvis Fosso-Kankeu, Vhahangwele Masindi, Johannes Maree and Bhekie Mamba (eds.) Customized Technologies for Sustainable Management of Industrial Wastewater: A Circular Economy Approach, (111–134) © 2025 Scrivener Publishing LLC

industries is posing a water availability security threat as its release into the environment significantly negatively impacts the quality of water sources for human consumption, industrial development, and the environment. Hence, the information discussed in this chapter is useful to water sector practitioners, industry, municipality regulators, academics, and intrapreneurs.

Keywords: Industrial wastewater, chemical methods, physical techniques, tailoring technology, water quality

3.1 Introduction

Industrial processes usually generate a large amount of wastewater, which is contaminated with potentially toxic or nutrient substances that may include dissolved ions and complexes, polar organics, emulsifiers, suspended colloids, etc. These aqueous soluble solids enter into the water during the treatment process such as extraction, absorption, sequestrating (absorption), washing (solids or liquids) or as products of the hydrolysis coupled reaction. In most cases, the contaminants will require more than one treatment process for their complete removal. Common types of water effluent include wastewater from mining and metallurgical operations, battery manufacturing dairy processing, brewing and related food industries, abattoirs, pharmaceutical, industrial chemical production, coal washing, textile manufacturing, food processing, automotive, military equipment factories, effluent streams from non-zero discharge water and wastewater treatment plants, and acid mine drainage (AMD).

The major constituents of metallurgical wastewater are base metal ions, which may include known associated and toxic heavy metal ions such as those of lead, mercury, and uranium. In addition to base metal ions, metallurgical wastewater tends to have high salinity due to high levels of sodium, magnesium, and other group I and II ions with sulfate, chloride, and nitrate as counter ions. Coal is washed with large volumes of water streams through the density medium separation (DMS) and releases wastewater that tends to have high levels of ferrous, sodium, magnesium, sulfate, and chloride ions. The dairy and brewing industry processes result in wastewater rich in soluble organics such as monosaccharides, soluble fats, and other inorganic contaminants. Hence, wastewater from dairy processing factories tends to also have high total dissolved solids (TDSs), biological oxygen demand (BOD), and chemical oxygen demand (COD). In brewing and dairy processing, abattoir wastewater has high BOD and COD apart from fats and other biodegradable organic components in its matrix. Wastewater from pharmaceutical industries has a diversified composition emanating from the use of a variety of chemical feedstocks and biologically active microorganisms. Textile manufacturing processes use mordant dyes and dye stabilizers, thereby producing wastewater with a high residual content of these chemicals and heavy metal ions such as aluminium, chromium and copper which are used as dye stabilizers. These dyes are toxic and reduce sunlight penetration for photosynthesis by aqueous plants and microorganism. Brine wastewater has a high content of group I salts that are the major components of membrane filtration and reverse osmosis used in water treatment. Acid mine drainage water is acidic and is generated from hydrolysis and oxidation of pyritic-bearing ores exposed to water and oxygen after mining.

Most industrial wastewater streams contain various pollutant components that may not necessarily be removed by a single-step process. In South Africa, the quality of industrial wastewater effluent allowed to be released into the environment is governed by three legislations, namely, the National Water Act 36 of 1998. This Act is used to regulate the issuance of Water Use Licenses, which allows the licensees to abstract specified amounts of water from ground and surface water sources for their business operations. The same Act is used to regulate the quality of wastewater, which can be released into the environment. The wastewater quality parameter limits are usually stated in the Water Use License, and, to achieve the stipulated quality parameters limits, the user may need to treat the wastewater stream by more than one process in a stepwise manner. The release of waste that could be either liquid, solid, sludge, or gaseous in South Africa is regulated by the National Environmental Management: Waste Act 59 of 2008. The wastewater quality parameters of the liquid stream are then stipulated in the permitting license known as the Waste Authorization License. This document stipulates the conditions of storage and handling of waste of a commercial facility and the quality of the waste effluent permissible to be released into the environment. The Mine Health and Safety Act 29 of 1996 is the Act used by the Department of Mineral Resources to issue integrated environmental management permits or licenses. Again, like the other Acts discussed previously, the Act among other things, regulates the quality of wastewater streams that a mining operation is allowed to release into the environment.

3.2 Stepwise Treatment of Industrial Wastewater Using Chemical Operations

There are several processes and engineering design considerations that may necessitate the construction of a stepwise process for treating industrial wastewater. In most cases, the design decision is made after failing to get a one-step treatment process that can remove or degrade various pollutant components to the required or regulated effluent quality parameters limits. Although, in some cases, a one step process may be available, process cost considerations maybe less favorable than to use a stepwise process. The stepwise treatment of some major industrial wastewater is discussed in the following sections. The following industrial wastewater streams from metallurgical operations, dairy food processing industries, abattoir, pharmaceutical, coal washing, textile manufacturing, food processing, automotive assembly, water and wastewater treatment, and AMD are discussed.

3.2.1 Stepwise Removal of Heavy Metal Ions from Metallurgical Wastewater Stream

3.2.1.1 Removal of Chromium and Other Heavy Metal Ions

Metallurgical operations generally involve the extraction of minerals from their ores and or the purification of metals that result in the production of high value composite materials such as alloys. Water is usually one of the major feed stocks of any metallurgical process used for cooling the final products. Wastewater from metallurgical processes is usually laden with dissolved metal ions and other inorganic ions such as CN^-(aq), SO_4^{2-}(aq), Cl^-(aq), and S^{2-}(aq). There are several reasons and targets for treating industrial wastewater. The treatment of industrial wastewater may be aimed at meeting regulatory wastewater quality parameters limits and producing water that meets process and or potable water quality. Heavy metals are the major contaminants of metallurgical wastewaters. For example, chromium ions are present in the wastewaters of several industries, including stainless steel manufacturing, protective coatings on metal, magnetic tapes, chrome plating, tanneries, textile dyes production, pigments and paint production, production of cement, paper, and rubber. The hexavalent chromium [Cr(VI)] ions are predominant in the wastewater, are the most toxic species of chromium, and are difficult to remove from aqueous systems in a single step. However, they can be removed completely from wastewater of the aforementioned processes in two steps: reduction and precipitation.

Reducing agent substances such as ferrous ions and sodium bisulfite or thiosulfate ions are used to reduce Cr(VI), as shown in Equations (3.1) to (3.3). Typical reduction reactions for Cr(VI) are as follows:

$$CrO_4^{2-}(aq) + 8H^+(aq) + 3Fe^{2+}(aq) \leftrightarrow Cr^{3+}(aq) + 3Fe^{3+}(aq) + 4H_2O(aq) \tag{3.1}$$

$$Cr_2O_7^{2-}(aq) + 14H^+(aq) + 6Fe^{2+}(aq) \leftrightarrow 2Cr^{3+}(aq) + 6Fe^{3+}(aq) + 7H_2O(l) \tag{3.2}$$

$$2Cr_2O_7^{2-}(aq) + 28H^+(aq) + 6S_2O_3^{2-}(aq) \leftrightarrow 4Cr^{3+}(aq) + 3S_4O_6^{2-}(aq) + 14H_2O(l) \tag{3.3}$$

A stepwise removal of heavy metal ions from industrial wastewater such as metallurgical or tunnel wastewater is shown in Figure 3.1.

After reduction, the alkalinization of the wastewater precipitates out Cr(III) ions as $Cr(OH)_3(s)$. The alkalinization will precipitate other heavy metal ions in the form of $M^{n+}(aq)$ to metal hydroxides, $M(OH)_n(s)$. For heavy metal ions found in low concentrations in wastewater, it is very difficult to separate the metal hydroxide sludge from the effluent, and electrocoagulation is usually a preferred removal technology.

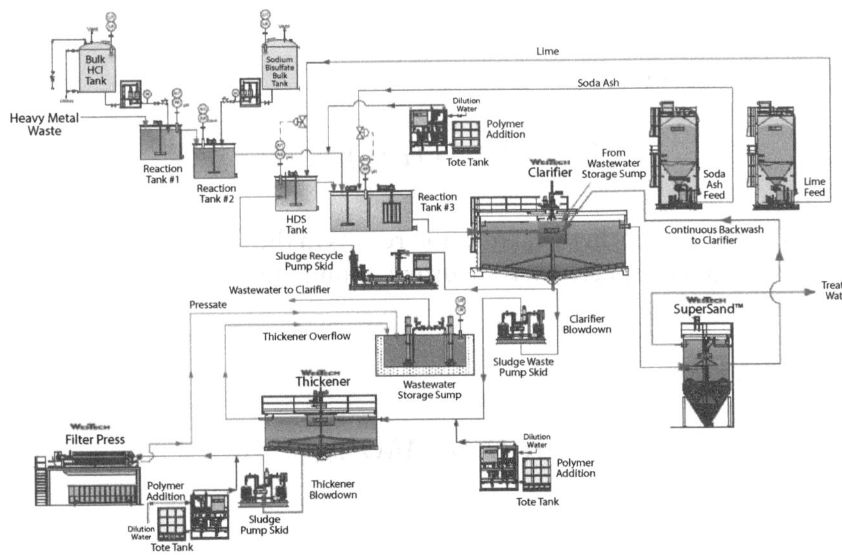

Figure 3.1 A stepwise removal of heavy metal ions from wastewater (https://www.westech-inc.com/solutions/metals-and-steel).

3.2.1.2 Removal of Heavy Metal Ions and Organic Pollutants in Wastewater by Electrocoagulation

Electrocoagulation is a process in which ionic coagulants are electrochemically generated *in situ* usually in a continuous manner for the flocculation of contaminants in water or wastewater [1]. Some of the common ionic coagulants that are used in water and wastewater treatment are Al^{3+} (aq)– and Fe^{3+} (aq)–based salts marketed as polyaluminum chloride and ferric chloride, respectively. In the electrochemical cell, the metal is made the anode to generate the respective coagulating ions (Al^{3+}/Fe^{3+}), whereas the hydroxyl (OH–) ions are generated at the inert metal electrode (cathode) according to Equations (3.4) and (3.5).

While electrocoagulation of solutes in the wastewater is induced by electrochemically generated ions, there are some pH-dependent redox reactions, which also play an important role, potentially generating the same ions, as shown in reactions (3.1) to (3.3) [2]:

Anodic reaction:

$$Al(0) \rightarrow Al^{3+} + 3e^{-} \tag{3.4}$$

Cathode reaction:

$$2H_2O + 2e^{-} \rightarrow H_2(g) + 2OH^{-} \tag{3.5}$$

The overall electrochemical reaction results in the formation of the most reactive coagulant species, tetrahydroxy aluminate, according to the following reaction:

At high pH values:

$$2Al + 6H_2O \rightarrow 2Al(OH)_4^{-} + 3H_2 \tag{3.6}$$

At low pH values:

$$Al^{3+} + 3H_2O \rightarrow Al((OH)_3 + 3H^{+} \tag{3.7}$$

When an iron or steel anode is used together with an inert reference, the ferric hydroxide coagulant is generated.

Anode reaction:

$$Fe \rightarrow Fe^{2+} + 2e^{-} \quad (3.8)$$

Cathode reaction:

$$2H_2O + 2e^{-} \rightarrow 2OH^{-} + H_2 \quad (3.9)$$

In a large-scale electrocoagulation process, the half-cell reactions (3.8) and (3.9) happen in an open and well-aerated pond and well-aerated and well-aerated reactor/pond to which the oxygen oxidizes the Fe^{2+} to Fe^{3+}:

$$4Fe^{2+} + 10H_2O + O_2 \rightarrow 4Fe(OH)_3 + 8H^{+} \quad (3.10)$$

If iron or aluminum electrodes are used, then the generated $Fe^{3+}(aq)$ or $Al(aq)^{3+}$ ions will immediately undergo further spontaneous reactions to produce corresponding hydroxides and/or polyhydroxides [3]. The common polyhydroxide Fe-based coagulants formed from the hydrolysis and other multi-metal complexes $Fe(OH)_3$ flocs are $Fe(H_2O)_6^{3+}$, $Fe(H_2O)_5(OH)^{2+}$, $Fe(H_2O)_4(OH)^{2+}$, $Fe_2(H_2O)_8(OH)_2^{4+}$, and $Fe_2(H_2O)_6(OH)_4^{4+}$. Ferric hydroxide is the most active coagulant species of iron. Similarly, Al^{3+} ions generated at the anode forms the Al(OH)3, which further hydrolizes amphoprotically to form various polydroxides that include the following:

$Al(H_2O)_6^{3+}$, $Al(H_2O)_5OH^{2+}$, $Al(H_2O)_4(OH)^{2+}$, $Al(OH)^{2+}$, $Al(OH)^{2+}$, $Al_2(OH)_2^{4+}$, $Al(OH)_4^{-}$, $Al_6(OH)_{15}^{3+}$, $Al_7(OH)_{17}^{4+}$, $Al_8(OH)_{20}^{4+}$, $Al_{13}O_4(OH)_{24}^{7+}$, and $Al_{13}(OH)_{34}^{5+}$.

The *in-situ*-generated coagulants clarify the wastewater during the electrocoagulation (EC) flocculation-coagulation process by various mechanisms, which include (i) surface complexation, (ii) electrostatic attraction, (iii) adsorption, and (iv) direct precipitation by the formation of the pollutant metal hydroxides [4].

Electrocoagulation has advantages when treating wastewater from metallurgical operations, such as low sludge volume generation, reduction of secondary pollution courtesy of the reactive species that are generated *in situ*, and the hydrogen gas generated at the cathode that serves as a flotation gas to help stabilise and thus remove coagulated pollutants as floc bubbles.

However, the efficiency of the electrocoagulation process is curtailed by passivation of the cathode, which lowers its efficiency to generate hydroxides ions. Moreover, there is need to replace the anode from time to time because it gets dissolved during the process.

While electrocoagulation of metallurgical wastewater efficiently removes metal ions, the counter anions such as SO_4^{2-}, Cl−, and NO_3^- may still persists in the effluent acids such as HCl, H_2SO_4, and HNO_3 that are used in the pickling and itching processes. Thus, the effluent is passed to a sludge thickening and separation holding tank where it is desalinated, and other residual to inorganic anions are removed by techniques such reverse osmosis and membrane filtration.

3.2.2 Stepwise Treatment of Wastewater Streams from Automotive Assembly Operations

The automotive industries assemble vehicles and manufacture spare parts. Large volumes of water are used in the assembly lines resulting in the release of large volumes of wastewater of complex matrix. The wastewater from automotive industry is generally classified as hazardous and thus regulated. These waste streams, therefore, require treatment before it is released into the environment. Wastewater is contaminated with high levels of TDSs in the form of heavy metal ions and complexes, counter ions that includes phosphates and sulfates, as well as biodegradable and persistent organic polymers. These contaminants are released into natural water, which would result in high BOD, COD turbidity, as well as incidence of persistence organics. The removal of these contaminants is usually achieved using a crystallizer process, as shown in Figure 3.2. However, distillate remains contaminated with volatile organic substances.

The presence of dissolved organics in the wastewater will adversely affect the receiving natural water bodies with high COD and BOD. To reduce the contamination, treatment processes like advanced oxidation processes (AOPs), electrochemical oxidation, electrooxidation, and electrocoagulation [5] should be used.

Most automotive companies aim at zero liquid discharge (ZLD) from their process streams due to the large volumes of water required per unit output. Hence, the concept strives to treat the wastewater to the highest quality to reuse in the car production cycle, thereby achieving ZLD. The concept of ZLD is well illustrated by integrating the coagulation treatment of wastewater from a car wash concerned, followed by the electrooxidation of residuals before water is reused for washing, as shown in Figures 3.3 and 3.4. The treatment process is

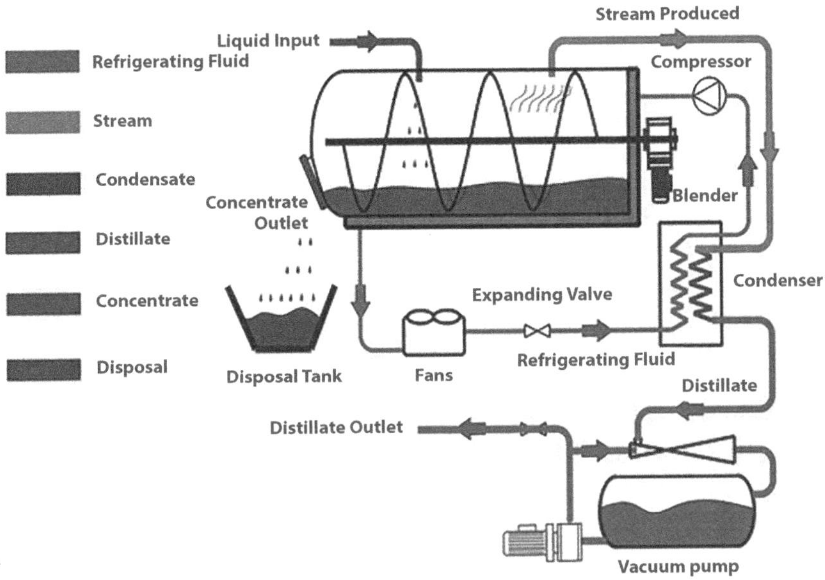

Figure 3.2 Crystallizer for automotive wastewater treatment (https://www.yasa.ltd/solidest-wastewater-crystallizer).

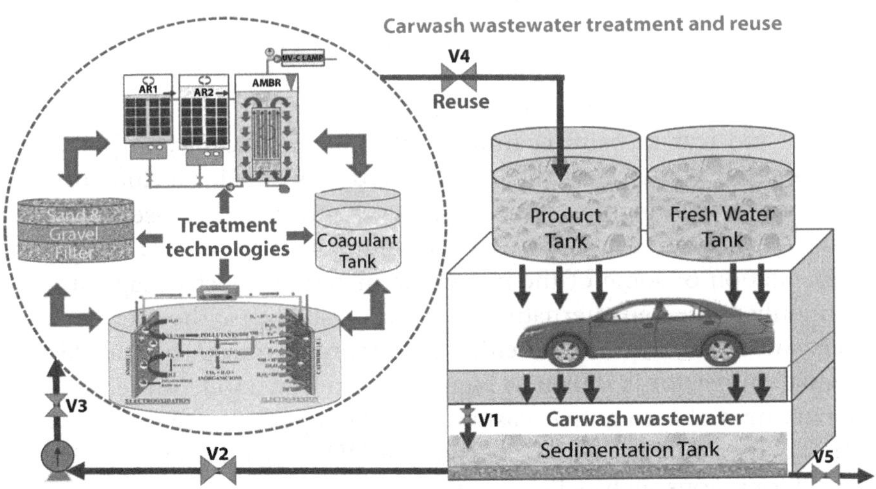

Figure 3.3 Carwash wastewater treatment and reuse [6].

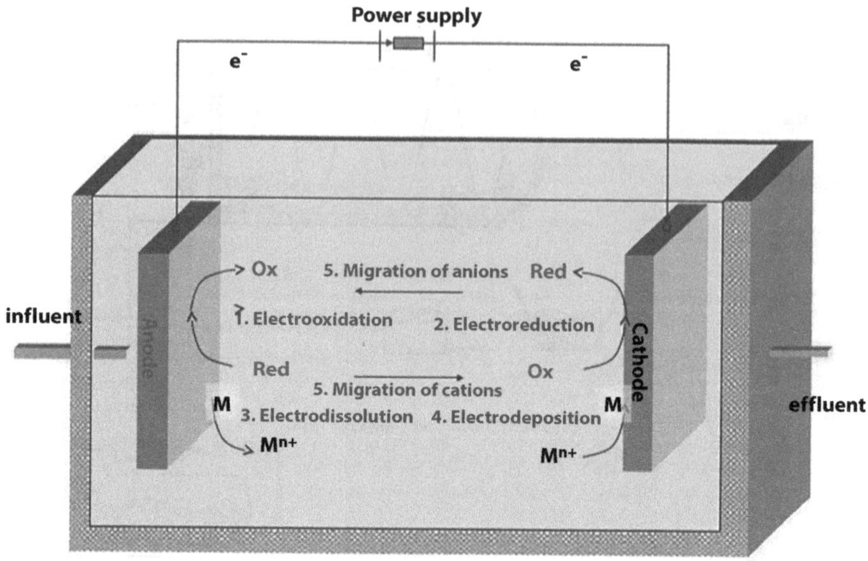

Figure 3.4 Electrooxidation reaction unit (https://www.yasa.ltd/electrooxidation-equipment).

cyclic and thus averts external discharge effluent. The species generated at each electrode depends on the electrolyte conditions, the make-up of the electrolyte, and the electrode materials used. For example, oxidants such as oxygen, chlorine, hydroxyl, or hypochlorite radicals are some of the species generated at the anode depending on the make-up of the electrolyte. These oxidants are capable of oxidising volatile organic pollutants in the distillate.

Wear and tear of paint coatings contribute significantly to pollutants in automotive wastewater. The major ingredients of paint include organic polymeric resins, pigments, additives, and solvents. The use of evapo-crystallizers to remove paint-derived pollutants may lead to the partial fragmentation of some of the polymeric materials. Electrocoagulation is commonly used as an alternative to evapo-crystallizers. The common sacrificial anode material used is either aluminum or iron. A typical electrocoagulation reactor is shown in Figure 3.5. An iron anode is preferred for the treatment of automotive wastewater at high pH because the generated Fe(III) plays a dual role as a coagulant [$Fe(OH)_3(s)$] and as a precipitating reagent for anions such as PO_4^{3-} ions.

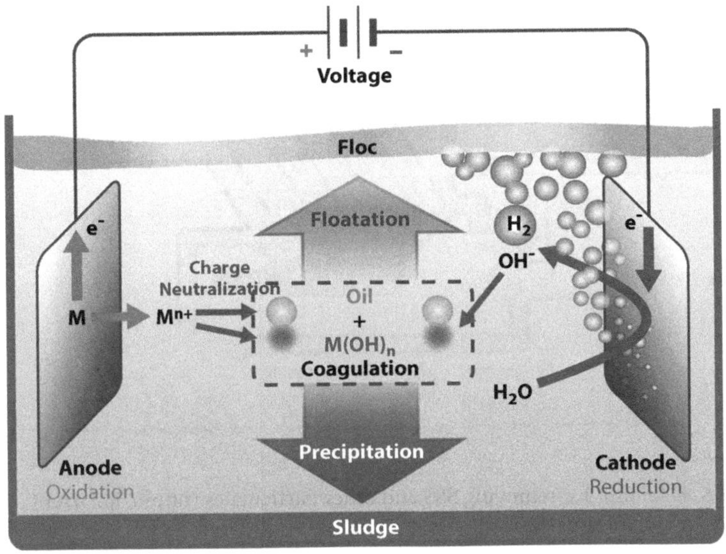

Figure 3.5 Mechanisms of electrocoagulation [7].

3.2.3 Stepwise Treatment of Wastewater Streams from Abattoir Processing Industries

The processing and handling of carcasses and related animal products from slaughterhouses and abattoirs especially in the beef industry use large volumes of water. Water is used to clean the slaughtering facility as well maintaining the general hygienic handling of the animal products. Consequently, huge volumes of wastewater streams are usual.

The general composition of wastewater released from abattoirs is significantly elevated levels of total organic carbon (TOC), which would cause a biochemical oxygen demand (BOD), COD, total nitrogen (TN), total suspended solids (TSSs), pH, ortho-phosphate (orto-PO_4), ortho–phosphorus pentoxide (orto-P_2O_5), potassium, (K), color, and turbidity (NTU) [8]. The diversity in the composition of pollutants in wastewater from abattoirs requires more than one treatment technique as a single process will not be adequate to clean the wastewater.

3.2.3.1 *Dissolved Air Floatation and Anaerobic Treatment*

Pretreatment techniques such as dissolved air floatation (DAF) are used to remove solids from blood and other fluids from the animals being slaughtered and carcasses being processed. These wastes would, otherwise, cause

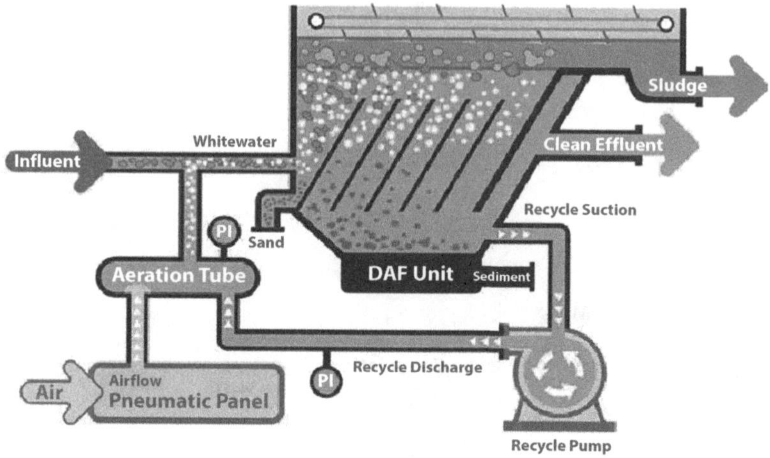

Figure 3.6 DAF unit for removing SSS and other particulates (https://ig7.ir/en/dissolved-air-flotation-daf/).

a high BOD/COD. To reduce TSS, the effluent is passed on to the flow anaerobic sludge blanket (UASB) reactor. The tandem treatment steps can remove BOD and COD to as low as 75% to 95%, as has been reported [8]. A typical DAF unit for removing fats, SSS, and other particulates is shown in Figure 3.6. The effluent free of particulates is then processed in the upflow anaerobic sludge blanket (UASB) reactor (Figure 3.7).

The upflow anaerobic sludge blanket (UASB) reactor for the treatment of wastewater from abattoirs is shown in Figure 3.7.

3.2.3.2 Physicochemical and Advanced Oxidation Processes for Abattoir Wastewater Treatment

To address the multi-challenges faced in treatment of abattoir wastewater to meet the stringent regulations in some countries, various hybrid treatment regime pertaining the combined biological and non-biological approaches for slaughterhouse wastewater have been devised to improve pollutant removal efficiency [10]. These involve pre-screening to remove solid particles greater than 2 mm and removal of fats and grease from the wastewater using the DAF. After the DAF treatment, the next physical process involves coagulation-flocculation and sedimentation. All these steps lead to the removal of large polymeric carbohydrates and glycoproteins, and these processes help reduce the COD and BOD levels [11]. Despite this integrated treatment, not all of the organic compounds found in abattoir

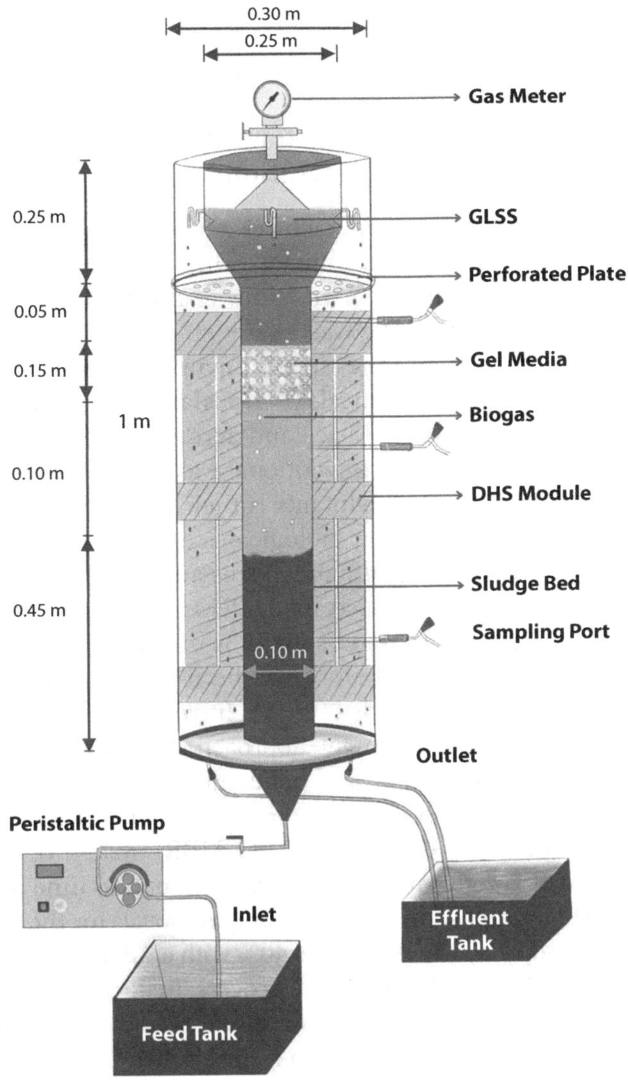

Figure 3.7 The upflow anaerobic sludge blanket (UASB) reactor for [9].

wastewater are resistant to biological treatment using aerobic bacteria, and some are not removed by physical processes.

The persistent organic pollutants that remain in the effluent and are resistant are passed on for further treatment by the AOPs. AOPs are an interesting complementary treatment option for primary or secondary treatment of slaughter wastewater, showing excellent overall treatment efficiencies for water reuse [11]. After treatment by AOP, the quality of the effluent is likely to meet the environmental quality standards.

3.2.4 Stepwise Treatment of Wastewater Streams from Pharmaceutical Operations

The pharmaceutical industry manufactures a broad range of products for use as medicine in humans and animals. The manufacturing processes that include chemical synthesis, extraction, formulation, and packaging produce wastewater streams. The wastewater streams may contain a high organic load if discharged into the aquatic environment. Varying concentrations of numerous pharmaceuticals (PHACs) have been recorded in the aqueous environment in South Africa and across the globe [12–14], and this indicates the inefficiency of conventional wastewater treatment plants. The risk associated with these pollutants is their ubiquity and persistence in the environment as well as their biological activity that may distress the sustainable development of aquatic organisms and the natural food chain that may include birds [15]. This has made PHACs emerging pollutants of concern, whose removal from water sources is mandatory.

Literature shows that a lot of work has been done in exploring effective alternative treatment methods [16–19]. In recent years, the focus has shifted toward a stepwise treatment approach encompassing integrated hybrid processes, to achieve improved pollutant removals. An almost complete removal of COD and BOD_5 from the wastewater was achieved by the electro-oxidation membrane bioreactor (EO-MBR) system operating at a recirculation rate of 160 mL/min and hydraulic retention time of 24 hours. Under the same conditions, the corresponding removals for the single MBR ranged from 60% to 87%, whereas it was 40% with a 2.5-fold biodegradability index (BOD_5/COD) for the single EO process. Reduced membrane fouling was also observed with the EO-MBR after a 28-day prolonged contact. Compared to the single MBR, lower concentrations of extracellular polymeric substances (EPSs) and proteinaceous compounds were observed in the sludge cake of EO-MBR. This was ascribed to the minimized fouling of the EO-MBR to the electrophilic interaction of the EO-generated •OH with the electron-rich moieties of EPS organic foulants.

Granular-activated carbon (GAC) filtration combined with ozonation (O_3) has also been utilized to degrade PHAC residues, among other contaminants in sewage treatment plants (STPs) [17]. The GAC/O_3 system was deployed to the effluent streams after the secondary stage that included biological treatment. Biomarker responses of rainbow trout exposed to GAC/O_3-treated effluents were evaluated. The GAC/O_3 treatment degraded from 87% to 95% of the PHACs and reduced biomarker responses due to unknown pollutants in STP effluents [17].

Marcelino and colleagues [18] studied a multistage ozone process combined with aerobic degradation for the elimination of the antibiotic amoxicillin in real wastewater. Effluent from the biological treatment was subjected to ozonation. Up to 99% amoxicillin removal was achieved with the single ozonation process, without complete mineralization. However, complete mineralization was achieved using a multistage treatment system composed of ozonation and biodegradation. In addition, at an O3 consumption of ~500 mg, >90% amoxicillin residue levels, >98% COD, and 90% toxicity were eliminated.

In another study, the performance of a heterogeneous Fenton hollow-fiber membrane (HFHFM) reactor in degrading PHACs was evaluated by Lan *et al.* [19]. A commercial micro-sized zeolite catalyst containing iron (Fe-ZSM5) was used in the heterogeneous Fenton-like process, which was coupled with membrane filtration. Through the design of the wastewater treatment system and a wide pH range, catalyst recyclability is possible. The oxidation of Ibuprofen (IBP) was initiated by the addition of H_2O_2 to the HFHFM reactor after the attainment of steady-state conditions (in terms of adsorption) and constant IBP concentration. More than 99% IBP removal was achieved after 180 min, regardless of the flux. A critical operating flux was established, below which membrane fouling was minimal. The authors were able to determine the critical operating flux and concluded that operating below the critical flux could help control membrane fouling. The results further showed that reversible fouling of the hydraulics might be eliminated by hydrodynamic backwashing [19].

The integrated adsorption-photocatalytic degradation system was demonstrated for the elimination of three antibiotic PHACs, *viz.*, sulfamethoxazole and ciprofloxacin antibiotics in water [20]. A synthesized Ferrocene-functionalized rGO-ZnO nanocomposite (Fc@rGO-ZnO) was used as the photocatalyst. The reaction solution containing the chosen antibiotics was agitated in the dark for 60 min before exposure to UV-C radiation in a photobioreactor. Agitation in the dark was done to achieve pre-adsorption/desorption equilibrium on the surface of the catalyst. The results demonstrated about 50% antibiotic elimination by the adsorption alone. However, higher removals of the three PHACS >95% were achieved after 180 min of photocatalytic treatment.

The aforementioned research results clearly show that improved pollutant removals were achieved in combined/hybrid systems compared to the single processes.

3.2.5 Stepwise Treatment of Wastewater Streams from Dairy Processing Industries

The dairy industry generates large volumes of high-nutrient and organic-loaded dairy processing sludge (DPS) [21]. DPS is classified as a nutrient-rich (P and N) biosolid and, therefore, can be spread on grassland or used as a secondary fertilizer supplement for crop production [22]. This is in line with one of the fundamental principles of sustainable agriculture and the circular economy, which emphasizes the reuse of inputs that would otherwise be regarded as waste. The three main treatment options for dairy wastewater are (a) onsite treatment in the factory WWTP, (b) onsite extraction of semisolids and special wastes by waste management, or (c) discharge to and subsequent treatment at a nearby STP [23]. Because of the high costs of the last two options and tightening regulatory discharge limits, most dairy processing adopts the onsite treatment option. The basic treatment regime is outlined in Figure 3.8.

Figure 3.8 shows that the dairy effluent treatment depends on the BOD. Typical dairy wastewater treatment produces a distinct form of DPS, which may be classified as (i) biochemically treated activated sludge or (ii) lime-treated (Figure 3.2) DAF sludge. The former constitutes stabilized sludge from secondary biological treatment, which may be aerobic, anaerobic, or a combination of both, whereas the latter is generated after chemical and DAF treatment of raw wastewater in the primary treatment stage [22]. Chakchouk *et al.* [24] studied the use of a combination of

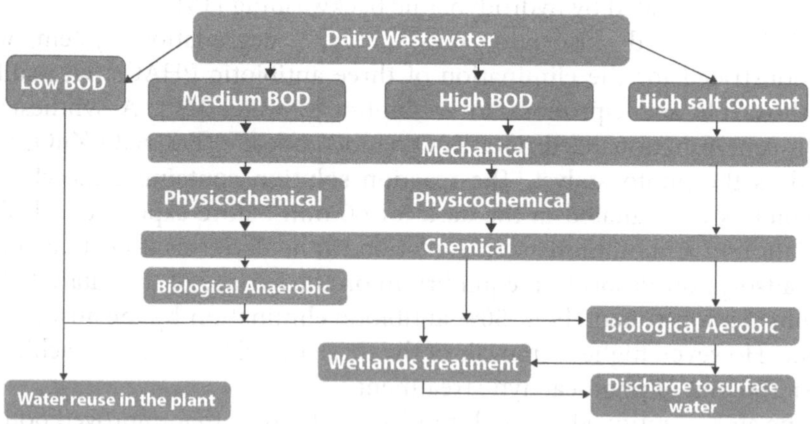

Figure 3.8 Basic dairy wastewater treatment techniques (adopted from Slavov [21]).

electrocoagulation and electrooxidation (EC-EO) systems for which both comprised cylindrical cells with aluminum plate and expanded platinized titanium electrodes, respectively. The current was supplied from a current generator and the dairy effluent was used as the electrolyte. After separate optimization, the two cells were coupled together, and the treatment was initiated. There was efficient removal of the dairy biosolids. The tandem removal process achieved removal indices of 100% turbidity, 90.4% color, and 66.4% COD after 21 min for dairy wastewater that had an initial COD of 3850 ppm. However, single EC and EO achieved COD eliminations of about 50% and 40%, respectively [24]. The results of the tandem/coupled cells were much higher than single processes, probably due to the complimentary nature of the combined processes. In recent years, enhanced removals of biological phosphorus from dairy effluents have been realized using the activated sludge (ACS) process without the addition of chemical agents. The process involves recirculating sludge through anaerobic and aerobic digesters [22].

3.2.6 Stepwise Treatment of Wastewater Streams from Food Processing Operations

The food and beverage industry uses substantial amounts of water as a raw material, resulting in production of wastewater. Faced with global water scarcity challenges, it is, therefore, crucial for the food industry to optimize water utilization *via* integrated recycling. Different treatment methods can be utilized depending on, among other factors, the type of food processing concerned and, thus, the wastewater characteristics and the level of contamination. The most common wastewater treatment technologies for food are highlighted in Figure 3.9.

Figure 3.9 shows that single or combined hybrid processes can be utilized to treat wastewater from the food industry. Most treatment processes target the removal of turbidity, COD, and BOD. The combination of reverse osmosis (RO) and membrane filtration (MF) (RO-MF) is the most effective to remove BOD from food processing wastewater, whereas the combined physicochemical and biological treatment (PSC-BIO) performs well for both COD and BOD eliminations (Figure 3.9). For example, a study investigating the efficiency of a PSC-BIO treatment system study, food wastewater with high concentrations of fat and oils was conducted [25]. The treatment process was made up of the primary stage consisting of screens and a neutralization reactor, a secondary aerobic-anaerobic system,

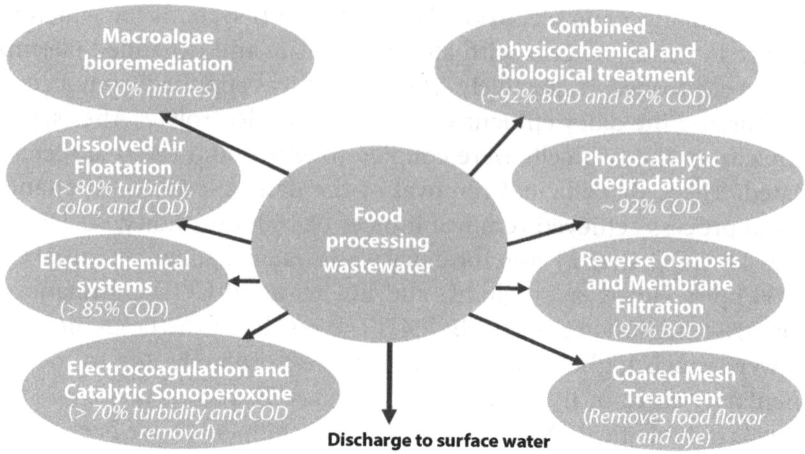

Figure 3.9 Common technology for food wastewater treatment [25].

and a coagulation-flocculation process. In another study, the removal efficiency of the combined electrocoagulation (EC) and catalytic sonoperoxone (US/H_2O_2) process was evaluated for treating wastewater from an olive plant, for which 75% elimination of TOC was achieved. The % removals of TOC from the food processing wastewater processes that were single ozonation (O_3) and catalytic ozonation ($catO_3$) were 68% ($catO_3/US$), 61% ($catO_3$), 25% (O_3/H_2O_2), 20% (O_3), 15% (H_2O_2), and 8% (US), respectively. The results indicate that the integrated $EC\text{-}US/H_2O_2$ process was the most effective for treating olive wastewater. In a related study, micro-ultra-nano-filtration (MF-UF-NF) was combined with reverse osmosis to treat olive wastewater containing phenol, and 60% COD removal was obtained [26]. NF alone was insufficient as the membranes were susceptible to free chlorine and suffered limited retention of univalent ions and salts. To purify water from food processing effluents, Hernández *et al.* [27] integrated UF with RO (UF-RO) and achieved overall efficiency greater than 95% in all the quality parameters. UF was chosen to decrease COD and total solids (TS) by removing divalent salts and colloidal species, whereas RO targeted the removal of monovalent salts and residual organic matter. In a separate study, the hybrid electrochemical and chemical oxidation (EC-CO) system was employed to treat sugar production wastewater [28]. Greater than 99% color and 98% COD were removed compared to a color removal efficiency of about 83% and an 81% COD lowering achieved by EC.

3.2.7 Stepwise Treatment of Wastewater Streams from Textile Manufacturing

The textile manufacturing industry consumes huge amounts of water and a broad range of synthetic dyes and chemicals. The latter includes detergents, desizers, stabilizing agents, inorganic salts, heavy metals, and degradable organics, all of which contribute to the pollution of the generated wastewater stream. Therefore, typically, textile effluents have a complex composition, with high organic load, unpleasant coloration, and extreme pH [29]. In addition to removing pollutants from textile wastewater streams, increasing research trends have shifted toward water-recovering solutions.

Several treatment technologies are available for treating textile wastewater, such as adsorption, biological methods, membrane separation, and AOPs [30]. However, due to the complex nature and variable composition of the wastewater streams [29], single processes are not very effective and most of them are costly, have high energy demand, and generate sludge [31]. Combined or hybrid processes have been explored in recent years.

Bioremediation has shown promising results in the remediation of textile effluents [32]. Advanced hybrid models of the technology include the sequencing batch reactor (SBR), membrane bioreactors (MBR), and hybrid bioreactors (HBR). SBR is an advancement on the traditional ACS process, which combines all process units in a single vessel. The process is cost-effective and easy to maintain [33]. In a typical SBR, multistage effluent treatment occurs. This involves feeding the wastewater into the reactor and mixing it with the biomass mechanically. Afterward, the effluent/biomass blend is exposed to aerated mechanical agitation to initiate aerobic biodegradation. The product-activated sludge is then passed to the settling stage and allowed to settle under anoxic conditions. In the final stage, treated effluent is drawn from the settling stage and disinfected [33]. In one study, a hybrid SBR-NF system achieved 98% COD and dye removal in textile effluent contaminated with dyes [34].

On the other hand, MBR combines membrane technologies with suspended microbial processes. The technique is highly efficient in eliminating refractory emerging pollutants that are resistant to the traditional ACS process [35]. However, membrane fouling is a major hindrance to industrial applications. Although washing and filter replacements can mitigate the problem, this increases the treatment cost. The most advanced bioremediation technology is the HBR, which encompasses the combination of two different processes. In a recent publication, Ayed and colleagues [36] combined coagulation and flocculation with an anaerobic-aerobic sequential biological process to treat textile wastewater. The authors recorded 76% COD reduction, 82% color removal, and 35% phenol removal.

Other studies have exploited the integration of AOPs as a post-treatment stage in the conventional biological ACS process for the remediation of textile effluents. Ma *et al.* [37] obtained > 90% color removal from textile effluents after treatment with a combined ACS-filtration-UVC/H_2O_2 system. In similar research, combining anaerobic acidification + ACS + O_3/zero valence iron (ZVI) removed 79% dissolved organic carbon (DOC) and 67% COD from the dyeing and finishing effluent streams [37], whereas coupling the photocatalytic membrane reactor (with ZnO) with anaerobic biofilm reactor yielded 99% COD and 93% color removals [38]. Overall, the integration of AOPs with biological processes resulted in improved pollutant removals compared to single processes [39]. However, limitations of some AOPs such as high energy demand and incomplete mineralization were observed in some studies [39].

References

1. Akbal, F. and Camcı, S., Treatment of Metal Plating Wastewater by Electrocoagulation. *Environ. Prog. Sustain. Energy*, *31*, 340–350, **2012**.
2. Barrera-Díaz, C.E., Balderas-Hernández, P., Bilyeu, B., Electrocoagulation: Fundamentals and Prospectives, in: *Electrochemical water and wastewater treatment*, pp. 61–76, Elsevier, Butterworth-Heinemann, 2018.
3. Mollah, M.Y.A., Morkovsky, P., Gomes, J.A.G., Kesmez, M., Parga, J., Cocke, D.L., Fundamentals, Present and Future Perspectives of Electrocoagulation. *J. Hazard. Mater.*, *114*, 199–210, **2004**.
4. Garcia-Segura, S., Eiband, M.M.S.G., de Melo, J.V., Martínez-Huitle, C.A., Electrocoagulation and Advanced Electrocoagulation Processes: A General Review about the Fundamentals, Emerging Applications and Its Association with Other Technologies. *J. Electroanal. Chem.*, *801*, 267–299, **2017**.
5. Viktoryová, N., Szarka, A., Hrouzková, S., Recent Developments and Emerging Trends in Paint Industry Wastewater Treatment Methods. *Appl. Sci.*, *12*, 10678, **2022**.
6. Espinoza-Montero, P.J., Martínez-Huitle, C.A., Loor-Urgilés, L.D., Technologies Employed for Carwash Wastewater Recovery. *J. Clean. Prod.*, 401, 136722, **2023**.
7. An, C., Huang, G., Yao, Y., Zhao, S., Emerging Usage of Electrocoagulation Technology for Oil Removal from Wastewater: A Review. *Sci. Total Environ.*, *579*, 537–556, **2017**.

8. Bustillo-Lecompte, C.F. and Mehrvar, M., Slaughterhouse Wastewater Characteristics, Treatment, and Management in the Meat Processing Industry: A Review on Trends and Advances. *J. Environ. Manage.*, *161*, 287–302, **2015**.
9. Arivalagan, A.L. and Stanislaus, A.R., Performance Evaluation of a Novel Upflow Anaerobic Sludge Blanket Reactor with Built-in Downflow Hanging Sponge Module for the Treatment of Municipal Sewage. *Environ. Sci. Pollut. Res.*, *29*, 89108–89120, **2022**.
10. Moreroa, M. and Basitere, M., The Challenges and Treatment of Abattoir Effluents: A South African Perspective. *Water Pract. Technol.*, *17*, 2598–2613, **2022**.
11. Bustillo-Lecompte, C. and Mehrvar, M., Slaughterhouse Wastewater: Treatment, Management and Resource Recovery, in: *Physico-chemical wastewater treatment and resource recovery*, pp. 153–174, **2017**.
12. Khan, A.H., Khan, N.A., Zubair, M., Shaida, M.A., Manzar, M.S., Abutaleb, A., Naushad, M., Iqbal, J., Sustainable Green Nanoadsorbents for Remediation of Pharmaceuticals from Water and Wastewater: A Critical Review. *Environ. Res.*, *204*, 112243, **2022**.
13. Caban, M. and Stepnowski, P., How to Decrease Pharmaceuticals in the Environment? A Review. *Environ. Chem. Lett.*, *19*, 3115–3138, **2021**.
14. Ohoro, C.R., Adeniji, A.O., Semerjian, L., Okoh, O.O., Okoh, A.I., Occurrence and Distribution of Pharmaceuticals in Surface Water and Sediment of Buffalo and Sundays River Estuaries, South Africa and Their Ecological Risk Assessment. *Emerg. Contam.*, *7*, 187–195, **2021**, doi: 10.1016/J.EMCON.2021.09.002.
15. Wooding, M., Rohwer, E.R., Naudé, Y., Determination of Endocrine Disrupting Chemicals and Antiretroviral Compounds in Surface Water: A Disposable Sorptive Sampler with Comprehensive Gas Chromatography – Time-of-Flight Mass Spectrometry and Large Volume Injection with Ultra-High Performance Liquid Chromatography–Tandem Mass Spectrometry. *J. Chromatogr. A*, *1496*, 122–132, **2017**, doi: 10.1016/j.chroma.2017.03.057.
16. Gharibian, S. and Hazrati, H., Towards Practical Integration of MBR with Electrochemical AOP: Improved Biodegradability of Real Pharmaceutical Wastewater and Fouling Mitigation. *Water Res.*, *218*, 118478, **2022**, doi: 10.1016/j.watres.2022.118478.
17. Beijer, K., Björlenius, B., Shaik, S., Lindberg, R.H., Brunström, B., Brandt, I., Removal of Pharmaceuticals and Unspecified Contaminants in Sewage Treatment Effluents by Activated Carbon Filtration and Ozonation: Evaluation Using Biomarker Responses and Chemical Analysis. *Chemosphere*, *176*, 342–351, **2017**, doi: 10.1016/j.chemosphere.2017.02.127.
18. Marcelino, R.B.P., Leão, M.M.D., Lago, R.M., Amorim, C.C., Multistage Ozone and Biological Treatment System for Real Wastewater Containing Antibiotics. *J. Environ. Manage.*, *195*, 110–116, **2017**, doi: 10.1016/j.jenvman.2016.04.041.

19. Lan, Y., Barthe, L., Azais, A., Causserand, C., Feasibility of a Heterogeneous Fenton Membrane Reactor Containing a Fe- ZSM5 Catalyst for Pharmaceuticals Degradation : Membrane Fouling Control and Long-Term Stability. *Sep. Purif. Technol., 231*, 115920, **2020**, doi: 10.1016/j.seppur.2019.115920.
20. Roy, N., Kannabiran, K., Mukherjee, A., Integrated Adsorption and Photocatalytic Degradation Based Removal of Ciprofloxacin and Sulfamethoxazole Antibiotics Using Fc@rGO-ZnO Nanocomposite in Aqueous Systems. *Chemosphere, 333*, 138912, **2023**, doi: 10.1016/j.chemosphere.2023.138912.
21. Slavov, A.K., General Characteristics and Treatment Possibilities of Dairy Wastewater-A Review. *Food Technol. Biotechnol., 55*, 14–28, **2017**, doi: 10.17113/ft b.55.01.17.4520.
22. Shi, W., Healy, M.G., Ashekuzzaman, S.M., Daly, K., Leahy, J.J., Fenton, O., Dairy Processing Sludge and Co-Products: A Review of Present and Future Re-Use Pathways in Agriculture. *J. Clean. Prod., 314*, 128035, 2021.
23. Britz, T.J., van Schalkwyk, C. and Hung, Y.T., Treatment of dairy processing wastewaters, in: *Handbook of industrial and hazardous wastes treatment*, pp. 673–705, CRC Press, Boca Raton, 2004.
24. Chakchouk, I., Elloumi, N., Belaid, C., Mseddi, S., Chaari, L., Kallel, M.A., Combined Electrocoagulation-Electrooxidation Treatment for Dairy Wastewater. *Braz. J. Chem. Eng., 34*, 109–117, **2017**, doi: 10.1590/0104-6632.20170341s20150040.
25. Shrivastava, V., Ali, I., Marjub, M.M., Rene, E.R., Soto, A.M.F., Wastewater in the Food Industry: Treatment Technologies and Reuse Potential. *Chemosphere, 293*, 133553, **2022**, doi: 10.1016/j.chemosphere.2022.133553.
26. Ochando-Pulido, J.M., Corpas-Martínez, J.R., Martinez-Ferez, A., About Two-Phase Olive Oil Washing Wastewater Simultaneous Phenols Recovery and Treatment by Nanofiltration. *Process Saf. Environ. Prot., 114*, 159–168, **2018**, doi: 10.1016/J.PSEP.2017.12.005.
27. Hernández, K., Muro, C., Ortega, R.E., Velazquez, S., Riera, F., Water Recovery by Treatment of Food Industry Wastewater Using Membrane Processes. *Environ. Technol. (U.K.), 42*, 775–788, **2021**, doi: 10.1080/09593330.2019.1645739.
28. Sahu, O., Rao, D.G., Gopal, R., Tiwari, A., Pal, D., Treatment of Wastewater from Sugarcane Process Industry by Electrochemical and Chemical Process: Aluminum (Metal and Salt). *J. Water Process Eng., 17*, 50–62, **2017**, doi: 10.1016/j.jwpe.2017.03.005.
29. Dhaouefi, Z., Toledo-Cervantes, A., Ghedira, K., Chekir-Ghedira, L., Muñoz, R., Decolorization and Phytotoxicity Reduction in an Innovative Anaerobic/Aerobic Photobioreactor Treating Textile Wastewater. *Chemosphere, 234*, 356–364, **2019**, doi: 10.1016/j.chemosphere.2019.06.106.

30. Shirazi, M.M.A., Bazgir, S., Meshkani, F.A., Novel Dual-Layer, Gas-Assisted Electrospun, Nanofibrous SAN4-HIPS Membrane for Industrial Textile Wastewater Treatment by Direct Contact Membrane Distillation (DCMD). *J. Water Process Eng.*, *36*, 101315, **2020**, doi: 10.1016/j.jwpe.2020.101315.
31. Fazal, T., Mushtaq, A., Rehman, F., Ullah Khan, A., Rashid, N., Farooq, W., Rehman, M.S.U., Xu, J., Bioremediation of Textile Wastewater and Successive Biodiesel Production Using Microalgae. *Renew. Sustain. Energy Rev.*, *82*, 3107–3126, **2018**.
32. Reddy, S. and Osborne, J., An Insight on the Advancements of Biological Technologies in the Bioremediation of Textile Effluents. *Urban Water J.*, *19*, 468–480, 2022.
33. Nancharaiah, Y.V., Sarvajith, M., Lens, P.N.L., Selenite Reduction and Ammoniacal Nitrogen Removal in an Aerobic Granular Sludge Sequencing Batch Reactor. *Water Res.*, *131*, 131–141, **2018**, doi: 10.1016/j.watres.2017.12.028.
34. Khosravi, A., Karimi, M., Ebrahimi, H., Fallah, N., Sequencing Batch Reactor/Nanofiltration Hybrid Method for Water Recovery from Textile Wastewater Contained Phthalocyanine Dye and Anionic Surfactant. *J. Environ. Chem. Eng.*, *8*, 103701, **2020**, doi: 10.1016/J.JECE.2020.103701.
35. Xiao, K., Liang, S., Wang, X., Chen, C., Huang, X., Current State and Challenges of Full-Scale Membrane Bioreactor Applications: A Critical Review. *Bioresour. Technol.*, *271*, 473–481, **2019**, doi: 10.1016/J.BIORTECH.2018.09.061.
36. Ayed, L., Ladhari, N., El Mzoughi, R., Chaieb, K., Decolorization and Phytotoxicity Reduction of Reactive Blue 40 Dye in Real Textile Wastewater by Active Consortium: Anaerobic/Aerobic Algal-Bacterial-Probiotic Bioreactor. *J. Microbiol. Methods*, *181*, 106129, **2021**, doi: 10.1016/j.mimet.2020.106129.
37. Ma, J., Chen, Y., Nie, J., Ma, L., Huang, Y., Li, L., Liu, Y., Guo, Z., Pilot-Scale Study on Catalytic Ozonation of Bio-Treated Dyeing and Finishing Wastewater Using Recycled Waste Iron Shavings as a Catalyst. *Sci. Rep.*, 8, 1, 7555, **2018**, doi: 10.1038/s41598-018-25761-6, /704/172/169/896 /639/638/77/887 /128 /140/146 Article.
38. Deveci, EU., Dizge, N., Yatmaz, H.C., Aytepe, Y., Integrated Process of Fungal Membrane Bioreactor and Photocatalytic Membrane Reactor for the Treatment of Industrial Textile Wastewater. *Biochem. Eng. J.*, *105*, 420–427, **2016**, doi: 10.1016/J.BEJ.2015.10.016.
39. Paździor, K., Bilińska, L., Ledakowicz, S., A Review of the Existing and Emerging Technologies in the Combination of AOPs and Biological Processes in Industrial Textile Wastewater Treatment. *Chem. Eng. J.*, *376*, 120597, **2019**, doi: 10.1016/j.cej.2018.12.057.

4

Trends on the Occurrence, Challenges, Migration, and Remediation of Emerging Contaminants in Aquatic Environments

Paki Israel Dikobe[1], Memory Tekere[1], Beauclair Nguegang[2] and Vhahangwele Masindi[1*]

[1]Department of Environmental Sciences, College of Agriculture and Environmental Sciences, University of South Africa (UNISA), Florida, South Africa
[2]Department of Chemical Sciences, Faculty of Science, University of Johannesburg, Auckland Park, Johannesburg, South Africa

Abstract
Water pollution problems have continued to increase not only in South Africa or Sub-Saharan Africa region but globally due to human activities. The presence of emerging contaminants (ECs) in surface water is mostly due to industrial development, agricultural practice and runoff and population growth, wastewater treatment facilities, landfill leachates, pharmaceutical companies, hospital wastes, and lifestyle waste. The ECs are unregulated substances that have entered the environment for as long as they have been produced through anthropogenic activities. Within the EC group, pharmaceuticals and personal care products are a group of the regular compounds employed in day-to-day urban activities. To date, regulatory approaches are inadequate to address and control these ECs, and the availability of such compounds in the environment and their potential detrimental effects has received an increasing attention from the scientific community and public at large. The accumulation of ECs in water and wastewater sources has raised various questions regarding the safety of potable water used for drinking, households, and other activities. The employed conventional water treatment processes can remove the organic pollutants; however, these ECs escape and ultimately present in the drinking water. There is, therefore, vital to design a robust approach to curtail ECs and their potential toxicity in drinking water sources. This book chapter reports

**Corresponding author*: emasinv@unisa.ac.za; masindivhahangwele@gmail.com

Elvis Fosso-Kankeu, Vhahangwele Masindi, Johannes Maree and Bhekie Mamba (eds.) Customized Technologies for Sustainable Management of Industrial Wastewater: A Circular Economy Approach, (135–192) © 2025 Scrivener Publishing LLC

the sources, classes, migration, effects, regulatory frameworks, and remediation of ECs in different spheres of the environment. Finally, future research avenues were also highlighted. This will guide future research and other areas of considerations.

***Keywords*:** Water pollution, emerging contaminants, wastewater treatment facilities, pharmaceuticals, anthropogenic activities, personal care products, water treatment processes, drinking water

4.1 Introduction

The scarcity of clean and fresh water globally and more specifically in South Africa (SA) has been a challenge that requires immediate remediation. Water contamination occurs easily because water is an universal solvent that can dissolve many substances [1, 2]. Water contamination and effectiveness of water treatment plants has become the most investigated subject in environmental research globally. Indeed, wastewater denotes any water with quality that is undesirably modified by natural causes and anthropogenic activities. The mixture of commercial, domestic, agricultural, and industrial activities, landfill, medical wastes, sewer outflow, storm water, or surface runoff has been identified as the principal sources of wastewater [3]. Municipal wastewater, mostly referred to as sewage, is generally transported in a clean or collective drain straight to a wastewater treatment plants (WWTPs) where it gets gradually cleaned. However, the occurrence and persistence of non-prescription and prescription ECs in aquatic environment streams have drawn worldwide attention in recent years. Although the health impact of these ECs on humans and aquatic species are still under investigation or unknown, various research studies have published the impact of some ECs. Various categories of the ECs are detected at high concentrations for both dam and drinking water. Despite political and government campaigns such as discharge effluent limits, Green and Blue Drop systems, and SANS 241, effluents discharged to the dams still necessitate more practical methods for the EC remediation. This, therefore, entails treatment processes, reduction, and removal of ECs before being discharged for human consumption. To alleviate water contamination challenges, numerous conventional water treatment unit processes such as flocculation, sedimentation, filtration, coagulation, and chlorination have been established [4]. Although these conventional unit processes have been proved suitable for the attenuation of inorganics contaminants from water, the ECs elude these unit process techniques and end up in the treated effluents and subsequently drinking water. Lately, adsorption is seen as an alternative to conventional mechanisms because it is cost-effective and ideal mechanism to reduce and remove the numerous ECs, primarily based on the type of adsorbent selected for the target adsorbate [5, 6]. Activated carbon (AC) has been selected as an ideal adsorbent, due to its well-developed porous structure and extensive surface area that renders it with effective and efficient adsorption capacities [7]. The carbon-based adsorption/AC has also been reported as a unique process for removing various categories of ECs from water [8]. This chapter presents a review on the occurrence and persistence of the selected ECs in aquatic environment and the removal by adsorption.

4.2 Emerging Contaminants in the Environment

The ECs are an enormous group of naturally or synthetic occurring chemical compounds or microorganisms, i.e., flame retardants, pharmaceuticals containing nonsteroidal anti-inflammatory drugs, fragrances, antibiotics and antifungal/antimicrobial agents, pesticides, food additives, hormones, artificial sweeteners, lipid regulators, ultraviolet (UV) filters, and other industrial products, with detrimental impact on human and environmental health. The previous extensive investigations of ECs during the past decades indicated their ubiquitous availability in the environment [9–12]. The ECs, such as endocrine-disrupting chemicals (EDCs), pesticides, personal care products (PCPs), and pharmaceutically active compounds (PhACs), have attracted worldwide attention because of their possible toxicity. The main sources of these compounds in the aqueous environment are agricultural, waste disposal sites, health care units, domestic, and industrial pathways [13–17]. Besides this, the current standard treatment plants do not completely remove the microbial metabolites and the ECs. Henceforth, this results in the introduction of the ECs into aqueous environment [17–19]. The main concern with ECs is the lack of information about their human health and ecological toxicity [20].

4.3 Classes of Emerging Contaminants

ECs of various categories can be recognized depending on their sources, effects in the ecosystem, chemical compositions, interaction, and mode of action. The ECs are very useful chemicals at the production stage; however, their metabolite and origination forms cause deleterious impact to human and the environment, resulting in concerns on their fate and impacts in the environment. The following section briefly narrates the various categories of ECs.

4.3.1 Personal Care Products

The PCPs are varied groups of synthetic organic compounds that include fungicides, antioxidants, pesticides, disinfectants, musk compounds, and preservatives employed as fragrances in some consumer products. They are generally used during manufacturing of products, i.e., soaps, sunscreens, toothpastes, lotions, and many cosmetics as active ingredients [21, 22]. For instances, frequently employed PCPs includes fragrances, i.e., polycyclic and nitro musks, sunscreens, UV filters such as methylbenzylidene camphor, antioxidants, and insect repellents such as N,N-diethyl-meta-toluamide, and applied as preservatives such as phenols and parabens to alter appearance, taste, odor, and touch [23, 24]. PCPs are not most essential product substances; however, their application in consumer goods makes them globally ubiquitous in the aquatic environment and poses doubtlessly complicated and ecological challenges that are no longer clearly understood [25–27]. During WTPs, some of those ECs are efficaciously eliminated; however, their fate remains dubious. In most instances, PCPs are applied externally on human skin or anatomy and not consumed through ingestion, resulting in high concentration of unchanged PCPs introduced into different environmental compartments in their unaltered forms [28, 29]. The PCPs are mostly consumed in larger quantities, resulting in high chances of damaging both the environment and human beings [30].

4.3.2 Pharmaceuticals

Pharmaceuticals are natural or synthesized chemical compounds employed for nutrition prevention, treatment, or diagnosis of diseases in animals and humans, also contributing to their lifespan and to enhance the growth rate and feeding efficiency of the animals [31, 32]. Pharmaceuticals are of various molecular formulas, chemical structures, and behavior, and their application varies according to the receiving organism: human, animal, or plants [31–33]. Richardson and Bowron [34] were first to report pharmaceutical as environmental pollutants; however, the negative ecological and environmental impact was later identified as "agents of subtle change" in late nineties [35]. Pharmaceuticals are categorized on the basis of their therapeutic uses, as per Table 4.1, as follows: antidepressant (e.g., venlafaxine (VFX)), antibiotics (e.g., sulfamethoxazole sodium (SMX)), anti-epileptic (e.g., carbamazepine (CBZ)), narcotic analgesics, (e.g., tremadol (TMD)), antiretroviral (e.g., FTC), anti-inflammatories and analgesics (e.g., ibuprofen (IBU), diclofenac (DCF), and paracetamol (ACT)), piperazine (e.g., monoacylglycerol lipase (CTZ)), and stimulants

Table 4.1 Major classes of pharmaceutical and their physicochemical properties.

Therapeutic groups	Acronyms	Molecular formula	Molecular weight (g/mol)	pKa	Log kow	Vapor pressure (mm Hg at 25°C)	Log P	Density (g/cm^3)	References
Stimulants	MET	$C_{10}H_{15}N$	149.23	9.99	2.07	5.4E-03	2.07	0.91	[38, 51, 52]
	1,7 DXN	$C_7H_8N_4O_2$	187.12	n/a	−2.45	n/a	1.57	1.6	
	MDMA	$C_{11}H_{15}NO_2$	193.24	9.9	2.15	2.27E-03	1.81	1.1	
	BZE	$C_{16}H_{19}NO_4$	289.33	n/a	1.3	1.1	2.71	1.3	
	COC	$C_{17}H_{21}NO_4$	303.35	8.7	2.30	1.91E-07	2.3	0.61	
	CAF	$C_8H_{10}N_4O_2$	194.20	10.40	−0.07	9.0E-07	−0.07	1.23	
Analgesic/Anti-Inflammatory	ACT	$C_8H_9NO_2$	151.17	9.40	0.46	6.29E-05	0.91	1.26	[51]
	DCF	$C_{14}H_{11}Cl_2NO_2$	296.15	4.20	4.51	6.14E-08	4.51	1.40	
	NAP	C14H14O3	230.26	4.15	3.18	1.89E-06	3.18	1.20	
Antibiotics	SMX	$C_{10}H_{10}N_3O_3S$	253.28	5.70	0.89	6.93E-08	0.89	1.50	[38, 51]
	TMP	$C_{14}H_{18}N_4O_3$	290.32	7.12	0.91	9.88E-09	0.91	1.30	
Antiretroviral	FTC	$C_8H_{10}FN_3O_3S$	247.25	2.65	−0.43	1.7E-08	−0.43	n/a	[38]
	EVZ	$C_{14}H_9ClF_3NO_2$	315.68	10.20	4.70	3.8E-07	4.60	n/a	
Narcotic Analgesics	TMD	$C_{16}H_{25}NO_2$	263.38	9.23	3.01	4.6E-07	1.34	1.01	[53]
	COD	$C_{18}H_{21}NO_3$	299.4	8.2	1.19	5.54E-09	1.39	1.32	

(Continued)

Table 4.1 Major classes of pharmaceutical and their physicochemical properties. (*Continued*)

Therapeutic groups	Acronyms	Molecular formula	Molecular weight (g/mol)	pKa	Log kow	Vapor pressure (mm Hg at 25°C)	Log P	Density (g/cm³)	References
Antidepressant	VFX	$C_{17}H_{27}NO_2$	277.41	9.50	3.20	2.46E-07	n/a	1.10	[54]
Piperazine	CTZ	$C_{21}H_{25}ClN_2O_3$	388.89	8.27	1.70	2.98E-11	2.80	1.20	[55]
Antiepileptics	CBZ	$C_{15}H_{12}N_2O$	236.27	13.90	2.45	1.84E-07	2.77	1.30	[51]
	BTA	$C_6H_5N_3$	119.12	8.37	1.44	2.5E-05	n/a	1.36	[55]
Pesticides	ATZ	$C_8H_{14}ClN_5$	215.68	1.60	2.61	2.89E-07	1.53	1.23	[56]
Depressant	MTQ	$C_{16}H_{14}N_2O$	250.29	2.4	2.50	0.9	2.50	1.2	[38]

(e.g., 1.3.7-trimetyl xanthine or caffeine (CAF)) [17, 36–38]. The frequent consumption of pharmaceuticals continues because of the high rate of preventing (instead of curing) diseases by consuming them as food additives and because of increasing or aging population. The inefficient municipal WWTPs highly contribute to a direct route of pharmaceuticals release to nature. The public, in general, is not aware or ignorant of the risk posed when carelessly discarding expired or unused drugs into toilets or sinks and ultimately introduced to water resources [39, 40].

High amount of these pharmaceutical products ends up in WWTPs because their metabolism in human, animal, and plants is incomplete. Other pathways are septic tanks, landfill site, showering or bathing, agricultural practices, industrial effluent, and urban wastewater [39, 41–43]. These classes of ECs are very stable and not easily removed by standard WWTPs because they are robust. Moreover, their chemical characteristics (non-volatile or polar nature) hinder the complete removal in WWTPs, resulting in their discharge to the environment [44]. Previously, in various countries like Brazil, Spain, USA, Holland, Canada, China, Germany, and even in SA, these PhACs have been discovered in different environmental water samples, i.e., groundwater, wastewaters, and surface water [37, 45, 46]. The elevated concentrations of veterinary and human pharmaceutical residues in different environmental compartment are of high concern because they are currently been detected above maximum permissible level in aquatic environment and sewage treatment plant [47] and potable water [48, 49], leading to their ubiquitous presence, resulting from their constant human consumption [50].

4.3.2.1 Different Type of Pharmaceutical Compounds

Various authors globally reported the presence of assorted pharmaceutical contaminants in river water and influents and effluents of WWTPs. These ECs are often categorized into various groups [57, 58].

4.3.2.1.1 Antibiotics

The discovery of first antibiotic substance by Sir Alexander Fleming in 1928 was opened door for a global manufacturing, commercialization, and usage of different type of antibiotics for various purposes, resulting to gradual accumulation in the environment [59–62]. From literature, if follows that the antibiotic consumption increased significantly from 21.1 to 34.8 billion in 76 countries within the period of 15 years (2000–2015) [63–65]. However, about 30%–90% of antibiotics are discarded into the environment as metabolites or in their natural forms [9, 61, 66, 67].

The antibiotics are employed to decrease mortality rates in various developing countries by minimizing fatalities from pathogenic diseases, thus resulting in high costs of antibiotic contamination [9, 66]. Antibiotics are also used for cultivation of aquatic organisms, i.e., to treat fish diseases [61, 68–70]. Once these compounds are metabolized by animals and humans, they yield the activated metabolites, hydroxylate forms, and conjugates [9, 71, 72].

The chemical structures of antibiotics compounds consist of cyclic components characterized by hexahydro pyrimidines, benzene rings, piperazine units, as well as morpholine, sulfonamides, and quinolone groups and have meta-stable properties [73–75]. Implying that, when antibiotics are introduced into the water resources, the enormous unknown consequences for various type of active chemical compounds with high reactive properties are introduced [66, 76, 77]. Antibiotics can also be introduced into the environment through the discarding of unused medicines by direct discharge as active pharmaceutical ingredients waste during production processes and excretion of residues after consumption [78–80]. A high concentration of mobile residues is also observed in urine and feces of individuals who consumes antibiotics, resulting in high concentrations of antibiotic residues in wastewater and which are also available in marine environments [77, 81, 82]. Various antibiotics degrades at different rates, i.e., fluoroquinolones and tetracycline are very persistent than penicillin, which implies that both fluoroquinolones and tetracycline will have better capacity to acquire in the water environment [83]. However, it is important to note that very minute amount from the total amounts presumably remains bioactive due to the binding capacity of many antibiotics to the particles [77, 84].

Antibiotics represent a group of ECs; then, their presence in water resources must be constantly monitored [85]. Ideally, antibiotics are designed and produced to eliminate bacteria and microbes; however, their side effects can be detrimental for human health and environment [83]. Previous studies reported that metabolites of antibiotics are persistent and are acquired in drinking and food supplies, including groundwater [85]. As previously highlighted, wastewater is a popular environmental sector influenced by these ECs, and modern decontamination techniques or approaches within the WWTPs do not entirely remove antibiotics contributing to the availability of the ECs in water resources [85], and they are also resistant to bacteria's in WWTPs [86]. As reported in literature, Bueno *et al.* [87] found ECs in concentration range of ng/L to hundreds ng/L in wastewater. Baralla *et al.* [88] reported the group-overall concentrations of antibiotics from 3,400 to 19,800 ng/L. Some of the ECs are ACT, NAP, IBU, DCF, and CAF [86, 87, 89].

4.3.2.1.2 Non-Steroidal Anti-Inflammatory Drugs

The non-steroidal anti-inflammatory drugs (NSAIDs) are another category of medication that consists of antipyretic, anti-inflammatory, and pain-relieving properties. The NSAIDs are mostly employed in both veterinary and human for pain relieve and long treatment of rheumatic diseases, and adverse side effects are inclusive of renal damage, cardiovascular, gastrointestinal, and liver complications [77, 90–93].

It has been previously reported that high consumption of NSAIDs leads to high concentration that are expressed in tens to hundreds of mg/L in the WWTP streams [38, 94]. IBU has been reported as dietary supplements and adulterants in food and are analyzed by quality control laboratories and regulatory agencies to ensure product quality and potency [57, 95, 96]. Moreover, it was highlighted that pharmaceuticals like DCF, NAP, ASA, IBU, ACT, and ketoprofen are present in potable water [58, 97, 98]. Due to their high mobility and persistent characteristic in the aqueous environment, the ECs are forever present in aquatic environment [99]. NSAIDs have a polar organic compound group that comprises one or two phenolic hydroxyl groups and carboxylic moieties [93, 100]. In acidic solution, the hydroxyl and carboxyl moieties are protonated because of the absence of ionic functional groups [101, 102].

ACT is mostly consumed for treating minor pains such as period pain and flu and for treatment conditions like muscle aches, backache, toothaches, arthritis, and headache [103, 104]. The mode of action of ACT is not yet understood; however, some studies revealed that ACT hampers the COX of the central nervous system [92]. ACT is used as veterinary medication for treatment of fermentation disorders, high concentration of acetone in the blood (acetonemia) in livestock, and treatment of pigs correlated with fever in poultry [103].

4.3.2.1.3 Antihypertensives

Antihypertensives are an additional group of pharmaceuticals mostly employed to reduce high blood pressure and contains beta blockers like atenolol, propranolol, and metoprolol. These antihypertensives, calcium channel blockers, angiotensin II receptor antagonists, diuretic furosemide, and beta blockers have constantly been discovered in tens of ng/L in various water matrices, i.e., drinking, ground, and surface water [55, 105]. Due to biodegradability resistance at lower concentrations, there are about 10% of antihypertensives removed from the water [106–108]. Other than that, Petrie *et al.* [109] and Evgenidou *et al.* [106] reported that CBZ metabolites are resistant in WWTPs, thus resulting in their presence in drinking water [97, 109]. Some of the sources for introduction of pharmaceuticals in water resources are direct discharge of untreated manufacturing effluents, municipality WWTPs, and indiscriminate dumping of unused or expired medicines [109].

4.3.2.1.4 Anticonvulsants/Antiepileptics

Anticonvulsants or antiepileptics drugs (AEDs) are assorted type of medicines consumed for the relief of generalized and partial seizures, epilepsy, neuralgia, and other mental disorders such as CBZ [38, 97, 109–112]. The latest-generation AEDs like retigabine, levetiracetam, gabapentin, rufinamide, lacosamide, and pregabalin have largely replaced first generation AEDs like barbiturates, valproic acid, and phenytoin, which were previously consumed to treat seizure disorders [112]. Because the severity of convulsions varies with patients, AEDs are administered to enhance medications efficacy at the accepting dosage range. Routine analytical testing of AEDs is also important in determining drug–drug interactions, assessing compliance and evaluating overdose [112]. Muñoz *et al.* [113] reported that CBZ can be observed in rivers receiving effluents from WWTPs in concentrations range of 1–3,000 ng/L due to its moderate persistence in water resources. Nkoom *et al.* [114] also reported that about 13% of CBZ that is used is excreted through feces and urine unchanged. It was reported that, in Daphnia, CBZ reduced population growth rate at 200 μg/L [115]. The combination of CBZ and diazepam initiates the effect of the phenotypic responses (like behavior) immediately after the organism is exposed to them, whereas different neuro-active compounds (NACs) such as fluoxetine needs about a week to induce the same response [112]. Other NACs can alter reproduction by delayed response in time for initial reproduction and thus negatively affecting population growth rates [112].

4.3.2.1.5 Hormones/Steroids

Hormones/steroids promote growth, osmoregulation, and stress responses in both human and veterinary pharmaceuticals [116, 117]. Their metabolism in both human and animal organism is incomplete, and they end up in water courses [117, 118]. Some of the synthetic steroids that promote hormonal functions or menopausal supplement for natural steroid hormones and birth control, i.e., estriol and estrone, have been detected in the water resources [117, 119]. Steroids consist of various groups of cell-signaling particles that have therapeutic efficacy in the treatment of diseases such as sexual dysfunction, inflammation, anemia, osteoporosis, and the reassembling of muscles post debilitating disease. Steroids are often consumed illicitly to boost athletic performance, increment strength, and lean muscle mass [120–122].

4.3.2.1.6 Antidepressants

Antidepressants (ADPs) are drugs mostly prescribed to treat conditions such as chronic pain conditions (off-label use), major depressive disorder, and anxiety disorders and to help manage addictions. The aftereffects of ADPs are dizziness, sexual dysfunction, dry mouth, weight gain, and headaches [38, 123]. ADPs are mostly combined with AEDs to control depression in patients with epilepsy [38]. ADP drugs aid as the synaptic junctive to inhibit their uptake of serotonin, noradrenaline, and other neurotransmitters [121] (e.g., dopamine). ADPs are discarded to WWTPs in their metabolized or parent form through urination. The increased consumption of ADPs includes high efficacy of ADPs for various health conditions and cultural acceptance of ADPs. ADPs can also diminish psychomotor and cognitive function, thus elevating the risk accident when driving or voluntary intoxication [124]. The presence of these contaminants in waste resources can cause detrimental health effects or corporeal [123]. Dolliver and Gupta [125] published that antibiotic contaminates the ground and surface water through leaching process and agricultural runoff. Consumption of antibiotics results in some of the health issues, such as dysfunction of hepatic, suppression of bone marrow, cardiac arrhythmia, and disruption of immune system, and affects food chain [123]. Lajeunesse *et al.* [126] and Metcalfe *et al.* [127], through their investigations, detected the ADPs and their metabolites in raw sewage and final effluent of WWTPs and surface water. Despite their efficacy in relieving depression, various side effects including sexual dysfunction, acute and chronic toxicity, hypoglycemia, and hindering or delaying the development of aquatic life form [123]. Increased consumption of these ADPs by children is mainly due to high threat of suicidal thinking and behavior in children and juvenile. People also undergo discontinuation syndrome, duplicating their previous depression when they cease consuming these ADPs [128, 129].

4.3.2.1.7 Antiretroviral (ARVs)

The ARVs are medications used worldwide for the treatment of people living with human immunodeficiency virus (HIV). In 2021, it was published that SA is among the leading countries consuming ARVs due to high number of people (around 7.7 millions) living with HIV [130]. Following the digestion of these ARVs throughout the human organisms, they are released into the different environmental compartment through urination or feces as revealed by their presence in various WTPs, raw sewage and final effluent, and surface water in most European countries, thus suggesting their presence in South African water resources [131]. The country implemented the biggest antiretroviral treatment (ART) program globally to reduce the HIV/AIDS pandemic. The fixed dosage combination (FDC) of antiretroviral drugs (ARVDs) was implemented as primary stage for treatment to HIV-infected patients additional to other mostly used ARVDs. The FDC is one pill containing the multiple drugs, commonly EVZ, emtricitabine (FTC), and tenofovir [132]. Suppose a daily consumption of a patient is 991 mg/day/person of HIV-ARVDs with almost mean range of 590–1996 [133] that would result to about 1,085 metric tons of HIV-ARV compounds ingested annually in SA. Due to the elevated numbers in SA of people absorbing and relying on ART, it has been regarded that this will lead to high availability of these compounds in the environment [130, 132–134]. The ubiquitous of these compounds in the environment may result to a detrimental and unique scenario, and these compounds might transform. The potential repercussions caused by the availability of these compounds in water resources within SA can be exacerbated by the water challenges and water scarcity in Sub-Saharan Africa and overall lower percentage rainfall, resulting in no or minimal reduction of concentration of these ARV compounds in the water resources. The sheer types of ARV compounds employed for HIV/AIDS treatment poses analytical challenges [130, 131, 135, 136]. These compounds are entry inhibitors, protease inhibitors, integrase strand transfer inhibitors, fusion inhibitors, and nucleoside and non-nucleoside reverse transcriptase, which are employed to treat HIV [135] and to prevent mother-to-child transmission [137]. The compounds include Lopinavir, Abacavir, Lamivudine, Didanosine, Zalcitabine, Abacavir, Tenofovir, EVZ, Stavudine, Nevirapine, Ritonavir, Zidovudine, and Indinavir [134, 135]. The consequences and impacts of ARVs in water resources and environmental relevance might not be immediately apparent as compared to other substances, i.e., antibiotics, which is drug resistance in the environment and water resources [135, 138]. The HIV is not waterborne disease and is only hosted by human, as is the case with other viruses or bacteria [134, 136].

4.3.2.1.8 Stimulants (Non-Amphetamine)

Stimulants are class of medications used to stimulate the central nervous system to produce hormones. They are used for various purposes including enhancing alertness, motivation, cognition, mood, attention, and physical performance; however, with various side effects such as anxiety, sleeplessness or inability to sleep, and tiredness. Among the stimulants, nicotine (NIC) is the most consumed due to the high number of smokers globally [134, 136].

4.3.3 Pesticides

The pesticides are fusion of substances intended for preventing and controlling undesirable types of animals or plants and destroying pest (vectors of animal or human disease). These substances are used to destroy insects or other organism harmful to plant or animal [139–142]. The mainly contributing sources of pesticides are as follows: domestic wastewater (runoff from lawns and gardens, improper cleaning, roadways, etc.) and agricultural runoff [143]. Pesticides are environmental pollutants of concern because they contaminate water, air, and soil because they seep into groundwater, enter surface through runoff, and transported to long distances [144, 145].

4.4 Sources of Emerging Contaminants

The uncontrolled discharge of EC waste through sources such as fertilizer industry, pharmaceutical industry, hormones, PCPs, antibiotics, pesticides, artificial sweeteners, clothing adhesives, illegal drugs, cleaning solvents, and antibacterial adds to their distribution in aquatic systems, hence enhancing the possibility of posing detrimental and deleterious effects to living organisms [146–149]. Major sources and the distribution pathways of ECs are highlighted in Figure 4.1. Their presence in various water bodies is affected by environmental conditions and natural mechanisms [150]. Sources such as pesticides, PCPs, and pharmaceuticals are the key contributors for ECs in aquatic environment, and this is primarily based on their high solubility and consumption [151]. The ECs are not biodegradable and are not completely attenuated using conventional wastewater treatment technologies [152], and it results in different ECs and their transformation products present in the final effluent to be discharged [153].

The level of concentration and number of various ECs available in surface water and ground water depends on sewer conditions and catchment characteristics, i.e., land use and population density [32]. The availability of ECs, such as nanoparticles, ECDs, medicine, per fluorinated chemicals, and cosmetic products in the ecosystem of the emerging contaminants, poses a big threat to the marine and individual organism's welfare and health [98]. The environmental presence and persistence of the ECs have irreparable

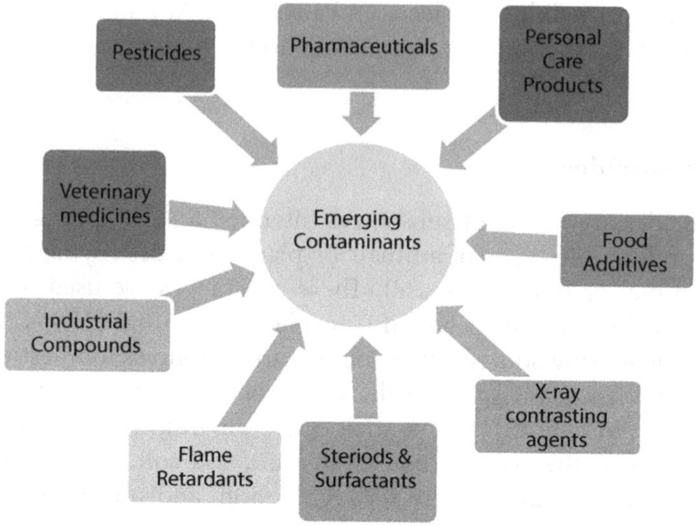

Figure 4.1 Pollutants of the emerging contaminants.

toxic effects, presumably resulting to the extinction of ecosystems [154]. Therefore, it is important to develop and regulate strict discharge limits for ECs and model advanced technologies for removal of ECs from the water treatment processes (WTPs). The unregulated effluent discharge in SA is due to the lack of strict legislation and insufficient information on the toxicity and fate of the ECs. Some countries have established interim aquatic guidelines for some ECs, and they include the Australian Drinking Water Guidelines 2011 [155], United States Environmental Protection Agency (USEPA) [11], World Health Organization (WHO) [156], European Union (EU) [157], and other global regulatory bodies, which have documented a list of inorganics contaminants with high ecological and human health effects.

4.4.1 Agricultural Practices

In some countries, about 70% of freshwater is consumed through irrigation and agricultural practices, i.e., it is anticipated that United States will undergo constant increase in the destined future [158, 159]. Farming does not only contribute to major contaminants, but it also leads to the reestablishment and rise of these contaminants via wastewater into water bodies. During farming, significant inputs of fertilizers and pesticides are consumed with main objective of producing best possible quantities of fiber and food [160–162]. Pesticides contain a broad various substance with chemical and physical properties, which can be used in agricultural activities to limit the dissemination of weeds, dangerous insects, and microorganism like bacteria and fungi. There are four types of pesticides: fungicides, insecticides, bactericides, and herbicides [163]. Pesticides reach the aquatic environment from agricultural sites through runoff and these contaminants may be bioaccumulated in vegetation, sediments, and living organisms, depending on the degree of substance solubility [140]. Some of the ECs from pesticides include, but not limited to, ATZ, Chlorpyrifos, Clomazone, Estriol, Carbendazim, Estrone, Difenoconazole, DCF, Triclosan, Progesterone, Ametryn, and CAF [163].

4.4.2 Wastewater Treatment Facilities

Wastewater is the water released from domestic, commercial, or industrial use. This water must be properly treated prior to being discharged into water bodies to avoid groundwater contamination. Some wide ranges of synthetic chemicals that can be regarded as ECs are used by humans in domestic settings and ultimately discharged with the resultant diverse human wastes. A conventional WWTPs cannot successfully eliminate such massive contaminant groups, and, therefore, final effluent is discharged into rivers and dams with such ECs [16, 38, 164]. The sources of EC in human waste are mainly PCPs and pharmaceuticals. Some of the ECs in pharmaceuticals detected from the human wastes are DCF, IBU, ACT, and CAF. Some ECs in PCPs includes disinfectants contaminants like triclocarban and triclosan; fragrances contaminants such as butylparaben tonalide, octinoxate methoxycinnamate, butylated hydroxyanisole, celestolide and phantolide, benzophenone, parabens, and galaxolide; UV filters; microplastics; butylated hydroxytoluene; and polydimethylsiloxane [16, 165]. For example, some of antibiotics (SMX, erythromycin, triclosan, and TMP) were detected in influents of WWTWs located in South California [166]. Other classes of pharmaceuticals, like ACT, NAP, COD, and CAF, were also detected WWTPs in Costa Rica [167]. Furthermore, pharmaceuticals conserve their therapeutic properties in order to remain persistent in the environment [164, 168]. Based on that, the physicochemical properties of pharmaceuticals as indicated on Table 4.1, such as volatility, high polarity, persistence, high lipophilicity, and adsorption, can influence their removal effectiveness during the treatment processes in WWTPs [153, 164]. The PCPs are a wide range of basic household chemicals that are used for cleaning, washing, and improving health, skin treatment, and beauty. Hair foods or care products, skin products, cleaning and washing products, body perfumes, and ornamental cosmetics are samples of PCPs. The main PCPs used in the whole world in high volumes are personal care and skincare products resulting in high daily usage and ultimately discharged constantly in larger volumes into the environment [169]. Consequently, the amount of ECs released into the environment is increasing enormously on the daily basis [16]. The ECs are mostly used for skin care products then ultimately introduced to the environment at concentrations that can damage human beings and public health [170].

4.4.3 Landfill Leachates

Leachates flowing from municipal solid waste (MSW) end in surface water and groundwater through infiltration [171–174]. The MSW is a major waste type containing daily byproducts from various industries [172, 175–178]. Previous studies have found a broad range of toxic contaminants in landfill leachate, including microplastics [8, 150], pesticides [179], polycyclic aromatic hydrocarbons (PAHs) [180], pharmaceuticals [181], pathogenic organisms [182], and phenolic compounds [183]. Pharmaceuticals are main collective of ECs, and its availability in potable water has caused notable attention concerning the risks of unfavorable effects on fauna, humans, and estrogenic [184, 185]. Some of pharmaceuticals are also employed in livestock, fishery, and farm animals to increase size of the mammals and to reduce sicknesses in some animals [40], later released to the environment as waste then ultimately being introduced to aquatic environment via run off during raining seasons.

4.4.4 Industrial Effluents and Pharmaceutical Companies

Wastewater from pharmaceutical companies contains more ECs than wastewater from others industries due to the presence of antibiotics residues. Antibiotics are consumed globally to eradicate various types of bacteria and virus; however, once consumed, the non-digestible part enters the water courses and other environmental compartments through urination and feces release [186, 187]. In addition to eradicate bacteria in human and animal, antibiotics are also widely employed in the aquacultural trade to treat fish diseases [68–70]. The wastes generated during manufacturing processes of feeding products for farm animals are reclaimed as organic fertilizers for plantation, ultimately released in the aquatic environment through runoff and comprise of antibiotics [68, 188], and are main contributors of high concentrations of antibiotics in the aquatic environment [189]. Carraro *et al.* [190] cited a high concentration of SMX (7,950 ng/L) in industrial effluent. Henceforth, antibiotic traces have been reported in wastewater from pharmaceutical institutions, substantially polluting the environment. The findings reported above illustrate that pharmaceutical production companies contribute notably to the presence of antibiotic in the aquatic system [191].

4.4.5 Hospital Wastes

Most antibiotic medications consumed by hospitalized patients are incompletely absorbed and metabolized by the body, resulting in some of excreted antibiotics flushed down the waste drain and discharged into hospital effluents, increasing concentrations of antibiotic residues [192, 193]. Due to the unregulated widespread consumption of antibiotics in medical facilities [190], antibiotic resistance genes are a cause for concern globally [194]. It has been reported that waste generated by hospitals contains substantial amount of antibiotics and microorganisms [195]. It has been reported by Carraro *et al.* [190] that the content of certain antibiotics such as Ciprofloxacin, Erythromycin, and Ofloxacin is elevated in hospital wastewater. Li *et al.* [196] also reported an extreme antibiotic concentration of 3,701 ng/L in hospital wastewater. Figures that significantly increased due to recent COVID -19 pandemic, resulting to a high number of patients worldwide and the consumption of antibiotics as well as the vaccination of against COVID-19, thus contributing to the release of ECs into different compartment [197, 198].

4.4.6 Lifestyle Waste

The septic tank is employed globally as the most customary and conventional on-site sewage management. Septic tanks are used in places with limited or non-admittance to public sewage systems [199]. However, they are not as effective and efficient as centralized WWTPs due to anaerobic conditions, poor placements, and undetected failures [200]. The leakage of these on-site treatment tanks is a main source behind regional and local contamination of surface and ground water in China, the USA, and many other countries [201–203]. According to a study and research conducted, about 25% of US households depend on septic tank systems. It is the main source of organic contaminants such as CAF, estradiol, nonylphenol, diethoxy carboxylates, nonylphenol mono and estrone, steroid hormones, and fluorescent whitening agents into the groundwater [200]. Various factors such as the septic tank density, maintenance of the septic tank, and improper construction and installation are main contributors to the likelihood of the septic tank to contaminate the groundwater. The geology of the site also plays an important role [204];

for instance, sandy aquifers have low organic carbon, low specific surface area, and high hydraulic conductivity. These characteristics make the infiltration or adsorption of contaminants easier when compared to other aquifers, so sandy aquifers are prone to septic tank contamination [205]. Some of the areas having highly transmissible aquifers and shallow groundwater tables are at utmost risk [206].

4.5 The Effects of the Emerging Contaminants

4.5.1 Effects on Human and Animal Health

Human consumption of some of the EDCs consequently interferes with the hormone system and possible side effects of cancer in human beings. It also changes the organism's growth and suppresses their reproduction, leading to their inhabitants' decline in reproduction and thereby resulting to population decline [207, 208]. In addition, EDCs can lead to acute or long-term effects such as endocrine disruption, cancers, immunotoxicity, neurology disorders, antimicrobial resistance, estrogenicity, and intersex [209, 210]. ATZ is a man-made herbicide from the triazine class, employed for controlling the broadleaf weeds in agricultural industries (i.e., the sorghum, sugar cane, and maize). Epidemiological studies published a correlation between elevated ATZ concentrations in potable water and low sperm motility and volume [211, 212], small-for-gestational-age [213], fetal gastroschisis [214], intrauterine growth retardation [215], and fetal growth defects, including restriction [216], spina, cryptorchidism, epispadias, and hypospadias; and increases in limb reductions (upper and lower) [217]; and chances of both prostate and breast cancer [211, 212].

The presence of atenolol and CBZ can disturb the developments of human embryonic stem cells [218]. CBZ is a therapeutic used as a mood-stabilizing and an anticonvulsant drug [58, 219–221]. It was previously reported that epilepsy patients who receive CBZ therapy during pregnancy delivered babies with urinary and cardiovascular tract anomalies and with a high rate of congenital anomalies such as neural tube defects [222]. It also affects development and normal growth in larvae and zebrafish embryos, accelerating the hatching rate, swim bladder appearance, body length, yolk sac absorption rate, and expression pattern of neural related genes in other fishes [223, 224]. DCF is analgesic and NSAID; it is used to relieve arthritis. Various authors [225–228] reported that DCF is toxic to scavenging birds and affects the metabolism function in fish. NAP is NSAID used to treat menstrual cramps, pain, and inflammatory diseases, i.e., fever, gout and rheumatoid arthritis. It affects the mRNA expression, causes thyroid disruption, causes renal and gastrointestinal effects in zebrafish, and accumulates in the bile of fishes [229–235]. CAF is a psychoactive drug that acts as stimulant and is predominantly found in food and beverages particularly coffee [236]. It causes panic disorders and anxiety among humans and thus leads to hepatocellular, endometrial, and colorectal cancer [22, 237–238]. It also induces lipid peroxidation and oxidative stress, affecting metabolic activity and energy reserves, and neurotoxic effects, affecting reproduction on terrestrial insects and aquatic species [239–241]. The leaves are used as herbicides due to their plant being touted to inhibit the development and growth of other neighboring plants [237]. It tastes recalcitrant bitter, and, based on that, it is mostly employed as an insect repellant and toxicant at high dose [242] and boost pollination to enhance reproduction [243]. According to Doepker *et al.* [244], it has been implicated to trigger behavioral, cardiovascular, and reproductive effects [244]. ACT is an analgesic and anti-inflammatory that is used to treat mild to moderate pain (from osteoarthritis, toothaches, headaches, backaches, menstrual periods, or flu/cold aches and pains) and to reduce fever. It can cause severe toxicity effects in fish such as Oryzias latipes, in bacterium such as Vibrio fischeri, and in the aquatic invertebrates such as Daphnia magna. It also affects reproduction system, growth, and endocrinal function in fish and nephrotoxicity in humans [245–247]. COD is an opiate, analgesic, and antidiarrheal drug and is antitussive. It is also known as methylmorphine. It causes obstructive sleep apnea because of opioid sensitivity, dizziness, nausea, vomiting, sedation, and lightheadedness and, sometimes, death in children [38]. TMP is used as the antibiotic, especially for bladder infections. It is highly toxic to algae, can disrupt thyroid hormone homeostasis, and can exert developmental and reproductive effects in fishes [246, 248]. FTC and EVZ are the antivirals used to cure HIV AIDS, and their presence of these antivirals poses a risk to fish, daphnia, and algae, resulting in ecosystem imbalance. They can also cause liver damage and lower immunity in tilapia fish [57, 249].

4.5.2 Effects on the Environment and Ecosystem

Once released, pharmaceuticals and personal care products (PPCPs) become toxic to various spheres of the environment if not properly managed because some are bonded with other contaminants, thus increasing their toxicity effects [63, 250]. The aquatic environment is area of interest because it is the most severely affected. There are plenty illnesses related to ECs, and about 700 ECs and their transformation products and metabolites are most likely introduced to the water resources, resulting in severe harm to the aquatic ecosystem. Aquatic animals are easily affected by the any minute variations in their surroundings, and the main affections include mutations in the genetic of aquatic life form. The noxious impacts of ECs include adapted immune system, feminization, changes in behavior, and varied sex ratios in aquatic species [109, 251–254]. During bioaccumulation process that happens by ingestion or by absorption through skin, the ECs stimulate the toxins in the anatomy of organisms. Bioaccumulation transpires at the base of food web concentration, due to direct exposure to ECs [255, 256]. The toxicity concentration elevates to state where it is deadly to the organism. Constant deceases of the genus lead to their extermination, leading to the imbalance in the ecosystem. Various alterations in the aquatic ecosystem cause the climatic changes [256]. Rahman *et al.* [257] described the health entailments of ECs in aquatic environments.

The ECs can be introduced in the aquatic food chain via various pathways, resulting in causing biomagnification [238, 255], and, as a result, increased compound concentration in organisms at the successive trophic concentrations can be obtained in the food chain [153]. Dodgen *et al.* [258] have conducted the study on the accumulation and uptake of EDCs/PPCPs, NAP, and DCF sodium in collards (Brassica oleracea) and lettuce (Lactuca sativa). The prolonged exposure of the ECs can provoke adverse effects on the aquatic environment [259]. The main concerns in relation to some of the following ECs (i.e., pesticides and antibiotics) are the development of bacterial resistance once introduced to the environment [260, 261] and the detrimental impact on the biodeterioration of plant components, which ultimately disarrange the basic food cycle in aquatic environment [150].

4.6 Variation of Emerging Contaminants in Aqueous Environments

As it can be seen in Table 4.2, ECs have been observed in different countries worldwide. Some of the obtained concentrations are very high and pose the high risk to the consumers as most of the DWTPs are not able to completely remove or reduce them. The highest COD concentrations were obtained in Lagos River [45], followed by Uhlava River and Po River [55]. Highest BZT concentrations were obtained in Besòs River [122], followed by Sha River [260] and then Ahar River [102]. The high ACT concentrations were obtained in St. Lawrence River, followed by Montmorency River and Chaudière River [15]. The high CAF concentrations were obtained at Jundiai River [262], followed by Huangpu River [263], and the low concentrations were detected in

Table 4.2 List of emerging contaminants detected in various rivers used as raw water sources for drinking water treatment plants.

Analytes	Concentration range (ng/L)	River names	Countries	References
ACT	0–17; 0–4; 5.0–270; 1.7–260; 6–150	Saigon; Ganges; St. Lawrence River; Montmorency River; Chaudière River	Vietnam; India; USA; Canada	[15, 55]
COD	0–1; 0–1; 2–1780	Uhlava; Po; Lagos	Czech Republic; Italy; Nigeria	[45, 55]
CAF	2148–15336; 0–37; 0–108; 0–31; 0–67; 0–90; 0–32; 0–18; 3–32.4	Jundiai; Yangtze; Huangpu; Malaren; Uhlava; Po; Saigon; Zurich; Ikpa	Brazil; China; Sweden; Czech Republic; Italy; Vietnam; Switzerland; Nigeria	[55, 262, 263]
DCF	27–364; 0–28	Jundiai; Uhlava;	Brazil; Czech Republic;	[55, 263]
Analyte	**Concentration range (ng/L)**	**River names**	**Areas**	**References**
NAP	7–145; 3–2120	Jundiai; Lagos	Brazil; Nigeria	[45, 263] [45, 263]
SMX	1.0–24; 0–4; 0–14; 0–7; 0–3; 0–6; 0–161; 0–2.8; 11–816	Delaware; Huangpu; Uhlava; Zurich; Po; Saigon; St. Lawrence River; Ikpa River; Ahar River	Pennsylvania; China; Czech Republic; Switzerland; Italy; Vietnam; USA; Nigeria; India	[55, 122, 262, 265, 269]

(*Continued*)

Table 4.2 List of emerging contaminants detected in various rivers used as raw water sources for drinking water treatment plants. (*Continued*)

Analytes	Concentration range (ng/L)	River names	Countries	References
CBZ	2–243; 15–659; 0–660; 0–0.04	Delaware; Jundiai; Nairobi; Ikpa	Pennsylvania; Brazil; Kenya; Nigeria	[79, 262, 263, 265]
ATZ	0–9; 0–3; 0–6; 0–74; 0–7; 0–0.51; 0–171; 0–0.15	Yangtze; Uhlava; Po; Huangpu; Saigon; Zurich; Parana; Owan	China; Czech Republic; Italy; China; Vietnam; Switzerland; Brazil; Nigeria	[55, 270]
Analyte	**Concentration range (ng/L)**	**River names**	**Areas**	**References**
TMD	0–7.20; 0–14.88; 0–45.95; 0–2.26; 0–4.32; 10.74–23.04; 0–852	Malaren; Uhlava; Zurich; Po; Saigon; St. Lawrence River; Lagos	Sweden; Czech Republic; Switzerland; Italy; Vietnam; Nigeria	[45, 54, 55]
VFX	0–1.57; 0–11.85; 0–4.59; 0–2.31; 0–1.02; 10.8–41.9	Malaren; Uhlava; Zurich; Po; Saigon; St. Lawrence River;	Sweden; Czech Republic; Switzerland; Italy; Vietnam; USA	[54, 55]
TMP	0.1–63; 0–388; 0–17; 4–223; 2–388	Delaware; Lagos,	Pennsylvania; Nigeria, Bangladesh; Serbia	[45, 265, 271]

(*Continued*)

Table 4.2 List of emerging contaminants detected in various rivers used as raw water sources for drinking water treatment plants. (*Continued*)

Analytes	Concentration range (ng/L)	River names	Countries	References
CTZ	0–20.26; 0–2.08; 0–12.62	Uhlava; Po; Saigon;	Czech Republic; Italy; Vietnam	[55]
1,7 DXN	0–116	Yangtze	China	[8]
Analyte	**Concentration range (ng/L)**	**River names**	**Areas**	**References**
EVZ	3-303; 134–354; 0–560; 0–33; 0–148	Hartbeespoort dam; Juskei; Kisumu; Mathare; Rietvlei Nature Reserve	South Africa; Kenya	[79, 267, 268, 272]
FTC	0–45; 0–8	Rhine; uMngeni	Germany; South Africa	[55, 267]
BTA	39–526; 1080–2856; 1–1732	Ahar River; Besòs River; Sha River	India; Spain; South China	[102, 122, 260]
COC	0.3–6		United Kingdom	[262]
BZE	0–92		United Kingdom	[262]
MDMA	0.5–9		United Kingdom	[262]

other rivers. The highest DCF concentrations were obtained in Jundiai River [262], followed by Uhlava River [263]. The highest NAP concentrations were obtained in Jundiai River [262], followed by Lagos River [45]. The highest

SMX concentrations were obtained in Ahar River [102], followed by St. Lawrence River [15], and the low concentrations were detected in other rivers. The same highest CBZ concentrations were obtained in both Nairobi [264] and Jundiai River [262], followed by Delaware River [265], and the low concentrations were detected at Ikpa River. The highest ATZ concentrations were obtained in Parana River [266], followed by Huangpu River [263], and the low concentrations were detected in other rivers. The high TMD concentrations were obtained at Lagos River [45], followed by Zurich River [55], St. Lawrence River [15], and Uhlava River [263], and the low concentrations were detected in other rivers. The highest VFX concentrations were obtained in St. Lawrence River [15], followed by Uhlava River [263], and the low concentrations were detected in other rivers. The highest TMP concentrations were obtained in Delaware River [265], followed by Lagos River [45], and the low concentrations were detected in other rivers. The highest CTZ concentrations were obtained in Saigon River [265], followed by Uhlava River [55], and the low concentrations were detected in Po River [55]. The highest 1,7 DXN concentrations were obtained in Yangtze River [8]. The highest EVZ concentrations were obtained in Kisumu River [264], followed by Juskei River, Hartbeespoort dam, and Rietvlei Nature Reserve River [267], and the low concentrations were detected in Mathare River [268]. The highest FTC concentrations were obtained in uMngeni River [267], followed by Rhine River [143].

4.7 Required Limits of Potable Water Quality Standards and Guidelines

The concentration of ECs in environment is regulated by various regulatory bodies, which include international regulator such as WHO and local regulator such as Department of Forestry Fisheries and Environment and Department of Water and Sanitation. These regulatory bodies regulate the EC concentration in an aquatic environment, which, when exceeded, may pose a threat to human and other organisms on exposure [53]. Perhaps due to insufficient data on behavioral and occurrence pattern, in SA, there are no regulatory guidelines to ECs. Table 4.3 lists the different guideline values of some EC concentrations in the aquatic environment proposed by various international regulatory bodies.

Table 4.3 Statutory guidelines given by international regulatory bodies [11, 155–157, 273].

Compounds	CAS number	U.S. EPA	Europe union	WHO	ADWG	Minnesota
Carbamazepine	298-46-4	NL	2000 μg/L	NL	NL	40 μg/L
Diclofenac	15307-86-5	NL	NL	NL	NL	NL
Naproxen	22204-53-1	NL	NL	NL	NL	NL
Caffeine	58-08-2	NL	NL	NL	NL	NL
Sulfamethoxazole	723-46-6	NL	NL	NL	NL	100 μg/L
Acetaminophen	103-90-2	NL	NL	NL	175 μg/L	200 μg/L
Naproxen	22204-53-1	NL	NL	NL	NL	NL
Efavirenz	154598-52-4	NL	NL	NL	NL	NL
Emtricitabine	143491-57-0	NL	NL	NL	NL	NL
Venlafaxine	93413-69-5	NL	NL	NL	NL	10 μg/L
Tramadol	27203-92-5	NL	NL	NL	NL	NL
Cetirizine	83881-51-0	NL	NL	NL	NL	NL
Trimethoprim	738-70-5	NL	120 μg/L	NL	NL	NL
Benzotriazole	95-14-7	NL	NL	NL	NL	20 μg/L
Benzolecgonine	519-09-5	NL	NL	NL	NL	NL
Atrazine	1912-24-9	3 μg/L	0.6 μg/L	100 μg/L	2.00 μg/L	3 μg/L
1,7 DXN	611-59-6	NL	NL	NL	NL	NL

N.B.: Not listed (NL)

4.8 Treatment of Emerging Contaminants

Various techniques have been designed and are currently applied to attenuate ECs in environment. Despite their ability to attenuate ECs to acceptable maximum permissible limit, there are various drawbacks associated with the existing technologies as reported in Table 4.4. As a result, research on the ideal application of additional treatment processes has paid significant attention to novel and innovative ways over the last few years [274].

Table 4.4 Shortcomings of various conventional drinking water treatment plants.

Method	Disadvantages
Chemical Oxidation (Ozone, Chlorination, Potassium Permanganate, etc.)	• Inability to completely remove ECs • Limited lifespan (ozone) • Formation of toxins • Release of toxic sludge • Inability to remove dyes • Not cost-effective
Adsorption	• Nondestructive technique • High capital cost • Non-selective process • Rapid saturation of adsorbents
Filtration	• Higher energy requirement • High capital cost • Membrane fouling • Limited flow rate
Photocatalysis	• High cost • Formation of secondary pollutants • Useful for concentrated effluents only • Air pollution
Electro Fenton Process	• Nondestructive technique • High investment • Non-selective process • Rapid saturation of adsorbents
Electrocoagulation	• Economical not sustainable • High maintenance cost • Discontinuous process • Inhibition of electrodes due to sludge deposition

4.8.1 Oxidation

4.8.1.1 Chlorination

Chlorine gas, hypochlorite, and chlorine dioxide are widely used for oxidation. The chlorine dioxide (ClO_2) is mainly used as a disinfectant instead of chlorine, due to its efficacy in pathogen deactivation and less yields of byproducts. During ClO_2 origination, chlorine is naturally organized into resulted ClO_2 solution as "contamination," which could accord the benefit of ClO_2 disinfection. The ClO_2 produces lesser amounts or fewer types of organo-halogenated byproducts as compared to chlorine [275]. The disproportionation reactions when using ClO_2 result in existence of other DBPs, like chlorites (ClO_2^-) and chlorates (ClO_3^-) in the purified water [276]. The ClO_2 displays a higher bactericidal capability and oxidation capacity in comparison with chlorine [277]. The ClO_2 is regularly applied as a disinfectant in water treatment; however, recent reports highlighted that it can also be used as the oxidant to remove the pharmaceutical residues like tetracyclines [278], DCF [279–282], and SMX [276, 283, 284] in wastewaters.

4.8.1.2 Potassium Permanganate Solution

Potassium permanganate ($KMnO_4$) is a very strong chemical oxidant globally used in WTPs, and it can be used to synthesize organic compounds [285]. In conventional WTPs, potassium permanganate mixture is dosed into raw water, at the mixing tank in together with coagulants. When used in filtration plants, the oxidant is typically dosed into raw water input to allow more contact duration upstream of filter process units. In all cases, this oxidant is added before the filtration. For better improved manganese discharge, expanded detention time is highly required [286]. McBeath and Graham [287] reported that ATZ was degraded by simultaneous hydroxyl radical coupled with permanganate mediated oxidation.

4.8.1.3 Ozonation

Ozonation cycle is commonly during wastewater and WTPs, specifically for disinfection and for removing harmful organic compounds. Catalytic ozonation has been recently used and has gained high attention [288]. During catalytic ozonation, the catalysts are used to create the hydroxyl radicals (OH^-) and promote ozone decay that can address the ozonation drawbacks. Due to its outstanding catalytic efficiency, simple formulation, and, generally, the iron availability, iron-based materials are mostly utilized catalysts for heterogeneous catalysts ozonation [289]. An important issue with ozonation the high capital cost in addition to its complexity in operation. Ozone is toxic chemical that can be harmful if inhaled in large doses. Furthermore, ozone can also corrodes the pipes and tank containing water to be treated [154]. Rout *et al.* [259] reported a very high removal efficiency (>95%) of DCF, CBZ, indomethacin, and TMP at an ozone dose of 5 mg/L.

4.8.1.4 *Adsorption*

The ECs can represent a problem in DWTPs because their removal cannot always be efficient with conventional treatments [290]. Numerous optional treatment methods have been considered to remove the ECs, including the membrane filtration or AC treatment. Adsorption has been globally employed to remove the ECs because it uses phase varying techniques, in which the adsorbates (contaminants) migrate from the aqueous to the solid phase (adsorbent) [5, 6, 150, 280]. AC is considerably used adsorbent for a wide-spectrum EC adsorption due to its larger specific surface area, high porosity, and the high concentration of surface interactions [291–295] and because it is rich in functional group, i.e., carbonyl, lactonic, quinone, carboxyl, and phenolic group [7, 294]. AC is classified as granular activated carbon (GAC) and powder activated carbon (PAC) based on the particle size as mesoporous (2–50 nm), microporous (2 ≥ 0.8 nm), and macroporous (≥50 nm), and this has been confirmed by the Webber-Morris model of inter-particle diffusion. Both GAC and PAC can effectively and efficiently remove ECs from wastewater. The mesoporous AC is the most ideal to remove ECs because it interferes less with the organic compounds for the adsorption active sites [7, 296–298]. Effective removal of the ECs in water is mostly related to the applied AC dosage [285, 299, 300]. Naddeo *et al.* [207] reported that the removal efficiency of 99% on selective ECs was achieved by using PAC. AC treatments shows great EC removal efficiency, and it is an attractive technology for upgrading WWTPs and DWTPs [301, 302].

4.8.1.5 *Filtration*

Filtration process is the final stage of clarification process taking place in a potable water treatment plant, whereby colloidal and suspended particles available in an aquatic suspension that passes through a porous medium are removed [303] and are also used to ensure that the water produced is fit and safe for human consumption [304]. The use of reverse osmosis

(RO) and nanofiltration (NF) is for the expansion of potable water from contaminated stream or river water, and reusing of the potable water then completely removes the ECs. The most prominent membrane processes that are capable to remove the ECs but are not yet effectively used are membrane distillation, forward osmosis (FO), and electrolyzed membrane (EM) [305]. When used in conjunction with the RO automation for the elimination of ECs, the percentage efficiency is greater than 99% [274]. In the study by Wintgens *et al.* [306], estrogenic pollutants were removed from landfill leachate while using membrane bioreactors (MBRs) and NF techniques. Kim *et al.* [307] reported the removal of PhACs (i.e., EDCs and pesticides) from water while using NF90. In Germany, Beier *et al.* [308] reported the removal of ciprofloxacin, carbamazepine, ibuprofen, and diclofenac, among others, from hospital wastewater using pretreatment MBR combined with advanced membrane system technologies, and the results confirmed that MBR technology succeeded by an advanced treatment for ECs removal was an ideal approach for treatment of hot spot wastewater effluents such as hospital wastewater.

4.8.1.6 Photocatalysis

Photocatalysis is an effective technique for the destruction of ECs [309], but the technique alone suffers from high rates of concoction of photogenerated e^-/h^+ charges and complex recovery of the catalysts from the water [310]. The pairing of photocatalysis and electrochemical oxidation is used as combined electrophotocatalytic processes (EPPs) for efficient operation. The EPPs can effectively and efficiently retard the concoction of e^-/h^+ charges by applying bias potential, which consequently yields much improved photonic efficiency [311]. During EPPs, the catalyst is localized on the anode, and the catalyst is conveniently recovered after treatment. Therefore, the coupling process offers collective benefits for the removal of ECs from water and reduces the shortcomings. More attention is given to the kind of light source (visible light and UV), the electrode materials used, and their effect on removing of ECs and pathogens [312].

4.8.1.7 Electro Fenton Process

The Electro Fenton Process (EFP) is perhaps the most recent AOPs and is employed to overpower the obstacles of traditional advances for toxin removal and is considered most cost-effective process [313]. The EFP connects the Fenton's cycle together with electrochemistry where it has been widely utilized for treating drug compounds and wastewater. During main stage, the Fenton reagents are dosed outward, and highly reactive actions of static terminal material are used, whereas, in the following stage, only H_2O_2 is introduced from external and Fe^{2+} is transmitted from iron anodes. Contrary to the conventional Fenton procedure, in EFP, the H_2O_2 is manufactured and carried in a more managed manner. Because there is no devastative reagent used in EFP, the methodology is sustainable for purifying wastewater [314]. The main limitation with Fenton process is to maximize the initiation and propagation steps while minimizing the termination steps [104]. The following conditions are required for effective process and to generate sufficient hydroxyl radicals: high quantities of H_2O_2 and Fe^{2+} at the range of 30–6,000 mmol/L and 18–410 mmol/L, respectively.

4.8.1.8 Electrocoagulation

Electrocoagulation is an electrochemical method or approach used for the purification of various kinds of contaminated water, and, due to its high efficiency and effectiveness in removing various stubborn contaminants, it has recently received high considerable attention. It has been effective in dealing with inorganic and organic contaminants without completely generating the byproduct wastes. In the past decade, the electrocoagulation has been used in most of research for treatment of various types of wastewater, ranging from highly contaminated industrial wastewater to polluted groundwater [315]. The technique is dependent on parameters such as voltage, inter electrode spacing, reaction time, electrode materials, and concentration of contaminants. The pH is significant in the removal of contaminants during electrocoagulation [316]. The electrocoagulation integrates three various technologies that are separately well established, but one of the limitations of the technique is that lack of information or data on how to assemble or integrate these technologies to produce the reliable electrocoagulation technology for industrial use. The other challenge is the high operating costs, resulting from high electricity consumption during wastewater treatment [315, 316].

4.9 Conclusions

This book chapter successfully highlighted the trends on the occurrence, challenges, migration, and remediation of emerging contaminants in aquatic environments. Specifically, water pollution problems have continued to increase not only in low- and middle-income countries or region but also globally due to human activities. The presence of emerging contaminants (ECs) in surface water is mostly due to industrial development, agricultural practice and runoff and population growth, wastewater treatment (WWT) facilities, landfill leachates, pharmaceutical companies, hospital wastes, and lifestyle waste along with cosmetics. In most regions, ECs are unregulated substances that mainly find their ways into the environment if not properly managed. Thenceforth, among these compounds, PPCPs are groups of regular compounds that are used on daily basis. To date, there are limited regulatory frameworks to address and control these ECs and their availability in different sphere of the environment and their potential eco-toxicological effects. The accumulation of ECs in water and wastewater sources poses risk to potable water. Conventional WTPs can remove the organic pollutants; however, these ECs escape and ultimately present in the drinking water. In drinking water treatment facilities, AC is used for the removal of ECs before water could be consumed. However, there are numerous avenues of consideration, and they are elucidated in the subsequent section.

4.10 Future Research Outlook

This comprehensive study highlighted several avenues of consideration with gross focus on the trends for the occurrence, challenges, migration, and remediation of emerging contaminants in aquatic environments, and they comprise the following:

- assessing and quantifying their eco-toxicological effects to living organisms including plants and animals;
- incorporating a wide range of contaminants into the regulatory frameworks in a quest to minimize their toxicological effects;
- utilizing AC, i.e., powdered or granular carbon, as a way of removing emerging contaminants from drinking water;
- assessing the efficacy of different disinfectants and unit processes in the removal of emerging contaminants;
- assessing potential metabolites of emerging contaminants in different spheres of the environment and their toxicological effects;
- performing techno-economic analysis (TEA) for the removal of emerging contaminants from aqueous solution.

Acknowledgments

The authors of this book chapter would like to thank the University of South Africa for extending their facilities toward the accomplishment of the objectives of this work.

References

1. Karimi-Maleh, H., Ayati, A., Ghanbari, S., Orooji, Y., Tanhaei, B., Karimi, F., Alizadeh, M., Rouhi, J., Fu, L., Sillanpää, M., Recent advances in removal techniques of Cr (VI) toxic ion from aqueous solution: A comprehensive review. *J. Mol. Liq.*, 329, 115062, 2021.
2. Karimi-Maleh, H., Ranjbari, S., Tanhaei, B., Ayati, A., Orooji, Y., Alizadeh, M., Karimi, F., Salmanpour, S., Rouhi, J., Sillanpää, M., Novel 1-butyl-3-methylimidazolium bromide impregnated chitosan hydrogel beads nanostructure as an efficient nanobio-adsorbent for cationic dye removal: Kinetic study. *Environ. Res.*, 195, 110809, 2021.
3. Tilley, E., Ulrich, L., Lüthi, C., Reymond, P., Zurbrügg, C., *Compendium of sanitation systems and technologies*, Swiss Federal Institute of Aquatic Science and Technology, Dubendorf, Switzerland, 2014.
4. Cesano, F., Cravanzola, S., Brunella, V., Damin, A., Scarano, D., From polymer to magnetic porous carbon spheres: combined microscopy, spectroscopy, and porosity studies. *Front. Mater.*, 6, 84, 2019.
5. Arslan, H., Eskikaya, O., Bilici, Z., Dizge, N., Balakrishnan, D., Comparison of Cr (VI) adsorption and photocatalytic reduction efficiency using leonardite powder. *Chemosphere*, 300, 134492, 2022.
6. Shahnaz, T., Priyan, V.V., Pandian, S., Narayanasamy, S., Use of Nanocellulose extracted from grass for adsorption abatement of Ciprofloxacin and Diclofenac removal with phyto, and fish toxicity studies. *Environ. Pollut.*, 268, 115494, 2021.
7. Patra, C., Gupta, R., Bedadeep, D., Narayanasamy, S., Surface treated acid-activated carbon for adsorption of anionic azo dyes from single and binary adsorptive systems: A detail insight. *Environ. Pollut.*, 266, 115102, 2020.
8. Sun, Y., Chen, Z., Wu, G., Wu, Q., Zhang, F., Niu, Z., Hu, H.-Y., Characteristics of water quality of municipal wastewater treatment plants in China: implications for resources utilization and management. *J. Cleaner Prod.*, 131, 1–9, 2016.
9. Okoye, C.O., Nyaruaba, R., Ita, R.E., Okon, S.U., Addey, C.I., Ebido, C.C., Opabunmi, A.O., Okeke, E.S., Chukwudozie, K.I., Antibiotic resistance in the aquatic environment: Analytical techniques and interactive impact of emerging contaminants. *Environ. Toxicol. Pharmacol.*, 96, 103995, 2022.

10. Tran, N.H. and Gin, K.Y.-H., Occurrence and removal of pharmaceuticals, hormones, personal care products, and endocrine disrupters in a full-scale water reclamation plant. *Sci. Total Environ.*, 599, 1503–1516, 2017.
11. *UNESCO, Project on "Emerging Pollutants in Wastewater Reuse in Developing Countries"* fully funded by the Swedish International Development Cooperation Agency (Sida) under the Programme Cooperation Agreement between UNESCO and Sweden for 2014-2017, SIDA, Stockholm, 2017.
12. Vishnu Priyan, V., Kumar, N., Narayanasamy, S., Development of Fe(3)O(4)/CAC nanocomposite for the effective removal of contaminants of emerging concerns (Ce(3+)) from water: An ecotoxicological assessment. *Environ. Pollut.*, 285, 117326, 2021.
13. Archer, E., Petrie, B., Kasprzyk-Hordern, B., Wolfaardt, G.M., The fate of pharmaceuticals and personal care products (PPCPs), endocrine disrupting contaminants (EDCs), metabolites and illicit drugs in a WWTW and environmental waters. *Chemosphere*, 174, 437–446, 2017a.
14. Boraah, N., Chakma, S., Kaushal, P., Attributes of wood biochar as an efficient adsorbent for remediating heavy metals and emerging contaminants from water: A critical review and bibliometric analysis. *J. Environ. Chem. Eng.*, 10, 107825, 2022.
15. Pulicharla, R., Proulx, F., Behmel, S., Serodes, J.B., Rodriguez, M.J., Occurrence and seasonality of raw and drinking water contaminants of emerging interest in five water facilities. *Sci. Total Environ.*, 751, 141748, 2021.
16. Rathi, B.S., Kumar, P.S., Show, P.L., A review on effective removal of emerging contaminants from aquatic systems: Current trends and scope for further research. *J. Hazard. Mater.*, 409, 124413, 2021.
17. Sarkar, D.J., Das Sarkar, S., Das, B.K., Praharaj, J.K., Mahajan, D.K., Purokait, B., Mohanty, T.R., Mohanty, D., Gogoi, P., Kumar, V.S., Behera, B.K., Manna, R.K., Samanta, S., Microplastics removal efficiency of drinking water treatment plant with pulse clarifier. *J. Hazard. Mater.*, 413, 125347, 2021.
18. Peralta Muniz Moreira, R., Cabrera Reina, A., Soriano Molina, P., Sánchez Pérez, J.A., Li Puma, G., Computational fluid dynamics (CFD) modeling of removal of contaminants of emerging concern in solar photo-Fenton raceway pond reactors. *Chem. Eng. J.*, 413, 127392, 2021.
19. Saidulu, D., Gupta, B., Gupta, A.K., Ghosal, P.S., A review on occurrences, eco-toxic effects, and remediation of emerging contaminants from wastewater: Special emphasis on biological treatment based hybrid systems. *J. Environ. Chem. Eng.*, 9, 105282, 2021.
20. Sharma, A., Kumar, V., Shahzad, B., Tanveer, M., Sidhu, G.P.S., Handa, N., Kohli, S.K., Yadav, P., Bali, A.S., Parihar, R.D., Dar, O.I., Singh, K., Jasrotia, S., Bakshi, P., Ramakrishnan, M., Kumar, S., Bhardwaj, R., Thukral, A.K., Worldwide pesticide usage and its impacts on ecosystem. *SN Appl. Sci.*, 1, 1446, 2019.

21. Lu, F., Jiang, Y., Wu, D., Zhou, J., Li, S., Zhang, J., Levels and profiles of polychlorinated dibenzo-p-dioxin and dibenzofurans in raw and treated water from water treatment plants in Shenzhen, China. *Environ. Pollut.*, 211, 233–240, 2016.
22. Nawaz, T. and Sengupta, S., Contaminants of Emerging Concern: Occurrence, Fate, and Remediation, in: *Advances in Water Purification Techniques*, pp. 67–114, 2019.
23. Brausch, J.M. and Rand, G.M., A review of personal care products in the aquatic environment: environmental concentrations and toxicity. *Chemosphere*, 82, 1518–1532, 2011.
24. Pietrogrande, M.C. and Basaglia, G., GC-MS analytical methods for the determination of personal-care products in water matrices. *TrAC, Trends Anal. Chem.*, 26, 1086–1094, 2007.
25. de García, S.O., Pinto, G.P., Encina, P.G., Mata, R.I., Consumption and occurrence of pharmaceutical and personal care products in the aquatic environment in Spain. *Sci. Total Environ.*, 444, 451–465, 2013.
26. Saroyan, H., Ntagiou, D., Samanidou, V., Deliyanni, E., Modified graphene oxide as manganese oxide support for bisphenol A degradation. *Chemosphere*, 225, 524–534, 2019.
27. Saroyan, H.S., Bele, S., Giannakoudakis, D.A., Samanidou, V.F., Bandosz, T.J., Deliyanni, E.A., Degradation of endocrine disruptor, bisphenol-A, on an mixed oxidation state manganese oxide/modified graphite oxide composite: A role of carbonaceous phase. *J. Colloid Interface Sci.*, 539, 516–524, 2019.
28. Ternes, T.A., Meisenheimer, M., McDowell, D., Sacher, F., Brauch, H.J., Haist-Gulde, B., Preuss, G., Wilme, U., Zulei-Seibert, N., Removal of pharmaceuticals during drinking water treatment. *Environ. Sci. Technol.*, 36, 3855–3865, 2002.
29. Yadav, D., Rangabhashiyam, S., Verma, P., Singh, P., Devi, P., Kumar, P., Hussain, C.M., Gaurav, G.K., Kumar, K.S., Environmental and health impacts of contaminants of emerging concerns: Recent treatment challenges and approaches. *Chemosphere*, 272, 129492, 2021.
30. Haarhoff, J., Coetzee, L., Ceronio, A., Msagati, T., Mamba, B., Gumbi, N., Nkambule, T., *Status quo report on the state of knowledge on persistent organic pollutants and contaminants of emerging concern. Draft Final Report to the Water Research Commission for WRC Project No. K5/2509*, Water Research Commission, Pretoria, 2015.
31. Chinnaiyan, P., Thampi, S.G., Kumar, M., Mini, K., Pharmaceutical products as emerging contaminant in water: relevance for developing nations and identification of critical compounds for Indian environment. *Environ. Monit. Assess.*, 190, 1–13, 2018.
32. Patel, M., Kumar, R., Kishor, K., Mlsna, T., Pittman Jr., C.U., Mohan, D., Pharmaceuticals of emerging concern in aquatic systems: chemistry, occurrence, effects, and removal methods. *Chem. Rev.*, 119, 3510–3673, 2019.

33. Jiang, X., Qu, Y., Mengmeng., Z., Li, W., Huang, J., Seasonal and spatial variations of pharmaceuticals and personal care products occurrence and human health risk in drinking water - A case study of China. *Sci. Total Environ.*, 694, 133711, 2019.
34. Richardson, M.L. and Bowron, J.M., The fate of pharmaceutical chemicals in the aquatic environment. *J. Pharm. Pharmacol.*, 37, 1–12, 1985.
35. Daughton, C.G. and Ternes, T.A., Pharmaceuticals and personal care products in the environment: agents of subtle change? *Environ. Health Perspect.*, 107, 907–938, 1999.
36. Kanakaraju, D., Glass, B.D., Oelgemöller, M., Titanium dioxide photocatalysis for pharmaceutical wastewater treatment. *Environ. Chem. Lett.*, 12, 27–47, 2014.
37. Rivera-Utrilla, J., Sánchez-Polo, M., Ferro-García, MA., Prados-Joya, G., Ocampo-Pérez, R., Pharmaceuticals as emerging contaminants and their removal from water. A review. *Chemosphere*, 93, 1268–1287, 2013.
38. Tran, N.H., Reinhard, M., Gin, K.Y.-H., Occurrence and fate of emerging contaminants in municipal wastewater treatment plants from different geographical regions-a review. *Water Res.*, 133, 182–207, 2018.
39. Khan, M.T., Shah, I.A., Ihsanullah, I., Naushad, M., Ali, S., Shah, S.H.A., Mohammad, A.W., Hospital wastewater as a source of environmental contamination: An overview of management practices, environmental risks, and treatment processes. *J. Water Process Eng.*, 41, 101990, 2021.
40. Richardson, S.D. and Kimura, S.Y., Emerging environmental contaminants: Challenges facing our next generation and potential engineering solutions. *Environ. Technol. Innovation*, 8, 40–56, 2017.
41. Naushad, M., Sharma, G., Alothman, Z.A., Photodegradation of toxic dye using Gum Arabic-crosslinked-poly (acrylamide)/Ni (OH) 2/FeOOH nanocomposites hydrogel. *J. Cleaner Prod.*, 241, 118263, 2019.
42. Stuart, M., Lapworth, D., Crane, E., Hart, A., Review of risk from potential emerging contaminants in UK groundwater. *Sci. Total Environ.*, 416, 1–21, 2012.
43. Tran, N.H., Ngo, H.H., Urase, T., Gin, K.Y., A critical review on characterization strategies of organic matter for wastewater and water treatment processes. *Bioresour. Technol.*, 193, 523–533, 2015.
44. Baker, D.R. and Kasprzyk-Hordern, B., Spatial and temporal occurrence of pharmaceuticals and illicit drugs in the aqueous environment and during wastewater treatment: new developments. *Sci. Total Environ.*, 454, 442–456, 2013.
45. Ebele, A.J., Oluseyi, T., Drage, D.S., Harrad, S., Abdallah, M.A.-E., Occurrence, seasonal variation and human exposure to pharmaceuticals and personal care products in surface water, groundwater and drinking water in Lagos State, Nigeria. *Emerging Contam.*, 6, 124–132, 2020.
46. Yan, Q., Gao, X., Huang, L., Gan, X.-M., Zhang, Y.-X., Chen, Y.-P., Peng, X.-Y., Guo, J.-S., Occurrence and fate of pharmaceutically active compounds in

the largest municipal wastewater treatment plant in Southwest China: Mass balance analysis and consumption back-calculated model. *Chemosphere*, 99, 160–170, 2014.

47. Schwaiger, J., Ferling, H., Mallow, U., Wintermayr, H., Negele, R., Toxic effects of the non-steroidal anti-inflammatory drug diclofenac: Part I: histopathological alterations and bioaccumulation in rainbow trout. *Aquat. Toxicol.*, 68, 141–150, 2004.
48. Kümmerer, K., Drugs in the environment: emission of drugs, diagnostic aids and disinfectants into wastewater by hospitals in relation to other sources–a review. *Chemosphere*, 45, 957–969, 2001.
49. Martin-Diaz, L., Franzellitti, S., Buratti, S., Valbonesi, P., Capuzzo, A., Fabbri, E., Effects of environmental concentrations of the antiepilectic drug carbamazepine on biomarkers and cAMP-mediated cell signaling in the mussel Mytilus galloprovincialis. *Aquat. Toxicol.*, 94, 177–185, 2009.
50. Nunes, B., Gaio, A., Carvalho, F., Guilhermino, L., Behaviour and biomarkers of oxidative stress in Gambusia holbrooki after acute exposure to widely used pharmaceuticals and a detergent. *Ecotoxicol. Environ. Saf.*, 71, 341–354, 2008.
51. Behera, S.K., Kim, H.W., Oh, J.-E., Park, H.-S., Occurrence and removal of antibiotics, hormones and several other pharmaceuticals in wastewater treatment plants of the largest industrial city of Korea. *Sci. Total Environ.*, 409, 4351–4360, 2011.
52. Mohan, M., Richardson, G., Gopan, G., Aghai, M.M., Bajaj, S., Galgamuwa, G.P., Vastaranta, M., Arachchige, P.S.P., Amorós, L., Corte, A.P.D., UAV-supported forest regeneration: Current trends, challenges and implications. *Remote Sens.*, 13, 2596, 2021.
53. Pivokonsky, M., Pivokonska, L., Novotna, K., Cermakova, L., Klimtova, M., Occurrence and fate of microplastics at two different drinking water treatment plants within a river catchment. *Sci. Total Environ.*, 741, 140236, 2020.
54. Glassmeyer, S.T., Furlong, E.T., Kolpin, D.W., Batt, A.L., Benson, R., Boone, J.S., Conerly, O., Donohue, M.J., King, D.N., Kostich, M.S., Nationwide reconnaissance of contaminants of emerging concern in source and treated drinking waters of the United States. *Sci. Total Environ.*, 581, 909–922, 2017.
55. Troger, R., Ren, H., Yin, D., Postigo, C., Nguyen, P.D., Baduel, C., Golovko, O., Been, F., Joerss, H., Boleda, M.R., Polesello, S., Roncoroni, M., Taniyasu, S., Menger, F., Ahrens, L., Yin Lai, F., Wiberg, K., What's in the water? - Target and suspect screening of contaminants of emerging concern in raw water and drinking water from Europe and Asia. *Water Res.*, 198, 117099, 2021.
56. Stevens-Garmon, J., Drewes, J.E., Khan, S.J., McDonald, J.A., Dickenson, E.R., Sorption of emerging trace organic compounds onto wastewater sludge solids. *Water Res.*, 45, 3417–3426, 2011.
57. Nannou, C., Ofrydopoulou, A., Evgenidou, E., Heath, D., Heath, E., Lambropoulou, D., Antiviral drugs in aquatic environment and wastewater

treatment plants: a review on occurrence, fate, removal and ecotoxicity. *Sci. Total Environ.*, 699, 134322, 2020.
58. Swartz, C., Genthe, B., Chamier, J., Petrik, L., Tijani, J., Adeleye, A., Coomans, C., Ohlin, A., Falk, D., Menge, J., *Emerging contaminants in wastewater treated for direct potable re-use: The human health risk priorities in South Africa*, WRC project, Pretoria, 2018.
59. Arregui, L., Ayala, M., Gómez-Gil, X., Gutiérrez-Soto, G., Hernández-Luna, C.E., Herrera de Los Santos, M., Levin, L., Rojo-Domínguez, A., Romero-Martínez, D., Saparrat, M.C., Laccases: structure, function, and potential application in water bioremediation. *Microb. Cell Fact.*, 18, 1–33, 2019.
60. Okeke, E.S., Chukwudozie, K.I., Nyaruaba, R., Ita, R.E., Oladipo, A., Ejeromedoghene, O., Atakpa, E.O., Agu, C.V., Okoye, C.O., Antibiotic resistance in aquaculture and aquatic organisms: a review of current nanotechnology applications for sustainable management. *Environ. Sci. Pollut. Res.*, 29, 69241–69274, 2022.
61. Okoye, C.O., Okeke, E.S., Okoye, K.C., Echude, D., Andong, F.A., Chukwudozie, K.I., Okoye, H.U., Ezeonyejiaku, C.D., Occurrence and fate of pharmaceuticals, personal care products (PPCPs) and pesticides in African water systems: A need for timely intervention. *Heliyon*, 8, 1–10, 2022.
62. Sodhi, K.K., Kumar, M., Balan, B., Dhaulaniya, A.S., Shree, P., Sharma, N., Singh, D.K., Perspectives on the antibiotic contamination, resistance, metabolomics, and systemic remediation. *SN Appl. Sci.*, 3, 1–25, 2021.
63. Chaturvedi, P., Shukla, P., Giri, B.S., Chowdhary, P., Chandra, R., Gupta, P., Pandey, A., Prevalence and hazardous impact of pharmaceutical and personal care products and antibiotics in environment: A review on emerging contaminants. *Environ. Res.*, 194, 110664, 2021.
64. Dinh, Q., Moreau-Guigon, E., Labadie, P., Alliot, F., Teil, M.-J., Blanchard, M., Eurin, J., Chevreuil, M., Fate of antibiotics from hospital and domestic sources in a sewage network. *Sci. Total Environ.*, 575, 758–766, 2017.
65. Song, L., Li, L., Yang, S., Lan, J., He, H., McElmurry, S.P., Zhao, Y., Sulfamethoxazole, tetracycline and oxytetracycline and related antibiotic resistance genes in a large-scale landfill, China. *Sci. Total Environ.*, 551, 9–15, 2016.
66. Koch, N., Islam, N.F., Sonowal, S., Prasad, R., Sarma, H., Environmental antibiotics and resistance genes as emerging contaminants: Methods of detection and bioremediation. *Curr. Res. Microb. Sci.*, 2, 100027, 2021.
67. Wang, Y.K., Ma, X.Y., Tang, L., Wang, X.C., Zhang, S., Ngo, H.H., Yu, M., Capability of shallow open-water unit for emerging contaminants attenuation and ecological safety improvement in a treated effluent polishing process. *J. Water Process Eng.*, 40, 101788, 2021.
68. Chen, J., Tong, T., Jiang, X., Xie, S., Biodegradation of sulfonamides in both oxic and anoxic zones of vertical flow constructed wetland and the potential degraders. *Environ. Pollut.*, 265, 115040, 2020.

69. Lulijwa, R., Rupia, E.J., Alfaro, A.C., Antibiotic use in aquaculture, policies and regulation, health and environmental risks: a review of the top 15 major producers. *Rev. Aquacult.*, 12, 640–663, 2020.
70. Meek, R.W., Vyas, H., Piddock, L.J.V., Nonmedical uses of antibiotics: time to restrict their use? *PLoS Biol.*, 13, e1002266, 2015.
71. Gao, N., Liu, C.-X., Xu, Q.-M., Cheng, J.-S., Yuan, Y.-J., Simultaneous removal of ciprofloxacin, norfloxacin, sulfamethoxazole by co-producing oxidative enzymes system of Phanerochaete chrysosporium and Pycnoporus sanguineus. *Chemosphere*, 195, 146–155, 2018.
72. Wang, Y., Yang, Y., Liu, X., Zhao, J., Liu, R., Xing, B., Interaction of microplastics with antibiotics in aquatic environment: distribution, adsorption, and toxicity. *Environ. Sci. Technol.*, 55, 15579–15595, 2021.
73. Mutiyar, P.K. and Mittal, A.K., Occurrences and fate of selected human antibiotics in influents and effluents of sewage treatment plant and effluent-receiving river Yamuna in Delhi (India). *Environ. Monit. Assess.*, 186, 541–557, 2014.
74. Renew, J.E. and Huang, C.-H., Simultaneous determination of fluoroquinolone, sulfonamide, and trimethoprim antibiotics in wastewater using tandem solid phase extraction and liquid chromatography–electrospray mass spectrometry. *J. Chromatogr. A*, 1042, 113–121, 10422004.
75. Xu, W., Zhang, G., Li, X., Zou, S., Li, P., Hu, Z., Li, J., Occurrence and elimination of antibiotics at four sewage treatment plants in the Pearl River Delta (PRD). *South China Water Res.*, 41, 4526–4534, 2007.
76. Nappier, S.P., Liguori, K., Ichida, A.M., Stewart, J.R., Jones, K.R., Antibiotic resistance in recreational waters: state of the science. *Int. J. Environ. Res. Public Health*, 17, 8034, 2020.
77. Waleng, N.J. and Nomngongo, P.N., Occurrence of pharmaceuticals in the environmental waters: African and Asian perspectives. *Environ. Chem. Ecotoxicol.*, 4, 50–66, 2022.
78. Golovko, O., Kumar, V., Fedorova, G., Randak, T., Grabic, R., Seasonal changes in antibiotics, antidepressants/psychiatric drugs, antihistamines and lipid regulators in a wastewater treatment plant. *Chemosphere*, 111, 418–426, 2014.
79. K'oreje, K., O., L., Ombaka, D., De Wispelaere, P., Okoth, M., Van Langenhove, H., Demeestere, K., Occurrence patterns ofpharmaceutical residues in wastewater, surface water and groundwater of Nairobi and Kisumu city, Kenya. *Chemosphere*, 149, 238–244, 2016.
80. Mehdi, Y., Létourneau-Montminy, M.-P., Gaucher, M.-L., Chorfi, Y., Suresh, G., Rouissi, T., Brar, S.K., Côté, C., Ramirez, A.A., Godbout, S., Use of antibiotics in broiler production: Global impacts and alternatives. *Anim. Nutr.*, 4, 170–178, 2018.
81. Felis, E., Kalka, J., Sochacki, A., Kowalska, K., Bajkacz, S., Harnisz, M., Korzeniewska, E., Antimicrobial pharmaceuticals in the aquatic

environment-occurrence and environmental implications. *Eur. J. Pharmacol.*, 866, 172813, 2020.
82. Kraemer, S.A., Ramachandran, A., Perron, G.G., Antibiotic pollution in the environment: from microbial ecology to public policy. *Microorganisms*, 7, 180, 2019.
83. Larsson, D.J., Antibiotics in the environment. *Ups. J. Med. Sci.*, 119, 108–112, 2014.
84. Boxall, A.B., Rudd, M.A., Brooks, B.W., Caldwell, D.J., Choi, K., Hickmann, S., Innes, E., Ostapyk, K., Staveley, J.P., Verslycke, T., Pharmaceuticals and personal care products in the environment: what are the big questions? *Environ. Health Perspect.*, 120, 1221–1229, 2012.
85. Manzetti, S. and Ghisi, R., The environmental release and fate of antibiotics. *Mar. Pollut. Bull.*, 79, 7–15, 2014.
86. Van Doorslaer, X., Dewulf, J., Van Langenhove, H., Demeestere, K., Fluoroquinolone antibiotics: an emerging class of environmental micropollutants. *Sci. Total Environ.*, 500, 250–269, 2014.
87. Bueno, M.M., Gomez, M., Herrera, S., Hernando, M., Agüera, A., Fernández-Alba, A., Occurrence and persistence of organic emerging contaminants and priority pollutants in five sewage treatment plants of Spain: two years pilot survey monitoring. *Environ. Pollut.*, 164, 267–273, 2012.
88. Baralla, G., Pinna, A., Tonelli, R., Marchesi, M., Ibba, S., Ensuring transparency and traceability of food local products: A blockchain application to a Smart Tourism Region. *Concurrency Comput. Pract. Exper.*, 33, e5857, 2021.
89. Magureanu, M., Mandache, N.B., Parvulescu, V.I., Degradation of pharmaceutical compounds in water by non-thermal plasma treatment. *Water Res.*, 81, 124–136, 2015.
90. Izadi, P., Izadi, P., Salem, R., Papry, S.A., Magdouli, S., Pulicharla, R., Brar, S.K., Non-steroidal anti-inflammatory drugs in the environment: Where were we and how far we have come? *Environ. Pollut.*, 267, 115370, 2020.
91. Nguegang, B., Ambushe, A.A., Municipal wastewater treatment technologies: Challenges and perspectives, in: *Development in wastewater treatment, Research and Processes.* Ed Maulin, P.S. pp. 333–362, 2024.
92. Fent, K., Weston, A.A., Caminada, D., Ecotoxicology of human pharmaceuticals. *Aquat. Toxicol.*, 76, 122–159, 2006.
93. Tyumina, E., Bazhutin, G., Cartagena Gómez, AP., Ivshina, I., Nonsteroidal anti-inflammatory drugs as emerging contaminants. *Microbiology*, 89, 148–163, 2020.
94. Lopez-Roldan, R., Rubalcaba, A., Martin-Alonso, J., Gonzalez, S., Marti, V., Cortina, J.L., Assessment of the water chemical quality improvement based on human health risk indexes: Application to a drinking water treatment plant incorporating membrane technologies. *Sci. Total Environ.*, 540, 334–343, 2016.
95. Agunbiade, F.O. and Moodley, B., Occurrence and distribution pattern of acidic pharmaceuticals in surface water, wastewater, and sediment of the

Msunduzi River, Kwazulu-Natal, South Africa. *Environ. Toxicol. Chem.*, 35, 36–46, 2016.
96. Matongo, S., Birungi, G., Moodley, B., Ndungu, P., Pharmaceutical residues in water and sediment of Msunduzi River, KwaZulu-Natal, South Africa. *Chemosphere*, 134, 133–140, 2015.
97. Proctor, K., Petrie, B., Lopardo, L., Munoz, D.C., Rice, J., Barden, R., Arnot, T., Kasprzyk-Hordern, B., Micropollutant fluxes in urban environment–A catchment perspective. *J. Hazard. Mater.*, 401, 123745, 2021.
98. Tijani, J.O., Fatoba, O.O., Babajide, O.O., Petrik, L.F., Pharmaceuticals, endocrine disruptors, personal care products, nanomaterials and perfluorinated pollutants: a review. *Environ. Chem. Lett.*, 14, 27–49, 2016.
99. Khetan, S.K. and Collins, T.J., Human pharmaceuticals in the aquatic environment: a challenge to green chemistry. *Chem. Rev.*, 107, 2319–2364, 2007.
100. Chopra, S. and Kumar, D., Ibuprofen as an emerging organic contaminant in environment, distribution and remediation. *Heliyon*, 6, e04087, 2020.
101. Mejía-García, A., Islas-Flores, H., Gómez-Oliván, L.M., SanJuan-Reyes, N., Ortega-Olvera, J.M., Hernández-Navarro, M.D., Overview of non-steroidal anti-inflammatory drugs as emerging contaminants, in: *Non-Steroidal Anti-Inflammatory Drugs in Water: Emerging Contaminants and Ecological Impact*, pp. 41–53, 2020.
102. Serra-Roig, M.P., Jurado, A., Díaz-Cruz, M.S., Vázquez-Suñé, E., Pujades, E., Barceló, D., Occurrence, fate and risk assessment of personal care products in river–groundwater interface. *Sci. Total Environ.*, 568, 829–837, 2016.
103. Demarchi, C.A., Michel, B.S., Nedelko, N., Ślawska-Waniewska, A., Dłużewski, P., Kaleta, A., Minikayev, R., Strachowski, T., Lipińska, L., Dal Magro, J., Preparation, characterization, and application of magnetic activated carbon from termite feces for the adsorption of Cr (VI) from aqueous solutions. *Powder Technol.*, 354, 432–441, 2019.
104. Zhu, Y., Zhu, R., Xi, Y., Zhu, J., Zhu, G., He, H., Strategies for enhancing the heterogeneous Fenton catalytic reactivity: A review. *Appl. Catal., B*, 255, 117739, 2019.
105. Ngqwala, N.P. and Muchesa, P., Occurrence of pharmaceuticals in aquatic environments: A review and potential impacts in South Africa. *S. Afr. J. Sci.*, 116, 1–7, 2020.
106. Evgenidou, E.N., Konstantinou, I.K., Lambropoulou, D.A., Occurrence and removal of transformation products of PPCPs and illicit drugs in wastewaters: a review. *Sci. Total Environ.*, 505, 905–926, 2015.
107. Hu, X., Hu, X., Tang, C., Wen, S., Wu, X., Long, J., Yang, X., Wang, H., Zhou, L., Mechanisms underlying degradation pathways of microcystin-LR with doped TiO2 photocatalysis. *Chem. Eng. J.*, 330, 355–371, 2017.
108. Luo, Y., Guo, W., Ngo, H.H., Nghiem, L.D., Hai, F.I., Zhang, J., Liang, S., Wang, X.C., A review on the occurrence of micropollutants in the aquatic environment and their fate and removal during wastewater treatment. *Sci. Total Environ.*, 473, 619–641, 2014.

109. Petrie, B., Barden, R., Kasprzyk-Hordern, B., A review on emerging contaminants in wastewaters and the environment: Current knowledge, understudied areas and recommendations for future monitoring. *Water Res.*, 72, 3–27, 2015.
110. Ambrósio, A.F., Soares-da-Silva, P., Carvalho, C.M., Carvalho, A.P., Mechanisms of action of carbamazepine and its derivatives, oxcarbazepine, BIA 2-093, and BIA 2-024. *Neurochem. Res.*, 27, 121–130, 2002.
111. Egbuna, C., Amadi, C.N., Patrick-Iwuanyanwu, K.C., Ezzat, S.M., Awuchi, C.G., Ugonwa, P.O., Orisakwe, O.E., Emerging pollutants in Nigeria: A systematic review. *Environ. Toxicol. Pharmacol.*, 85, 103638, 2021.
112. Rivetti, C., Campos, B., Barata, C., Low environmental levels of neuro-active pharmaceuticals alter phototactic behaviour and reproduction in Daphnia magna. *Aquat. Toxicol.*, 170, 289–296, 2016.
113. Muñoz, N., Manalastas, R., Pitisuttithum, P., Tresukosol, D., Monsonego, J., Ault, K., Clavel, C., Luna, J., Myers, E., Hood, S., Safety, immunogenicity, and efficacy of quadrivalent human papillomavirus (types 6, 11, 16, 18) recombinant vaccine in women aged 24–45 years: a randomised, double-blind trial. *Lancet*, 373, 1949–1957, 2009.
114. Nkoom, M., Lu, G., Liu, J., Yang, H., Dong, H., Bioconcentration of the antiepileptic drug carbamazepine and its physiological and biochemical effects on Daphnia magna. *Ecotoxicol. Environ. Saf.*, 172, 11–18, 2019.
115. Hansen, J., Sato, M., Kharecha, P., Beerling, D., Berner, R., Masson-Delmotte, V., Pagani, M., Raymo, M., Royer, D.L., Zachos, J.C., Target atmospheric CO2: Where should humanity aim? arXiv preprint arXiv:0804.1126, 2008.
116. Chen, J., Liu, Y.-S., Deng, W.-J., Ying, G.-G., Removal of steroid hormones and biocides from rural wastewater by an integrated constructed wetland. *Sci. Total Environ.*, 660, 358–365, 2019.
117. Ojoghoro, J., Scrimshaw, M., Sumpter, J., Steroid hormones in the aquatic environment. *Sci. Total Environ.*, 792, 148306, 2021.
118. Iancu, V.-I., Radu, G.-L., Scutariu, R., A new analytical method for the determination of beta-blockers and one metabolite in the influents and effluents of three urban wastewater treatment plants. *Anal. Methods*, 11, 4668–4680, 2019.
119. Bhandari, R.K., Deem, S.L., Holliday, D.K., Jandegian, C.M., Kassotis, C.D., Nagel, S.C., Tillitt, D.E., Vom Saal, F.S., Rosenfeld, C.S., Effects of the environmental estrogenic contaminants bisphenol A and 17α-ethinyl estradiol on sexual development and adult behaviors in aquatic wildlife species. *Gen. Comp. Endocrinol.*, 214, 195–219, 2015.
120. Amin, M.M., Bina, B., Ebrahimi, A., Yavari, Z., Mohammadi, F., Rahimi, S., The occurrence, fate, and distribution of natural and synthetic hormones in different types of wastewater treatment plants in Iran. *Chin. J. Chem. Eng.*, 26, 1132–1139, 2018.
121. Fan, J.-J., Wang, S., Tang, J.-P., Zhao, J.-L., Wang, L., Wang, J.-X., Liu, S.-L., Li, F., Long, S.-X., Yang, Y., Bioaccumulation of endocrine disrupting

compounds in fish with different feeding habits along the largest subtropical river, China. *Environ. Pollut.*, 247, 999–1008, 2019.

122. Williams, M., Kookana, R.S., Mehta, A., Yadav, S., Tailor, B., Maheshwari, B., Emerging contaminants in a river receiving untreated wastewater from an Indian urban centre. *Sci. Total Environ.*, 647, 1256–1265, 2019.
123. Bhushan, S., Rana, M.S., Raychaudhuri, S., Simsek, H., Prajapati, S.K., Algae- and bacteria-driven technologies for pharmaceutical remediation in wastewater, in: *Removal of toxic pollutants through microbiological and tertiary treatment*, pp. 373–408, Elsevier, The Netherlands, 2020.
124. Karki, A.J., Cappelli, P., Dirks, C., Pekar, H., Hellenas, K.E., Rosen, J., Westerberg, E., New efficient methodology for screening of selected organic micropollutants in raw- and drinking water from 90 Swedish water treatment plants. *Sci. Total Environ.*, 724, 138069, 2020.
125. Dolliver, H. and Gupta, S., Antibiotic losses in leaching and surface runoff from manure-amended agricultural land. *J. Environ. Qual.*, 37, 1227–1237, 2008.
126. Lajeunesse, A., Smyth, S., Barclay, K., Sauvé, S., Gagnon, C., Distribution of antidepressant residues in wastewater and biosolids following different treatment processes by municipal wastewater treatment plants in Canada. *Water Res.*, 46, 5600–5612, 2012.
127. Metcalfe, C.D., Chu, S., Judt, C., Li, H., Oakes, K.D., Servos, M.R., Andrews, D.M., Antidepressants and their metabolites in municipal wastewater, and downstream exposure in an urban watershed. *Environ. Toxicol. Chem.*, 29, 79–89, 2010.
128. Gabriel, M. and Sharma, V., Antidepressant discontinuation syndrome. *Cmaj*, 189, E747–E747, 2017.
129. Wilson, E. and Lader, M., A review of the management of antidepressant discontinuation symptoms. *Ther. Adv. Psychopharmacol.*, 5, 357–368, 2015.
130. Akullian, A., Vandormael, A., Miller, J.C., Bershteyn, A., Wenger, E., Cuadros, D., Gareta, D., Bärnighausen, T., Herbst, K., Tanser, F., Large age shifts in HIV-1 incidence patterns in KwaZulu-Natal, South Africa. *Proc. Natl. Acad. Sci.*, 118, e2013164118, 2021.
131. Wood, T.P., Basson, A.E., Duvenage, C., Rohwer, E.R., The chlorination behaviour and environmental fate of the antiretroviral drug nevirapine in South African surface water. *Water Res.*, 104, 349–360, 2016.
132. Nkomo, P., Davies, N., Sherman, G., Bhardwaj, S., Ramokolo, V., Ngandu, N.K., Noveve, N., Ramraj, T., Magasana, V., Singh, Y., How ready are our health systems to implement prevention of mother to child transmission Option B+? *South. Afr. J. HIV Med.*, 16, 1–5, 2015.
133. Swanepoel, C., Bouwman, H., Pieters, R., Bezuidenhout, C., Presence, concentrations and potential implications of HIV-anti-retrovirals in selected water resources in South Africa. *Water Res. Commission. WRC Rep.*, 2144, 14, 2015.

134. Wood, T.P., Duvenage, C.S., Rohwer, E., The occurrence of anti-retroviral compounds used for HIV treatment in South African surface water. *Environ. Pollut.*, 199, 235–243, 2015.
135. Peng, X., Ou, W., Wang, C., Wang, Z., Huang, Q., Jin, J., Tan, J., Occurrence and ecological potential of pharmaceuticals and personal care products in groundwater and reservoirs in the vicinity of municipal landfills in China. *Sci. Total Environ.*, 490, 889–898, 2014.
136. Schoeman, C., Dlamini, M., Okonkwo, O.J., The impact of a Wastewater Treatment Works in Southern Gauteng, South Africa on efavirenz and nevirapine discharges into the aquatic environment. *Emerging Contam.*, 3, 95–106, 2017.
137. Mofenson, L.M., Prevention in neglected subpopulations: prevention of mother-to-child transmission of HIV infection. *Clin. Infect. Dis.*, 50, S130–S148, 2010.
138. Adeola, A.O. and Forbes, P.B., Antiretroviral drugs in African surface waters: prevalence, analysis, and potential remediation. *Environ. Toxicol. Chem.*, 41, 247–262, 2022.
139. Bilal, M., Iqbal, H.M., Barceló, D., Persistence of pesticides-based contaminants in the environment and their effective degradation using laccase-assisted biocatalytic systems. *Sci. Total Environ.*, 695, 133896, 2019.
140. Gani, K.M. and Kazmi, A.A., Contamination of emerging contaminants in Indian aquatic sources: first overview of the situation. *J. Hazard. Toxic Radioact. Waste*, 21, 04016026, 2017.
141. Primel, E.G., Caldas, S.S., Marube, L.C., Escarrone, A.L.V., An overview of advances in dispersive liquid–liquid microextraction for the extraction of pesticides and emerging contaminants from environmental samples. *Trends Environ. Anal. Chem.*, 14, 1–18, 2017.
142. Prasad, R., Bhattacharyya, A., Nguyen, Q.D., Nanotechnology in sustainable agriculture: recent developments, challenges, and perspectives. *Front. Microbiol.*, 8, 1014, 2017.
143. Boudh, S. and Singh, J.S., Pesticide contamination: environmental problems and remediation strategies, in: *Emerging and eco-friendly approaches for waste management*, pp. 245–269, 2019.
144. Saleh, I.A., Zouari, N., Al-Ghouti, M.A., Removal of pesticides from water and wastewater: Chemical, physical and biological treatment approaches. *Environ. Technol. Innovation*, 19, 101026, 2020.
145. Close, M.E., Humphries, B., Northcott, G., Outcomes of the first combined national survey of pesticides and emerging organic contaminants (EOCs) in groundwater in New Zealand 2018. *Sci. Total Environ.*, 754, 142005, 2021.
146. Arman, N.Z., Salmiati, S., Aris, A., Salim, M.R., Nazifa, T.H., Muhamad, M.S., Marpongahtun, M., A Review on Emerging Pollutants in the Water Environment: Existences. *Health Eff. Treat. Processes*, 13, 3258, 2021.

147. Gogoi, A., Mazumder, P., Tyagi, V.K., Chaminda, G.T., An, A.K., Kumar, M., Occurrence and fate of emerging contaminants in water environment: a review. *Groundwater Sustainable Dev.*, 6, 169–180, 2018.
148. Kim, S., Chu, K.H., Al-Hamadani, Y.A., Park, C.M., Jang, M., Kim, D.-H., Yu, M., Heo, J., Yoon, Y., Removal of contaminants of emerging concern by membranes in water and wastewater: A review. *Chem. Eng. J.*, 335, 896–914, 2018.
149. Xie, H., Wang, X., Chen, J., Li, X., Jia, G., Zou, Y., Zhang, Y., Cui, Y., Occurrence, distribution and ecological risks of antibiotics and pesticides in coastal waters around Liaodong Peninsula, China. *Sci. Total Environ.*, 656, 946–951, 2019.
150. Rodriguez-Narvaez, O.M., Peralta-Hernandez, J.M., Goonetilleke, A., Bandala, E.R., Treatment technologies for emerging contaminants in water: A review. *Chem. Eng. J.*, 323, 361–380, 2017.
151. Bai, X., Lutz, A., Carroll, R., Keteles, K., Dahlin, K., Murphy, M., Nguyen, D., Occurrence, distribution, and seasonality of emerging contaminants in urban watersheds. *Chemosphere*, 200, 133–142, 2018.
152. Al-Jaseem, Q., Almasoud, F.I., Ababneh, A.M., Al-Hobaib, A.S., Radiological assessment of water treatment processes in a water treatment plant in Saudi Arabia: Water and sludge radium content, radon air concentrations and dose rates. *Sci. Total Environ.*, 563-564, 1030–1036, 2016.
153. Majumder, A., Gupta, B., Gupta, A.K., Pharmaceutically active compounds in aqueous environment: A status, toxicity and insights of remediation. *Environ. Res.*, 176, 108542, 2019.
154. de Oliveira, M., Frihling, B.E.F., Velasques, J., Magalhães Filho, F.J.C., Cavalheri, P.S., Migliolo, L., Pharmaceuticals residues and xenobiotics contaminants: occurrence, analytical techniques and sustainable alternatives for wastewater treatment. *Sci. Total Environ.*, 705, 135568, 2020.
155. Charles, K.J., Nowicki, S., Bartram, J.K., A framework for monitoring the safety of water services: from measurements to security. *npj Clean Water*, 3, 36, 2020.
156. WHO, *<WHO 2017 LIMITS.pdf>*, American Water Work Association, Denver, USA, 2017.
157. Directive (EU) 2020/2184, Directive (EU) 2020/2184 of the European Parliament and of the Council of 16 December 2020 on the quality of water intended for human consumption, 2020.
158. Cheng, W., Li, J., Wu, Y., Xu, L., Su, C., Qian, Y., Zhu, Y.-G., Chen, H., Behavior of antibiotics and antibiotic resistance genes in eco-agricultural system: a case study. *J. Hazard. Mater.*, 304, 18–25, 2016.
159. Dieter, C.A., Maupin, M.A., Caldwell, R.R., Harris, M.A., Ivahnenko, T.I., Lovelace, J.K., Barber, N.L., Linsey, K.S., Estimated use of water in the United States in 2015, US Geological Survey, 2018.
160. Brusseau, M. and Artiola, J., Chemical contaminants, in: *Environmental and pollution science*, pp. 175–190, Elsevier, The Netherlands, 2019.

161. Snow, D.D., Cassada, D.A., Bartelt–Hunt, S.L., Li, X., D'Alessio, M., Zhang, Y., Zhang, Y., Sallach, J.B., Detection, occurrence and fate of emerging contaminants in agricultural environments. *Water Environ. Res.*, 88, 913–929, 2016.
162. Yang, Y., Ok, Y.S., Kim, K.H., Kwon, E.E., Tsang, Y.F., Occurrences and removal of pharmaceuticals and personal care products (PPCPs) in drinking water and water/sewage treatment plants: A review. *Sci. Total Environ.*, 596-597, 303–320, 2017.
163. López-Doval, J.C., Montagner, C.C., de Alburquerque, A.F., Moschini-Carlos, V., Umbuzeiro, G., Pompêo, M., Nutrients, emerging pollutants and pesticides in a tropical urban reservoir: Spatial distributions and risk assessment. *Sci. Total Environ.*, 575, 1307–1324, 2017.
164. Khasawneh, O.F.S. and Palaniandy, P., Occurrence and removal of pharmaceuticals in wastewater treatment plants. *Process Saf. Environ. Prot.*, 150, 532–556, 2021.
165. Kumar, R., Sarmah, A.K., Padhye, L.P., Fate of pharmaceuticals and personal care products in a wastewater treatment plant with parallel secondary wastewater treatment train. *J. Environ. Manage.*, 233, 649–659, 2019.
166. Phonsiri, V., Choi, S., Nguyen, C., Tsai, Y.-L., Coss, R., Kurwadkar, S., Monitoring occurrence and removal of selected pharmaceuticals in two different wastewater treatment plants. *SN Appl. Sci.*, 1, 1–11, 2019.
167. Ramírez-Malule, H., Quiñones-Murillo, D.H., Manotas-Duque, D., Emerging contaminants as global environmental hazards. A bibliometric analysis. *Emerging Contam.*, 6, 179–193, 2020.
168. Lee, D. and Choi, K., Comparison of regulatory frameworks of environmental risk assessments for human pharmaceuticals in EU, USA, and Canada. *Sci. Total Environ.*, 671, 1026–1035, 2019.
169. Kim, E., Jung, C., Han, J., Her, N., Park, C.M., Jang, M., Son, A., Yoon, Y., Sorptive removal of selected emerging contaminants using biochar in aqueous solution. *J. Ind. Eng. Chem.*, 36, 364–371, 2016.
170. Jianlong, W., Shizong, W., Removal of pharmaceuticals and personal care products (PPCPs) from wastewater: a review. *J. Eviron. Manag.*, 182, 620–640, 2016.
171. Hamid, H., Li, L.Y., Grace, J.R., Review of the fate and transformation of per- and polyfluoroalkyl substances (PFASs) in landfills. *Environ. Pollut.*, 235, 74–84, 2018.
172. Nika, M.C., Ntaiou, K., Elytis, K., Thomaidi, V.S., Gatidou, G., Kalantzi, O.I., Thomaidis, N.S., Stasinakis, A.S., Wide-scope target analysis of emerging contaminants in landfill leachates and risk assessment using Risk Quotient methodology. *J. Hazard. Mater.*, 394, 122493, 2020.
173. Qi, C., Huang, J., Wang, B., Deng, S., Wang, Y., Yu, G., Contaminants of emerging concern in landfill leachate in China: A review. *Emerging Contam.*, 4, 1–10, 2018.

174. Ramakrishnan, A., Blaney, L., Kao, J., Tyagi, R.D., Zhang, T.C., Surampalli, R.Y., Emerging contaminants in landfill leachate and their sustainable management. *Environ. Earth Sci.*, 73, 1357–1368, 2015.
175. K'oreje, K., Vergeynst, L., Ombaka, D., De Wispelaere, P., Okoth, M., Van Langenhove, H., Demeestere, K., Occurrence patterns of pharmaceutical residues in wastewater, surface water and groundwater of Nairobi and Kisumu city, Kenya. *Chemosphere*, 149, 238–244, 2016.
176. Gallen, C., Drage, D., Eaglesham, G., Grant, S., Bowman, M., Mueller, J., Australia-wide assessment of perfluoroalkyl substances (PFASs) in landfill leachates. *J. Hazard. Mater.*, 331, 132–141, 2017.
177. Victoria, R.P., Amila, O.D., Spency, C., Brown, S.J., Catingan, S.D., Smith, J.E., Roy, J.W., Organic contaminants of emerging concern in leachate of historic municipal landfills. *Environ. Pollut.*, 276, 116474, 2021.
178. Hölzle, I., Contaminants in landfill soils–Reliability of prefeasibility studies. *Waste Manage. (Oxford)*, 63, 337–344, 2017.
179. Bandala, E.R., Liu, A., Wijesiri, B., Zeidman, A.B., Goonetilleke, A., Emerging materials and technologies for landfill leachate treatment: A critical review. *Environ. Pollut.*, 291, 118133, 2021.
180. Ates, H. and Argun, M.E., Fate of PAHs under subcritical and supercritical conditions in landfill leachate: removal or formation? *Chem. Eng. J.*, 414, 128762, 2021.
181. Lu, M.-C., Chen, Y.Y., Chiou, M.-R., Chen, M.Y., Fan, H.-J., Occurrence and treatment efficiency of pharmaceuticals in landfill leachates. *Waste Manage. (Oxford)*, 55, 257–264, 2016.
182. Ye, M., Sun, M., Chen, X., Feng, Y., Wan, J., Liu, K., Tian, D., Liu, M., Wu, J., Schwab, A.P., Feasibility of sulfate-calcined eggshells for removing pathogenic bacteria and antibiotic resistance genes from landfill leachates. *Waste Manage. (Oxford)*, 63, 275–283, 2017.
183. Boonyaroj, V., Chiemchaisri, C., Chiemchaisri, W., Yamamoto, K., Enhanced biodegradation of phenolic compounds in landfill leachate by enriched nitrifying membrane bioreactor sludge. *J. Hazard. Mater.*, 323, 311–318, 2017.
184. Hu, Q. and Zhang, Z., Application of Dubinin–Radushkevich isotherm model at the solid/solution interface: A theoretical analysis. *J. Mol. Liq.*, 277, 646–648, 2019.
185. Liu, A., Goonetilleke, A., Egodawatta, P., *Role of rainfall and catchment characteristics on urban stormwater quality*, Springer, Sinagpore, 2015.
186. Hubeny, J., Harnisz, M., Korzeniewska, E., Buta, M., Zieliński, W., Rolbiecki, D., Giebułtowicz, J., Nałęcz-Jawecki, G., Płaza, G., Industrialization as a source of heavy metals and antibiotics which can enhance the antibiotic resistance in wastewater, sewage sludge and river water. *PLoS One*, 16, e0252691, 2021.
187. Castiglioni, S., Davoli, E., Riva, F., Palmiotto, M., Camporini, P., Manenti, A., Zuccato, E., Mass balance of emerging contaminants in the water cycle of a

highly urbanized and industrialized area of Italy. *Water Res.*, 131, 287–298, 2018.
188. Naquin, A., Shrestha, A., Sherpa, M., Nathaniel, R., Boopathy, R., Presence of antibiotic resistance genes in a sewage treatment plant in Thibodaux, Louisiana, USA. *Bioresour. Technol.*, 188, 79–83, 2015.
189. Grehs, B.W., Linton, M.A., Clasen, B., de Oliveira Silveira, A., Carissimi, E., Antibiotic resistance in wastewater treatment plants: Understanding the problem and future perspectives. *Arch. Microbiol.*, 203, 1009–1020, 2021.
190. Carraro, E., Bonetta, S., Bertino, C., Lorenzi, E., Bonetta, S., Gilli, G., Hospital effluents management: chemical, physical, microbiological risks and legislation in different countries. *J. Environ. Manage.*, 168, 185–199, 2016.
191. Kotwani, A., Joshi, J., Kaloni, D., Pharmaceutical effluent: a critical link in the interconnected ecosystem promoting antimicrobial resistance. *Environ. Sci. Pollut. Res.*, 28, 32111–32124, 2021.
192. Guo, X., Tang, N., Lei, H., Fang, Q., Liu, L., Zhou, Q., Song, C., Metagenomic analysis of antibiotic resistance genes in untreated wastewater from three different hospitals. *Front. Microbiol.*, 12, 709051, 2021.
193. Rozman, U., Duh, D., Cimerman, M., Turk, SS., Hospital wastewater effluent: hot spot for antibiotic resistant bacteria. *J. Water Sanit. Hyg. Dev.*, 10, 171–178, 2020.
194. Zhang, S., Huang, J., Zhao, Z., Cao, Y., Li, B., Hospital wastewater as a reservoir for antibiotic resistance genes: a meta-analysis. *Front. Public Health*, 8, 574968, 2020.
195. Rowe, W.P., Baker-Austin, C., Verner-Jeffreys, D.W., Ryan, J.J., Micallef, C., Maskell, D.J., Pearce, G.P., Overexpression of antibiotic resistance genes in hospital effluents over time. *J. Antimicrob. Chemother.*, 72, 1617–1623, 2017.
196. Li, J., Cheng, W., Xu, L., Strong, P., Chen, H., Antibiotic-resistant genes and antibiotic-resistant bacteria in the effluent of urban residential areas, hospitals, and a municipal wastewater treatment plant system. *Environ. Sci. Pollut. Res.*, 22, 4587–4596, 2015.
197. Chen, Z., Guo, J., Jiang, Y., Shao, Y., High concentration and high dose of disinfectants and antibiotics used during the COVID-19 pandemic threaten human health. *Environ. Sci. Eur.*, 33, 1–4, 2021.
198. Teymoorian, T., Teymourian, T., Kowsari, E., Ramakrishna, S., Direct and indirect effects of SARS-CoV-2 on wastewater treatment. *J. Water Process Eng.*, 42, 102193, 2021.
199. Hinga, M., *The Effect of Septic Tanks Sewage Disposal System Distances on Borehole Water Quality in Ongata Rongai, Kajiado County, Kenya*, University of Nairobi, *Eur. Int. J. Sci. Technol.*, 6, 1–11, 2016.
200. Swartz, C.H., Reddy, S., Benotti, M.J., Yin, H., Barber, L.B., Brownawell, B.J., Rudel, R.A., Steroid estrogens, nonylphenol ethoxylate metabolites, and other wastewater contaminants in groundwater affected by a residential septic system on Cape Cod, MA. *Environ. Sci. Technol.*, 40, 4894–4902, 2006.

201. Conn, K.E., Lowe, K.S., Drewes, J.E., Hoppe-Jones, C., Tucholke, M.B., Occurrence of pharmaceuticals and consumer product chemicals in raw wastewater and septic tank effluent from single-family homes. *Environ. Eng. Sci.*, 27, 347–356, 2010.
202. Flanagan, K., Dixon, B., Rivenbark, T., Griffin, D., An integrative GIS approach to analyzing the impacts of septic systems on the coast of Florida, USA. *Phys. Geogr.*, 41, 407–432, 2020.
203. Lu, Y., Tang, C., Chen, J., Sakura, Y., Impact of septic tank systems on local groundwater quality and water supply in the Pearl River Delta, China: case study. *Hydrol. Processes Int. J.*, 22, 443–450, 2008.
204. Yates, M.V., Septic tank density and ground-water contamination. *Groundwater*, 23, 586–591, 1985.
205. Ball, W., Buehler, C., Harmon, T., Mackay, D., Roberts, P., Characterization of a sandy aquifer material at the grain scale. *J. Contam. Hydrol.*, 5, 253–295, 1990.
206. Godfrey, E., Woessner, W.W., Benotti, M.J., Pharmaceuticals in on-site sewage effluent and ground water, western Montana. *Groundwater*, 45, 263–271, 2007.
207. Naddeo, V., Secondes, M.F.N., Borea, L., Hasan, S.W., Ballesteros Jr., F., Belgiorno, V., Removal of contaminants of emerging concern from real wastewater by an innovative hybrid membrane process–UltraSound, Adsorption, and Membrane ultrafiltration (USAMe®). *Ultrason. Sonochem.*, 68, 105237, 2020.
208. Yadav, S., Jadeja, N.B., Dafale, N.A., Purohit, H.J., Kapley, A., Pharmaceuticals and personal care products mediated antimicrobial resistance: future challenges, in: *Pharmaceuticals and personal care products: waste management and treatment technology*, pp. 409–428, Elsevier, 2019.
209. Pereira, L.C., de Souza, A.O., Bernardes, M.F.F., Pazin, M., Tasso, M.J., Pereira, P.H., Dorta, D.J., A perspective on the potential risks of emerging contaminants to human and environmental health. *Environ. Sci. Pollut. Res.*, 22, 13800–13823, 2015.
210. Venkidasamy, B., Subramanian, U., Samynathan, R., Rajakumar, G., Shariati, M.A., Chung, I.-M., Thiruvengadam, M., Organopesticides and fertility: where does the link lead to? *Environ. Sci. Pollut. Res.*, 28, 6289–6301, 2021.
211. Graymore, M., Stagnitti, F., Allinson, G., Impacts of atrazine in aquatic ecosystems. *Environ. Int.*, 26, 483–495, 2001.
212. Lladó, J., Lao-Luque, C., Ruiz, B., Fuente, E., Solé-Sardans, M., Dorado, A.D., Role of activated carbon properties in atrazine and paracetamol adsorption equilibrium and kinetics. *Process Saf. Environ. Prot.*, 95, 51–59, 2015.
213. Ochoa-Acuña, H., Frankenberger, J., Hahn, L., Carbajo, C., Drinking-water herbicide exposure in Indiana and prevalence of small-for-gestational-age and preterm delivery. *Environ. Health Perspect.*, 117, 1619–1624, 2009.

214. Waller, S.A., Paul, K., Peterson, S.E., Hitti, J.E., Agricultural-related chemical exposures, season of conception, and risk of gastroschisis in Washington State. *Am. J. Obstet. Gynecology*, 202, 241. e241–241. e246, 2010.
215. Munger, R., Isacson, P., Hu, S., Burns, T., Hanson, J., Lynch, C.F., Cherryholmes, K., Van Dorpe, P., Hausler Jr., W.J., Intrauterine growth retardation in Iowa communities with herbicide-contaminated drinking water supplies. *Environ. Health Perspect.*, 105, 308–314, 1997.
216. Chevrier, C., Limon, G., Monfort, C., Rouget, F., Garlantézec, R., Petit, C., Durand, G., Cordier, S., Urinary biomarkers of prenatal atrazine exposure and adverse birth outcomes in the PELAGIE birth cohort. *Environ. Health Perspect.*, 119, 1034–1041, 2011.
217. Davis, J., Environmental atrazine exposure and congenital malformations in New York State: an ecologic study, in: *Proceedings of the Teratology Society 45th Annual Meeting, St. Pete Beach, Florida. Birth Defects Res A Clin Mol Teratol*, p. 926, 2005.
218. Estrada-Arriaga, E.B., Cortés-Muñoz, J.E., González-Herrera, A., Calderón-Mólgora, C.G., de Lourdes Rivera-Huerta, M., Ramírez-Camperos, E., Montellano-Palacios, L., Gelover-Santiago, S.L., Pérez-Castrejón, S., Cardoso-Vigueros, L., Assessment of full-scale biological nutrient removal systems upgraded with physico-chemical processes for the removal of emerging pollutants present in wastewaters from Mexico. *Sci. Total Environ.*, 571, 1172–1182, 2016.
219. Lee, S.-H., Kim, K.-H., Lee, M., Lee, B.-D., Detection status and removal characteristics of pharmaceuticals in wastewater treatment effluent. *J. Water Process Eng.*, 31, 100828, 2019.
220. Mladinic, M., Zeljezic, D., Shaposhnikov, S.A., Collins, A.R., The use of FISH-comet to detect c-Myc and TP 53 damage in extended-term lymphocyte cultures treated with terbuthylazine and carbofuran. *Toxicol. Lett.*, 211, 62–69, 2012.
221. Nassiri Koopaei, N. and Abdollahi, M., Health risks associated with the pharmaceuticals in wastewater. *DARU J. Pharm. Sci.*, 25, 9, 2017.
222. Matalon, S., Schechtman, S., Goldzweig, G., Ornoy, A., The teratogenic effect of carbamazepine: a meta-analysis of 1255 exposures. *Reprod. Toxicol.*, 16, 9–17, 2002.
223. Stamatelatou, K., Frouda, C., Fountoulakis, M., Drillia, P., Kornaros, M., Lyberatos, G., Pharmaceuticals and health care products in wastewater effluents: the example of carbamazepine. *Water Sci. Technol. Water Supply*, 3, 131–137, 2003.
224. Zhang, Y., Geng, J., Ma, H., Ren, H., Xu, K., Ding, L., Characterization of microbial community and antibiotic resistance genes in activated sludge under tetracycline and sulfamethoxazole selection pressure. *Sci. Total Environ.*, 571, 479–486, 2016.

225. Alkimin, G., Soares, A.M., Barata, C., Nunes, B., Evaluation of ketoprofen toxicity in two freshwater species: Effects on biochemical, physiological and population endpoints. *Environ. Pollut.*, 265, 114993, 2020.
226. Bedia, J., Peñas-Garzón, M., Gómez-Avilés, A., Rodriguez, J.J., Belver, C., A review on the synthesis and characterization of biomass-derived carbons for adsorption of emerging contaminants from water. *J. Carbon Res.*, 4, 1–53, 2018.
227. Diniz, M., Salgado, R., Pereira, V., Carvalho, G., Oehmen, A., Reis, M., Noronha, J., Ecotoxicity of ketoprofen, diclofenac, atenolol and their photolysis byproducts in zebrafish (Danio rerio). *Sci. Total Environ.*, 505, 282–289, 2015.
228. Fernandez, M.E., Ledesma, B., Román, S., Bonelli, P.R., Cukierman, A.L., Development and characterization of activated hydrochars from orange peels as potential adsorbents for emerging organic contaminants. *Bioresour. Technol.*, 183, 221–228, 2015.
229. Boyd, G.R., Zhang, S., Grimm, D.A., Naproxen removal from water by chlorination and biofilm processes. *Water Res.*, 39, 668–676, 2005.
230. El-Sheikh, A.H., Qawariq, R.F., Abdelghani, J.I., Adsorption and magnetic solid-phase extraction of NSAIDs from pharmaceutical wastewater using magnetic carbon nanotubes: Effect of sorbent dimensions, magnetite loading and competitive adsorption study. *Environ. Technol. Innovation*, 16, 100496, 2019.
231. Wojcieszyńska, D. and Guzik, U., Naproxen in the environment: its occurrence, toxicity to nontarget organisms and biodegradation. *Appl. Microbiol. Biotechnol.*, 104, 1849–1857, 2020.
232. Hernández-Pérez, A., Noonin, C., Söderhäll, K., Söderhäll, I., Environmental concentrations of sulfamethoxazole increase crayfish Pacifastacus leniusculus susceptibility to White Spot Syndrome Virus. *Fish Shellfish Immunol.*, 102, 177–184, 2020.
233. Hu, F., Dong, F., Yin, L., Wang, H., Zheng, M., Fu, S., Zhang, W., Effects of sulfamethoxazole on the growth, oxidative stress and inflammatory response in the liver of juvenile Nile tilapia (Oreochromis niloticus). *Aquaculture*, 543, 736935, 2021.
234. Jansomboon, W., Boontanon, S.K., Boontanon, N., Polprasert, C., Da, C.T., Monitoring and determination of sulfonamide antibiotics (sulfamethoxydiazine, sulfamethazine, sulfamethoxazole and sulfadiazine) in imported Pangasius catfish products in Thailand using liquid chromatography coupled with tandem mass spectrometry. *Food Chem.*, 212, 635–640, 2016.
235. Zhao, H., Wang, Y., Guo, M., Liu, Y., Yu, H., Xing, M., Environmentally relevant concentration of cypermethrin or/and sulfamethoxazole induce neurotoxicity of grass carp: involvement of blood-brain barrier, oxidative stress and apoptosis. *Sci. Total Environ.*, 762, 143054, 2021.
236. Zhu, X., Wang, X., Wang, L., Fan, X., Li, X., Jiang, Y., Biodegradation of lincomycin in wastewater by two-level bio-treatment using chloroperoxidase

and activated sludge: degradation route and eco-toxicity evaluation. *Environ. Technol. Innovation*, 20, 101114, 2020.
237. Ohoro, C., Adeniji, A., Okoh, A., Okoh, O., Distribution and chemical analysis of pharmaceuticals and personal care products (PPCPs) in the environmental systems: A review. *Int. J. Environ. Res. Public Health*, 16, 3026, 2019.
238. Parida, V.K., Saidulu, D., Majumder, A., Srivastava, A., Gupta, B., Gupta, A.K., Emerging contaminants in wastewater: A critical review on occurrence, existing legislations, risk assessment, and sustainable treatment alternatives. *J. Environ. Chem. Eng.*, 9, 105966, 2021.
239. Acero, J.L., Benitez, F.J., Real, F.J., Teva, F., Coupling of adsorption, coagulation, and ultrafiltration processes for the removal of emerging contaminants in a secondary effluent. *Chem. Eng. J.*, 210, 1–8, 2012.
240. Appa, R., Mhaisalkar, V.A., Naoghare, P.K., Lataye, D.H., Adsorption of an emerging contaminant (primidone) onto activated carbon: kinetic, equilibrium, thermodynamic, and optimization studies. *Environ. Monit. Assess.*, 191, 1–16, 2019.
241. Bruton, T., Alboloushi, A., De La Garza, B., Kim, B.-O., Halden, R.U., Fate of caffeine in the environment and ecotoxicological considerations, in: *Contaminants of emerging concern in the environment: Ecological and human health considerations*, pp. 257–273, ACS Publications, 2010.
242. Lee, C.S., Lee, Y.C., Chiang, H.M., Abrupt state change of river water quality (turbidity): Effect of extreme rainfalls and typhoons. *Sci. Total Environ.*, 557-558, 91–101, 2016.
243. Wright, P.E. and Dyson, H.J., Intrinsically disordered proteins in cellular signalling and regulation. *Nat. Rev. Mol. Cell Biol.*, 16, 18–29, 2015.
244. Doepker, C., Lieberman, H.R., Smith, A.P., Peck, J.D., El-Sohemy, A., Welsh, B.T., Caffeine: friend or foe? *Annu. Rev. Food Sci. Technol.*, 7, 117–137, 2016.
245. Guiloski, I.C., Ribas, J.L.C., Piancini, L.D.S., Dagostim, A.C., Cirio, S.M., Fávaro, L.F., Boschen, S.L., Cestari, M.M., da Cunha, C., de Assis, H.C.S., Paracetamol causes endocrine disruption and hepatotoxicity in male fish Rhamdia quelen after subchronic exposure. *Environ. Toxicol. Pharmacol.*, 53, 111–120, 2017.
246. Nogueira, A.F., Pinto, G., Correia, B., Nunes, B., development, E., locomotor behavior, biochemical, and epigenetic effects of the pharmaceutical drugs paracetamol and ciprofloxacin in larvae and embryos of Danio rerio when exposed to environmental realistic levels of both drugs. *Environ. Toxicol.*, 34, 1177–1190, 2019.
247. Xia, L., Zheng, L., Zhou, J.L., Effects of ibuprofen, diclofenac and paracetamol on hatch and motor behavior in developing zebrafish (Danio rerio). *Chemosphere*, 182, 416–425, 2017.
248. Oliver, M., Kudłak, B., Wieczerzak, M., Reis, S., Lima, S.A., Segundo, M.A., Miro, M., Ecotoxicological equilibria of triclosan in Microtox, XenoScreen YES/YAS, Caco2, HEPG2 and liposomal systems are affected by the

occurrence of other pharmaceutical and personal care emerging contaminants. *Sci. Total Environ.*, 719, 137358, 2020.
249. Horn, S., Vogt, T., Gerber, E., Vogt, B., Bouwman, H., Pieters, R., HIV-antiretrovirals in river water from Gauteng, South Africa: Mixed messages of wastewater inflows as source. *Sci. Total Environ.*, 806, 150346, 2022.
250. Oluwole, A.O., Omotola, E.O., Olatunji, O.S., Pharmaceuticals and personal care products in water and wastewater: A review of treatment processes and use of photocatalyst immobilized on functionalized carbon in AOP degradation. *BMC Chem.*, 14, 1–29, 2020.
251. Rajamohan, N. and Rajasimman, M., Biosorption of Selenium using activated plant based sorbent–Effect of variables, isotherm and kinetic modeling. *Biocatal. Agric. Biotechnol.*, 4, 795–800, 2015.
252. Rajamohan, N., Rajasimman, M., Dilipkumar, M., Parametric and kinetic studies on biosorption of mercury using modified Phoenix dactylifera biomass. *J. Taiwan Inst. Chem. Eng.*, 45, 2622–2627, 2014.
253. Sarojini, G., Babu, S.V., Rajamohan, N., Kumar, P.S., Rajasimman, M., Surface modified polymer-magnetic-algae nanocomposite for the removal of chromium-equilibrium and mechanism studies. *Environ. Res.*, 201, 111626, 2021.
254. Talib, A. and Randhir, T.O., Managing emerging contaminants: status, impacts, and watershed-wide strategies. *Exposure Health*, 8, 143–158, 2016.
255. Shahid, M.K., Kashif, A., Fuwad, A., Choi, Y., Current advances in treatment technologies for removal of emerging contaminants from water – A critical review. *Coord. Chem. Rev.*, 442, 213993, 2021.
256. Zenker, A., Cicero, M.R., Prestinaci, F., Bottoni, P., Carere, M., Bioaccumulation and biomagnification potential of pharmaceuticals with a focus to the aquatic environment. *J. Environ. Manage.*, 133, 378–387, 2014.
257. Rahman, M., Yanful, E., Jasim, S., Occurrences of endocrine disrupting compounds and pharmaceuticals in the aquatic environment and their removal from drinking water: Challenges in the context of the developing world. *Desalination*, 248, 578–585, 2009.
258. Dodgen, L., Li, J., Parker, D., Gan, J., Uptake and accumulation of four PPCP/EDCs in two leafy vegetables. *Environ. Pollut.*, 182, 150–156, 2013.
259. Rout, P.R., Zhang, T.C., Bhunia, P., Surampalli, R.Y., Treatment technologies for emerging contaminants in wastewater treatment plants: A review. *Sci. Total Environ.*, 753, 141990, 2021.
260. Zhang, Q.-Q., Ying, G.-G., Pan, C.-G., Liu, Y.-S., Zhao, J.-L., Comprehensive evaluation of antibiotics emission and fate in the river basins of China: source analysis, multimedia modeling, and linkage to bacterial resistance. *Environ. Sci. Technol.*, 49, 6772–6782, 2015.
261. Juliano, C. and Magrini, G.A., Cosmetic ingredients as emerging pollutants of environmental and health concern. A mini-review. *Cosmetics*, 4, 11, 2017.
262. Inam, E., Offiong, N.-A., Kang, S., Yang, P., Essien, J., Assessment of the occurrence and risks of emerging organic pollutants (EOPs) in Ikpa River

Basin freshwater ecosystem, Niger Delta-Nigeria. *Bull. Environ. Contam. Toxicol.*, 95, 624–631, 2015.
263. de Sousa, D.N.R., Mozeto, A.A., Carneiro, R.L., Fadini, P.S., Spatio-temporal evaluation of emerging contaminants and their partitioning along a Brazilian watershed. *Environ. Sci. Pollut. Res. Int.*, 25, 4607–4620, 2018.
264. K'Oreje K, O., Okoth, M., Van Langenhove, H., Demeestere, K., Occurrence and treatment of contaminants of emerging concern in the African aquatic environment: Literature review and a look ahead. *J. Environ. Manage.*, 254, 109752, 2020.
265. Vilimanovic, D., Andaluri, G., Hannah, R., Suri, R., Ronald MacGillivray, A., Occurrence and aquatic toxicity of contaminants of emerging concern (CECs) in tributaries of an urbanized section of the Delaware River Watershed. *AIMS Environ. Sci.*, 7, 302–319, 2020.
266. Sposito, J.C., Montagner, C.C., Casado, M., Navarro-Martín, L., Solórzano, J.C.J., Piña, B., Grisolia, A.B., Emerging contaminants in Brazilian rivers: occurrence and effects on gene expression in zebrafish (Danio rerio) embryos. *Chemosphere*, 209, 696–704, 2018.
267. Rimayi, C., Odusanya, D., Weiss, J.M., de Boer, J., Chimuka, L., Contaminants of emerging concern in the Hartbeespoort Dam catchment and the uMngeni River estuary 2016 pollution incident, South Africa. *Sci. Total Environ.*, 627, 1008–1017, 2018.
268. Wood, T.P., Du Preez, C., Steenkamp, A., Duvenage, C., Rohwer, E.R., Database-driven screening of South African surface water and the targeted detection of pharmaceuticals using liquid chromatography-high resolution mass spectrometry. *Environ. Pollut.*, 230, 453–462, 2017.
269. Benson, R., Conerly, O.D., Sander, W., Batt, A.L., Boone, J.S., Furlong, E.T., Glassmeyer, S.T., Kolpin, D.W., Mash, H.E., Schenck, K.M., Human health screening and public health significance of contaminants of emerging concern detected in public water supplies. *Sci. Total Environ.*, 579, 1643–1648, 2017.
270. Ogbeide, O., Tongo, I., Ezemonye, L., Risk assessment of agricultural pesticides in water, sediment, and fish from Owan River, Edo State, Nigeria. *Environ. Monit. Assess.*, 187, 1–16, 2015.
271. Enyoh, C.E., Verla, A.W., Qingyue, W., Ohiagu, F.O., Chowdhury, A.H., Enyoh, E.C., Chowdhury, T., Verla, E.N., Chinwendu, U.P., An overview of emerging pollutants in air: Method of analysis and potential public health concern from human environmental exposure. *Trends Environ. Anal. Chem.*, 28, e00107, 2020.
272. K'oreje, K.O., Demeestere, K., De Wispelaere, P., Vergeynst, L., Dewulf, J., Van Langenhove, H., From multi-residue screening to target analysis of pharmaceuticals in water: development of a new approach based on magnetic sector mass spectrometry and application in the Nairobi River basin, Kenya. *Sci. Total Environ.*, 437, 153–164, 2012.

273. Merrett, H., Lance, C., Bath, A., Catchment Surveillance–Use of the ADWG Observational Monitoring Approach, *Australian Water Association Oz water conference*, Adelaide, Australia, 2011.
274. Dhangar, K. and Kumar, M., Tricks and tracks in removal of emerging contaminants from the wastewater through hybrid treatment systems: a review. *Sci. Total Environ.*, 738, 140320, 2020.
275. Van Haute, S., López-Gálvez, F., Gómez-López, V.M., Eriksson, M., Devlieghere, F., Allende, A., Sampers, I., Methodology for modeling the disinfection efficiency of fresh-cut leafy vegetables wash water applied on peracetic acid combined with lactic acid. *Int. J. Food Microbiol.*, 208, 102–113, 2015.
276. Willach, S., Lutze, H.V., Eckey, K., Löppenberg, K., Lüling, M., Terhalle, J., Wolbert, J.-B., Jochmann, M.A., Karst, U., Schmidt, T.C., Degradation of sulfamethoxazole using ozone and chlorine dioxide-compound-specific stable isotope analysis, transformation product analysis and mechanistic aspects. *Water Res.*, 122, 280–289, 2017.
277. Hassenberg, K., Geyer, M., Mauerer, M., Praeger, U., Herppich, W.B., Influence of temperature and organic matter load on chlorine dioxide efficacy on Escherichia coli inactivation. *LWT Food Sci. Technol.*, 79, 349–354, 2017.
278. Wang Pei, W.P., He YiLiang, H.Y., Huang ChingHua, H.C., Reactions of tetracycline antibiotics with chlorine dioxide and free chlorine. *Water Res.*, 45, 1838–1846, 2011.
279. Wang, Y., Liu, H., Xie, Y., Ni, T., Liu, G., Oxidative removal of diclofenac by chlorine dioxide: Reaction kinetics and mechanism. *Chem. Eng. J.*, 279, 409–415, 2015.
280. Alnajrani, M.N. and Alsager, O.A., Removal of antibiotics from water by polymer of intrinsic microporosity: Isotherms, kinetics, thermodynamics, and adsorption mechanism. *Sci. Rep.*, 10, 794, 2020.
281. Antoniadou, M., Falara, P.P., Likodimos, V., Photocatalytic degradation of pharmaceuticals and organic contaminants of emerging concern using nanotubular structures. *Curr. Opin. Green Sustainable Chem.*, 29, 100470, 2021.
282. Rayaroth, M.P., Aravind, U.K., Aravindakumar, C.T., Degradation of pharmaceuticals by ultrasound-based advanced oxidation process. *Environ. Chem. Lett.*, 14, 259–290, 2016.
283. Jagannathan Madhavan, J.M., Jayaraman Theerthagiri, J.T., Dhandapani Balaji, D.B., Salla Sunitha, S.S., Choi MyongYong, C.M., Muthupandian Ashokkumar, M.A., Hybrid advanced oxidation processes involving ultrasound: an overview, *Molecules*, 24, 1–18, 2019.
284. Yabalak, E., Al-Nuaimy, M.N.M., Saleh, M., Isik, Z., Dizge, N., Balakrishnan, D., Catalytic efficiency of raw and hydrolyzed eggshell in the oxidation of crystal violet and dye bathing wastewater by thermally activated peroxide oxidation method. *Environ. Res.*, 212, 113210, 2022.

285. Yu, F., Li, Y., Han, S., Ma, J., Adsorptive removal of antibiotics from aqueous solution using carbon materials. *Chemosphere*, 153, 365–385, 2016.
286. Piezer, K., Li, L., Jeon, Y., Kadudula, A., Seo, Y., The Application of Potassium Permanganate to Treat Cyanobacteria-Laden Water: a Review. *Process Saf. Environ. Prot.*, 148, 400–414, 2021.
287. McBeath, S.T. and Graham, N.J.D., *In-situ* electrochemical generation of permanganate for the treatment of atrazine. *Sep. Purif. Technol.*, 260, 118252, 2021.
288. Wang, J. and Chen, H., Catalytic ozonation for water and wastewater treatment: Recent advances and perspective. *Sci. Total Environ.*, 704, 135249, 2020.
289. Wang, J. and Bai, Z., Fe-based catalysts for heterogeneous catalytic ozonation of emerging contaminants in water and wastewater. *Chem. Eng. J.*, 312, 79–98, 2017.
290. Simazaki, D., Kubota, R., Suzuki, T., Akiba, M., Nishimura, T., Kunikane, S., Occurrence of selected pharmaceuticals at drinking water purification plants in Japan and implications for human health. *Water Res.*, 76, 187–200, 2015.
291. Aydin, S., Aydin, M.E., Beduk, F., Ulvi, A., Removal of antibiotics from aqueous solution by using magnetic Fe3O4/red mud-nanoparticles. *Sci. Total Environ.*, 670, 539–546, 2019.
292. Elias, M.A., Hadibarata, T., Sathishkumar, P., Modified oil palm industry solid waste as a potential adsorbent for lead removal. *Environ. Chem. Ecotoxicol.*, 3, 1–7, 2021.
293. Melliti, A., Srivastava, V., Kheriji, J., Sillanpää, M., Hamrouni, B., Date Palm Fiber as a novel precursor for porous activated carbon: optimization, characterization and its application as Tylosin antibiotic scavenger from aqueous solution. *Surf. Interfaces*, 24, 101047, 2021.
294. Reza, M.S., Yun, C.S., Afroze, S., Radenahmad, N., Bakar, M.S.A., Saidur, R., Taweekun, J., Azad, A.K., Preparation of activated carbon from biomass and its' applications in water and gas purification, a review. *Arab J. Basic Appl. Sci.*, 27, 208–238, 2020.
295. Rizzo, L., Malato, S., Antakyali, D., Beretsou, V.G., Đolić, M.B., Gernjak, W., Heath, E., Ivancev-Tumbas, I., Karaolia, P., Ribeiro, A.R.L., Consolidated vs new advanced treatment methods for the removal of contaminants of emerging concern from urban wastewater. *Sci. Total Environ.*, 655, 986–1008, 2019.
296. Budimirović, D., Veličković, Z.S., Djokić, V.R., Milosavljević, M., Markovski, J., Lević, S., Marinković, A.D., Efficient As (V) removal by α-FeOOH and α-FeOOH/α-MnO2 embedded PEG-6-arm functionalized multiwall carbon nanotubes. *Chem. Eng. Res. Des.*, 119, 75–86, 2017.
297. Othmani, A., John, J., Rajendran, H., Mansouri, A., Sillanpää, M., Chellam, P.V., Biochar and activated carbon derivatives of lignocellulosic fibers towards adsorptive removal of pollutants from aqueous systems: Critical study and future insight. *Sep. Purif. Technol.*, 274, 119062, 2021.

298. Patra, C., Medisetti, R.M.N., Pakshirajan, K., Narayanasamy, S., Assessment of raw, acid-modified and chelated biomass for sequestration of hexavalent chromium from aqueous solution using Sterculia villosa Roxb. shells. *Environ. Sci. Pollut. Res.*, 26, 23625–23637, 2019.
299. Gopinath, A. and Kadirvelu, K., Strategies to design modified activated carbon fibers for the decontamination of water and air. *Environ. Chem. Lett.*, 16, 1137–1168, 2018.
300. Korotta-Gamage, S.M. and Sathasivan, A., A review: Potential and challenges of biologically activated carbon to remove natural organic matter in drinking water purification process. *Chemosphere*, 167, 120–138, 2017.
301. Garg, D., Kumar, S., Sharma, K., Majumder, C., Application of waste peanut shells to form activated carbon and its utilization for the removal of Acid Yellow 36 from wastewater. *Groundwater Sustain. Dev.*, 8, 512–519, 2019.
302. Ling, Y., Klemes, M.J., Xiao, L., Alsbaiee, A., Dichtel, W.R., Helbling, D.E., Benchmarking micropollutant removal by activated carbon and porous β-cyclodextrin polymers under environmentally relevant scenarios. *Environ. Sci. Technol.*, 51, 7590–7598, 2017.
303. Brandt, M., Johnson, K., Elphinston, A., Ratnayaka, D., Chemical storage, dosing and control, in: *twort's water supply*, 2017.
304. Yi, J., Lee, J., Jung, H., Park, P.-K., Noh, S.H., Reduction of bacterial regrowth in treated water by minimizing water stagnation in the filtrate line of a gravity-driven membrane system. *Environ. Eng. Res.*, 24, 17–23, 2019.
305. Nghiem, L.D. and Fujioka, T., Removal of emerging contaminants for water reuse by membrane technology, in: *Emerging membrane technology for sustainable water treatment*, pp. 217–247, 2016.
306. Wintgens, T., Gallenkemper, M., Melin, T., Occurrence and removal of endocrine disrupters in landfill leachate treatment plants. *Water Sci. Technol.*, 48, 127–134, 2003.
307. Kim, T.-U., Amy, G., Drewes, J., Rejection of trace organic compounds by high-pressure membranes. *Water Sci. Technol.*, 51, 335–344, 2005.
308. Beier, S., Köster, S., Veltmann, K., Schröder, H., Pinnekamp, J., Treatment of hospital wastewater effluent by nanofiltration and reverse osmosis. *Water Sci. Technol.*, 61, 1691–1698, 2010.
309. Domínguez-Espíndola, R.B., Bruguera-Casamada, C., Silva-Martínez, S., Araujo, R.M., Brillas, E., Sirés, I., Photoelectrocatalytic inactivation of Pseudomonas aeruginosa using an Ag-decorated TiO2 photoanode. *Sep. Purif. Technol.*, 208, 83–91, 2019.
310. Brugnera, M.F., Miyata, M., Zocolo, G.J., Gonçalves Tessaro, L.L., Fujimura Leite, C.Q., Boldrin Zanoni, M.V., A promising technology based on photoelectrocatalysis against Mycobacterium tuberculosis in water disinfection. *Environ. Technol.*, 42, 743–752, 2021.
311. Zhao, C., Wang, Y., Wang, X., Dionysiou, D.D., Treatment of contaminants of emerging concern and pathogens using electrophotocatalytic processes: A review. *Curr. Opin. Green Sustainable Chem.*, 32, 100527, 2021.

312. Girgis, E., Adel, D., Tharwat, C., Attallah, O., Rao, K.V., Cobalt ferrite nanotubes and porous nanorods for dye removal. *Adv. Nano Res.*, 3, 111, 2015.
313. Dewil, R., Mantzavinos, D., Poulios, I., Rodrigo, M.A., New perspectives for Advanced Oxidation Processes. *J. Environ. Manage.*, 195, 93–99, 2017.
314. Kushwaha, A., Goswami, S., Hans, N., Goswami, L., Devi, G., Deshavath, N.N., Yadav, M.K., Lall, A.M., An insight into biological and chemical technologies for micropollutant removal from wastewater, in: *Fate and transport of subsurface pollutants*, pp. 199–226, 2021.
315. Moussa, D.T., El-Naas, M.H., Nasser, M., Al-Marri, M.J., A comprehensive review of electrocoagulation for water treatment: Potentials and challenges. *J. Environ. Manage.*, 186, 24–41, 2017.
316. Garcia-Segura, S., Eiband, M.M.S., de Melo, J.V., Martínez-Huitle, C.A., Electrocoagulation and advanced electrocoagulation processes: A general review about the fundamentals, emerging applications and its association with other technologies. *J. Electroanal. Chem.*, 801, 267–299, 2017.

5

An Update on the Progress, Trends and Challenges of Drinking Water Treatment and Provision

Maphanga Donald[1], Mapula Lucey Moropeng[1], Vhahangwele Masindi[2]* and Beauclair Nguegang[3]

[1]Department of Chemical, Metallurgy and Materials Engineering, Tshwane University of Technology, Pretoria West, Pretoria, South Africa
[2]Department of Environmental Sciences, School of Agriculture and Environmental Sciences, University of South Africa (UNISA), Florida, South Africa
[3]Department of Chemical Sciences, Faculty of Science, University of Johannesburg, Auckland Park, Johannesburg, South Africa

Abstract

Safe drinking water requires proper and effective treatment followed by sound monitoring and maintenance of leftover chlorine within the drinking water distribution networks. The main quest is to suppress potential regrowth of microbial fractions and secondary contamination that prevails in the distribution network. A body of knowledge has been developed; however, there is a need to distill the progress, trends, and challenges of drinking water treatment and provision. Therefore, this literature review seeks to highlight the processes of drinking water treatment, disinfectants that are used for drinking water treatment, chlorination as a way of keeping the distribution network sterile, laboratory and field experiments and assays, disinfection challenges, simulation of chlorine demand and decay using different models, their weaknesses, and challenges thereof. Benefits of combining experimental results with simulation were also highlighted. The effect of bulk and wall chlorine decay has been unpacked among other contributing factors. The importance of modeling in understanding chlorine decay and demand has been emphasized because conducting water samples every time the water quality changes is a time-consuming and labor-intensive task. This book chapter gave an

**Corresponding author*: emasinv@unisa.ac.za; masindivhahangwele@gmail.com

Elvis Fosso-Kankeu, Vhahangwele Masindi, Johannes Maree and Bhekie Mamba (eds.) Customized Technologies for Sustainable Management of Industrial Wastewater: A Circular Economy Approach, (193–222) © 2025 Scrivener Publishing LLC

update on the progress, trends, and challenges of drinking water treatment and provision. Findings reported in this book chapter will aid in the development of robust and more effective systems and technologies for the treatment and provision of drinking water in low- and middle-income countries.

Keywords: Drinking water treatment, drinking water provision, chlorine demand and decay, modeling, first-order model, simulation, chlorine-reactive substances, water distribution network

5.1 Raw Water

Unprocessed or feed water refers to water that is not purified and enriched with minerals, bacteria, pathogens, cations (K^+, Na^+, Ca^+, Mg^+, Fe^{2+}, and more), and anions (Cl^-, NO_3^-, SO_4^{2-}, and more) [1]. The feed water can be found as a product of precipitated water from rainfall, water stored under the landscape, unfiltered water wells, and from water flowing sources such as rivers and lakes [2]. The contaminants that are found in the feed water are tempered with the regulation of water pH, taste, odor, and turbidity and generate unsafe water for consumption [3]. Consumption of drinking water that is microbiologically and chemically polluted has been causing severe health damage on consumers such as infants, young children, elderly, or unwell people or those who are immune compromised [3]. Feed water is further processed in drinking water treatment plant to critically control the water quality and produce safe water for consumers [4].

5.2 Drinking Water Treatment Process

The processes of treating drinking water involve the combination of physical, chemical, and biological methods to improve the water quality for consumers [5]. A water treatment plant is equipped with units that perform pre-disinfection, coagulation, flocculation, sedimentation, filtration, and post-disinfection. The pre-disinfection phase is when the dosing of chlorine or other disinfectant is added into the feed water to remove the pathogens, bacteria, microbes, and other chlorine-reactive contaminants present in the water [5]. A coagulant in a form of chemical is then added to attract the negatively and positively charged colloids that are dissolved and suspended in the water [6]. The added chemical removes the dissolved organic and inorganic matters [7].

The flocculation phase employs the slow stirring of the treated water to allow the inter-particle collision and to form large flogs that could be physically filtered. In the sedimentation process, the dense-suspended particles or flocks are allowed to settle through the gravitational force. The settled particles are removed by passing them into a filtration medium (Slow Sand Filters) at a rate of 4–24 $min^{-1}/10\ dm^2$. A post-disinfection process is the last step in the water treatment plant to ensure that 99.9% of the inorganic and organic matters that managed to survive during the purification phases are destroyed and inactivated completely [8–10]. Chlorine is widely used in the post-disinfection stages to react with all the pathogens present in the final water; however, the chlorine residual is not effectively controlled due to the interchanging of bulk and wall chlorine demand [11].

5.3 Functionalities of Drinking Water Treatment Process

To safeguard the health of end-users, drinking water should be of a specified quality to ensure that it is safe for consumption without posing an intermediate or long-term harm to the consumers [12]. The drinking water purification processes are designed and cultivated to prohibit the deterioration of final water quality by ensuring that the water purification process is effective [11]. The inter-link of the water treatment plant units are designed to ensure that if pathogens managed to survive the first stages of the treatment process; therefore, they should be eliminated in the next stage until the final step [2, 13].

5.4 Final Water and Challenges

Different processes including physical, chemical and biological that occurs in the water treatment plants dictate the water quality that would reach the consumer's tap [14]. Drinking water treatment and post-chlorination processes do not destroy all biomass present in raw water sufficiently and microorganisms in final water still persists in lower concentrations to further consume the chlorine residual [15, 16]. Some of the contributing factors that affect the final water quality integrity are biofilm growth [16], nitrification [6], leaching substances from network materials, internal pipe corrosion [17], scale formation on the pipe surfaces [18], and the formation of disinfectant by-products that further consume the chlorine residual [17].

Scale formation on the drinking water conduits is being caused by the solid material deposits during post-treatment, upstream corrosion, and deposits [18]. The effect of low flow of water, low chlorine residual levels, the presence of nutrients in water, and microorganisms that are able to form biofilm in bulk water collectively enables a conducive environment to develop biofilm on the wall surface of the pipes in the water treatment and distribution systems [19]. Consumer demand of final water determines how long the water could be stagnant within the plumbing system and thus leads to change in microbes due to regrowth [10].

5.5 Distribution Water Challenges

Fresh water is being treated through the drinking water treatment plants and pumped into the water distribution system for domestic use [11]. The design of water distribution system consists of pipes, pumping system, fittings, facilities for water storages, and installed devices that are used to avoid backflow [18]. Water quality in post-treatment depends mostly on the characteristics of the distribution system and the conditions in which it is being operated [7]. The quality of the tap water is solely dependent on the pipes that flows through [18]. Depending on the flows and other factors, biological substances establish their community in the distribution network, and this leads to the formation of biofilms and microbiological regrowth [12]. Also, to a certain extent, microorganisms establish their community in the biofilms because they derive or acquire nutrients from the biofilms and moving water [20]. Microorganisms that are attached to the biofilm of the municipal water distribution pipes are a provocative factor in supplying safe water [21]. Specifically, the biofilms in water distribution network promote a conducive living environment for bacteria, archaea, fungus, and protozoa [22]. The presence of bacteria on the biofilm in distribution system produces soluble microbial products, such as proteins that accelerate disinfectant decay and concomitantly contribute to the formation of disinfectant by-products [7].

Pipe corrosion and taste odor in the distribution system are mostly influenced by the presence of biofilm in the distribution network [21]. Bacteria self-attach on the surface of the suspended particles in such a way that the disinfectant could not reach [15]. When hydraulic properties such as flow velocity or back flush of the system are disturbed, the suspended particles are being released to the bulk water [23, 24]. There is insufficient knowledge regarding the composition of species and ecology of biofilms within distribution system, because it is challenging and costly to get samples from underground pipes [25]. A low number of suspended particles can still exist in the water distribution system due to being partially removed from the raw water treatment and the contaminants that are found in reservoirs or pipes [22, 26].

The final water mainly carries along little concentrations of organic carbon, inorganic nutrients, and chlorine residuals [22]. Disinfection by-products (DBPs) are formed when chlorine is reacting with inorganic and organic matters during the transportation of water within the distribution system [21]. The contributing factors toward disinfection byproducts are mostly the water pH, temperature, contact time, type, and concentration of chlorine [27]. The bulk and wall chlorine decay in water distribution systems are very complex and vary from one distribution system network to the other [28]. These factors are being characterized by the age of the system, type of pipes, biomass on the wall surface, hydrodynamics, and fluid flow of the network [29]. It is critical to understand the factors that contributes to the chlorine decay in the affected distribution network and best model the bulk chlorine decay before the wall decay studies [30].

5.6 Types of Disinfectants and Oxidants

There are several water disinfectants or chemicals that are dosed in the water to destroy pathogens that are harmful for consumers. Chemical oxidants are disinfecting techniques that are used to oxidize natural organic and inorganic matters in water.

5.6.1 Chlorine Gas

A greenish-yellow chlorine gas is widely used in the process of disinfecting water for both primary and secondary disinfection. Employing chlorine to water ensures the effectiveness role of nearly killing all the pathogenic microorganisms found in the water [3]. The production of chlorine gas involves passing of a brine solution through electrolysis in a diaphragm, membrane, or mercury cell, as shown below:

$$2NaCl_{(aq)} + 2H_2O_{(g)} \rightarrow 2NaOH_{(aq)} + Cl_{2(g)} + H_{2(g)} \tag{5.1}$$

The brine in Equation 5.1 decomposes to sodium hydroxide, chlorine gas, and hydrogen gas. The process of chlorination is not costly as compared to ultraviolet (UV) or ozone disinfection methods, and the chlorine dosages can be easily controlled [31]. Chlorine gas is harmful when is being inhaled at a concentration of between 3 5 mg/L and 5 mg/L, and it is fatal when a human being is exposed to a 4 mg/L for an hour. Chlorine gas in water treatment can be applied during the pre-chlorination, re-chlorination, post-chlorination, and de-chlorination stages [32]. The application of chlorine in water is to control microorganisms, remove ammonia (NH_3), control the water taste and odor, reduce watercolor, destroy organic matters, and oxidize hydrogen sulfide (H_2S), iron (Fe^{2+}), and manganese (Mn^{2+}). Chlorine gas is highly reactive and effective in water disinfection; however, the reactivity is pH-dependent [33].

5.6.2 Chlorine Dioxide

Chlorine dioxide (ClO_2) is a water disinfectant that is being used worldwide to inhibit the regrowth of microorganisms and bacteria in municipal water. It is also used as a pre-oxidant to extract water taste, odor, color, phenol, iron (Fe^{2+}), and manganese (Mn^{2+}) during the pre-disinfectant stages of water treatment [34]. The production of chlorine dioxide involves the reaction of sodium chlorite with acid, as shown in Equation 5.2:

$$5NaClO_2 + 4HCl \rightarrow 4ClO_2 + 5NaCl + 2H_2O \tag{5.2}$$

Chlorine dioxide gas is highly volatile, and, hence, it can be only produced onsite in small quantities using special reactors. For safety and health purposes, chlorine dioxide gas dosages in large quantities are prohibited and can be used at the point of use. The instability of chlorine dioxide is much better as compared to ozone, and the residual does not last long [35].

There are less chlorinated compounds generated after reacting with natural organic matters (NOMs), and its effectiveness is independent of water pH. Chlorine dioxide is efficient in eliminating the contaminants that are harmful to consumers and requires 4 h of reaction time in low dosage [36]. It is 99% more efficacious in destroying viruses as compared to bacteria. Chlorine dioxide is a strong oxidizing agent, does not react with the ammonia available in the water, and has no effect in producing trihalomethanes (THMs) [33].

5.6.3 Sodium Hypochlorite Solution

Sodium hypochlorite (NaOCl) is mostly employed as a bleaching agent for paper or textiles bleaching and as a water disinfection. Chlorine concentration in the solution varies between 10% and 15% as free chlorine, and the addition of hypochlorite in water produces the hypochlorous acid as shown below [34]:

$$NaOCl + H_2O \rightarrow HOCl + NaOH^- \quad (5.3)$$

The hypochlorous formed in Equation 5.3 is defined as a free available chlorine (FAC) and is further separated to hydrochloric acid (HCl) and oxygen (O) as an effective oxidizing agent. The efficacy and reactivity of sodium hypochlorite are similar as compared to those of chlorine gas; however, chlorine in the hypochlorous acid quickly loses its effectiveness during the process of solution storing [37].

5.6.4 Calcium Hypochlorite

Calcium hypochlorite [Ca $(ClO)_2$] is a water disinfectant that have the similar effects as sodium hypochlorite and calcium hypochlorite mostly available in solid with 70% of chlorine than liquid sodium hypochlorite [38]. The formation of chlorine from calcium hypochlorite is shown as in Equation 5.4:

$$Ca(ClO)_2 + 2H_2O \rightarrow 2HOCl + Ca(OH)_2 \quad (5.4)$$

Compared to chlorine gas and sodium hypochlorite, calcium hypochlorite is user-friendly and has an exceptional stability to protect its strength gradually when stored in a dry place [33]. The hypochlorous acid produced in Equation 5.4 has the same chlorine reactivity as that of sodium hypochlorite with 99% of destroying all contaminants in the water [39].

5.6.5 Chloramines

The excess of FAC with the reaction of ammonia generates the formation of chloramines at a theoretical pH of 8.4 as shown in Equation 5.5 [8]:

$$NH_{3(aq)} + HOCl \rightarrow NH_2Cl + H_2O \quad (5.5)$$

Chloramines perform a crucial role in the distribution system by ensuring that residual protection is maintained. Compared to chlorine, chloramines have a higher stability and produce less of halogenated by-products [33].

The chloramines react with other microorganisms and inhibit the survival of bacteria; however, the reaction rate is very slow as compared to chlorine [3].

5.6.6 Ozonation

Ozone is a water disinfectant that has been used widely in wastewater and water reuse applications. Liquid oxygen is passed through UV light to produce ozone molecule, as shown below [40]:

$$3O_2 \xrightarrow{hv} 2O_3 \tag{5.6}$$

The formed ozone gas is a highly reactive oxidant; destroys organic matters, viruses, and bacteria; and eliminates water odor or taste. Ozone is a better water disinfectant as compared to chlorine; there is no formation of disinfectant by-products and residuals. The ozonation process produces a multitude of carbonyl disinfection byproducts. However, it is a very expensive technique and requires a sophisticated operating equipment [40].

5.6.7 Ultraviolet Light (UV)

Ultraviolet light is widely used to disinfect wastewater, drinking water, and aquaculture. The UV light kills the biological microorganism components by breaking the chemical bonds in DNA, RNA, and proteins [33].

5.6.8 Photocatalysis

Photocatalytic disinfection in water disinfection is being used widely for having exceptional benefits of using sunlight during the process, and a solid catalyst like (TiO_2) titanium dioxide is used [41].

5.7 Role of Chlorine in Water Treatment

The significance of using chlorine in water treatment plants will be broadly highlighted in this section.

5.7.1 Aqueous Chlorine Chemistry

Gaseous chlorine Cl_2 or hypochlorite is mostly used during chlorination processes in water treatment; the following equation describes the hydrolysis of chlorine in water [38]:

$$Cl_2 + H_2O \rightarrow HOCl + H^+ + Cl^- \tag{5.7}$$

The hypochlorous (HOCl) acid produced from reaction equation 5.7 is a weak acid that rapidly dissociates in aqueous solution [38]:

$$HOCl \rightarrow H^+ + OCl^- \tag{5.8}$$

5.7.2 Organic Compounds

Organic nitrogen compounds are found in wastewater that are being originated from domestic use [42]. Chlorine reacts with olefins, oxygenated moieties (alcohols, aldehydes, ketones, and acids), sulfur-containing moieties, nitrogen-containing moieties, and aromatic compounds [38]. The reaction of chlorine and organic nitrogen produces N-chloro compounds or chloramines that have a high stability, and, in that case, the chloramines cannot be effectively oxidized during the chlorination process, and the formation of DBP occurs [42].

5.7.3 Inorganic Compounds

The most valuable properties of free chlorine is that of being a very strong chemical oxidant, and the reactiveness of free chlorine solely rely on the water pH [42]. In the process of aqueous chlorination, hypochlorous acid reacts with ammonia, halides, other anionic inorganic compounds (SO_3^{2-}, CN^-, NO_2^-, or sulfide), As (II), Fe (II), and Mn (II). The hypochlorous acid oxidizes all the inorganic compounds to their respective oxidized compounds, such as Fe(II) being oxidized to Fe(III) [38].

The inorganic compounds found in wastewater treatment are complex chemicals and rapidly consume 90% of free chlorine. The overall chlorine-reactive compounds are being represented by the following reaction [42]:

$$Cl_2 + H_2O + (\text{Chlorine demand}) \rightarrow 2HCl + (\text{Oxidized compounds}) \quad (5.9)$$

However, hydrochloric acid generated from Equation 5.9 continue to react with the alkalinity of the water as indicated in the following equation [42]:

$$2HCl + H_2O + CaCO_3 \rightarrow CaCl_2 + CO_2 + 2H_2O \quad (5.10)$$

The above reaction indicates that chlorination of water can result in a pH shift provided that the alkalinity of the solution is insufficient to buffer the reaction [42].

5.8 Effects of Chlorine as a Post-Disinfectant

The commonly used disinfectant worldwide is chlorine for being cost-effective and having the capacity to prolong residual chlorine within the water distribution system [43]. Chlorine reacts with bromide ion and dissolved organic matter in the system; hence, possible health-threatening disinfection products like THMs and haloacetic acids (HAAs) could be produced [44]. Controlling desired chlorine residual in the system requires booster doses (re-chlorination) to generate a higher dosage at the inlet, and this will be reduced at the exit point; however, a high dose of chlorine produces taste or odor at the upstream and complaints from consumers arise [35]. The chlorine dose should be sufficient enough to inactivate pathogens and the residual should be minimum to limit generation of disinfectant by-products and prevent bacterial growth [35, 45, 46].

5.9 Regulatory Requirements

According to World Health Organization (WHO), the required concentrations of THMs, HAAs, and haloacetonitrile are 0.3 mg/L, 0.02 mg/L, and 0.002 mg/L, respectively [43]. In South Africa, the allowed concentrations of organic compounds such as THMs are 0.2 mg/L and 10 mg/L for total organic carbon; for inorganic compounds like iron and sulfate, it was found to be 0.2 mg/L and 0.4 mg/L, respectively, to mention few [47]. Maintaining chlorine residual in the distribution is necessary to prevent water contamination, and the recommended concentration of chlorine

Table 5.1 Health-related inorganic chemical parameters in South Africa [47].

Determinant	Value (mg/L)	Determinant	Value (mg/L)
Aluminum (Al)	0.3	Lead	0.02
Ammonia	1.0	Manganese	0.1
Antinomy	0.01	Mercury	0.001
Arsenic	0.01	Nickel	0.15
Chromium	0.1	Sodium	0.2
Copper	1.0	Zinc	5.0
Iron	0.2	Chloride	200
Cyanide	0.05	Fluoride	1.0
Sulfate	400	Selenium	0.02
Nitrate	10	Nitrite	10

residual by WHO is between 0.2 and 1.0, whereas, in South Africa, it is found to be 0.005 mg/L [17, 47]. Table 5.1 summarizes the maximum allowed concentrations of inorganic compounds that are regulated by the South African water-governing bodies.

5.10 Chlorine Decay

5.10.1 Chlorine Degradation in Water Distribution Network

Chlorine degradation is the process in which chlorine is being consumed by reacting with chlorine-reactive substances that include bacteria regrowth, cell inactivation, and chemicals [2, 48]. Furthermore, chlorine degradation is accelerated in the presence of pipe deposits that are formed during biomass growth and corrosion. Prediction of chlorine decay modeling in large water distribution systems is challenging because it depends on the water flow in the whole treatment system [23]. The chlorine decay model consists of two reaction mechanisms: the reaction of chlorine with bulk water and pipe wall [35]. Chlorine decay model considers the reaction rate of chlorine with bulk water (k_b) and the reaction rate of chlorine with the pipe walls (k_w) [49].

5.10.1.1 *Bulk Decay Reaction Mechanism*

The reaction rate of bulk degradation should be differentiated from wall degradation by quantifying chlorine decomposition of the water source under controlled environment [39]. Bulk decay rate is being determined by timeously tracing the chlorine concentration using glass bottles that are already filled with the sample water. Inverse relationship between the bulk decay coefficient and initial chlorine concentration was discovered in the laboratory analysis [32, 50]. This illustrates that, if chlorine dose is being increased in a system, then, the bulk reaction rate decreases leaving more chlorine residual in the water. However, the more chlorine leftover that exceeds 0.5 mg/L, the more it is highly chronic for consumption as declared by the WHO, thus calling for the optimization and control bulk chlorine decay [17, 51].

5.10.1.2 *Wall Reaction Mechanism*

It is necessary to recognize the interaction between biofilm (biomass) formation and the chlorine demand for the reaction prior to model chlorine decay [32]. After determining the bulk decay rate from laboratory tests, the wall decay rate is established through a calibration process that determines the difference of chlorine consumed within the distribution system and the chlorine consumed through bulk decay tests [52]. However, the chlorine decay coefficient at the wall solely rely on the conditions of the pipe flowing in and also the pipe materials [12]. The surface-to-volume ratio (S/V) also contributes to the chlorine decay within the pipe walls. When increasing the S/V, so does the inner pipe surface in contact with chlorine residuals increase [53].

5.11 Chlorine Decay Models

Insufficient knowledge on the kinetics of chlorine reaction in water distribution is a challenge, and the following reaction describes the foundation of chlorine decay models [51]:

$$\mathrm{Cl} + \mathrm{reactants} \rightarrow \mathrm{products} \tag{5.11}$$

In the above reaction, Cl represents chemical chlorine residual, and reactants describe the chlorine demand species to form DBPs as represented by products [51].

5.11.1 Zero-Order Model

Zero-order reaction kinetic model takes place when the reaction rate in the system remains constant and the limiting reactant is not chlorine. Zero-order model is used mostly in chlorine decay models that involves corrosion of cast iron pipes [54]. The following equation describes the zero-order reaction kinetics of chlorine in water:

$$Cl(t) = Cl_0 - k.t \tag{5.12}$$

Here, chlorine content at time t(h) is being represented by Cl(t) in (mg/L), whereas Cl_0 represents the initial chlorine dose (mg/L), and k is the decay rate (mg/L·h) [55].

5.11.2 First-Order Model

The first-order reaction kinetic model takes place when the chlorine concentration is the limiting reactant when reacting with chlorine-reactive substances in both wall and bulk decay rate. For first-order model, the chlorine degradation rate relies mostly on the initial chlorine dose concentration and not depending on the detected chlorine demand [49, 52]. Chlorine concentration relies on the mass transfer rate from bulk flow to the pipe walls, and it is in equivalent proportion with the chlorine residual. Film resistance model represents the mass transfer process when chlorine is being transferred from the bulk flow to the pipe walls at a proportional rate to the slope of chlorine decay rate [54, 56]. The first-order reaction rate model and the overall decay rate are presented in Equations 5.13 and 5.14:

$$Cl(t) = Cl_0 e^{-kt} \tag{5.13}$$

$$K = k_w + k_b \tag{5.14}$$

where chlorine content at any given time is being represented by Cl(t) (mg/L), Cl_0 represents the initial chlorine dose (mg/L), and K represents the first order over all decay rate coefficient of chlorine (mg/L·h), whereas k_w and k_b represent the wall and bulk decay rate coefficients, respectively (mg/L·h) [51].

5.11.3 Second-Order Model

The second-order model kinetics includes a two-component irreversible reaction during chlorine reaction with organic matter. The two-component second-order model constitutes for both chlorine and chlorine-reactive substance through the irreversible reaction; the following reaction describes the second-order chlorine decay model [56]:

$$Cl(t) = \frac{C_0(1-R)}{1-R.e^{-u.t}} \tag{5.15}$$

Here, the chlorine content at time (t) is being represented by Cl(t) (mg/L), C_0 is the initial chlorine content, and t represents the reaction time (min). R represents the model parameter that is dimensionless and needs to be estimated, and u represents the reaction rate(1/min) [20, 57].

5.11.4 The Nth-Order Model

$$Cl(t) = \left(k.t(n-1) + \left(\frac{1}{C_0}\right)^{(n-1)}\right)^{\frac{-1}{n-1}} \tag{5.16}$$

The Cl(t) represents the chlorine concentrations (mg/L) at time t (min), whereas n is the reaction order model, C_0 is the initial chlorine dose (mg/L), and k is the bulk decay coefficient (1/min).

5.11.5 Determining the Bulk Reaction Order of the Samples

The measurements of chlorine residual lab test and their respective time intervals should be plotted graphically to determine the relationship. A zero-reaction-order relationship is when the measured residual chlorine and time plot are linear. A first-reaction-order relationship is when the measure chlorine residual takes a natural log or In(C) against time is linear. A second-reaction-order relationship is when the measured chlorine as (1/C) plotted against time and thus gives a straight line [49].

5.12 The Limitations of Traditional Chlorine Decay Models

Modeling first-order chlorine decay is easy and possible due to only defining a singular parameter [bulk decay rate (k)] through analytical experiments and assuming that chlorine reacts with a singular substance [39]. First-order chlorine decay models are dependent on the chlorine concentration as a limiting factor; hence, this models has not produced accurate results because chlorine reacts with multitudes of chlorine-reactive substances [12]. The nth-order and parallel-order chlorine decay models were proposed to improve the accuracy of first-order models by accommodating the reaction in both slow and fast decay phases [39]; however, this models did not produce accurate results that were fitting with the analytical experiments and not accumulating the effect of the rapid chlorine dose well; even though parallel-first-order model was the best, it did not yield enough results to fit the analytical data [50].

These models are less effective and do not produce accurate results for drinking water due to the assumption that the content of available chlorine is not sufficient to react with the chlorine-reactive substances completely [15, 58, 59]. The second-order chlorine degradation models are based on the effect of chlorine reacting with another unknown chlorine constituent. The limitation of second-order chlorine degradation model is that the content of free chlorine should not be the same as the concentration of the singular chlorine-reactive substance. In most cases, chlorine reacts with more than one chlorine-reactive substance; hence, this model is also not sufficient to define the chlorine decay in both slow and fast decay phases. However, second-order chlorine decay models are better when compared to first-order chlorine decay models [32, 60, 61].

5.13 Experimental Approaches

5.13.1 Bulk Chlorine Decay Using Analytical Methods

5.13.1.1 *Effect of Water Indexes (Dissolved Organic Carbon, UV_{254}, Ammonia, and EEM)*

Water quality indexes play a significant role in chlorine decay modeling; Wang *et al.* [2] analyzed the effect of dissolved organic carbon (DOC), UV_{254}, ammonia, and UV absorption (UV_{254}), pre-ozonation using ferric sulfate and aluminum sulfate as coagulants, and synthesized water to investigate the effect of pre oxidation especially in the essence of removing DOC [2]. In the study of Wang *et al.* [2], the chlorine decay established from the chlorine consumption in a period of 8 h found that the treatment system that included lower pre-ozonation and coagulant concentration of ferric sulfate resulted in the best chlorine stability and THMs below 0.06 mg/L followed by the ferric sulfate coagulant; however, the DOC, UV_{254}, EEM, and ammonia did not constitute to chlorine decay. The study of Wang *et al.* [2] strongly suggested that pre-ozonation assists in stabilizing chlorine residual.

Bu *et al.* [62] have investigated the effect of using coagulation and pre-ozonation to remove the NOM and determine the formation of DBPs. In their study, it was found that the addition of ozone after the coagulation dosage has increased the removal of NOM by achieving a less UV_{254} that was below 0.023 cm^{-1} and the DOC was below 2.31 mg/L, whereas the coagulant dose was set to 7 mg/L. However, the formation of DBPs was established after the addition of sodium hypochlorite (NaOCl). The coagulation after ozone has performed better results, because the coagulant after ozone stabilizes the formation of DBPs especially for water that contains bromine, and the chlorine dosage had no impact on forming DBPs [62]. It is critical to perform pre-ozonation as a pre-treatment technique that will stabilize the chlorine residual after chlorination [63].

5.13.1.2 *Effect of Natural Organic Matters Toward Chlorine Decay*

Krishna *et al.* [64] studied the effect of chlorine tablet-dosed toward the bacterial community that includes opportunistic pathogens and nitrifies. In their study, the selected bacterial groups were Methyl bacterium, Hyphomicrobium, and unclassified Rhizobia. Samples were collected from the entry and exit points of the downstream of a reservoir where chlorine tablets were dosed timeously. It was found in their studies that the bacterial group dominated 93.8% of bacterial genera in the downstream of the distribution pipelines, and E. coli was not detected in the period of the study [64]. The dosages of chlorine tablets had a very positive impact toward increment of total chlorine to the ratio of total ammoniacal nitrogen, notwithstanding the slightest changes of total chlorine residual. The total chlorine residuals and status of nitrification did not have an impact on the excess of bacterial community [64].

5.13.1.3 *Effect of Temperature and pH*

Powel *et al.* [50] studied the factors that contribute to chlorine degradation within distribution systems. It was revealed that the bulk degradation coefficient of chlorine decay has shown a dependence on temperature, initial chlorine dose, and total organic matters. There was no effect of pH toward bulk chlorine decay coefficient observed because the pH variance was minimum [50]. Zhang and Andrews [65] have investigated the effect of pH using samples that had three pH values (6.6, 7.6, and 8.6), and there was no statistical variation in their findings, and the pseudo-first-order bulk decay chlorine coefficients were found to be ranging between 0.0018 and 0.0022 (1/h) [65].

5.13.2 Wall Chlorine Decay Using Analytical Methods

5.13.2.1 *The Effect of Biofilm*

Biofilm formation on the pipe walls accelerates the chlorine degradation in the water distribution network. Xu *et al.* [10] assessed the effect of biofilms growth toward chlorine degradation and TTHM formation that prevails during physical and chemical water treatment processes. Pipe loop system of poly vinyl chloride pipes was used to resemble the distribution system, and the system was checked after every 2 weeks. The biofilm development in the pipe loops allowed enough time to occur, then the pipes were removed from the system, and organic-free chlorine solution buffer was added to the pipes for a period of 48 h. It was found that the chlorine decay in the pilot study was higher than that in the real distribution systems. Biofilms have a great impact in chlorine decay within distribution system by generating toxic disinfection byproducts [10]. Chen *et al.* [29] investigated the chemical and biological threat during the expansion of water distribution pipe renewal. The sole concerned parameters were the DBPs and the potential for DBPs (DBPsFP) over a period of 2 years. It was found that the total bacterial count in the new pipelines ranged from 1.38 to 9.9×10^8 per cm^2. The potential for DBPs slightly varied at both front and back ends. Stainless steel pipes were found to have less biofilm growth as compared to ductile and galvanized pipes. The total DBPsFP of biofilms were found to be the highest as compared to the DBPs [29].

5.14 Simulations and Mathematical Estimates

5.14.1 Bulk Chlorine Decay Rate Using Simulations and Mathematical Estimates

5.14.1.1 Effect of Temperature on Bulk Chlorine Decay Rate (NOM and THMs)

Monteiro *et al.* [9] have investigated the influence of varying temperature toward chlorine degradation rates for raw and final water from various locations. The model was based on the variation of natural NOM and to establish the effect of temperature toward fast and slow decay rates using two reactant models (2R). It was found that, in both raw and final water, the bulk chlorine degradation coefficients in fast and slow decay phases were extremely impacted by the temperature variations. The slope of Arrhenius equation for the slow decay phase was found to be the highest and highly sensitive toward temperature as compared to that of the fast chlorine decay phase [9]. Furthermore, Abokifa *et al.* [44] investigated the influence of temperature variation the formation of THMs using a multiple species reactive transport model. The temperature varied from 0°C to 30°C; it was found that temperature has an influence on the formation of bacterial regrowth, whereas temperature increment affected various parallel processes. Temperature increment from 0°C to 25°C activates the exponential growth of bacterial community, faster consumption of chlorine residuals, and the rapid depletion of chlorine reaction with biomass that generates higher concentrations of THMs [44].

5.14.1.2 Influence of Hydraulic Conditions

Jamwal and Kumar [60] investigated the influence of flow rate and initial chlorine levels to predict the chlorine decay coefficients using a first-order model. The initial chlorine dose (IC) was at 3 mg/L and 6 mg/L, whereas the velocity was ranging from 0.03 m/s to 0.12 m/s. The finding revealed a moderate bulk chlorine degradation coefficient at IC of 3 mg/L, which was a double compared to the bulk decay coefficient for initial chlorine dose of 6 mg/L. This implies that the chlorine consumption in the low velocity phase was due to the effect of fast and slow chlorine-reactive substances that were present in the pipe [60].

Additionally, Abokifa *et al.* [66] investigated the effect of flow velocity toward the formation of biofilm using a multiple species reactive transport model. It was found that, when the flow velocity increases, the enhancement of mass transfer of solutes from the bulk water to the biofilm interface was established. At elevated shear stress rate, the biomass gets detached from the biofilm or pipe wall to the bulk water. A Blasius equation was established to predict the shear stress under turbulent flow conditions, and it was found that the shear stress increases with flow velocity to a power above 1 ($n > 1$) [66]. Van der Walt *et al.* [11] investigated the effect of water age in Sedibeng water distribution system using EPANET software. The model order of the reaction kinetics was first order. It was found that the maximum water age within the system is 160 h in a 70-km-long pipe network. The summer chlorine decay rate was found to be 0.052 (1/h) and 0.027 (1/h) during winter seasons. The simulated chlorine model was used to optimize the disinfection systems and a switch to using ClO_2 as a disinfectant was demonstrated to stop the re-chlorination processes. However, the use of ClO_2 as their disinfectant will require a capital of extra 100,000 US dollars [11].

5.14.1.3 The Initial Chlorine Dose Effect on Bulk Decay Rate

In controlling the leftover chlorine in water network distribution, bulk chlorine degradation models and simulations are continuously being developed and optimized to control bulk chlorine decay according to the review of Li *et al.* [7]. Furthermore, Tiruneh *et al.* [12] have studied the chlorine-reactive substances based on chlorine demand and total chlorine using mathematical estimates; in their studies, it was found that instantaneous chlorine decay happen within 5 min, and the bulk reaction rate relationship with initial chlorine dose was found to be inverse using mathematical estimates that were validated by the analytical results [12].

5.15 The Rate Constant of Chlorine Decay with the Wall of Water Pipe

5.15.1 Effect of Hydraulic Conditions

Cong *et al.* [51] determined the reaction rate constant of chlorine decay at the wall by running pilot study experiments using various flow velocities that were ranging between 0.1 and 0.5 m/s; the temperatures were varying from 15°C, 20, to 30°C; the initial chlorine dose concentration was ranging from 0.30 mg/L to 0.40 mg/L; and the total organic carbon (TOC) concentration was ranging from 3.5 mg/L to 5.5 mg/L. It was found that the chlorine decay coefficients of both bulk water were almost the same as the constant of the overall decay, implying that wall decay could be approximated to zero, illustrating that there was no mass transfer with the pipe wall because the molecular diffusivity of the model was laminar flow [51].

Additionally, Jamwal and Kumar [60] determined the effect of flow velocity and initial chlorine levels to predict the chlorine degradation coefficients at the wall using a first-order model. The flow rate within the pipe loop ranged from 0.02 m/s for laminar flow to 0.12 m/s for turbulent flow. It was found that the wall degradation coefficient was influenced by both initial chlorine dose and flow velocity. The transition of flow regime from laminar to turbulent resulted in an increase of wall decay coefficient by 66% due to the influence of biofilm [60]. Goodarzi *et al.* [45] have investigated the effect of temperature toward flow within a chlorine contact tank using solute transport model. Reynolds-averaged Navier–Stokes modeling equations coupled with turbulence closure model were used to define the flow properties in a three-dimensional (3D) advection-diffusion interface. Small temperature and inflow change have impacted the chlorine concentration within the tank. Their study revealed that the chlorine degradation in the tank was linearly dependent on the temperature, because, when the temperature increases, the chlorine depletion rate is also increasing; this was because of the effect that temperature has on the reaction rate as described by the Arrhenius equation [45].

5.16 Tools for Simulations and Mathematical Estimates

This section will provide the available software packages and numerical estimates to perform chlorine decays assays.

5.16.1 Integrated Chlorine Decay Mathematical Models Derived from Traditional Models

The integrated chlorine decay models are finite reactant models that are grounded on the basis of chlorine-reactive finite components that include single reactant (SR) second-order model, two reactants (2R) second-order model, and variable rate coefficient model [67]. These models produce accurate results as compared to the traditional chlorine decay models. However, the challenge arise when measuring the concentration of the finite chlorine-reactive substances because it cannot be measured directly, implying that the model parameters can be produced from fitting the experimental data to the models, and this is complex to an extent that other model parameters are being omitted for simplicity purposes [2, 28]. Some of the integrated chlorine decay models are shown in Table 5.2.

Table 5.2 A list of few proposed integrated chlorine decay models in the literature.

Model	Differential equation	Integrated equation	Adjustable parameters
nth-order	$dCl/dt = -k.C^n$	$Cl(t) = \left(k.t(n-1) + \left(\frac{1}{C_0}\right)^{(n-1)} \right)^{\frac{-1}{n-1}}$	k, n
Limited first order	$dCl/dt = -k(C - C^*)$	$Cl(t) = C^* + (C_0 - C^*)e^{-k.t}$	k, C^*
Parallel first order	$dCl/dt = -k_1C_1,$ $dCl/dt = -k_2C_2$ $C_{1,0} = C_0.x$ $C_{2,0} = C_0(1-x)$	$Cl(t) = C_0.x.e^{-k1.t} + C_0(1-x)e^{-k2.t}$	k_1, k_2, x
Two reactant first-order model [54]	$dCl/dt = \frac{2\,k_w k_f}{R(k_w k_f)} C$	$Cl(t) = Cl_0.e^{\frac{2\,k_w k_f}{R(k_w k_f)} t}$	$k_w, k_{f,} R$

5.17 Software Packages for Chlorine Decay Simulations

The use of software packages to predict and optimize the water quality models in network pipes or reservoirs has gained more attention to improve the performance and efficiency of the hydraulic systems [67]. Mathematical, physics, and chemistry concepts that constitute modeling of chlorine decay are being incorporated in the software packages to graphically visualize the water quality models within the distribution systems and water treatment systems.

5.17.1 EPANET Software

The Environmental Protection Agency Network (EPANET) software is a modeling tool that is being used worldwide to model and simulate the water quality models in the water distribution systems. The modeling tool was designed to comprehend the water quality within distributions systems, and it can be used to simulate and analyze variety of water quality analysis. Researchers use EPANET software to design, optimize, and develop the water infrastructure to optimally and cost-effectively improve the hydraulic and water quality. EPANET multi-species extension (EPANET-MSX) and EPANET real-time extension (EPANET-RTX) are the integrated modeling tools optimized from the traditional EPANET tool to improve the objectives of the software by accommodating even the integrated chlorine decay models [35, 54, 68].

5.17.2 COMSOL Multiphysics Software

COMSOL Multiphysics software is a multi-platform modeling tool that is designed to solve, analyze, and simulate finite multiphysics elements. This software enables a conventionally physics based user interfaces that are coupled with applications to solve ordinary and partial differential matrix equalities [69]. This software is used to simulate the effect of fluid flow dynamics, reaction mechanism, and chlorine transport in water distribution systems for both traditional and integrated chlorine decay models [45].

5.17.3 AQUASIM Software

AQUASIM software is a water quality modeling software that solves the application of ordinary differential equations to optimize the estimation of model parameter values. The software fits the experimental data with the estimated parameters by varying the estimated model parameters. This software is used widely in the simulation of chlorine decay reaction kinetics and tetrahalomethane formation [32, 52].

5.17.4 Other Modeling Software

Other chlorine decay and water quality modeling software packages include and not limited to the following: The H2Onet software is used to model and analyze water quality in water distribution systems and other modeling software [32].

5.18 Challenges of Simulations

Modeling chlorine decay rate constants at various temperature is quite a challenge because the model does not follow the Arrhenius model in many cases; the reaction kinetics requires simplification; and if it is oversimplified and applied, then, some model parameters are being omitted; this further contradicts the efficiency and validity of the model [8]. Other modeling parameters require further recalibration to compensate the effect of water flow changes [67]. The traditional first-order and second-order reaction rates having a singular reaction rate coefficient insufficiently interpret the bulk decay precisely. Integrated mathematical models have been developed to mitigate the inefficiency of the traditional models; however, the integrated models are complex to an extent that other modeling parameters are being omitted to simplify the model, and these models depend on types of water being treated, for example, reclaimed water models are different from fresh water models. Simulating water quality models using EPANET has its limits because the traditional EPANET software can only solve water quality model that follows the traditional reaction rate model. Hence, the extensions of multi-species and real-time application tools are being used to mitigate the limits; however, these extensions require coding approaches that are not user friendly. The role of biofilm modeling is a challenge, because the water distribution pipes are embedded below the ground for years and reaching a sample for biofilm analysis on the pipe wall is a major challenge, and pilot systems were developed to investigate the growth of biofilm. However, pilot studies are costly, and they need more time to develop the biofilm along the pipelines, and such biofilm constituents are season- and flow condition–dependent; this further delay the process.

5.19 Conclusion and Avenues for Future Research

5.19.1 Conclusion

This book chapter successfully highlighted an update on the progress, trends, and challenges of drinking water treatment and provision. The essence of chlorine in drinking water treatment was highlighted. The importance of chlorine demand and decay modeling in enhancing water quality was meticulously elucidated. Specifically, it was highlighted that modeling involves various reaction rate models with numerous parameters specific to the water being modeled. However, due to the complexity and cost involved in chlorine decay modeling, it is important to establish simplicity and clear boundaries in the depth of the models. Chlorine decay within the distribution system is influenced by numerous factors that vary based on the hydraulic systems in place. Extensive research has been conducted on chlorine decay, but it remains a complex field. Furthermore, it noted to be beneficial to categorize water quality models based on the classification of water types. This categorization can streamline the chlorine decay modeling process and reduce the required time. Performing chlorine decay assays through laboratory analysis of temporal chlorine decay profiles aims to enhance the efficiency of chlorine residuals in the water distribution network system. However, conducting water samples every time the water quality changes is a time-consuming and labor-intensive task. Chlorine decay experiments typically require more than 24 h to yield accurate results. The new approach in optimizing the water distribution network is to perform regression-based chlorine decay models that will predict the sufficient chlorine dose to be simulated in a water quality and solute transport model to determine the depletion of chlorine with distance. The combination of regression chlorine-based water quality and solute transport models were not studied in comparison before. The bulk chlorine decay in laboratory experiments has been extensively studied relying mostly on the effect of organic matters.

5.19.2 Avenues for Future Research

After the scanning of the existing body of knowledge, the following areas were highlighted as areas of investigation:

- Explore the efficacy of using other disinfectants as post-chlorination or -disinfection instead of chlorine which is the main chemical.

- Calibrate the model again raw water of different compositions and extremities, and this will allow for more in-depth understanding of required chemicals.
- More studies that constitute more extensive work on the effect of inorganic matters and then develop models that could predict the concentration of inorganic constituents within the system.
- Water quality models have been researched extensively and validated by only experimental and pilot studies; however, incorporating water quality models with pilot studies and solute transport models simultaneously will highlight some of the missing gaps in chlorine demand and decay modeling.
- Various mathematical and computational models have been researched extensively and remain complex; however, comparison of at least five models each to filter out, improve, and classify the best models according to the water type of modeling.

Acknowledgments

The authors of this manuscript would like to express their sincere gratitude to the Tshwane University of Technology and University of South Africa for extending their facilities toward the fulfillment of the objectives of this project.

References

1. Vingerhoeds, M.H., *et al.*, Sensory of drinking water produced by reverse osmosis membrane filtration followed by remineralisation. *Water Res.*, 94, 42–51, 2016.
2. Wang, Y.H., *et al.*, Quantifying chlorine-reactive substances to establish a chlorine decay model of reclaimed water using chemical chlorine demands. *Chem. Eng.*, 356, 791–798, 2019.
3. Krishna, K.B.C. and Sathasivan, A., Biofilm growth on PVC and HDPE pipes impacts chlorine stability in the recycled water. *J. Environ. Chem. Eng.*, 8, 104476, 2020.

4. Masindi, V. and Dunker, L.C., STATE OF WATER AND SANITATION IN SOUTHAFRICA, Council For Scientific And Industrial Research(CSIR), 2016.
5. De Boer, S. and Gonazalez-Rodriguez, J., Benchmarking tertiary water treatments for the removal of micropollutants and pathogens based on operational and sustainability criteria. *J. Water Process Eng.*, 46, 102587, 2022.
6. Lemus Peres, M.F. and Rodriguez Susa, M., Exopolymeric substances from drinking water biofilms:Dynamics of production and relation with disinfection byproducts. *Water Res.*, 116, 304–315, 2017.
7. Li, A., *et al.*, Disinfectant residual stability leading to disinfectant decay and by-product formation in drinking water distribution systems: A systematic review. *Water Res.*, 153, 335–348, 2019.
8. Liu, B., Reckhow, D.A., Li, Y., A two-site chlorine decay model for the combined effects of pH, water distribution temperature and in-home heating profiles using differential evolution. *Water Res.*, 54, 47–57, 2014.
9. Monteiro, L., *et al.*, Modelling chlorine residual decay as influenced by temperature. *Water Environ.*, 3, 29, 331–337, 2015.
10. Xu, J., *et al.*, Role Of drinking water biofilms on residual chlorine deccay and trihalomethane formation : An experimental and modeling study. *Sci. Total Environ.*, 642, 516–525, 2018.
11. Van der Walt, C., Chirwa, E., Traut, D., Calibration of an EPANET model by using monitoring points in the bulk distribution of Sedibeng Water, for predicting disinfectant decay. *European Symposium on Computer Aided Process Engineering*, 2021.
12. Tiruneh, A.T., *et al.*, A mathematical model for variable chlorine decay rates in water distribution systems. *Modell. Simul. Eng.*, 2019, 1–12, 2019.
13. Sun, Y.X., *et al.*, Changes in the components and biotoxity of dissolved organic matter in a municipal wastewater reclamation reverse osmosis system. *Environ. Technol.*, 17, 37, 2149–2156, 2016.
14. Marta, O. and Silva, M.T., Occurrence and removal of organic micropollutants: An overview of the watch list of EU Decision 2015/495. *Water Res.*, 94, 257–279, 2016.
15. Saidan, M., *et al.*, Evaluation of factors affecting bulk chlorine decay kinetics for Zai water supply system in Jordan, Case Study. *Environ. Prot. Eng.*, 43, 4, 223–231, 2017.
16. Srivastava, S. and Bhargava, A., Biofilms and human health. *Biotechnol. Lett*, 38, 1, 1–22, 2015.
17. WHO., *Guidelines for drinking-water quality*, 4th ed., Geneva, 2011.
18. Zhang, Y., Gang, L., Knibbe, W., Potential impacts of changing supply-water quality on drinking water distribution: A review. *Water Res.*, 116, 135–148, 2017.
19. Abe, Y., Cohesiveness and hydrodynamic properties of young drinking water biofilms. *Water Res.*, 46, 4, 1155–1166, 2012.

20. Monteiro, L., *et al.*, Modeling of chlorine decay in drinking water supply systems using EPANET MSX. *Procedia Eng.*, 70, 1192–1200, 2014.
21. Nono, D., *et al.*, The effects of hydraulic performance and operation interventions on water quality: The case of Phakalane water distribution system in Botswana. *Phys. Chem. Earth*, 108, 48–59, 2018.
22. Kelly, J., *et al.*, Temporal Variations in the Abundance and Composition of Biofilm Communities Colonizing Drinking Water Distribution Pipes. *PLoS One*, 5, 9, 1–11, 2014.
23. Kowalska, B., Kowalski, D., Musz-Pomorska, A., Chlorine decay in water distribution systems. *Environ. Prot. Eng.*, 32, 2, 1–12, 2006.
24. Rubulis, J., Verberk, J., Vreeburg, J., Chemical and Microbial Compositin of loose deposits in drinking water distribution systems. *Environ. Eng.*, 685–702, 2008.
25. Parsek, M.R. and Singh, P.K., Bacterial biofilms: An emerging link to disease pathogenesis. *Annu. Rev. Microbiol.*, 57, 677–701, 2003.
26. Dang, P. and Jayaretne, A., Ice Pigging-A better way to clean water mains. Yarra Valley Water, Australia, 2014.
27. Chowdhury, S., Models for predicting disinfection byproduct (DBP) formation in drinking waters: A chronological review. *Sci. Total Environ.*, 407, 14, 4189–4206, 2009.
28. Monteiro, L., *et al.*, Assessment of current models ability to describe chlorine decay and appraisal of water spectroscopic data as model inputs. *Environ. Eng.*, 143, 1, 1–9, 2017.
29. Chen, H., *et al.*, Formation of biofilms from new pipelines at both ends of the drinking water distribution system and comparison of disinfection by-products formation potential. *Environ. Res.*, 182, 109150, 2020.
30. Angelouds, A. and S.T.a.F.R, Disinfection Kinetics in CFD Modelling of Solutetransport in Contact Tanks. *Book Of proceedings*, 2014.
31. Khayyat, A., Study of point of use treatment methods for the disinfection of drinking water in Nepal, in: *Mechanical Engineering*, Master Thesis, Massachusetts Institute of Technology, 2000.
32. Kowalska, B., Holota, E., Kowalski, D., Simulation of chlorine concentration changes in a real water supply network using EPANET 2.0 and WATERGEMS software packages, in: *Urban water systems and floods 2*, p. 184, 2018.
33. Ishaq, M.S., *et al.*, Disinfection Methods, in: *Photocatalysis-Applications and Attributes*, vol. 10, p. 572, 2018.
34. Gomes, I.B., Simoes, M., Simoes, L.C., The effects of sodium hypochlorite against selected drinking water-isolated bacteria in planktonic and sessile states. *Sci. Total Environ.*, 565, 40–48, 2016.
35. Hua, P., Vasyukova, E., Uhl, W., A variable reaction rate model for chlorine decay in drinking water due to the reaction with dissolved organic matter. *Water Res.*, 75, 109–122, 2015.
36. Hallam NB, W.J., Foster, C.F., Powell, J.C., Spencer, I., The decay of chlorine associated with the pipe wall in water. *Water Res.*, 36, 3479–3488, 2002.

37. Rodrigues, A., *et al.*, Evaluation of active chlorine releasing of sodium hypochlorite during seven days, stored at different temperatures. *Rev. Sul Brasileria de Odontologia (RSBO)*, 10, 2, 143–148, 2013.
38. Debordea, M. and Guntena, U., Reactions of chlorine with inorganic and organic compounds during water treatment—Kinetics and mechanisms: A critical review. *Water Res.*, 42, 13–51, 2008.
39. Cong, L., *et al.*, Chlorine decay and trihalomethane formation following ferrate(VI) preoxidation and chlorination of drinking water. *Environ. Sustainability*, 187, 413–420, 2017.
40. Morrison, M.C., *et al.*, Ozone disinfection of waterbone pathogens and their surrogates: A crritical review. *Water Res.*, 214, 118206, 2022.
41. Dalrymple, O.K., *et al.*, A review of the mechanisms and modeling of photocatalytic disinfection. *Appl. Catal. B*, 98, 27–38, 2010.
42. Pretorius, W.A. and Pretorius, P.C., *A guide for the design of chloramine disinfection facilities for purified sewage effluent*, The University Of Pretoria: Water Research, 2001.
43. Alexandrou, L., Meehan, J.B., Jones, O.A.H., Regualtory and emerging disinfection by-products in recycled waters -Review. *Sci. Total Environ.*, 637-638, 1607–1616, 2018.
44. Abokifa, A., *et al.*, Investigating the Role of Biofilms in Trihalomethane Formation in Water Distirbution Systems with a Multicomponent Model. *Water Res.*, 104, 25, 2019.
45. Goodarzi, D., Abolfathi, S., Borzooeic, S., Modelling solute transport in water disinfection systems: Effects of temperature gradient on the hydraulic and disinfection efficiency of serpentine chlorine contact tanks. *J. Water Process Eng.*, 37, 11, 2020.
46. Kophaei, A.J. and Sathasiva, A., Chlorine decay prediction in bulk water using the parallel second order model:An analytical solution development. *Chem. Eng.*, 171, 232–241, 2011.
47. Mamba, B.B., Rietveld, L.C., Verberk, J., SA Drinking Water Standards under the microscope, 2008.
48. Clark, R.M., The USEPA's distribution system wate quality modelling program: a historical perspective. *Water Environ.*, 29, 3, 320–330, 2015.
49. Garcia-Avila, F., *et al.*, Relationship between chlorine decay and temperature in the drinking water. *Methods X*, 7, 1–10, 2020.
50. Powel, J.C., *et al.*, Factors which control bulk chlorine decay rates. *Water Res.*, 34, 1, 117–126, 2000.
51. Cong, L., Yang, Y.J., Jieze, Y., Second-order Chlorine Decay and Trihalomethanes Formation in a Pilot-Scale Water Distribution Systems. *Water Environ. Res.*, 84, 656, 2012.
52. Fisher, I., Kastl, G., Sathasivan, A., New model of chlorine-wall reaction for simulating chlorine concentration in drinking water distribution systems. *Water Res.*, 125, 427–437, 2017.

53. Wen, L., Kiene, L., Levi, Y., Chlorine demand of biofilms in water distribution systems. *Water Res.*, 33, 3, 827–835, 1999.
54. Monteiro, L., Carneiro, J., Covas, D., Modelling chlorine wall decay in a full-scale water supply system. *Urban Water*, 17, 8, 754–762, 2020.
55. Ralph, H., *et al.*, *General Chemistry: Principles and moder applications*, 9th ed., N, Pearson Education, Upper Saddle River, N.J, 2007.
56. Boccelli, D., *et al.*, A reactive species model for chlorine decay and THM formation under rechlorination conditions. *Water Res.*, 37, 2654–2666, 2003.
57. Robert, M. and Clark, M., Chlorine Demand and TTHM formation Kinetics: A second-Order Model. *Environ. Eng.*, 124, 1, 1–10, 1996.
58. Phillip M, R.J., Soon, T.K., Dragan, S., Hong, B., A variable rate coefficient chlorine decay model. *Environ. Sci. Technol.*, 43, 2, 408–414, 2009.
59. Rossman, L., Clark, R., Grayman, W., Modeling chlorine residuals in drinking - water distribution systems. *Environ. Eng.*, 120, 4, 803–820, 1994.
60. Jamwal, P. and Kumar, M.S.M., Effect of flow velocity on chlorine decay inwater distribution network: a pilot loop study. *Curr. Sci.*, 111, 48, 1–6, 2016.
61. Kastl., G.J., F., I.H., Jegatheesan, V., Evaluation of chlorine decay kinetics expressions for drinking water distribution systems modelling. *Water SRT-Aqua*, 48, 6, 219–226, 1999.
62. Bu, F., *et al.*, The combination of coagulation and ozonation as a pre-treatment of ultrafiltration in water treatment. *Chemosphere*, 231, 349–356, 2019.
63. Fan, B., *et al.*, Characterization of dissolved organic matter and membrane fouling in coagulation-ultrafiltration process treating micro-polluted surface water. *J. Environ. Sci.*, 75, 317–324, 2019.
64. Krishna, K.B.C., *et al.*, Bacterial community including opportunistic pathogens dynamics in a chlorine tablet-dosed chloraminated service reservoir and reticulation pipes. *Int. Biodeterior. Biodegrad.*, 153, 9, 2020.
65. Zhang, H. and Andrews, S.A., Catalysis of copper corrosion products on chlorine decay and HAA formation in simulated distribution systems. *Water Res.*, 46, 2665–2673, 2012.
66. Abokifa, A., *et al.*, Water quality modeling in the dead end sections of drinking water distribution networks. *Water Res.*, 89, 107–117, 2016.
67. Fisher, I., Kastl, G., Sathasiva, A., A suitable model of combined effects of temperature and initial condition on chlorine bulk decay in water distribution systems. *Water Res.*, 46, 3293–3303, 2012.
68. EPANET, Application for Modeling Drinking Water Distribution System, 2022. cited 2022 20 August 2022.
69. Comsol, *The Transport of Diluted Species Interface*, 2022, cited 2022 24 March 2022. Accessed from www.doc.comsol.com

Part II

TREATMENT OF INDUSTRIAL WASTEWATER USING SUSTAINABLE TECHNOLOGIES THAT ARE EFFECTIVE AND AFFORDABLE

6

A Comprehensive Assessment of the Chemical-Based Technologies for Waste(Water) Treatment

Linda L. Sibali[1], Zolani Dyosi[1], Beauclair Nguegang[2] and Vhahangwele Masindi[1]*

[1]Department of Environmental Sciences, College of Agriculture and Environmental Sciences, University of South Africa (UNISA), Florida, South Africa
[2]Department of Chemical Sciences, Faculty of Science, University of Johannesburg, Auckland Park, Johannesburg, South Africa

Abstract
Wastewater treatment is a critical process for safeguarding public health and the environment because it is one of the vital sources of survival for all living things, including drinking, industrial use, and environmental conservation. Chemical technology plays a significant role in water treatment by employing various chemical processes like ion exchange, coagulation, flocculation, adsorption, disinfection, and advanced oxidation processes to remove contaminants and impurities from water sources. The integration of innovative chemical materials into water treatment processes promises improved efficiency and sustainability in tackling the complex issue of wastewater treatment. This present article gives a comprehensive review on water treatment process based on chemical technology, addressing the reasons why chemical technology treatment processes are often preferred over biological processes. It further establishes the technical expertise required and equipment and machinery involved in chemical technology process and identifies the recent chemical materials used in wastewater treatment plants.

Keywords: Chemical technology, water treatment, coagulation, flocculation, advanced oxidation processes

**Corresponding author*: emasinv@unisa.ac.za; masindivhahangwele@gmail.com

Elvis Fosso-Kankeu, Vhahangwele Masindi, Johannes Maree and Bhekie Mamba (eds.) Customized Technologies for Sustainable Management of Industrial Wastewater: A Circular Economy Approach, (225–256)

6.1 Introduction

A greater understanding of freshwater scarcity has been made possible by population growth worldwide. This is due to the ensuing rise in demand for drinking water, as well as knowledge on the disease caused by usage of raw water that are intertwined to climate change's consequences, such as droughts and seawater incursion, and natural disasters [1, 2]. Several methods for water purification systems have been developed and put into use to guarantee the availability of safe drinking water. The creation and application of efficient wastewater treatment technologies has also been spurred by the knowledge that improperly treated wastewater may pollute sources of drinking water, bio-accumulate in the food chain, and result to detrimental effects on aquatic ecosystems and the environmental in general [2, 3].

Prior to the industrial revolution, river and stream pollution was not a big issue. There was a widespread belief that the sheer amount of water in these bodies could liquefy and disperse any waste materials, making them harmless [4]. To lower the hazards connected with water pollution, it is reported that smaller amounts of garbage/waste/effluents were released into waterbodies, and it was believed that the size of the receiving waterways would provide natural toxic dilution or breakdown [5]. However, multiple illness outbreaks during the eighteenth century including a significant cholera outbreak guided researchers to believe that polluted water may be a source of disease transmission. The germ hypothesis had not yet gained widespread acceptance, and the mechanics of disease transmission were still largely unknown [6]. During the London cholera epidemic, researchers suggested that the public water system and not the city's smoggy air was the disease's true cause [6, 7]. This finding disproved accepted wisdom and emphasized how crucial water treatment is toward maintaining public health. As time went on, worries regarding enduring protracted hazards as opposed to sudden acute threats started to influence how people thought about water purification. The priorities for public health have changed from focusing on diseases brought on by aquatic microbes to possible health problems brought on by human-induced water-related contaminants such as pesticide residues and industrial pollutants like heavy metals [8]. Other sources include detergents, oils, greases, phenols, polychlorinated biphenyls, halogenated aromatic hydrocarbons, formaldehyde, and polybrominated biphenyls [9]. The industrial heavy metals pollutants include titanium, vanadium, chromium, manganese, iron, cobalt, nickel, copper, zinc, arsenic, molybdenum, cadmium, lead, and mercury [10]. Public health impact as a result of water treatment has also become apparent.

Chemical treatment technologies are commonly applied to remove toxic pollutants like dye, toxic metals, and odor from industry effluents or wastewaters [11, 12]. Advances in automated monitoring and analytical techniques for detecting contaminants at lower levels have had an impact on the evaluation of water and wastewater treatment procedures, including the establishment of water quality standards by organizations/institutions like the World Health Organization [13, 14]. The chemical approaches for water purification require technical expertise to be used effectively. Chemicals, such as coagulants and flocculants, are added during the coagulation and flocculation processes to help remove colloidal matter and suspended particles [15–17]. The generated flocs and particulate particles are separated from one another and removed using filtering systems, sedimentation basins, and clarifiers [18]. Pathogens and bacteria are inactivated or eliminated via disinfection techniques like chlorination and ozonation [19]. Sophisticated oxidation techniques are used to decompose organic pollutants and eliminate taste and odor components. These methods include the use of ultraviolet (UV) radiation and advanced oxidants such as hydrogen peroxide (Fenton process) [20]. Chemical technologies and biological processes are the two primary methods for treating water. Although each approach has advantages, in most applications, chemical technology treatment approaches are frequently chosen over biological ones due to the speed and efficiency of operation, control and predictability, broad spectrum of contaminant removal, flexibility and adaptability, less space and infrastructure requirements, dependability, and cost-effectiveness of chemical technology treatment methods.

This review work sought to give comprehensive information on chemical technology treatment processes utilized in water treatment to improve understanding of the benefits and applications of chemical technology over biological technology, the technical know-how required to put it into effect, the equipment and machinery required, and the most recent breakthroughs in chemical materials. This information can assist water treatment specialists, researchers, and legislators in selecting and implementing the best water treatment strategies based on their specific needs and goals. It can also serve as a basis for new research and development in the field of chemical technology for water treatment.

6.2 Overview of Chemical Treatment Technologies

6.2.1 Water Treatment Processes Based on Chemical Technology

Chemical technology processes employ a wide range of techniques and procedures to remove pollutants, contaminants, and impurities from water

Table 6.1 Chemical-based water treatment technology.

Chemical process	Operating principle	Target pollutants/ contaminants
Coagulation	Particle destabilization to encourage aggregation during flocculation	Particle sizes between 0.01 µm and 1 µm (colloidal particles)
Flocculation	Encouraging the growth of smaller particles into bigger ones to improve sedimentation-based elimination	Total suspended solids
Ion Exchange	Process wherein in-solution ions of a different species (i.e., contaminants) displace ions of a specific species from an insoluble exchange material	Ammonia, metals, and organic substances in wastewater Drinking water: Calcium and magnesium-related hardness, as well as a smaller degree of iron, aluminum, and other cations (water softening)
Adsorption	The act of building up pollutants in solution on an adsorbent	Organic compounds and heavy metal ions
Disinfection	Usage of potent oxidizers to destroy bacteria and viruses, such as ultraviolet light, ozone, chlorine compounds, and hydrogen peroxide	Bacteria and viruses
Advanced Oxidation	Dispense with refractory organic pollutants	Recalcitrant pollutants, BOD, and COD

sources. An overview of the chemical processes based on chemical technology used in water and wastewater treatment are shown in Table 6.1.

6.2.1.1 *Chemical Precipitation (Coagulation and Flocculation)*

In chemical precipitation, coagulants are used as part of the fundamental chemical treatment process of coagulation in order to agglomerate and destabilize suspended particles in water. Figure 6.1 shows a simple illustration of coagulation, flocculation, and sedimentation. After coagulation, flocculation step follows, and it involves gently combining water and flocculants to enhance the aggregation of destabilized particles. Because they promote the formation of larger flocs, which aid in the removal of suspended particles, colloids, and microorganisms, high-molecular-weight polymers are frequently utilized as flocculants [21–23]. Lime [$Ca(OH)_2$], polyaluminum chloride (PAC), $Al_2(SO_4)_3$.14H2O (aluminum sulfate), $FeCl_3.6H_2O$ (ferric chloride), ($FeSO_4.7H_2O$) ferrous sulfate, and organic polyelectrolytes are some of coagulants employed as flocculation aids during the treatment of wastewater [24–26]. Polyacrylamides (PAMs), polyethylene oxide (PEO), and chitosan are typical flocculants [27]. As a result of the coagulants' ability to remove electrical charges from the particles, bigger flocs of the combined particles may readily be separated by procedures like sedimentation or filtering [28]. Figure 6.1 illustrates the schematic diagram of coagulation-flocculation and sedimentation

The quantity, type, and quality of the control approach used all affect how much removal may be achieved utilizing chemically accelerated sedimentation. According to established protocols, Jar tests are frequently carried out to establish the appropriate chemical dose and coagulant requirements for a certain water or wastewater media [29, 30]. The coagulant is poured

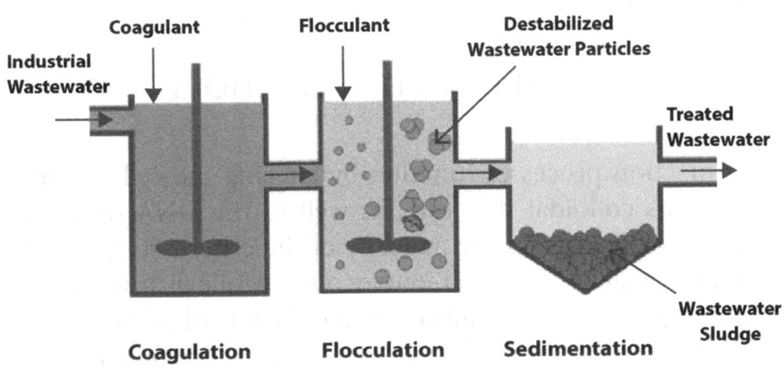

Figure 6.1 Schematic diagram of coagulation-flocculation and sedimentation [16].

into the rapid-mix basin's water for 20 s to 30 s to enable adequate interaction between the coagulant and the suspended particulates and to distribute the chemical uniformly [29–31]. The fine microflocs are then gently stirred for 15 min to 30 min in a flocculation basin using paddle- or turbine-type mixers to produce bigger floc particles that will be removed in the next sedimentation or filtration process [4]. Planning for the coagulation phase should primarily take into account the retention periods in the rapid-mixing basin, velocity gradient, i.e., the G values, type of coagulant, and the dose [18, 24, 32]. High levels of dissolved organic materials with a high molecular weight in raw water for consumption (such that present in lake water) improved the removal of ionizable medicines to removal levels of 30%–50% [33, 34]. According to several recent research using wastewater, chemical coagulation-flocculation typically performs poorly in the removal of pharmaceuticals (most are not removed at all) but can remove up to 50%–70% of highly lipophilic chemicals [4, 35].

6.2.1.2 *Disinfection*

For water to be safe for human consumption, germs and other microbes must be removed through the process of disinfection. The most widely used chemical disinfectants include chlorine, chlorine dioxide, ozone, and UV radiation. Chlorine-based substances, such chlorine gas, sodium hypochlorite, or calcium hypochlorite, are efficient at rendering a variety of microbes inactive [36]. Controlling taste and odor concerns is very helpful when using chlorine dioxide [37]. Alternative disinfection techniques include the use of ozone and UV radiation; the former effectively disinfects and oxidizes surfaces, whereas the latter eliminates bacteria by exposing surfaces to UV light [38]. Chlorine gas quickly hydrolyzes into hydrochloric acid (HCl) and hypochlorous acid (HOCl) when it is dissolved in wastewater, as illustrated in the chemical equation 6.1:

$$Cl_2 + H_2O <=> H^+ + Cl^- + HOCl \tag{6.1}$$

The disinfection processes function by altering the cell's permeability, the protoplasm's colloidal structure, as well as the DNA or RNA of the organism. Chlorine satisfies the majority of the criteria for an ideal disinfectant, i.e., it is widely available; removes odors while disinfecting; guarantees the homogeneity of the solution; cannot be absorbed by other organic matter; is corrosion-resistant; is non-toxic to either people or animals; is safe to carry, handle, store, and use; has excellent solubility; and has greater stability, with efficient great dilution at room temperature. Consequently, due to worries that disinfection by-products (organo-chlorines) may form, especially the use of various disinfection techniques based on O_3, O_3/H_2O_2, and UV radiation has developed, particularly in wastewater due to the high organic matter concentration [39–42].

The complex chemical process of disinfection is influenced by a number of variables, including the type of disinfectant used, the cyto-chemical makeup, physical state of the pathogens, and bioactivity as well as the media characteristics, including temperature (which affects the rate constant), pH (which affects how agents are distributed relative to one another), and electrolyte and other component concentrations, which might affect how a disinfectant functions [39, 40, 43]. When constructing this treatment phase, care must be taken to take into account the interaction duration of the chlorine in the water with reference to the pH, taste-producing compounds, bacterial quality, temperature, together with other important aspects [44, 45]. In addition to being easily accessible, affordable, and effective at low concentrations, chlorine has the added advantages of producing residual concentrations if the dose used is high enough to inhibit regrowth in the drinking water system [46, 47]. In contrast, O_3 and UV are uneconomical water treatment methods because it involves the use of strong oxidants; however, they have several environmental merits over the use of chlorine. Ozonation, for example, produces an effluent that is saturated with oxygen, which is reported to lower the risk of anoxia in the downstream region [48, 49]. The potential advantage of these methods over chlorination is further highlighted by the reaction that some refractory chemicals, like PPCPs of rising environmental engineering interest, can have with ozone and UV radiation [50].

6.2.1.3 Adsorption

Contaminants are removed from water through the process of adsorption by sticking to the surface of an adsorbent substance (Figure 6.2). Activated carbon has been a common adsorbent that may remove chemicals, organic molecules and impurities that give bad taste or smell. For specialized pollutant removal, other adsorbents such zeolites, clays, and synthetic resins are also used [51].

Adsorption is mostly employed in the treatment of drinking water to get rid of substances that give it an unpleasant odor, taste, or color. This process generally follows the biological water purification stage or subsequent tertiary treatment stages and aids in eliminating dissolved organic materials from wastewater [53]. As adsorbents, several granular materials

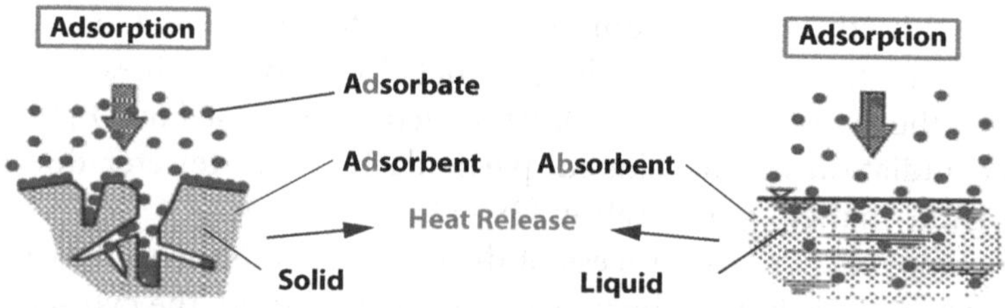

Figure 6.2 Representation of adsorption mechanism [52].

are employed, including synthetic polymers, activated carbon, silica-based compounds, and activated alumina [54]. Due to the greater expense of the other absorbents, by far, the most common absorbent used is activated carbon (activated carbon). Granular-activated carbon (GAC) and powdered-activated carbon (PAC) are the two terms for the two sizes of both of which are composed mostly of 0.1-mm (200-mesh)–diameter particles used in water treatment [55, 56]. Because activated carbon contains many capillaries and a larger surface area accessible for physical adsorption than other possible adsorbents (700–1,600 m^2/g for GAC and 1,400–2,200 m^2/g for PAC), it can effectively adsorb these target chemicals [4]. Whereas GAC systems have a greater initial expense, PAC treatment may be simply added to plants that already exist. The running expenses of GAC systems, i.e., the expense of refilling the used AC, are cheaper than those of PAC components, and they are able to achieve higher levels of removal [57].

There are two different types of activated carbon contactors: fixed-bed and expanded-bed. The primary flaw of the fixed-bed method, headloss buildup in the contactor, is eliminated by the expanded-bed architecture [58]. However, by oxidizing the adsorbed organic material off the surface in a furnace kept at an elevated temperature depending on the operating pressure (250°C–800°C), GAC may be renewed without considerable capacity loss [59]. The amount of organics that activated carbon or another adsorbent can absorb depends on the temperature, the amount of adsorbate present, as well as the adsorbate's properties (such as solubility, chemical make-up, and polarity). An adsorption isotherm, a mathematical function, may be used to calculate the quantity per mass of adsorbent of each organic that may be adsorbed. The isotherm models (Freundlich and Langmuir) are generally fitted with the experimental data to provide description of experimental isotherms [60]. The amount that can be absorbed, i.e., adsorbate to adsorbent (mg/g), and the concentration of impurities need to be eliminated.

6.2.1.4 Advanced Oxidation Processes (AOPs)

Powerful oxidizing agents or reactive oxygen species are used in advanced oxidation processes (AOPs) to break down and mineralize organic contaminants in water. AOPs include processes like ozonation (ozone treatment), Fenton's reaction (hydrogen peroxide and ferrous iron), and photochemical reactions (UV/H_2O_2 and UV/TiO_2). Recalcitrant organic compounds that are resistant to conventional treatment techniques can be broken down by AOPs. Table 6.2 shows the major mechanisms for organics removal from wastewater treatment by different AOPs.

Hydroxyl radicals (OH), which are produced in AOPs, are able to transform a range of contaminants and have a non-selective method of action that cannot be eliminated by traditional treatment [61]. The creation of these $OH^{\cdot}$ radicals in the treated water is possible using a number of methods, including photocatalytic oxidation, O_3, catalytic, UV, and ultrasonic [62]. Application of AOPs is majorly in situations with low volumes (cost) and low organic loadings because of the high operating costs of these technologies. Additionally, AOP-treated wastewaters could need additional biological or polishing treatments due to the fact that rather than complete mineralization of the contaminants, partial degradation, or merely transformation is typically accomplished [63].

Since the start of this century, there have been a growing number of scientific publications that discuss pharmaceuticals removal from wastewater through AOPs [64]. The removal of developing pollutants like PPCPs during water treatment has been the subject of extensive investigation in recent years. In an assessment of recent research in this area, Klavarioti *et al.* (2009) cover photolysis, ozonation, Fenton treatment, sonolysis, UV/H_2O_2, moist air oxidation, and electrolysis [64].

6.2.1.5 Ion Exchange Water Treatment Process

Ion exchange (IE) is a chemical water treatment technique where the water being treated and a solid resin exchange ions. It is frequently used to get rid of certain ions such heavy metals, fluoride, sulfate, and nitrate. Desirable ions are released in exchange for the undesired ions that the IE resin selectively adsorbs from the water [66]. The most typical use of IE in drinking water treatment is for water softening [67]. During this step of treatment, Na^+ produced by an IE solid replace the calcium and magnesium ions that were previously present in the aqueous phase. Ions that can be exchanged are present in the molecular structures on the surface of the IE resins. Table 6.3 shows various types of IE resins, functional groups, operating conditions target ions and regeneration reagents.

Table 6.2 Major mechanisms for organics removal for wastewater treatment by different AOPs.

AOP types	Oxidant for advanced oxidation	Other occurring mechanisms
O_3	OH^-	Direct O_3 oxidation
Ultrasonic irradiation	OH^-	Acoustic cavitation generates transient high temperatures (≥ 5,000 K) and pressures (≥ 1,000 atm) and produces H^+ and HO_2^-, besides OH^-
O_3/H_2O_2	OH^-	Direct O_3 oxidation H_2O_2 oxidation
Heat/persulfate	SO_4^-	Persulfate oxidation
Fe(II)/persulfate	SO_4^-	Persulfate oxidation Iron coagulation Iron sludge-induced adsorption
O_3/UV	OH^-	UV photolysis UV photolysis
UV/TiO_2	OH^-	UV photolysis
UV/persulfate	SO_4^-	Persulfate oxidation UV photolysis
UV/ H_2O_2	OH^-	H2O2 oxidation
OH^-/persulfate	SO_4^-/ OH^-	Persulfate oxidation
Fenton reaction	OH^-	Iron coagulation Iron sludge-induced adsorption
Photo-Fenton reaction	OH^-	Iron coagulation Iron sludge-induced adsorption UV photolysis

*source [65]

Table 6.3 Various categories of ion exchange resins.

Resin type	Functional groups	Ions removed	Regeneration	Operating range
Strong base anionic resin	Quaternary amine $RN(CH_3)_3^+$	NO_3^-, SO_4^{2-}, $C1O_4^-$, $HAsO_4^{2-}$, SeO_3^{2-}, etc.	Regenerated with NaOH or NaCl	1 to 13
Weak base anionic resin	Tertiary amine $RN(CH_3)_2H^+$	NO_3^-, SO_4^{2-}, $C1O_4^-$, $HAsO_4^{2-}$, SeO_3^{2-}, etc.	Regenerated with NaOH, or Ca(OH)2	<6
Strong acid cationic resin	Sulfonate RSO_3^-	Total hardness, Mg^{2+}, Ra^{2+}, Ba^{2+} Pb^{2+}, etc.	Regenerated with HCl or NaCl	1 to 14
Weak acid cationic resin	Carboxylate $RCOO^-$	Carbonate hardness, Mg^{2+}, Ra^{2+}, Ba^{2+}, Pb^{2+}, etc.	Regenerated with HCl	>7

*Source [68]

The right choice of IE resin and working conditions relies on the characteristics of the raw water, the requirements for water quality, the ease of operation, and financial factors [69]. The system must be properly chosen, constructed, managed, and maintained (including regeneration) in order to produce successful outcomes [70]. To reduce exchange with other ion species, the chosen resin should preferably also be selective. The original IE materials were porous, naturally occurring compounds (zeolites). These materials have now been superseded by resins made of synthetic organic ions, which have significantly larger IE capability. These polymeric materials' structural architectures are created with varied reactive groups, giving them the characteristics needed for the numerous cation and anion exchange applications [71]. Figure 6.3 shows a simple representation of IE column.

Ion exchange is employed in wastewater treatment, in addition to its employment in the production of drinking water, to replace harmful ions from the solid phase with aqueous pollutants like nitrogen and heavy metals. Using this method, metals including copper, zinc, copper, chromium, nickel, and cadmium can be eliminated. The methods and techniques used

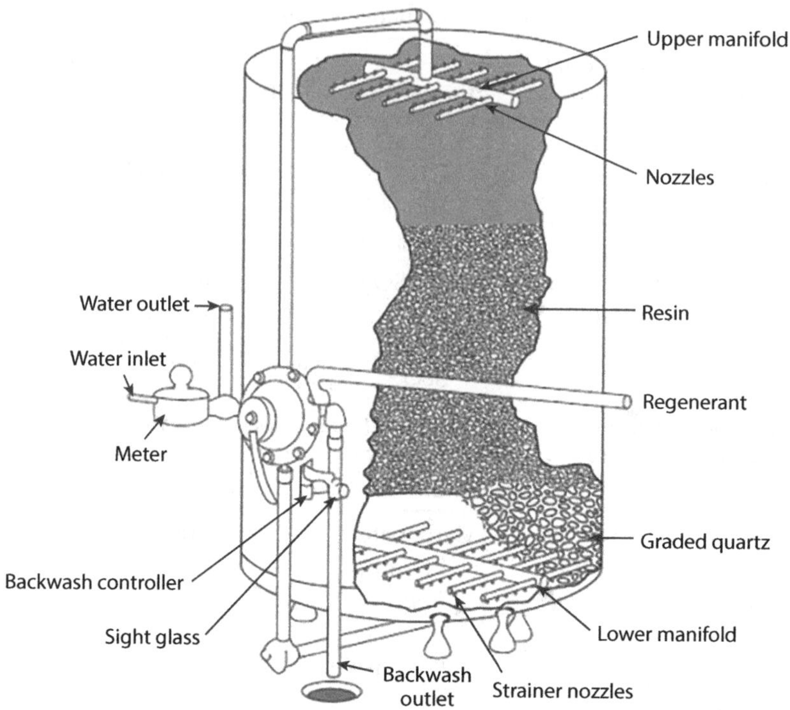

Figure 6.3 A typical ion exchange resin column.

in IE systems are comparable to adsorption systems [72]. Once breakthrough occurs, the column bed is backwashed to revive the resin [73]. NaOH solution is used to regenerate strong base exchangers, whereas strong acid cation exchangers employ H_2SO_4 or HCl solution [74]. Due to the complicated operating processes (including pre-treatment and regeneration) and uncertainties regarding the lifespan of the IE resins, the IE processes has only found constrained usage in wastewater treatment [75]. Designing IE units and choosing IE resins must take critical products' recovery into mind, such as heavy metals, as this will greatly impact on the economics of the treatment phase [4, 76].

6.3 Advantages of Chemical Technology Treatment Processes over Biological Processes

Chemically based water treatment technologies offer a wide range of options for resolving various water quality challenges. The particular pollutants present, the targeted treatment outcomes, and the features of the water supply all influence the choice and mix of these methods. Chemical technologies and biological processes are the two primary methods for treating water. Although each approach has advantages, in some applications chemical technology treatment approaches are frequently chosen over biological ones because of its efficiency, versatility, reliability, and control [77, 78]. Table 6.4 presents the general advantages of chemical treatment based-technologies.

Table 6.4 Advantages of chemical treatment based-technologies.

Efficiency
The efficacy of chemical technology treatment techniques in accomplishing water treatment goals is well established. They are more efficient than biological processes due to a number of considerations. • **Rapid Treatment:** In chemical technology, physical and chemical processes can happen more quickly, resulting in speedier treatment. This is especially helpful in instances where quick treatment is required, including emergencies or high-demand circumstances. • **High Removal Efficiency:** Many contaminants, including suspended particles, dissolved materials, and particular pollutants, can be successfully removed using chemical treatment procedures. Higher removal efficiencies are achieved when chemicals like coagulants, flocculants, and oxidants react fast with pollutants to precipitate or destroy them. • **Consistent Performance:** Processes using chemical technology deliver consistent performance independent of changes in the quality of the influent water or the operational environment. Contrarily, biological processes may be more susceptible to changes in environmental factors like temperature and pH, which may have an impact on their functionality and treatment effectiveness.
Versatility
• **Treatment Spectrum:** Chemical procedures have the ability to remove a variety of pollutants, including microorganisms, organic compounds, heavy metals, suspended solids, and emerging ones. While chemical technology procedures can manage a wider variety of pollutants, biological processes are often more suited for the decomposition of organic materials and the removal of nutrients.

(*Continued*)

Table 6.4 Advantages of chemical treatment based-technologies. (*Continued*)

• **Tailored Chemical Treatment:** By choosing the right chemicals, doses, and treatment schedules, chemical treatment may be customized to address certain water quality issues. Customization is possible due to this flexibility depending on the particular pollutants present and the intended treatment objectives. • **Complementary to Biological Processes:** To improve the effectiveness of therapy, chemical technology processes can be combined with biological processes. For instance, water can be pretreated using chemical coagulation to remove particulates and lighten the burden on future biological treatment units.
Control
Processes utilizing chemical technology provide users more control over treatment settings, facilitating simpler tinkering and optimization. Key aspects of control include the following: • **Dosage Optimization:** Chemical dosages can be adjusted based on real-time water quality data and desired treatment outcomes. This flexibility enables operators to optimize treatment efficiency, minimize chemical usage, and reduce the formation of disinfection by-products. • **Reaction Kinetics:** Understanding the kinetics of chemical reactions allows for precise control over the reaction time, ensuring sufficient contact time for effective contaminant removal. This control is particularly important in the case of rapid reactions or when dealing with contaminants that require specific reaction conditions. • **Process Monitoring:** Chemical treatment processes can be easily monitored using online sensors and analytical techniques, allowing for real-time monitoring of key parameters such as pH, oxidation-reduction potential (ORP), and turbidity. This monitoring facilitates immediate adjustments and ensures optimal treatment performance.
Reliability
Chemical technology processes are generally more reliable and less affected by external factors compared to biological processes. This reliability stems from several factors: • **Independence from Environmental Conditions:** Chemical processes are less reliant on specific environmental conditions such as temperature, pH, and microbial activity, which can affect the performance of biological processes. This independence allows chemical technology to deliver consistent treatment performance under varying operating conditions.

(*Continued*)

Table 6.4 Advantages of chemical treatment based-technologies. (*Continued*)

• **Robustness:** Chemical treatment processes can handle fluctuating influent water quality and load variations more robustly. They are less susceptible to upsets caused by sudden changes in flow rate, pollutant concentration, or toxicity levels. • **Reduced Sludge Production:** Biological processes often generate a significant amount of sludge, which requires proper management and disposal. Chemical technology processes typically produce less sludge, reducing the complexity and cost associated with sludge handling. • **Cost-Effectiveness:** The cost-effectiveness of water treatment is a critical consideration for many water treatment plants. Chemical technology processes can offer cost advantages in various scenarios. • **Capital and Operating Costs:** In certain cases, the capital and operating costs associated with chemical technology processes can be lower compared to biological processes. This cost advantage may arise from simplified process design, reduced space requirements, and lower energy consumption. • **Time Efficiency:** Chemical technology processes can treat water more rapidly, reducing the required treatment time and associated costs. This is particularly valuable in scenarios where rapid treatment is essential, such as emergency response situations. • **Flexibility in Chemical Selection:** Chemical treatment processes offer flexibility in selecting chemicals based on availability, cost, and treatment objectives. This flexibility allows for optimization of chemical usage and cost savings through the selection of cost-effective chemicals and dosages.

6.3.1 Limitations and Challenges

Chemical-based water treatment technology methods have some novel advantages over biological techniques; however, they also have drawbacks and hurdles that need to be resolved. Chemical reagent usage can result in the production of sludge and other by-products, which calls for careful treatment and disposal. Certain coagulants and oxidants, for example, may pose environmental or health risks and require cautious management. The cost of chemicals and the energy needs for some processes, including AOPs, may also make widespread application economically difficult [79, 80]. It is crucial to take into account these aspects and incorporate chemical treatment procedures within a thorough and sustainable framework for wastewater treatment that also incorporates biological procedures and other cutting-edge treatment techniques [81].

6.4 Overview on Technical Expertise

This section examines the technical knowledge required to use chemical technology for water treatment in an efficient manner. The discussion emphasizes the significance of comprehending chemical processes, kinetics, and thermodynamics in water treatment as well as the information and abilities needed for chemical process design, operation, and optimization.

6.4.1 Technical Expertise Required in Chemical Technology Water Treatment Process

6.4.1.1 Expertise in Chemical Process Design

A comprehensive strategy for implementing chemical technology in water treatment is developed through chemical process design [82]. It requires expertise in developing a comprehensive strategy for implementing chemical technology in water treatment that is a multifaceted task that demands expertise in various key areas; (i) safely handling and storing chemicals while taking into account their characteristics and potential risks; (ii) recognizing the compatibility and probable interactions between various substances; (iii) troubleshooting process problems, figuring out what causes inefficiencies or failures, and putting remedial actions in place; and (iv) routinely inspecting and maintaining equipment to guarantee peak performance and avoid breakdowns.

6.4.1.2 Expertise in Chemical Process Operation

Expertise in monitoring and managing process parameters, such as pH, temperature, chemical dosing rates, and reaction periods, is required for the efficient functioning of chemical technology in the chemical water treatment processes [83]. This expertise include the following: (i) safely handling and storing chemicals while taking into account their characteristics and potential risks; (ii) recognizing the compatibility and probable interactions between various substances; (iii) troubleshooting process problems, figuring out what causes inefficiencies or failures; and (iv) putting remedial actions in place and routinely inspecting and maintaining equipment to guarantee peak performance and avoid breakdowns.

6.4.1.3 Expertise in Chemical Process Optimization

The most efficient chemical treatments and dose rates must be determined by conducting laboratory research and pilot-scale investigations. This helps in (i) examining and evaluating data on water quality to pinpoint areas in need of development and improvement, (ii) using statistical approaches and data analysis methods to spot trends and patterns in the performance of water treatment systems, (iii) putting process improvements and changes into practice to improve the efficacy and efficiency of therapy, and (iv) assessing the financial viability of various treatment alternatives and enhancing the process' cost-effectiveness [83, 84].

6.4.1.4 Technical Expertise in Understanding Chemical Reactions, Kinetics, and Thermodynamics

A thorough grasp of chemical processes, kinetics, and thermodynamics is essential for the effective use of chemical technology in the treatment of water. This information makes it possible to (i) choose the right compounds and doses depending on how they react with certain pollutants; (ii) forecast and manage response rates to provide adequate interaction time for treatment outcomes; (iii) analyze reaction routes and mechanisms to enhance therapeutic procedures; (iv) take into account thermodynamic concepts while designing and running treatment procedures, such as equilibrium conditions and solubility; and (iv) knowledge on reaction kinetics analysis to establish dose rates and reaction times necessary for successful separation of pollutants.

Professionals in the water treatment industry can choose chemicals, develop processes, and optimize them by having a solid understanding of chemical interactions, kinetics, and thermodynamics. It aids in the prediction of chemical and contaminant behavior, evaluation of the viability of various treatment options, and assurance of the intended therapeutic effects [28].

6.5 Overview on Equipment and Machinery

This section gives an overview of the main machinery and equipment used in chemical technology procedures for water treatment. It discusses

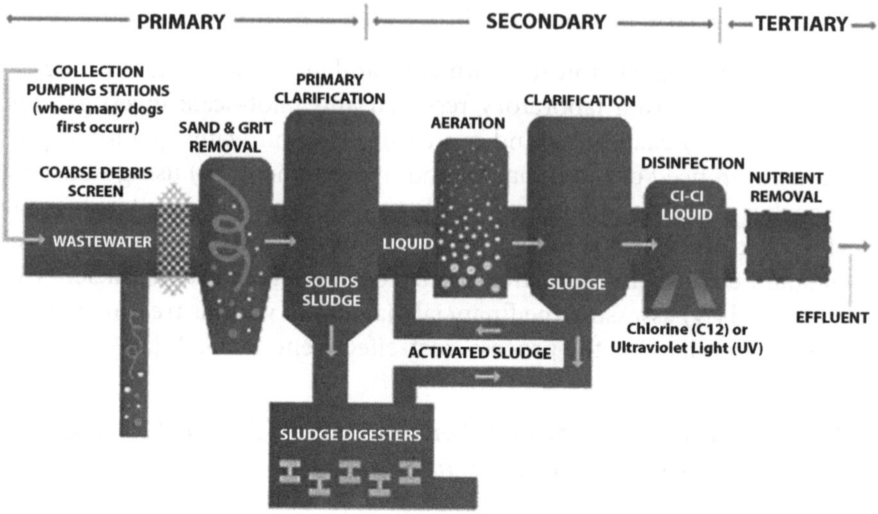

Figure 6.4 Combination of various equipment and machinery used in chemical water treatment process, adopted from [85].

these components' goals, roles, and its interactions. Figure 6.4 provides an illustration on how the equipment and machineries are connected. These elements are essential to the treatment process at various phases, such as mixing, reaction, separation, and disinfection. In order to assure proper chemical water treatment process.

6.5.1 Equipment and Machinery in Chemical Technology Processes for Water Treatment

6.5.1.1 Pumps

Pumps are crucial components in the chemical water treatment process because they transfer fluids like water and other contaminants [86]. For efficient flow of water from one unit operation to another, they provide the hydraulic power needed. There are two most common types of pumps. These include centrifugal and positive displacement. The centrifugal pumps are often used in water treatment plants due to their high efficiency, simplicity, and low maintenance requirements. They achieve this by converting rotational energy into kinetic energy, which they subsequently employ to give the fluid pressure and velocity. Positive displacement pumps, on the other hand, provide a fixed volume of fluid with each cycle, making them perfect for applications needing precise flow control [86].

6.5.1.2 *Mixers*

In chemical treatment of water, mixers facilitate the desired chemical interactions and reactions during the treatment process by ensuring that chemicals are blended and combined properly [87]. They promote reactant contact, enhance uniform chemical dispersion, and boost mass transfer effectiveness. Various mixers that can be used include mechanical agitators, impellers, propellers, and static mixers are among the different types of mixers that are regularly employed. The choice of mixer is influenced by a number of variables, including the desired mixing intensity, the fluid's viscosity, and the specific chemical processes occurring [87].

6.5.1.3 *Reactors*

Reactors are used to provide optimal chemical blending and mixing throughout the treatment process in order to promote the essential chemical reactions [28]. They enhance the efficiency of mass transfer, help in uniform chemical dispersion, and encourage interaction between the reactants. A variety of mixers, such as mechanical agitators, impellers, propellers, and static mixers, are used. Numerous factors, such as the required mixing intensity, the fluid's viscosity, and the specific chemical reactions involved, determine the best mixer to use.

6.5.1.4 *Filters*

In order to separate and remove solid particles and other contaminants from water during a chemical water treatment process, filters are crucial components. They are all meant to collect and hold larger particles, colloids, and suspended solids. Sand filters, activated carbon filters, membrane filters, and other types of filters are used in the purification of water. When it comes to removing suspended particles, sand filters employ a bed of sand, whereas activated carbon filters use adsorption to get rid of specific pollutants and organic molecules. High-efficiency filtration is provided by membrane filters, such as microfiltration, ultrafiltration, and reverse osmosis membranes, by keeping particles according to their size and molecular weight [28].

6.5.1.5 *Disinfection Systems*

Disinfection is a crucial stage in the treatment of water because it gets rid of dangerous bacteria and guarantees the safety of the water. Chemical and physical disinfection are just two of the techniques used by disinfection systems. Utilizing chemicals like chlorine, chlorine dioxide, ozone, or UV radiation as disinfectants is known as chemical disinfection. In terms of efficacy, disinfection by-products, and residual disinfectant, each approach has advantages and disadvantages. The desired level of disinfection, the target microorganisms, and the specific water quality criteria are among the elements that influence the choice of the disinfection system [28].

6.5.2 Integration of Equipment and Machinery

Effective water treatment is ensured by the coordination of the machinery and equipment mentioned above (section 6.5.1). The water moves through a series of unit activities, each of which contributes to the overall treatment process in a different way. For instance, pumps are in charge of moving water from one component to another, guaranteeing a steady flow throughout the procedure. Chemicals are blended properly with the use of mixers, guaranteeing even distribution and fostering efficient reactions in the reactors. Reactors create the conditions for chemical reactions to take place, which makes it easier to remove impurities. Clarified water is the consequence of the pollutants and suspended particles being removed by filters. Last but not least, disinfection systems get rid of dangerous bacteria, making sure the water is cleaned to the standards needed for safe use or disposal. To preserve water purity, achieve the best treatment efficiency, and adhere to regulatory standards, these components are carefully integrated. Considerations for design include things like hydraulic performance, residency time, mixing intensity, chemical dosing, and contact time for disinfection. To ensure that it operates well, to reduce downtime, and to enhance treatment effectiveness, it must be properly maintained and monitored.

6.6 Overview Recent Chemical Materials Used in Wastewater Treatment Plants

To increase the effectiveness of the treatment operations and meet the necessary water quality criteria, wastewater treatment facilities use a variety of chemical substances. Recent developments in chemical substances have increased pollutant removal, improved solid-liquid separation, and better disinfection. The development of chemicals used in wastewater treatment plants is explained in this section, with a focus on chemical-based water treatment technologies; these include coagulants and flocculants for solid-liquid separation, the use of AOPs with substances like hydrogen peroxide and ozone, the use of adsorbents and IE resins for contaminant removal, and improvements in disinfectants for microbial control.

6.6.1 Recent Chemical Materials Used in Wastewater Treatment Plants

6.6.1.1 Coagulants and Flocculants for Solid-Liquid Separation

In wastewater treatment facilities, coagulants and flocculants are crucial chemical components that make solid-liquid separation possible. In contrast to flocculants, which encourage the formation of bigger flocs, coagulants destabilize colloidal particles and cause them to aggregate into larger structures. Among the most recent developments in coagulants and flocculants are the following:

- Coagulants and flocculants made of synthetic polymers, such as PAC, PAM, and polydiallyldimethylammonium chloride, are more effective than conventional inorganic coagulants and require lower dosages [88, 89].
- Coagulants based on microbes: Chemical coagulants can be replaced with microbial-based solutions that are more environmentally friendly and sustainable [90, 91].
- Coagulation and flocculation composites: Improved coagulation and flocculation performance in composite materials, such as iron-based nanoparticles and hybrid materials, suggests the possibility of more effective solid-liquid separation [92].

6.6.1.2 Advanced Oxidation Processes (AOPs) Utilizing Ozone and Hydrogen Peroxide

Refractory organic compounds and emerging pollutants are removed in wastewater treatment facilities using AOPs. To oxidize and destroy pollutants, AOPs produce extremely reactive hydroxyl radicals (•OH). Recent developments consist of the following:

- Hydrogen peroxide (H_2O_2) as an AOP precursor: H_2O_2 is frequently employed in conjunction with catalysts or UV irradiation to create hydroxyl radicals, allowing for effective degradation of a variety of organic molecules [93].

- Ozone (O_3) is a strong oxidant that may be employed alone or in conjunction with other AOPs as an AOP oxidant. It has benefits including quick response kinetics, extensive pollutant removal capability, and no toxic disinfection by-products [94].
- Integration of AOPs with other treatment processes: Activated carbon adsorption and membrane filtration are two examples of other treatment processes that may be used with AOPs to improve overall treatment effectiveness and get rid of persistent organic pollutants [95].

6.6.1.3 Adsorbents and Ion Exchange Resins for Contaminant Removal

Ion exchange resins and adsorbents are essential for eliminating certain pollutants from wastewater. In these materials, recent developments include the following:

- Activated carbon–based adsorbents: For the adsorption of organic molecules, such as medications, insecticides, and industrial pollutants, activated carbon is frequently utilized. Modified activated carbon materials with improved selectivity and adsorption capability are a recent discovery [96, 97].
- Nanomaterial-based adsorbents: Due to their large surface area and adsorption capacities, nanomaterials including graphene oxide, carbon nanotubes, and magnetic nanoparticles have showed promise in wastewater treatment [98–100].
- Ion exchange resins: Ion exchange resins are used to remove certain ions and impurities in a selective manner. Improved selectivity, higher capacity, and the creation of innovative resin materials for the removal of developing pollutants like per- and polyfluoroalkyl substances are among the advancements in IE resins [101].

6.6.1.4 *Advancements in Disinfectants for Microbial Control*

Disinfection is a crucial stage in the treatment of wastewater because it gets rid of dangerous germs and protects the public's health. Improvements in disinfectants recently include the following:

- Chlorine alternatives: Alternative disinfectants including chlorine dioxide (ClO_2), chloramines, and UV irradiation have grown in favor as a result of worries about disinfection by-products. These substitutes effectively inhibit microbial growth while reducing the production of disinfection by-products [102–104].
- Advanced disinfection techniques: Electrochemical disinfection, ozonation, and sophisticated UV systems, such medium-pressure UV and UV-LED, have all been developed as disinfection methods. These systems offer better disinfection performance and energy efficiency [105–107].
- Combined disinfection approaches: Diverse disinfectants or methods have been used in integrated disinfection systems to generate synergistic effects and improve microbial inactivation while reducing reliance on a single disinfectant [108].

6.7 Conclusions

Chemical technology processes employ a wide range of techniques and procedures to remove pollutants, contaminants, and impurities from aqueous medium. The effectiveness of water treatment, pollutant removal based on solid-liquid separation, and microbiological control have all greatly increased as a result of recent advances in chemical materials used in wastewater treatment facilities. AOPs with hydrogen peroxide and ozone, adsorbents, and IE resins, as well as improvements in disinfectants, have all been used to create more efficient and long-lasting water treatment procedures. Continuous advancements in chemical materials research and development have increased wastewater treatment efficiency, minimized negative environmental effects, and raised requirements for water purity.

Acknowledgment

The authors of this book chapter would like to express their sincere gratitude to the University of South Africa for extending their facilities toward the completion of this project.

References

1. He, C., *et al.*, Future global urban water scarcity and potential solutions. *Nat. Commun.*, 12, 1, 4667, 2021.
2. Salehi, M., Global water shortage and potable water safety; Today's concern and tomorrow's crisis. *Environ. Int.*, 158, 106936, 2022.
3. Edokpayi, J.N., Odiyo, J.O., Durowoju, O.S., Impact of wastewater on surface water quality in developing countries: a case study of South Africa. *Water Qual.*, 10, 66561, 10–5772, 2017.
4. Yargeau, V., Water and wastewater treatment: chemical processes, in: *Metropolitan Sustainability*, pp. 390–405, Elsevier, Cambridge, 2012.
5. Siddiqua, A., Hahladakis, J.N., Al-Attiya, W.A.K., An overview of the environmental pollution and health effects associated with waste landfilling and open dumping. *Environ. Sci. Pollut. Res.*, 29, 39, 58514–58536, 2022.
6. Tulchinsky, T.H., John Snow, cholera, the broad street pump; waterborne diseases then and now, in: *Case Stud. Public Health*, p. 77, 2018.
7. Manisalidis, I., Stavropoulou, E., Stavropoulos, A., Bezirtzoglou, E., Environmental and health impacts of air pollution: a review. *Front. Public Health*, 8, 14, 2020.
8. Boelee, E., Geerling, G., van der Zaan, B., Blauw, A., Vethaak, A.D., Water and health: From environmental pressures to integrated responses. *Acta Trop.*, 193, 217–226, 2019.
9. Kumar, V., Sonkar, M., Pooja, Shukla, S.K., Recent advancement on bioaugmentation strategies for process industry wastewater (PIWW) treatment, in: *Water Remediat*, pp. 189–209, 2018.
10. Briffa, J., Sinagra, E., Blundell, R., Heavy metal pollution in the environment and their toxicological effects on humans. *Heliyon*, 6, 9, 1–26, 2020.
11. Anekwe, I.M.S., Adedeji, J., Akpasi, S.O., Kiambi, S.L., Available Technologies for Wastewater Treatment, in: *Wastewater Treat*, p. 33, 2022.
12. Adane, T., Adugna, A.T., Alemayehu, E., Textile industry effluent treatment techniques. *J. Chem.*, 2021, 1–14, 2021.
13. Zulkifli, S.N., Rahim, H.A., Lau, W.-J., Detection of contaminants in water supply: A review on state-of-the-art monitoring technologies and their applications. *Sens. Actuators B Chem.*, 255, 2657–2689, 2018.

14. Zainurin, S.N., *et al.*, Advancements in monitoring water quality based on various sensing methods: a systematic review. *Int. J. Environ. Res. Public Health*, 19, 21, 14080, 2022.
15. Estrada, J.M., Kraakman, N.B., Muñoz, R., Lebrero, R., A comparative analysis of odour treatment technologies in wastewater treatment plants. *Environ. Sci. Technol.*, 45, 3, 1100–1106, 2011.
16. Teh, C.Y., Budiman, P.M., Shak, K.P.Y., Wu, T.Y., Recent advancement of coagulation–flocculation and its application in wastewater treatment. *Ind. Eng. Chem. Res.*, 55, 16, 4363–4389, 2016.
17. Lee, C.S., Robinson, J., Chong, M.F., A review on application of flocculants in wastewater treatment. *Process Saf. Environ. Prot.*, 92, 6, 489–508, 2014.
18. Arab, M., *et al.*, A soft-sensor for sustainable operation of coagulation and flocculation units. *Eng. Appl. Artif. Intell.*, 115, 105315, 2022.
19. Lanrewaju, A.A., Enitan-Folami, A.M., Sabiu, S., Swalaha, F.M., A review on disinfection methods for inactivation of waterborne viruses. *Front. Microbiol.*, 13, 991856, 2022.
20. Marcinowski, P.P., Bogacki, J.P., Naumczyk, J.H., Cosmetic wastewater treatment using the Fenton, Photo-Fenton and H2O2/UV processes. *J. Environ. Sci. Health., Part A*, 49, 13, 1531–1541, 2014.
21. Chen, F., *et al.*, Characteristics and mechanism of chitosan in flocculation for water coagulation in the Yellow River diversion reservoir. *J. Water Process Eng.*, 34, 101191, 2020.
22. Kurniawan, S.B., *et al.*, What compound inside biocoagulants/bioflocculants is contributing the most to the coagulation and flocculation processes? *Sci. Total Environ.*, 806, 150902, 2022.
23. Zhao, J., Li, B., Wang, A., Ge, W., Li, W., Floc formation and growth mechanism during magnesium hydroxide and polyacrylamide coagulation process for reactive orange removal. *Environ. Technol.*, 43, 3, 424–430, 2022.
24. Ho, Y.-C., Chua, S.-C., Chong, F.-K., Coagulation-flocculation technology in water and wastewater treatment, in: *Handbook of Research on Resource Management for Pollution and Waste Treatment*, pp. 432–457, IGI Global, Pennsylvania, 2020.
25. Bratby, J., *Coagulation and Flocculation in Water and Wastewater Treatment*, IWA publishing, London, 2016.
26. Zarei Mahmoudabadi, T., Ehrampoush, M.H., Talebi, P., Fouladi-Fard, R., Eslami, H., Comparison of poly ferric chloride and poly titanium tetrachloride in coagulation and flocculation process for paper and cardboard wastewater treatment. *Environ. Sci. Pollut. Res.*, 28, 27262–27272, 2021.
27. Qian, J.W., Xiang, X.J., Yang, W.Y., Wang, M., Zheng, B.Q., Flocculation performance of different polyacrylamide and the relation between optimal dose and critical concentration. *Eur. Polym. J.*, 40, 8, 1699–1704, 2004.
28. Schutte, F. and Focke, W., Handbook for the operation of water treatment works, in: *Water Res. Comm. Water Inst. South. Afr. TT26506*, 2006.

29. Zainal-Abideen, M., Aris, A., Yusof, F., Abdul-Majid, Z., Selamat, A., Omar, S., II, Optimizing the coagulation process in a drinking water treatment plant–comparison between traditional and statistical experimental design jar tests. *Water Sci. Technol.*, 65, 3, 496–503, 2012.
30. Balls, M., *Relationships Between Floc Properties and NOM Removal Using a Moorland Water Source*, University College London, London, 2014, PhD Thesis, UCL.
31. Tchobanoglous, G., Burton, F.L., Stensel, H.D., Wastewater engineering. *Management*, 7, 1, 4, 1991.
32. Lappo, S., *The Removal of Residual Concentrations of Phosphorus from Wastewater by Means of Coagulation and Flotation*, University of Oulu, Finland, 2016, Master's Thesis, S. Lappo.
33. Kooijman, G., De Kreuk, M.K., Houtman, C., Van Lier, J.B., Perspectives of coagulation/flocculation for the removal of pharmaceuticals from domestic wastewater: A critical view at experimental procedures. *J. Water Process Eng.*, 34, 101161, 2020.
34. Vieno, N., Tuhkanen, T., Kronberg, L., Removal of pharmaceuticals in drinking water treatment: effect of chemical coagulation. *Environ. Technol.*, 27, 2, 183–192, 2006.
35. Treviño-Reséndez, J., Medel, A., Mijaylova, P., Meas, Y., Coagulation/flocculation and electrocoagulation methods for oily wastewater treatment, in: *Advanced Technologies in Wastewater Treatment*, pp. 173–212, Elsevier, The Netherland, 2023.
36. Jalali Milani, S. and Nabi Bidhendi, G., A review on the potential of common disinfection processes for the removal of virus from wastewater. *Int. J. Environ. Res.*, 16, 1, 9, 2022.
37. Xu, M.-Y., *et al.*, Chlorine dioxide-based oxidation processes for water purification: A review. *J. Hazard. Mater.*, 436, 129195, 2022.
38. Hussain, K., Khan, N.A., Vambol, V., Vambol, S., Yeremenko, S., Sydorenko, V., Advancement in Ozone base wastewater treatment technologies: Brief review. *Ecol. Quest.*, 33, 2, 7–19, 2022.
39. Pandian, A.M.K., Rajamehala, M., Singh, M.V.P., Sarojini, G., Rajamohan, N., Potential risks and approaches to reduce the toxicity of disinfection by-product–A review. *Sci. Total Environ.*, 822, 153323, 2022.
40. Von Gunten, U., Ozonation of drinking water: Part II. Disinfection and by-product formation in presence of bromide, iodide or chlorine. *Water Res.*, 37, 7, 1469–1487, 2003.
41. Criquet, J. and Allard, S., Influence of bromide and iodide on the formation of disinfection by-products in drinking water treatment, in: *Comprehensive Analytical Chemistry*, pp. 117–138, Elsevier, The Netherland, 2021.
42. Gao, Z.-C., *et al.*, Effect of UV wavelength on humic acid degradation and disinfection by-product formation during the UV/chlorine process. *Water Res.*, 154, 199–209, 2019.

43. Kali, S., *et al.*, Occurrence, influencing factors, toxicity, regulations, and abatement approaches for disinfection by-products in chlorinated drinking water: A comprehensive review. *Environ. Pollut.*, 281, 116950, 2021.
44. Kumar, J.K. and Pandit, A.B., *Drinking Water Disinfection Techniques*, Taylor and Francis, Boca Raton, 2012.
45. Organization, W.H., *Guidelines for Drinking-Water Quality*, vol. 1, World Health Organization, Geneva, 2004.
46. Crider, Y.S., Tsuchiya, M., Mukundwa, M., Ray, I., Pickering, A.J., Adoption of point-of-use chlorination for household drinking water treatment: a systematic review. *Environ. Health Perspect.*, 131, 1, 016001, 2023.
47. Cheung, P.C., A historical review of the benefits and hypothetical risks of disinfecting drinking water by chlorination (Updated and Revised). *J. Environ.*, 8, 73–146, 2017.
48. Joseph, C.G., Farm, Y.Y., Taufiq-Yap, Y.H., Pang, C.K., Nga, J.L., Puma, G.L., Ozonation treatment processes for the remediation of detergent wastewater: A comprehensive review. *J. Environ. Chem. Eng.*, 9, 5, 106099, 2021.
49. Lim, S., Shi, J.L., von Gunten, U., McCurry, D.L., Ozonation of organic compounds in water and wastewater: A critical review. *Water Res.*, 213, 118053, 2022.
50. Mouele, E.S.M., *et al.*, A critical review on ozone and co-species, generation and reaction mechanisms in plasma induced by dielectric barrier discharge technologies for wastewater remediation. *J. Environ. Chem. Eng.*, 9, 5, 105758, 2021.
51. Saravanan, A., *et al.*, Effective water/wastewater treatment methodologies for toxic pollutants removal: Processes and applications towards sustainable development. *Chemosphere*, 280, 130595, 2021.
52. Samer, M., Biological and chemical wastewater treatment processes. *Wastewater Treat. Eng.*, 150, 212, 2015.
53. Sharma, S. and Bhattacharya, A., Drinking water contamination and treatment techniques. *Appl. Water Sci.*, 7, 3, 1043–1067, 2017.
54. Gupta, A., *et al.*, A review of adsorbents for heavy metal decontamination: Growing approach to wastewater treatment. *Materials*, 14, 16, 4702, 2021.
55. Ng, C.A., *et al.*, Optimization of membrane bioreactors by the addition of powdered activated carbon. *Bioresour. Technol.*, 138, 38–47, 2013.
56. Yu, S., Wang, J., Zhao, Z., Cai, W., Simultaneous coupling of fluidized granular activated carbon (GAC) and powdered activated carbon (PAC) with ultrafiltration process: A promising synergistic alternative for water treatment. *Sep. Purif. Technol.*, 282, 120085, 2022.
57. Pivokonsky, M., *et al.*, Current knowledge in the field of algal organic matter adsorption onto activated carbon in drinking water treatment. *Sci. Total Environ.*, 799, 149455, 2021.
58. Tahmasebi-Boldaji, R., Hatamipour, M.-S., Khanahmadi, M., Sadeh, P., Najafipour, I., Ultrasound-assisted packed-bed extraction of hypericin from

Hypericum perforatum L. and optimization by response surface methodology. *Ultrason. Sonochem.*, 57, 89–97, 2019.
59. Larasati, A., Fowler, G.D., Graham, N.J., Insights into chemical regeneration of activated carbon for water treatment. *J. Environ. Chem. Eng.*, 9, 4, 105555, 2021.
60. Metcalf, L., Wastewater engineering: treatment and reuse. Metcalf & Eddy Inc. *McGraw-Hill Inc. N. Y. Mohammed ElBably MA 2016 Technol. Domest. Wastewater Treat. Reuse Options Appl. Dev. Ctries. JSM Env. Sci. Ecol.*, 4, 3, 1033, 2003.
61. Cheng, M., *et al.*, Hydroxyl radicals based advanced oxidation processes (AOPs) for remediation of soils contaminated with organic compounds: a review. *Chem. Eng. J.*, 284, 582–598, 2016.
62. Rekhate, C.V. and Srivastava, J.K., Recent advances in ozone-based advanced oxidation processes for treatment of wastewater-A review. *Chem. Eng. J. Adv.*, 3, 100031, 2020.
63. Amor, C., Marchão, L., Lucas, M.S., Peres, J.A., Application of advanced oxidation processes for the treatment of recalcitrant agro-industrial wastewater: A review. *Water*, 11, 2, 205, 2019.
64. Klavarioti, M., Mantzavinos, D., Kassinos, D., Removal of residual pharmaceuticals from aqueous systems by advanced oxidation processes. *Environ. Int.*, 35, 2, 402–417, 2009.
65. Deng, Y. and Zhao, R., Advanced oxidation processes (AOPs) in wastewater treatment. *Curr. Pollut. Rep.*, 1, 167–176, 2015.
66. Strathmann, H., Ion-exchange membrane processes in water treatment. *Sustain. Sci. Eng.*, 2, 141–199, 2010.
67. Tang, C., Merks, C.W.A.M., Albrechtsen, H.-J., Water softeners add comfort and consume water–comparison of selected centralised and decentralised softening technologies. *Water Supply*, 19, 7, 2088–2097, 2019.
68. *The National Academies Press*, https://nap.nationalacademies.org/ (accessed Jul. 23, 2023).
69. Zagorodni, A.A., *Ion Exchange Materials: Properties and Applications*, Elsevier, The Netherlands, 2006.
70. Rathi, B.S. and Kumar, P.S., A review on sources, identification and treatment strategies for the removal of toxic Arsenic from water system. *J. Hazard. Mater.*, 418, 126299, 2021.
71. Sodha, V., Shahabuddin, S., Gaur, R., Ahmad, I., Bandyopadhyay, R., Sridewi, N., Comprehensive review on zeolite-based nanocomposites for treatment of effluents from wastewater. *Nanomaterials*, 12, 18, 3199, 2022.
72. Trujillo, E.M., Spinti, M., Zhuang, H., Immobilized biomass: a new class of heavy-metal selective ion exchangers, in: *Ion Exchange Technology*, pp. 225–270, CRC Press, London, 2021.
73. Fiskum, S.K., Pease, L.F., Peterson, R.A., Review of ion exchange technologies for cesium removal from caustic tank waste. *Solvent Extr. Ion Exch.*, 38, 6, 573–611, 2020.

74. Höll, W.H., CARIX process—A novel approach to desalination by ion exchange, in: *Ion Exchange Technology*, vol. pp, pp. 271–313, CRC Press, 2021.
75. Huang, X., Guida, S., Jefferson, B., Soares, A., Economic evaluation of ion-exchange processes for nutrient removal and recovery from municipal wastewater. *npj Clean Water*, 3, 1, 7, 2020.
76. Hubicki, Z. and Kołodyńska, D., Selective removal of heavy metal ions from waters and waste waters using ion exchange methods. *Ion Exch. Technol.*, 7, 193–240, 2012.
77. Gupta, V.K., Ali, I., Saleh, T.A., Nayak, A., Agarwal, S., Chemical treatment technologies for waste-water recycling—an overview. *RSC Adv.*, 2, 16, 6380–6388, 2012.
78. Crini, G. and Lichtfouse, E., Advantages and disadvantages of techniques used for wastewater treatment. *Environ. Chem. Lett.*, 17, 145–155, 2019.
79. Cheng, S., *et al.*, Developing a UV dosing strategy for UV/chlorine process towards the trade-off between pharmaceuticals and personal care products degradation and disinfection by-products formation. *J. Clean. Prod.*, 385, 135584, 2023.
80. Srivastav, A.L., Patel, N., Chaudhary, V.K., Disinfection by-products in drinking water: Occurrence, toxicity and abatement. *Environ. Pollut.*, 267, 115474, 2020.
81. Gedda, G., *et al.*, Introduction to conventional wastewater treatment technologies: limitations and recent advances. *Mater. Res. Found*, 91, 1–36, 2021.
82. Hydrex - Water treatment chemical programs, in: *Veolia*, https://www.veoliawatertechnologies.com/en/services/water-treatment-chemical-programs (accessed Jul. 23, 2023).
83. Water Handbook - Monitoring And Control Of Water Treatment, in: *Veolia*, https://www.watertechnologies.com/handbook/chapter-36-monitoring-and-control-water-treatment (accessed Jul. 23, 2023).
84. *Chapter 2 Water Quality Monitoring, Standards and Treatment*, https://www.fao.org/3/X5624E/x5624e05.htm (accessed Jul. 23, 2023).
85. KSB, A Look at Wastewater Treatment Processes, in: *Empowering Pumps and Equipment*, Feb. 27, 2017, https://empoweringpumps.com/ksb-wastewater-treatment-process/ (accessed Jul. 23, 2023).
86. *The Ultimate Guide to Chemical Pumps: Types, Applications, and Safety Tips*, Mar. 19, 2023, https://www.rotechpumps.com/what-is-a-chemical-pump/ (accessed Jul. 23, 2023).
87. Wastewater Treatment Mixers, in: *Dynamix Agitators Inc*, https://dynamix-inc.com/industries/wastewater-treatment-industry/ (accessed Jul. 23, 2023).
88. Abujazar, M.S.S., Karaağaç, S.U., Amr, S.S.A., Alazaiza, M.Y., Bashir, M.J., Recent advancement in the application of hybrid coagulants in coagulation-flocculation of wastewater: A review. *J. Clean. Prod.*, 345, 131133, 2022.

89. Ibrahim, A., Yaser, A.Z., Lamaming, J., Synthesising tannin-based coagulants for water and wastewater application: A review. *J. Environ. Chem. Eng.*, 9, 1, 105007, 2021.
90. Ali, M.E.M., Moniem, S.M.A., Hemdan, B.A., Ammar, N.S., Ibrahim, H.S., Innovative polymeric inorganic coagulant-flocculant for wastewater purification with simultaneous microbial reduction in treated effluent and sludge. *South Afr. J. Chem. Eng.*, 42, 127–137, 2022.
91. Zhang, B., Mao, X., Tang, X., Tang, H., Shen, Y., Shi, W., Effect of modified microbial flocculant on membrane fouling alleviation in a hybrid aerobic granular sludge membrane system for wastewater reuse. *Sep. Purif. Technol.*, 290, 120819, 2022.
92. Jin, Y., *et al.*, Iron-based technology coupling moderate preoxidation with hybrid coagulation for highly effective removal and moderate growth inhibition of Oscillatoria in drinking water treatment plants. *J. Environ. Chem. Eng.*, 10, 3, 107723, 2022.
93. Farzanehsa, M., Vaughan, L.C., Zamyadi, A., Khan, S.J., Comparison of UV-Cl and UV-H2O2 advanced oxidation processes in the degradation of contaminants from water and wastewater: A review. *Water Environ. J.*, 1–12, 2023.
94. Khandelwal, A. and Baral, S.S., Advanced Oxidation Processes: Fundamental, Technologies, Applications and Recent Advances, *Sustain. Water Treat. Adv. Technol. Interv.*, pp. 1–25, New Jersey, USA, 2022.
95. Żyłła, R., Boruta, T., Gmurek, M., Milala, R., Ledakowicz, S., Integration of advanced oxidation and membrane filtration for removal of micropollutants of emerging concern. *Process Saf. Environ. Prot.*, 130, 67–76, 2019.
96. Turan, K., Canlidinç, R.S., Kalfa, O.M., Preconcentration of trace amount Cu (II) by solid-phase extraction method using activated carbon-based ion-imprinted sorbent. *Turk. J. Chem.*, 46, 2, 550–566, 2022.
97. Nguyen, X.C., Kang, J.-H., Bang, G., Kim, K.-M., Lee, C.-H., Pelletized activated carbon-based CO-selective adsorbent with highly oxidation-stable and aggregation-resistant Cu (I) sites. *Chem. Eng. J.*, 451, 138758, 2023.
98. Kumari, S., Kaur, M., Dhania, G., Carbon Nanomaterials for Wastewater Treatment, in: *Modern Nanotechnology: Volume 1: Environmental Sustainability and Remediation*, pp. 255–277, Springer, Switzerland, 2023.
99. Rasheed, T., Magnetic nanomaterials: Greener and sustainable alternatives for the adsorption of hazardous environmental contaminants. *J. Clean. Prod.*, 362, 132338, 2022.
100. Banerjee, P., Das, P., Mukhopadhayay, A., Ghosh, A.K., Graphene Oxide-Carbon Nanotube Composites for Wastewater Treatment,. *Carbon Nanotub. Green Environ.*, pp. 19–50, Palm Bay, USA, 2022.
101. Li, H., *et al.*, A recent overview of per-and polyfluoroalkyl substances (PFAS) removal by functional framework materials. *Chem. Eng. J.*, 452, 139202, 2023.

102. Pelayo, D., Rivero, M.J., Santos, G., Gómez, P., Ortiz, I., Techno-economic evaluation of UV light technologies in water remediation. *Sci. Total Environ.*, 868, 161376, 2023.
103. Deng, L., Wang, T., Shen, J., Tan, C., Hu, J., Singh, R.P., Comparison of UV/chloramine disinfection of methylamine water in the absence and presence of bromide: Halonitromethanes formation, toxicity alteration, and reaction mechanisms. *J. Environ. Chem. Eng.*, 11, 3, 109749, 2023.
104. Dalamarinis, P., Smith, B., Fusselman, S., Acid Restimulation in Legacy Wolfcamp Wells Utilizing Chlorine Dioxide (ClO2). An Operator Case Study of Reservoir Conductivity and near Wellbore Fracture System Reactivation, in: *SPE/AAPG/SEG Unconventional Resources Technology Conference*, URTeC, p. D011S005R001, 2023.
105. Wavelenghts, G.A., UV-LED water disinfection for small-scale decentralized systems. *Sci. Total Environ.*, 493, 185–196, 2022.
106. Pelayo, D., Rivero, M.J., Santos, G., Gómez, P., Ortiz, I., Techno-Economic Evaluation of UV Light Technologies for the Simultaneous Disinfection and Water Remediation, in: *Available SSRN 4261978*.
107. Li, Y., *et al.*, Degradation of nano-sized polystyrene plastics by ozonation or chlorination in drinking water disinfection processes. *Chem. Eng. J.*, 427, 131690, 2022.
108. Ping, Q., Yan, T., Wang, L., Li, Y., Lin, Y., Insight into using a novel ultraviolet/peracetic acid combination disinfection process to simultaneously remove antibiotics and antibiotic resistance genes in wastewater: Mechanism and comparison with conventional processes. *Water Res.*, 210, 118019, 2022.

7

Treatment of Flue Gas Desulfurization Wastewater from Power Stations Using Freeze Crystallization

A.L. Tau[1] and J.P. Maree[2*]

[1]Department of Chemical, Metallurgical and Materials Engineering, Faculty of Engineering and the Built Environment, Tshwane University of Technology, Pretoria, South Africa
[2]Institute for Nanotechnology and Water Sustainability (iNanoWS), College of Science, Engineering and Technology, University of South Africa, Florida Science Campus, Johannesburg, South Africa

Abstract
The wet flue gas desulfurization (FGD) system in thermal power plants, designed for emissions control, consumes large volumes of water. This process generates waste gypsum and large volumes of wastewater with high total dissolved solid (TDS) concentrations, reaching 50,000 mg/L. Eskom's Kusile Power Station uses a wet FGD system to manage SO_2 and other greenhouse gas emissions. This study examined the waste disposal, as well as the electrical and chemical costs associated with three FGD wastewater management options. The high TDS concentrations result from SO_2, Cl, and boron absorption from coal during the FGD process, along with the dissolution of Na, K, Mg, Mn, and Ca from limestone during its reaction with acidic gases from the combustion chamber. Most of the $CaSO_4$ in the solution crystallized as gypsum due to its low solubility.

For wastewater from a 4,800-MW plant, disposal at a toxic waste facility costs R9.7 million per month. Pre-treating the wastewater with lime or Na_2CO_3 followed by evaporation raised the cost to R11.41 million monthly. In contrast, freeze crystallization, which does not require chemical pre-treatment, had a much lower cost of R960,000 per month. Freeze crystallization was found to be the most economical option, using less energy than evaporation (330 kJ/kg for freezing vs. 2,260 kJ/kg

**Corresponding author*: maree.jannie@gmail.com

Elvis Fosso-Kankeu, Vhahangwele Masindi, Johannes Maree and Bhekie Mamba (eds.) Customized Technologies for Sustainable Management of Industrial Wastewater: A Circular Economy Approach, (257–310) © 2025 Scrivener Publishing LLC

for evaporation) and requiring 100 kWh/kg ice. The TDS in the feed was 50,000 mg/L, reduced to less than 3,000 mg/L in the melted ice.

***Keywords*:** Pipe freeze crystallization, flue gas desulfurization, coal-fired power stations, salt and clean water recovery, reverse osmosis cooling, modeling

Remark

This chapter is largely based on the Master's thesis of Asser Tau. A summarized version of the thesis was also published in the following Proceedings:

AL Tau, JP Maree, A Adeniyi, and MS Onyango (2022) Treatment of Flue Gas Desulphurisation effluent with Freeze-crystallization, International Institute of Biological Chemical and Environmental Engineering (IICBEE), Johannesburg, South Africa, Nov. 28-29, ISBN: 978-989-9121-14-0, https://doi.org/10.17758/IICBE4.C1122223, p60–75.

Abbreviations and Acronyms

ED	Electrodialysis
FGD	Flue gas desulfurization
IX	Ion exchange
MD	Membrane distillation
MED	Multi-effect distillation
MEE	Multi-effect evaporation
MSF	Multistage flash
RO	Reverse osmosis
ROC	Reverse osmosis-cooling
SSHE	Scraped surface heat exchanger
TDS	Total dissolved solids
TUT	Tshwane University of Technology
TSS	Total suspended solids
TVC	Thermal vapor compression
VCD	Vapour compression distillation
WFGD	Wet flue gas desulfurization
ZLD	Zero liquid discharge

7.1 Introduction

7.1.1 Background

Eskom uses approximately 2% of South Africa's freshwater resources. To comply with emissions regulations, Eskom requires enhanced water optimization technologies. While coal remains the primary energy source, it also leads to considerable air pollution. Power plants generate coal ash, wastewater, and release gaseous emissions, including NOx and SOx, into the atmosphere. Sulfur emissions, which contribute to acid rain and ozone layer depletion, are subject to strict environmental regulations. Eskom aims to adhere to the zero liquid effluent discharge (ZLED) policy. At the Kusile coal-fired power plant, a wet flue gas desulfurization (FGD) system has been implemented to lower SO_2 emissions [1].

The management of the currently available water sources is of utmost importance both to the government and industrial consumers in the manufacturing sectors such as the power industry. During thermal power plant operations integrated with wet FGD for emissions control, the latter requires large volumes of water. The process releases both solid gypsum and liquid waste, and wastewater that cannot be directly used elsewhere in the power station. Consequently, adequate management of these waste streams is critical to sustainable power plant operations. Kusile Power Station is a pulverized coal-fired power station with integrated wet FGD technology. The power station currently manages wastewater by disposal in a toxic waste site as a temporary measure, and, for the long term, the power station intends to treat the wastewater through evaporation. The evaporative desalination method is energy intensive [2, 3]. Other common management methods of saline streams are disposal to evaporation ponds and ash dams. The disposal cost to toxic waste disposal sites is as high as R2,000/m^3.

According to recent environmental legislation contained in Government Notice 636, Regulation 5 of 2013 [4], the hazardous wastes discarded to landfill sites should not contain more than 40% water. During rainy seasons, the evaporation ponds and ash dams overflow, resulting in the hyper-saline wastewater spilling into the environment. Also, this disposal method has a negative environmental impact with the potential of groundwater contamination as the liners may be worn out with time causing the stored wastewater to seep into the ground. The current treatment and disposal methods of the wastewater are thus insufficient in managing the wastes and also minimizing direct water utilization from limited water resources. This necessitates the need to investigate more environmentally friendly, improved water management, and cost-effective wastewater treatment methods that will assist the power industry to comply with the set ZLED policy and sustain limited water sources.

7.1.2 Description of the Wet Flue Gas Desulfurization System

Wet limestone FGD systems include unit operations such as limestone storage, limestone dosage, spray tower absorbers, dewatering of gypsum, and wastewater treatment [5]. Flue gas flows upward through the absorber where it is contacted with calcium carbonate slurry [6]. Calcium sulfite hemihydrate ($CaSO_3{\cdot}½H_2O$) formed when SO_2 from the flue gas is neutralized with $CaCO_3$. Calcium sulfite hemihydrate is oxidized to form calcium sulfate dihydrate ($CaSO_4{\cdot}2H_2O$), i.e., gypsum, by bubbling air through the slurry in the bottom contact reactor [7]. As the slurry is recycled, it becomes more concentrated due to soluble compounds in the limestone, e.g., Na^+, K^+, Mg^{2+}, Ca^{2+}, and some heavy metals and in flue gas (e.g., SO_2, HCl, and B). The Ca^{2+} and SO_4^{2-} concentrations are controlled by the solubility of gypsum that crystallizes at a solid's concentration of approximately 15%. A portion of the gypsum-slurry is bled off and dewatered with a horizontal vacuum belt filter. Solid gypsum and saline wastewater with a total dissolved solid (TDS) of 50,000 mg/L is produced. In order to achieve zero liquid discharge (ZLD), the wastewater needs to be treated in a cost-effective way. The purpose of this investigation is to identify such a cost-effective process. The Kusile Power Station is expected to produce about 900,000 tons of gypsum per annum [7].

Figure 7.1 shows a counterflow, open-spray tower where the slurry of limestone is contacted with the flue gas. The limestone slurry is pumped through banks of spray nozzles to form fine droplets [8].

During the operation of the FGD scrubber, flue gas enters the FGD absorber module and moves upward, where it comes into contact with a $CaCO_3$ slurry sprayed from the absorber's upper section. SO_2 is absorbed from the flue gas into the slurry and is neutralized by the $CaCO_3$, resulting in the formation of $CaSO_3{\cdot}½H_2O$. The slurry collects in a reaction tank at the base of the scrubber, allowing enough time for the reactions to complete, while air is introduced to oxidize $CaSO_3{\cdot}½H_2O$ into gypsum [7], and the slurry is then recycled. Wet FGD technologies, therefore, require significant water volumes, approximately 160 m^3/h/unit while on load, to form the limestone slurry (4 m^3/h/unit on load), which must be treated. As SO_2 is removed from the flue gas, solid gypsum and other compounds

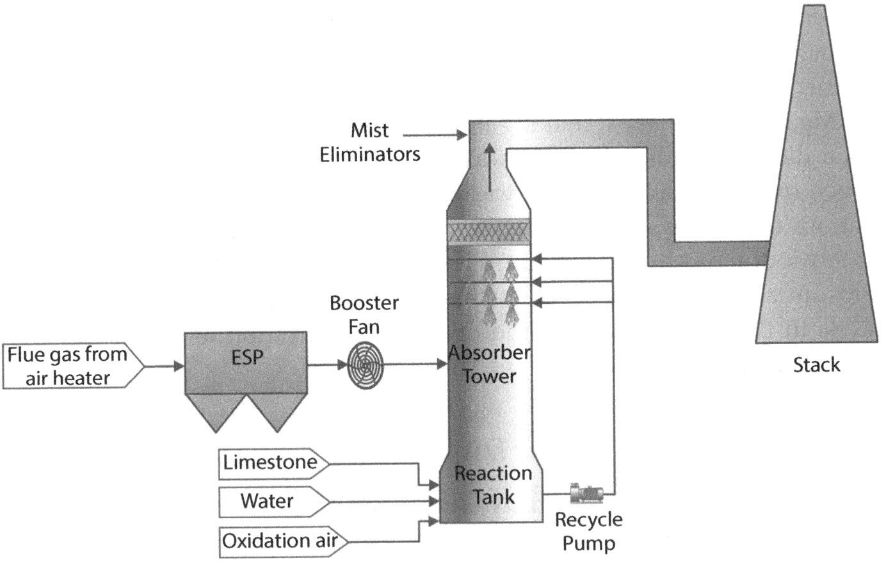

Figure 7.1 Process schematic of flue gas cleaning by electrostatic precipitation and desulfurization [5, 6].

accumulate in the reaction tank as gypsum slurry, increasing over time. When the solids content reaches a set point (around 15%), part of the gypsum slurry is extracted and sent to a dewatering system.

Dewatering is performed by feeding the gypsum slurry onto a horizontal vacuum belt filter, which generates wastewater that must be managed to meet ZLED requirements and produces solid gypsum (see Figure 7.2 for dewatered gypsum on a horizontal vacuum belt filter). Kusile Power

Figure 7.2 Gypsum on a horizontal vacuum belt filter [9].

Station is expected to generate approximately 900,000 tons of gypsum annually [7].

The wastewater from the FGD process contains high concentrations of Cl, Mg, Ca, and metals, making it unsuitable for reuse within the station. Total dissolved solids in FGD wastewater can reach up to 50,000 mg/L, which includes silica [5]. This wastewater may also contain high concentrations of sulfates, copper leachate, as well as $MgSO_4$ and $Al_2(SO_4)$, commonly found in wastewater from FGD systems. The soluble salts in the raw gypsum can be controlled by washing it before drying [7]. The pollutant levels in this wastewater depend on the concentration of impurities and metals present in the coal and limestone used [5].

7.1.3 Current Wastewater Disposal

Table 7.1 indicates three key facts: (i) the current cost for discharging FGD wastewater is R2,000 per ton; (ii) a total volume of 9 670 m^3 was disposed of at a toxic waste disposal facility between 1 January 2019 and 30 April 2019 (Pers. Comm, Eskom Rep, 2019); and (iii) with a flow rate of 6.63 m^3/h for 6 units, the total cost reaches R9.7 million per month.

FGD wastewater is transported and disposed of at a hazardous waste site, costing R9,670,000 per month. A three-stage treatment facility was developed to recover clean water from wastewater, comprising pre-treatment, evaporative concentration, and crystallization stages. During

Table 7.1 Waste transport and disposal costs [9].

Parameter	Value
Volume disposed of in 4 months with 3 units (m^3)	**9,670.00**
Flow rate disposed for 3 units (m^3/h)	3.31
Flow rate disposed for 6 units (m^3/h) (Eskom figure)	6.63
Transport cost (R/m^3)	1,000.00
Disposal cost (R/m^3)	1,000.00
Cost (R/m^3)	2,000.00
Cost (R/month)	9,670 000

the pre-treatment phase, Ca is removed through lime and soda-ash flocculation with ferric chloride, followed by clarification. This stage aims to reduce scale formation on the heat exchangers. The resulting water still has Ca concentration of 1,500 mg/L and Mg concentration of 320 mg/L. Some Ca remains because it is needed as a seeding slurry in the evaporator or concentrator. The pH of the clarified water is adjusted to between 7 and 8 to minimize further precipitation downstream before it flows into the brine concentrator tanks [5]. The FGD wastewater is introduced into the heat exchanger at 37°C and heated to 96°C. Water vapor is extracted from the top of the vessel, which is equipped with baffles to minimize solid carryover [5].

7.1.4 Objectives

The objectives established for this study are as follows:

- Gain a clear understanding with regards to the operations of a coal-fired power station.
- Identify the sources of impurities that lead to the poor water quality of FGD wastewater.
- Compare the costs associated with effluent management across the following scenarios: (i) disposal at hazardous waste sites; (ii) treatment involving softening and evaporation; and (iii) freeze crystallization.

7.2 Literature Review on Freeze Crystallization

This review focuses on wastewater and saline wastewater generation and its treatment in the electrical power industry with a strong emphasis on coal fired power stations. The review examines water usage in the electrical power industry and current technologies that are used to treat, reuse, and recycle the water. It discusses how wastewater and saline wastewater are generated in the power industry and what technologies are currently used to handle the wastewater and saline wastewaters. It also reviews what has been done in the area of freeze crystallization as applied to water treatment. It identifies the gap as to the importance of the treatment and beneficiation of saline wastewater and also emphasizes the importance of this application for the beneficiation of wastewater produced in the power industry. Beneficiation of saline wastewater generated by the coal powered industry may be a great approach to offset the high cost of saline wastewater treatment. Combining reverse osmosis with freeze crystallization as an integrated process may deliver both potable water and beneficiation of saline wastewater.

7.2.1 Basic Theory

7.2.1.1 Operating Principle

The key stages in freeze crystallization of saline water include (i) pre-cooling the feed wastewater, (ii) cooling and forming ice in a heat exchanger, and (iii) separating the ice crystals from the concentrated brine stream [10, 11]. Freeze-crystallization technology not only recovers high-purity ice but also effectively recovers salts. Rapid crystal formation can lead to impure ice due to the faster growth rate [3]. The principle of salt rejection during the phase change from liquid to solid underlies the freeze-crystallization process, which serves as an alternative method for recovering dissolved salts and water as ice. The phenomenon governing salt rejection involves the formation of small ice crystal lattice structures that release salt ions during phase change rather than incorporating them into the lattice [11–13].

This process can generate fresh water from brackish sources and industrial saline wastewaters through freezing and cooling [3]. Freeze crystallization effectively treats brackish water by harvesting and melting ice crystals to produce fresh water. Ice crystal growth during freezing results in solute rejection [2]. Freezing involves transferring cold energy into water to create solid ice or a semi-solid ice slurry. Crystallization isolates impurities present in the solution [14]. As contaminants concentrate in the mother liquor during slow freezing, nearly pure water can be achieved by separating and melting the ice crystals. The remaining mother liquor will have a higher concentration of salt, which crystallizes at the eutectic temperature.

Freeze crystallization offers several advantages compared to other methods that the technology competes with, including evaporation ponds and distillation. Its advantages include lower theoretical energy requirements, flexibility with varying salinity levels, and reduced potential for scaling, corrosion, and precipitation issues [10, 11, 15, 16]. The energy requirements for freeze crystallization are lower than those for evaporative desalination because the latent heat of ice is 334 kJ/kg, compared to 2 340 kJ/kg for the latent heat of vaporization of water. This means that freezing requires mechanical power, whereas evaporation necessitates heat input [2, 17].

The energy needed to produce one kilogram of pure water from ice is about one-seventh of that required to produce the same amount from vapor condensation.

Whereas evaporative desalination involves heating from 25°C to temperatures above 100°C (due to the boiling point elevation from dissolved salts), freeze crystallization requires lowering temperatures from 25°C to below 0°C [18]. The cost of desalination by freezing can be reduced from $1.85/m^3 for thermal desalination to $0.93/m^3 [10]. Cost comparisons at a power cost of ZAR0.30/kWh indicate that thermal desalination consumes ZAR214/m^3, whereas freeze crystallization consumes only ZAR36/m^3, highlighting the energy efficiency of freeze crystallization [10, 18]. The efficiency of the freeze-crystallization technology in removing dissolved solids, such as salts, can be evaluated using Equation 7.1.

$$\eta = \frac{1 - C_s \times V_s}{C_0 \times V_0} \tag{7.1}$$

where V_s is the melted ice crystals volume, C_o is the initial concentration of a specific component in the feed solution (i.e., in the saline wastewater), V_0 is the initial total volume of the feed solution, and C_s is the concentration of a specific component present in the ice phase.

7.2.1.2 Supersaturation

Solubility of pollutants in most solutions undergoing crystallization serves as the first process step in the crystallization process that gives information about the rate of supersaturation. Supersaturation being the driving force in the crystallization process refers to the deferential concentration of the solute's chemical potential in a supersaturated state rather than that in the saturated solution. A supersaturated solution can be achieved using either of the methods such as cooling, addition of anti-solvent, chemical reactions, and evaporation of the solvent with cooling being the most employed method. Cooling for achieving a supersaturated solution whereby heat is removed from a saturated solution is applied to solutions whose solubility is a function of temperature [19–21]. Thus, supersaturation is a thermodynamic driving force for crystallization defined as a deviation from thermodynamic equilibrium.

Figure 7.3 shows the plot of a typical solubility or super solubility, whereby the achievement of supersaturation according to the following stages can be reached [22]:

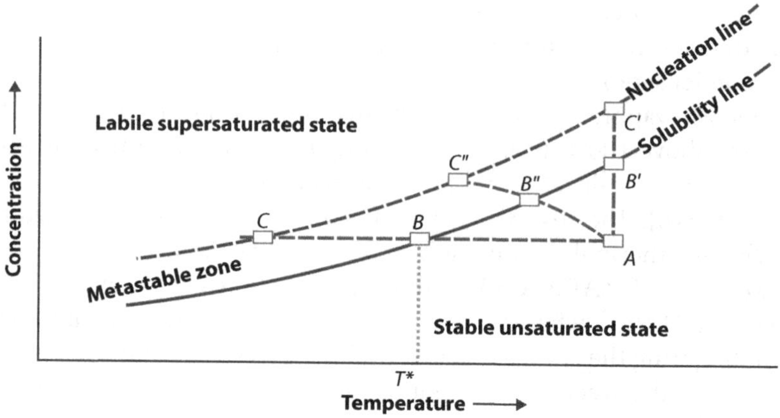

Figure 7.3 Plots for typical solubility/super-solubility [22].

- Cooling of the saturated solution as expressed by line AC
- Evaporation of the solvent thus concentrating the saturated solution expressed by line AC'
- Combining the processes of cooling and evaporation as expressed by line AC".

Other techniques for generating supersaturation conditions for crystallization process conditions in most solutions are described by Figure 7.4 [23]:

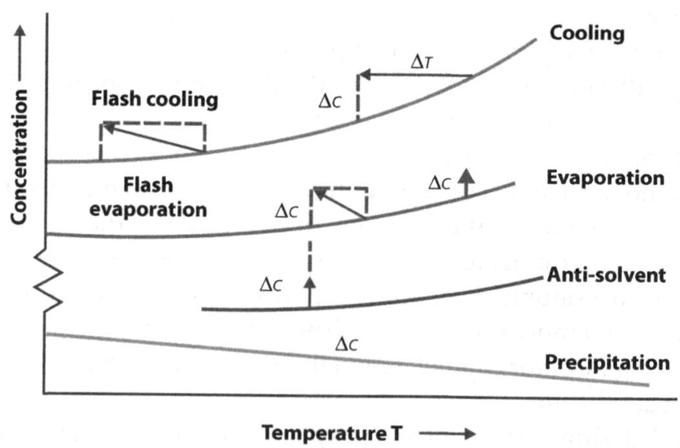

Figure 7.4 Techniques for generation supersaturation conditions in crystallization from most solutions [23].

The absolute supersaturation that is the driving force that governs the crystallization is expressed as the concentration differential between supersaturated solution, C, and the saturated solution C^* [23]:

$$\Delta C = C - C^* \tag{7.2}$$

where the concentration is expressed as kg/m^3.

The state of the supersaturation can be further described by quantities such as super-cooling and supersaturation, expressed by Equations 7.3 and 7.4 [22]:

$$\text{Relative supersaturation: } \sigma = \frac{\Delta C}{C^*} \tag{7.3}$$

where ΔC is the concentration driving force (as $\Delta C = C - C^*$), and C^* is the dissolution concentration at equilibrium (and $C > C^*$).

$$\text{Super-cooling: } \Delta T = T - T^* \tag{7.4}$$

where T is the temperature for dissolution, and T^* is the actual temperature (and $T^* > T$).

Supersaturation ratio, S, can be expressed as the quotient of supersaturated solution concentration (C) and the of saturated solution concentration (C^*). This can be expressed mathematically by Equation 7.5 [23]:

$$S = \frac{C}{C^*} \tag{7.5}$$

Also, relative supersaturation, σ, can again be expressed as $\sigma = S - 1$.

The super-cooling expression is typically applied in melt-crystallization, where supersaturation is represented as the temperature difference between the actual temperature and the melting temperature. This equation is also relevant in cooling crystallization, with solubility temperature replacing the melting temperature. The supersaturation of a moderately soluble solid, as a function of concentration and temperature, can be illustrated in Figure 7.5 [24].

Supersaturation in freeze crystallization can be regulated by the temperature of the primary refrigerant in use. In order to achieve this, the

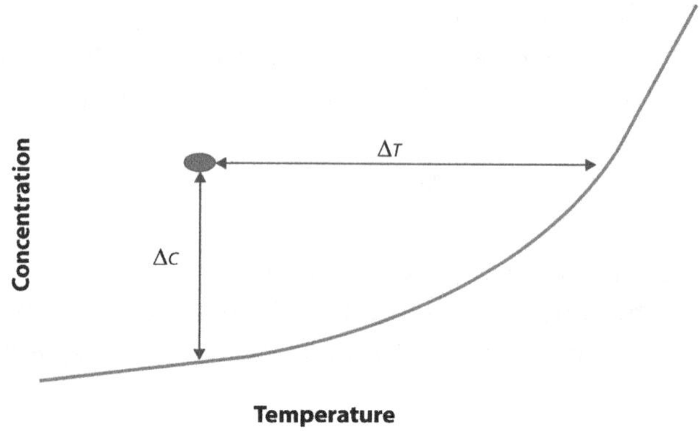

Figure 7.5 Moderately soluble solute supersaturation [24].

solution is super-cooled at the equilibrium temperature or by varying the operating pressure, which, in turn, shifts the equilibrium temperature. On the other hand, the temperature of the solution and the growing crystals of ice from the solution, which releases the heat of crystallization affects the supersaturation [25].

Supersaturation as a process result with two phenomena of phase change: nucleation and crystal growth. These two phenomena compete for solute as a function of their availability on the level of supersaturation, with their relative magnitude being the main factors that dictate the size distribution of the crystal product. There is a non-linear correlation between the nucleation and supersaturation; meanwhile, the rate of growth shows linear correlation with the supersaturation [25–28].

7.2.1.3 Crystallization

During crystallization, crystalline material is generated from a liquid, gas, or amorphous solid. The operating principle of crystallization as a process is a function of limited solubility of solute, thus dissolved in a solvent at a specific condition such as temperature and pressure. When these conditions are varied to a point whereby the solubility is low, crystalline solid will be formed under these new conditions. The crystals are generated with higher internal irregular structure with the basis, which is known as crystal lattice. In the now formed lattice, foreign molecules are not found imbedded in the crystal structure; thus, this makes the solid product of high level of purity. Crystallization processes generate supersaturated solution or melt [25, 28].

Many industries such as detergents, fertilizers, food, pharmaceutical, mineral processing, and waste effluents make use of crystallization as part of the processes that they undertake in order to yield their products. Crystallization is one of the long existing unit operations aimed at production of large volumes of components such as sodium chloride (NaCl), sodium sulfate (Na_2SO_4), aluminium sulfate [$Al_2(SO_4)_3$], and sucrose ($C_{12}H_{22}O_{11}$) with annual production rates of 10^8 tons on a world basis. Crystallization is a preferred means purification of many organic solutions than distillation; this is because the enthalpy of crystallization for many organic liquids is, in general, lower than enthalpy of vaporization. Temperatures close to ambient temperatures can be used in crystallization process, thus making the energy requirements much lower [29].

Depending on the type or extent of crystallization required, many different types of crystallizers are in use by most industries. For forced circulation, draft-tube baffled and fluidized bed crystallizers are mostly dominating. In most cases, crystallization through evaporation and cooling is executed from solution through continuous operation of the crystallizer. In heat and mass transfer, fluid and particle mechanics govern the operation of crystallization. Crystallization can be used in desalting seawater, fruit juice concentration, and other various industrial processes in order to recover valuable components in wastewater and to remove any foreign material not needed in water.

7.2.1.3.1 Crystal Nucleation

Once supersaturated conditions have been achieved in the solution, nucleation can be initiated, whereby a stable solid phase is formed. The nucleation process thus forms the new crystalline component from the solution. Nucleation can be defined as a process that involves a series of atomic or molecular components of a reactant phase re-organized into a nucleus of large product phase adequate to be able to grow irreversibly to a size that

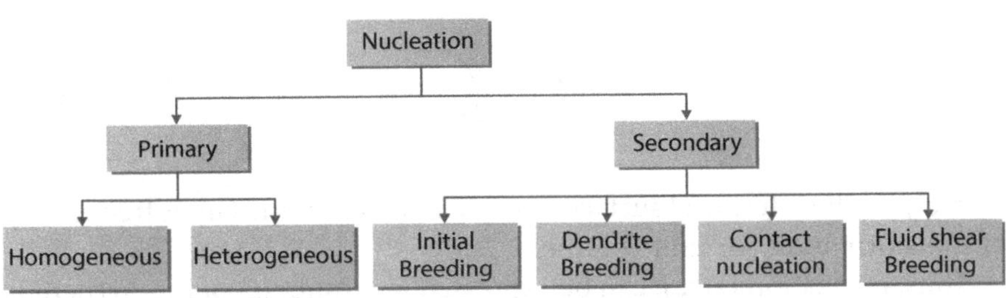

Figure 7.6 Mechanisms of crystal nucleation.

is large under microscopic viewing. The nucleation process can be a complex process as the creation of crystals in a supersaturated solution may be inhibited because of nuclei being generated by different mechanisms. Thus, most of the nucleation classification methods differentiate between primary and secondary nucleation [29]. The mechanisms of crystal nucleation are described in Figure 7.6.

7.2.1.3.2 Primary Nucleation

Primary nucleation refers to the formation of crystals in the early phase with no added crystal or with no effect or influence on the process should there be crystals in system. It is the initial generation of solid crystals from a solution or melt that is supersaturated, which is free from suspended particles of the material to be crystallized. This is classified into homogenous and heterogeneous nucleation [30–32]. In heterogeneous nucleation, primary nucleation is activated by suspended particles of foreign substances, and, in homogeneous nucleation, nucleation is initiated spontaneously. The energy involved in the formation of a solid phase and generation of arbitrary spherical crystals with radius r in a supersaturated fluid can be described by Equation 7.6 [29].

$$\Delta G = 4\pi^2\sigma + (4\pi / 3)r^3\Delta G_v \tag{7.6}$$

where ΔG is the overall excess free energy involved due to formation of a crystalline body, σ is the interfacial tension among crystals and their surroundings in a saturated solution, ΔG_v is the phase change's free energy change per unit volume, $4\pi r^2\sigma$ represents the surface contribution that is positive and is proportional to r^2 and $(4\pi / 3)$, and $r^3\Delta G_v$ represents the volume contribution and is negative and proportional to r^3.

The components of the crystallization should constitute stable clusters of critical size before forming into crystals. This can be better defined by Equation 7.7.

$$r_c = \frac{2\gamma}{\Delta G_v} \tag{7.7}$$

where γ is the interfacial surface tension and ΔG_v is the Gibbs free energy of transformation per unit volume.

Thus, nucleation rate, J, can be expressed as a function of the Arrhenius reaction rate equation as defined by Equation 7.8 [29]:

$$J = F\exp\left[-16\pi\delta^3V^2 / 3k^3T^3\left(\ln S^2\right)\right] \tag{7.8}$$

where F *is* the pre-exponential factor, v is the molar volume, k is the Boltzmann constant, and S is the supersaturation ratio.

Heterogeneous nucleation is initiated by foreign nuclei or surfaces that are available in solution and result in being significant at low supersaturation levels. Thus, heterogeneous nucleation can be referred to as generation of new particles from a solution or melt by the existence of foreign material that is insoluble [24]:

The rate of heterogeneous nucleation can be expressed by Equation 7.9 [33]:

$$B_{het}^0 = A_{het}\exp\left[\frac{-16\pi\sigma^3v^2f(\phi)}{3k^3T^3\left(\ln S^2\right)}\right] \tag{7.9}$$

where $f(\varphi)$ accounts for the reduced energy barrier to nucleation because of a foreign solid phase and $f(\varphi)$ is a geometric factor that can be defined by Equation 7.10:

$$f(\phi) = \left[\frac{(2+\cos\varphi)(1-\cos(\phi))}{4}\right] \tag{7.10}$$

where $f(\varphi)$ is the geometric factor that ranges between 0 and 1, and φ is the wetting or contact angle between foreign particles and deposit of crystalline material and ranges between 0°C and 180°C.

7.2.1.3.3 Secondary Nucleation

Secondary nucleation takes place in a supersaturated solution that is in this state due to the availability of parent crystals. Nucleation is thus initiated by being in contact with other existing crystals or seeds. Secondary nucleation consists of different types of processes, including initial bleeding, polycrystalline bleeding, macro-abrasion, dendritic, fluid shear, and contact nucleation. However, among these secondary nucleation types, contact nucleation is used in the majority of the industrial crystallizers. The generation of nuclei is initiated by contact between growing crystals and wall surfaces of the container, stirrer, and existing crystals [27, 30].

In the continuous crystallization process, secondary nucleation is mostly applied and is dependent on the level of supersaturation, degree of agitation, and the level of solids present. Thus, the rate of secondary nucleation can be expressed as a functional dependence on magma density, supersaturation, and the rate of stirring as shown by Equation 7.11 [24]:

$$BN_{T}{}^{Jhn}{}_{s\ sec} \tag{7.11}$$

where M_T is the magma density; ΔC is the degree of supersaturation; N_s is the rate of stirring; and j, l, and b denote the order dependence on secondary nucleation on the density of suspension, agitation, and supersaturation density, respectively.

7.2.1.3.4 Freezing Point Depression

The addition of a small amount of solute to a liquid solvent lowers the overall temperature of the mixture to T_f, where the pure solvent begins to solidify. This temperature is lower than the freezing point of the pure solvent, T_m, and the difference between these two temperatures due to the solute addition is known as freezing point depression, described by Equation 7.12 [19]:

$$\Delta T = T_m - T_f \tag{7.12}$$

If the solvent is pure water, then the solid that forms will be pure ice. The relationship between solute concentration and freezing point can be plotted using the simplified freezing point equation, which can be expressed as follows:

$$\Delta T = -\frac{RT_m^2}{\Delta H^{fus}\left(T_m\right)}\ln\left(\gamma_1 x_1\right) \tag{7.13}$$

In this equation, R is the universal gas constant, ΔH^{fus} represents the heat of fusion for water (6,025 J/mol), γ_1 is the water activity coefficient, and x_1 denotes the water mole fraction in the solution [19].

7.2.1.3.5 Seeding

Induction of crystallization in a supersaturated solution by adding small particles with their intention of acting as templates for growth and/or secondary nucleation is mostly used in secondary nucleation crystallization practices. This method is known as seeding and assists in controlling the size of particles and particle size distribution, whereby the crystal size of the seed is perceived to affect the secondary nucleation. Due to higher contact probabilities and the energy associated with collision, large seeds result in more secondary nuclei than small seeds do. It is also known that, as contact probability and energy of collision increase due to an increase in the total seed surface area, a link between this parameter and the creation of secondary nuclei cannot be avoided [21, 27, 34].

Addition of a certain amount of seed crystals having a uniform size, L_s, in the crystallizer at the temperature of saturation assists in the regulation of the process of crystallization. At low supersaturation levels, the process can be carried out within the metastable zone. The way in which seed crystals are being produced, seeding policy, and quantity of seed and loadings influence the process of crystallization in a seeded solution. Meanwhile, the seed crystal size distribution is influenced by the mass and size of the seed crystals. The size of the seeds is preferably small, and narrow seed size distribution is preferred. Agglomeration, sometimes, takes place with most materials and results in the size uniformity becoming less. The seed mass required can be calculated by making use of Equation 7.14 [21, 30].

$$\frac{M_c}{M_s} = \left(\frac{L_c}{L_s}\right)^3 \tag{7.14}$$

where M_c is the mass of the crystals produced, M_s is the mass of the seeds produced, L_c is the final crystal size, and L_s is the final seed size.

7.2.1.3.6 Crystal Growth

Crystal growth refers to the increase in crystal size resulting from the deposition of material on crystal surfaces in a supersaturated solution. During this process, molecules are transferred to the crystal surface, and the heat of crystallization is removed as it is released. This growth is associated with the mass increase of formed crystals or their dimensional characteristics, which are controlled by non-equilibrium thermodynamic conditions [31].

The growth process is governed by factors such as the overall rate of crystallization, temperature, supersaturation, and crystal content [35]. Crystal growth involves several processes that lead to the enlargement of crystals. These processes include the transportation of atoms through the solution, attachment of atoms to the crystal surface, movement of atoms across the surface, and the attachment of atoms to edges and kinks.

For crystals to grow in a solution, the dissolved solids must diffuse through the fluid-crystal interface, where they are adsorbed onto the crystal surface and integrated into the crystal lattice. At the same time, the heat of crystallization must be dissipated into the surrounding solution, although this heat dissipation is often considered negligible due to its minimal effect on crystal growth. Crystal growth can be understood as a process governed by diffusion-reaction theory, comprising both diffusion and surface integration steps, as illustrated in Figure 7.7.

Considering the crystal growth as a diffusion-controlled process, crystal growth rate can be described by Equation 7.15 on a first-order surface reaction [36]:

$$\frac{dm}{dt} = \frac{DA}{X + \frac{D}{k_r}} \tag{7.15}$$

where k_r is the first-order reaction rate constant, m is the mass of crystal, D is the solute diffusivity, X is the effective film thickness, and A is the surface area.

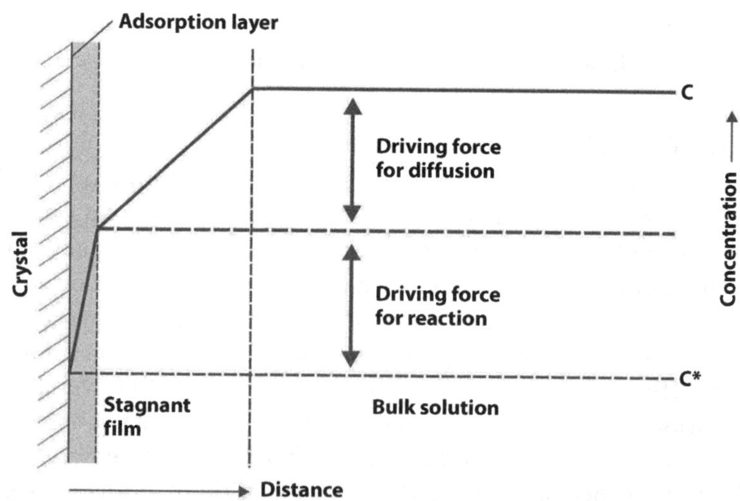

Figure 7.7 Growing crystal at the solution interface [36].

In a crystallization process, whereby the crystal growth is a function of mass transfer, the surface interaction occurs at a fast rate, and diffusive-convective transport of growth units from the bulk of the solvent to the surface of the crystal can be expressed by Equation 7.16 [36]:

$$m = k_d\left(C - C_l\right) \tag{7.16}$$

where C is the solute bulk concentration, $\dot{m}$ is the mass flux density, $C_{I\,is}$ the interfacial or surface concentration of solute, and k_d is the mass transfer coefficient.

The effect of saturation of solute on the crystal size, the growth rate, and the nucleation rate can be graphically represented as shown in Figure 7.8.

7.2.1.3.7 Ice Crystallization

At the equilibrium point of freezing, an energy barrier must be surpassed through the generation of a substantial driving force before crystallization in the solution can begin. The continuous removal of heat below 0°C without initiating phase change highlights the presence of this energy barrier. This principle underpins the concepts of undercooling or super-cooling, which induce a thermodynamic state of instability. As a result, sub-microscopic water aggregates eventually form an interface (nucleus, embryo, or seed) that promotes the significant transformation from liquid to solid. The degree of undercooling dictates when ice nucleation begins.

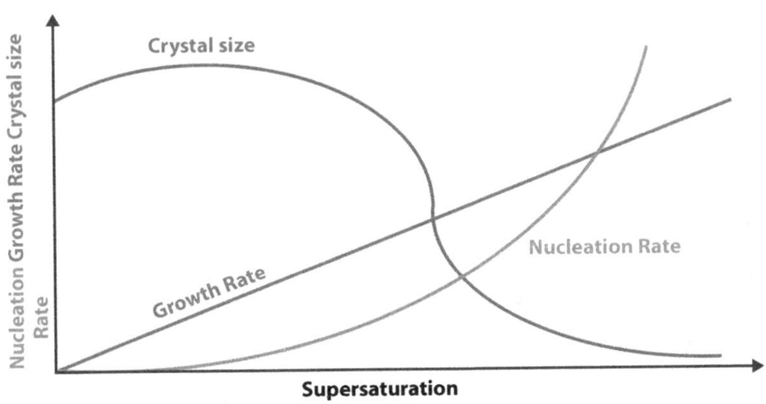

Figure 7.8 Effect of saturation of solute on the crystal size, the growth rate, and the nucleation rate [10].

A stable nucleus must form for phase change to occur, where liquid molecules arrange themselves in the crystal configuration [37].

The freezing process of pure water can be illustrated using a temperature-time profile, as shown in Figure 7.9. Below the freezing point, T_f, which equals the melting point, T_m, the crystallization induction occurs between points A to B or B′, representing the initial thermal event where pure water or solution undergoes undercooling. The activation energy for nucleation relates to non-equilibrium conditions within a metastable zone. At point B or B′, nucleation occurs once the required mass of nuclei is achieved, leading to the rapid release of latent heat faster than the system can dissipate it. This drives the temperature from point C (0°C) to C′ at T_f. Crystal growth in pure water continues along the time line from C to D. As the heat from ice is removed, the temperature decreases from D to E once crystallization completes. The C-D line remains at a constant temperature because the partially frozen mixture does not cool further until crystallization concludes. T_f represents the freezing time, from nucleation onset at point C to the full removal of latent heat at point D. Freeze concentration occurs during the freezing process of an aqueous solution, where water separates as ice crystals at point C′D′ [38].

7.2.1.4 Solubility, Phase Diagrams, and Metastable Zone

7.2.1.4.1 Solubility of Sodium Sulfate (Na_2SO_4)

Solubility is the maximum quantity of a substance that can be fully dissolved in a specific amount of solvent at a given temperature. This concept is fundamental in fields such as chemistry, physics, food science, pharmaceuticals, and biological sciences. The equilibrium solubility of a substance

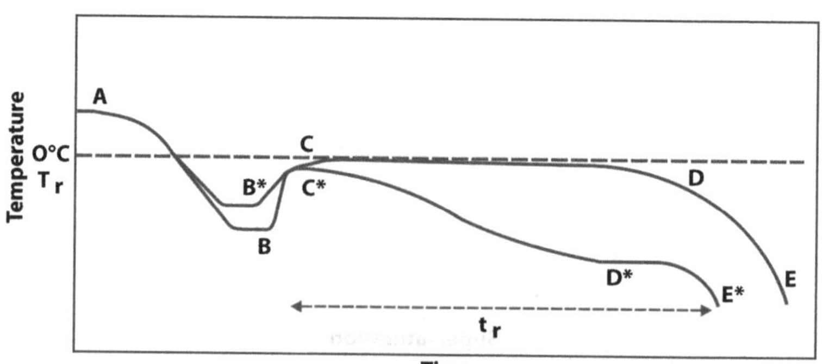

Figure 7.9 Crystallization process of ice from pure water and solution [38].

refers to the concentration of the solute in its saturated solution when the solution is in equilibrium with the pure solid solute [15, 39].

Saline wastewater, to be treated by reverse osmosis, may be composed of unwanted components such as sodium sulfate (Na_2SO_4) at various concentrations. Solid sodium sulfate can exist in the form of anhydrous Na_2SO_4, heptahydrate, $Na_2SO_4 \cdot 7H_2O$, and decahydrate, $Na_2SO_4 \cdot 10H_2O$ [40, 41]. As the temperature of a solution decreases, the solubility of each of these hydrated forms also decreases with $Na_2SO_4 \cdot 10H_2O$ crystallizing out of the solution at temperatures lower than 30°C, and the remaining saline wastewater mother liquor will be composed approximately of 30% (w/w) of sodium. The Na_2SO_4-water binary phase system can be described by the phase diagram as shown in Figure 7.10 [19]. The regions where various stable hydrates of sodium sulfate based on the Na_2SO_4 concentration in solution and at the mixture temperature can be seen on the phase diagram. At a mixture temperature of 32.27°C and concentration of 33.13% (w/w) Na_2SO_4, a transition point exists where a stable crystalline form of Na_2SO_4 varies by becoming hydrated from Na_2SO_4 to $Na_2SO_4 \cdot 10H_2O$.

7.2.1.4.2 Phase Diagrams

The use of phase diagrams offers benefits in the crystallization process, whereby operating conditions can be determined and predictions made of theoretical amounts of solid products and liquid residual solution exist after equilibrium is achieved. On a thermodynamic phase diagram, phases present at conditions of interest with certain composition, temperature,

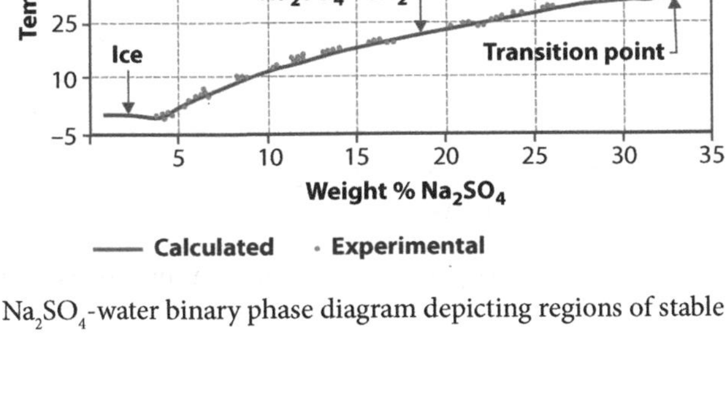

Figure 7.10 Na_2SO_4-water binary phase diagram depicting regions of stable phases [19].

and pressure can be obtained by making use of such a diagram. In most cases, the pressure effect is insignificant and the phase diagrams are commonly formulated by making use of variables such as temperature and composition. A typical binary electrolytic aqueous system phase diagram that portrays eutectic behavior can be described as shown by Figure 7.11.

The phase diagram for multi-component ternary systems can be described by Figure 7.12, which shows a ternary system for three components, namely, A, B, and C, whereby the operating conditions can be established in order to expect certain quantities of solids. The presence of multiple solutes in a solution has an impact on the solubilities and eutectic points, making it more complex for the processing of multicomponent solutions than for binary solutions.

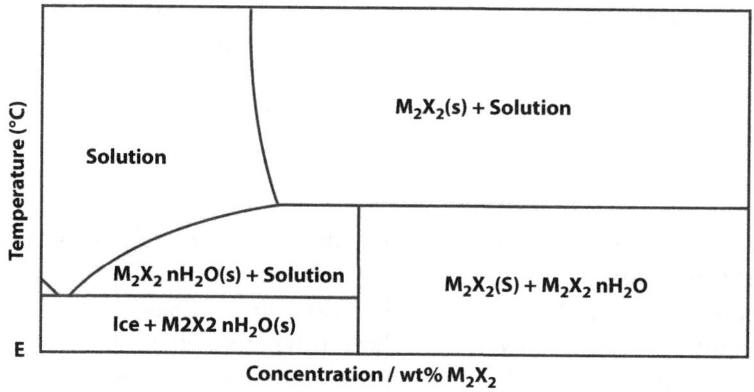

Figure 7.11 Aqueous system phase diagram [24].

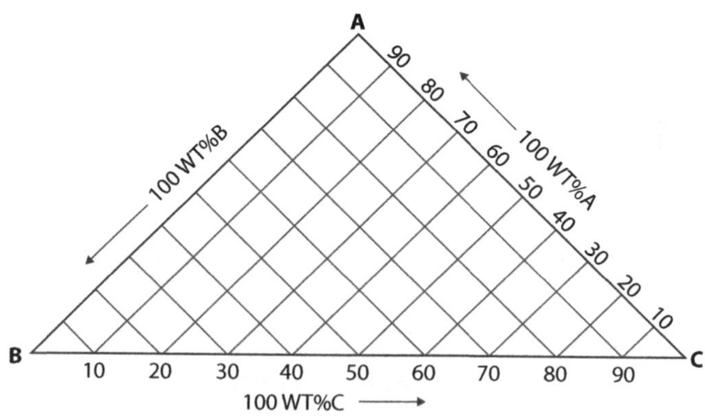

Figure 7.12 Multicomponent ternary system phase diagram consisting of component A, B, and C.

7.2.1.4.3 Metastable Zone

The region on a phase diagram that indicates supersaturation where the spontaneous nucleation has a lower chance of taking place is known as a metastable zone. In order to initiate crystal growth while operating at this region, addition of seeds promotes the growth [35]. The metastable zone can also be defined as the area between concentration solubility (solid-phase and liquid-phase thermodynamic equilibrium) and the concentration at which the first nuclei are detected (super-solubility). Thus, it involves a solution that is at supersaturation as a function of solute or solvent with no reaction that leads to formation of solid phases. However, this takes place in an area that lies between the line of solubility or line of freezing and the metastable limit. The width of this area is named the metastable zone width and is a property related to kinetics, which is dependent on the temperature of the solution in a process, rate of cooling, history of the solution, content of impurities, and mixing [26].

The metastable zone involves nucleation kinetics, and its width is a function of numerous factors associated with the crystallization process such saturation concentration, existence of crystal seeds, rate of cooling, impurity content, and the level of agitation. A typical metastable zone of the crystallization process can be described by Figure 7.13. In order for an industrial scale batch crystallization process to produce desirable crystalline products, the metastable zone should be noted and followed [42].

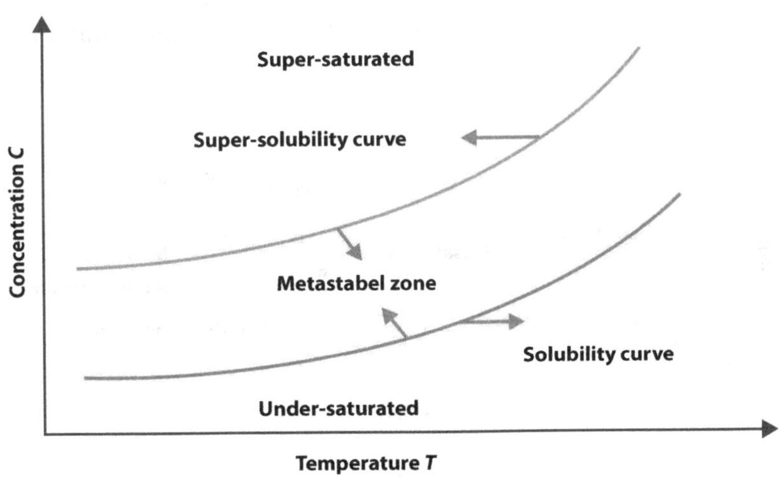

Figure 7.13 Typical metastable zone of a crystallization process [42].

The operating regions of the crystallization process that yield better separation and purified products can be recommended by making use of the metastable zone.

7.2.2 Techniques in Freeze Crystallization

The main types of freeze crystallization are indirect, direct, eutectic, and vacuum freeze crystallization. The indirect method, used on cold surfaces, is divided into progressive and suspension freeze crystallization. Direct contact freezing crystallization involves refrigerant directly interacting with the saline solution for heat exchange, whereas indirect freezing uses refrigerant without direct contact. Progressive freezing works by immersing the saline solution gradually in a tube in contact with a cold refrigerant. Vacuum freeze crystallization reduces the temperature of the saline solution by vaporizing part of the water, leading to ice formation [10, 11, 43].

Eutectic freeze crystallization allows for simultaneous recovery of salt and pure water at the eutectic point of the saltwater system. This eutectic point is the lowest temperature of crystallization for a given solution composition, where both solvent and solute crystallize from the wastewater. The eutectic composition and temperature vary depending on the salt concentration of the wastewater or brine (Figure 7.14) [24].

During freeze crystallization, the amount of ice formed from brine depends on factors like the solute concentration in the mother liquor, heat transfer coefficient, and temperature difference between the refrigerant and the saline water. This relationship is defined as $Q = m_b.Cp.dT + m_iL$, where Q represents the cooling capacity of the refrigeration unit, m_i is the

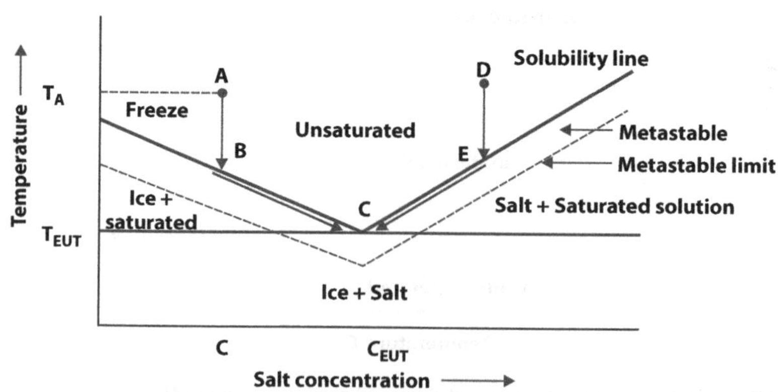

Figure 7.14 Binary system phase diagram [24].

ice mass fraction, m_b is the brine mass, L is the ice latent heat of fusion (333 kJ/kg), and dT is the temperature difference of the brine [40].

Additionally, brine's electrical conductivity is used to assess salt removal efficiency by freeze crystallization, calculated as salt removal (%) = $(C_f - C_i)/C_f \times 100$, where C_i is the initial conductivity of the brine and C_f is the conductivity of the melted ice [2]. This study explored freeze crystallization as a solution to treat FGD wastewater at Kusile Power Station.

7.2.2.1 Direct Contact Freeze-Crystallization

Direct contact freeze crystallization (Figure 7.15) uses a refrigerant that interacts directly with the solution being frozen, creating a cooling effect that leads to crystallization [10]. By controlling supersaturation, nucleation and crystal growth occur on the basis of the standard principles of freeze crystallization [29]. The refrigerant needs a boiling point below −4°C and must be non-toxic, immiscible with water, and chemically stable. Operating at around −5°C, this method consumes minimal energy, as refrigerant vaporization efficiently produces the cooling required to crystallize solutes in the process solution. The refrigerant is injected under pressure through nozzles and vaporizes upon entering the crystallizer, releasing cold energy that results in ice crystal formation.

This method offers several advantages, such as producing high-quality ice with minimal impurities by carefully controlling refrigerant mixing. Its energy efficiency is a key benefit, as it operates without moving parts, keeping power usage and maintenance costs low. The process yields a high

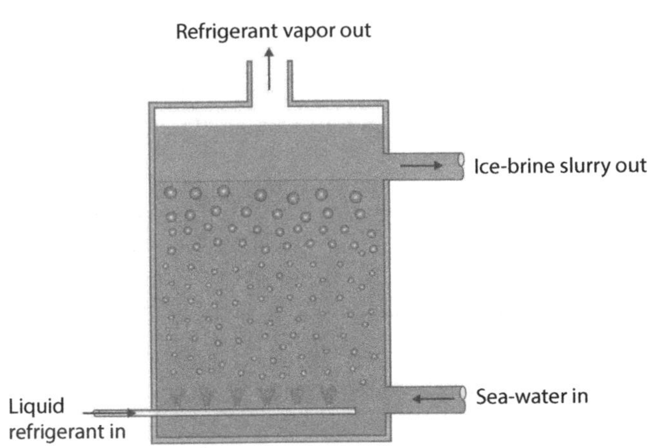

Figure 7.15 Direct freeze-crystallization process unit [15].

volume of crystals at a low driving force, making it both compact and efficient for large-scale operations. Direct contact freeze crystallization stands out for its cost-effectiveness, requiring simple equipment while providing significant results in crystal production and brine purification [44].

7.2.2.2 *Indirect Contact Freeze Crystallization*

In indirect contact freeze crystallization, the saline wastewater being treated remains separate from the refrigerant, and ice crystals are formed via mechanical refrigeration or other cooling methods [15]. This process uses physical barriers, such as heat exchanger walls, to transfer energy from the refrigerant to the wastewater, facilitating ice formation on the surfaces. The refrigerant cools the saline water through the heat exchanger walls, with ice crystal growth occurring on these cooled surfaces [45]. The general process flow is outlined in Figure 7.16, where seawater is pumped through a heat exchanger, cooled in a freezing chamber, and later separated into ice and saline wastewater. The ice is melted in a melter using heat from the compressed refrigerant, whereas the remaining saline wastewater is returned to the system [42].

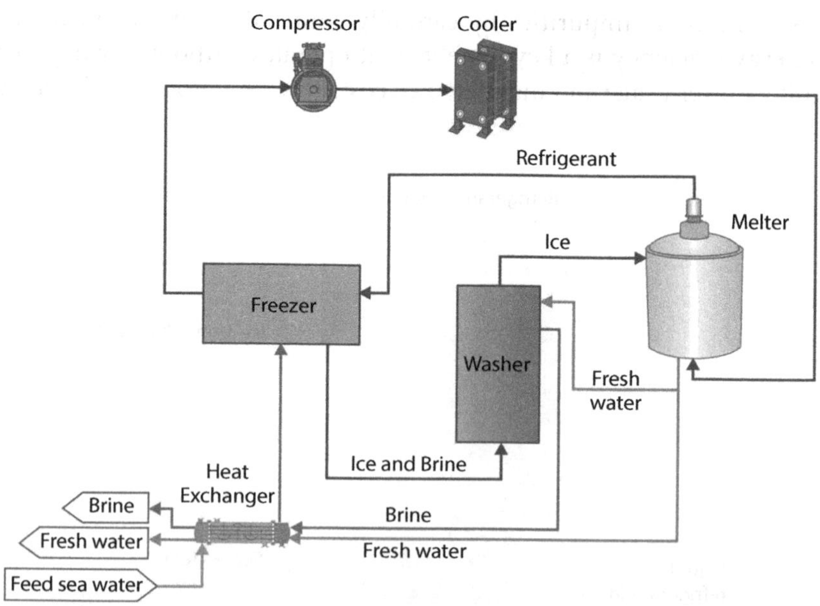

Figure 7.16 Indirect freeze-crystallization process [42].

This method can be categorized into suspension freeze crystallization and progressive freeze crystallization. Suspension freeze crystallization, often used in food processing, generates ice crystals in stages and is characterized by isothermal growth, although it has disadvantages like limited crystal size and lower purity [17]. Progressive freeze crystallization forms a single layer of large ice crystals on cooled surfaces, enabling easier separation and higher purity, although it is less productive than suspension freeze crystallization. While progressive freeze crystallization produces purer ice, the entrapment of impurities in the crystal structure poses a challenge [46].

7.2.2.3 *Vacuum Freeze Desalination*

During the freezing process, cooled saline water begins to freeze, causing the dissolved salts to be excluded from the forming ice crystals, as illustrated in Figure 7.17. The cooled saline water is then sprayed into a vacuum chamber at a pressure of 0.4 kPa, where some of the water evaporates as vapor. This evaporation further removes heat, prompting the formation of ice crystals. The ice that forms floats atop the saline wastewater, allowing for washing with fresh water and subsequent melting, resulting in fresh water being discharged from the ice melter, as noted by Clayton [47].

The vacuum freeze-crystallization process comprises four main components: the freezer, washer, melter, and heat removal system. In the freezer,

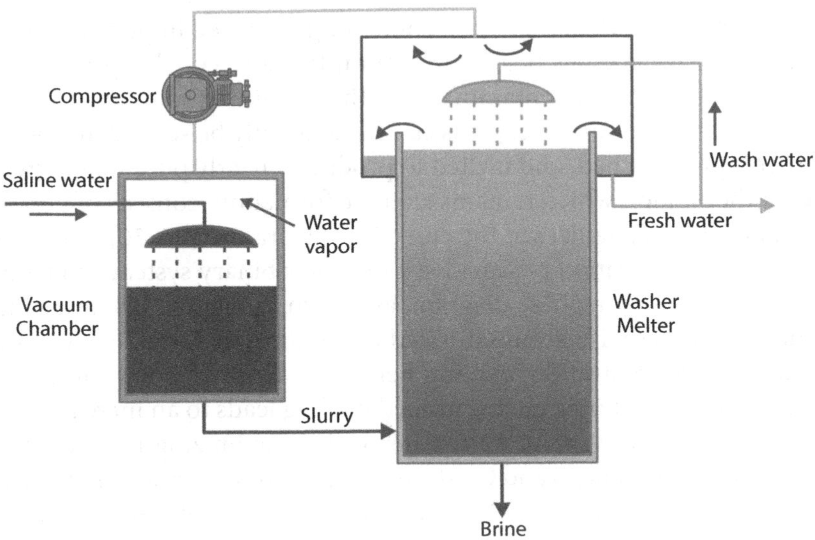

Figure 7.17 Vacuum freeze desalination [47].

ice crystals and vapor are formed simultaneously. The heat generated during the freezing process is typically directed to the melter, where it aids in melting the ice crystals. The size of the ice crystals is crucial, as smaller crystals complicate the washing process. The formed ice crystals are pumped as a slurry into the washer, where separation from the saline water occurs. In the melter, the ice from the washer is melted using heat transferred from the saline wastewater in the freezer, facilitated by discharging refrigerant into the melter, which promotes effective heat exchange [48].

Processes for vacuum freezing have been developed as a function of vacuum freezing vapor compression. This includes processes such as vacuum-freezing ejector absorption, vacuum freezing vapor absorption, vacuum-freezing multiple phase transformation, and vacuum-freeze high pressure ice-melting [42].

7.2.2.4 Eutectic Freeze Crystallization

Eutectic freeze crystallization is a technique used to segregate organic or inorganic aqueous solutions into solidified ice and solutes by operating at the eutectic point [34, 49]. The eutectic point is a specific point on a phase diagram of a saltwater solution where equilibrium is maintained among ice crystals, salt, and a solution with a defined concentration. At this point, the corresponding concentration and temperature are referred to as the eutectic concentration and eutectic temperature, respectively.

The fundamental mechanism of eutectic freeze crystallization relies on the principle that, when a solution containing dissolved impurities is gradually frozen, ice crystals begin to form on the surface. This process concentrates the dissolved contaminants in the remaining solution, known as the mother liquor. The ice crystals can subsequently be separated from the mother liquor, washed, and melted to produce a nearly pure water stream. The mother liquor, which remains after ice formation, consists of pure salt and water that crystallizes at the eutectic temperature [20, 50]. The eutectic freeze-crystallization process for a saltwater binary system can be better illustrated using a phase diagram, as shown in Figure 7.18. In a binary saline wastewater with an unsaturated concentration A that is lower than the eutectic concentration, cooling below the freezing point initiates ice crystal formation at point B. Continued cooling leads to an increase in salt concentration in the saline wastewater while maximizing ice crystal formation between points B and C. At the eutectic point, further cooling leads to saturation of salt concentration, resulting in the simultaneous formation of both ice and salt.

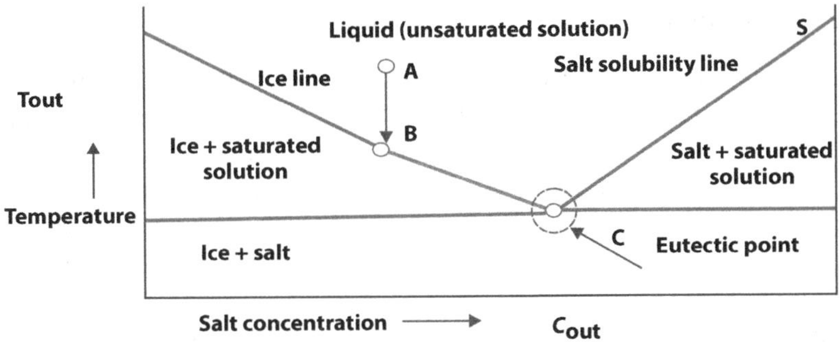

Figure 7.18 Saltwater mixture phase diagram [15].

Figure 7.19 shows a binary eutectic phase diagram for an aqueous Na_2SO_4 system. In a batch process, cooling the dilute solution from point A to point B results in ice crystallization. Further cooling concentrates the solution until it reaches point C, the eutectic point. Maintaining a slight level of supersaturation in the solution is essential to initiate crystallization, allowing the cooling curve to remain just below the equilibrium line. As the solution becomes more concentrated, cooling from point D to point E will lead to sodium sulfate dihydrate crystallization, continuing until point C is reached. Operating at the eutectic point enables simultaneous crystallization of ice and salt as long as the concentration of the residual solution remains uniform.

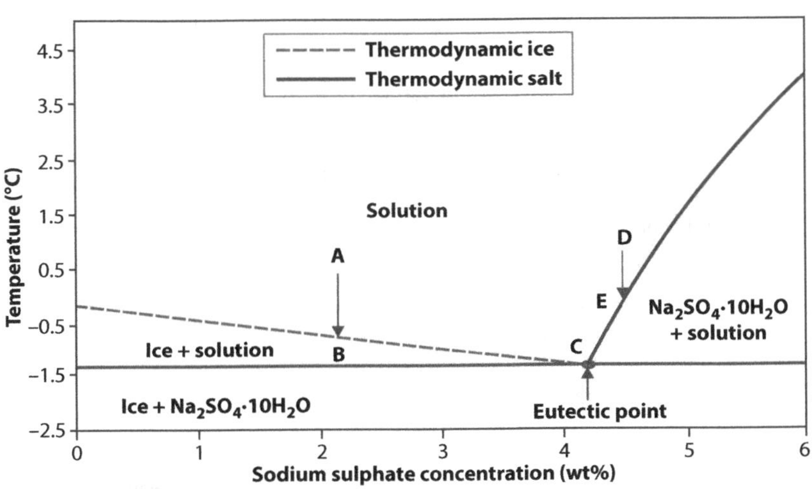

Figure 7.19 Phase diagram for a binary aqueous sodium sulfate system [19].

7.2.2.5 *Secondary Refrigerant Freezing*

In secondary refrigerant freezing process, saline water is in contact with an immiscible refrigerant. Formation of ice crystals occurs as a result of insoluble refrigerant evaporation. Butane with a boiling point of –0.5°C at atmospheric pressure, relatively near the freezing point of water, has been used as the first hydrocarbon refrigerant in this freezing process. The operating principle of the secondary refrigerant freezing process is described by Figure 7.20 [51].

The comparison of costs associated with saline wastewater treatment by the freeze-desalination and evaporation-desalination techniques, as outlined in the previous sections, without recovery of energy, is shown in Table 7.2.

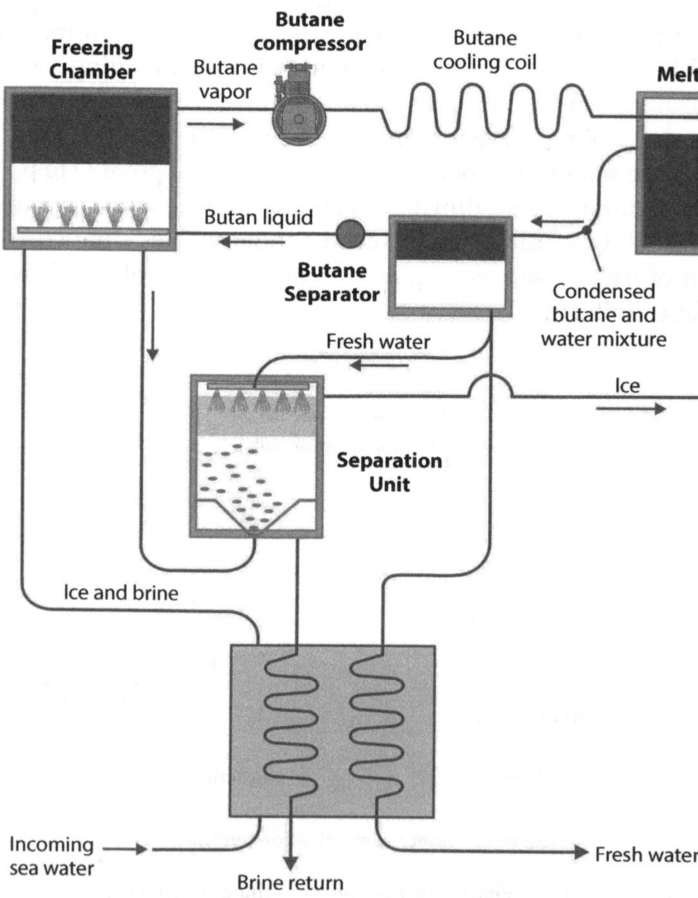

Figure 7.20 Process flow diagram of the secondary refrigerant freezing process [51].

Table 7.2 Comparison of freeze crystallization and evaporation without recovery of energy [40]. Temp, temperature.

Parameter	Freeze	Evaporate
Change in Temp. *d*T	25.0	75.0
Energy to reduce Temp. to 0°C (MJ/m³)	104.50	313.50
Energy to freeze or evaporate 1 m³ of water (MJ/m³)	333	2,257.0
Energy to freeze or evaporate 1 m³ of water (kWh/m³)	121.53	714.03
Price (R/kWh)	0.30	0.30
Cost to freeze or desalinate 1 m³ (R/m³)	36.46	214.21
Cost ratio	1	5.90

7.2.2.6 HybridICE Freeze-Crystallization Technology

The HybridICE technology is a type of freeze-crystallization technique that uses the phase change of water from liquid to solid to recover fresh water from saline sources through continuous crystallization on a cooled scrapped surface [40, 41]. Typical equipment for the HybridICE freeze-crystallization process is illustrated in Figure 7.21.

7.2.2.6.1 Description of the Process

The HybridICE unit is an innovative freeze-crystallization technology designed to efficiently recover fresh water from saline sources by leveraging the phase change of water from liquid to solid. At the core of this process is a refrigeration unit that recycles a primary refrigerant, specifically R404A, between the refrigeration unit and the scraped surface heat exchanger (SSHE). In the SSHE, saline wastewater undergoes indirect cooling, allowing for the formation of ice crystals that yield pure water. The continuous circulation of saline wastewater through a closed circuit

Figure 7.21 HybridICE freeze-crystallization pilot plant at TUT [40, 41].

enhances the overall efficiency of the system, facilitating the steady recovery of pure water in the form of ice crystals from the concentrated saline solution.

The configuration of the SSHEs is important to the process, consisting of two units operating in series to enhance nucleation and crystal ripening in the first and second heat exchangers, respectively. Each SSHE operates at distinct temperatures to promote the growth of ice crystals effectively, with the second heat exchanger maintained at a lower operating temperature than the first. The first heat exchanger uses a counter-current setup, whereas the second makes use of a parallel flow design. Once formed, a slurry composed of ice and saline wastewater, characterized by a known ice fraction, is transferred via a centrifugal pump to an ice filter. Notably, this separation process does not involve the addition of fresh water for washing the ice crystals.

Within the filter, the slurry experiences recrystallization, flotation, and agglomeration of ice crystals at the surface of the mother liquor. As ice continues to accumulate at the ice-saline wastewater interface, a column of ice cake develops. The attachment of saline wastewater to the ice crystals generates interfacial tension, which is mitigated through suction induced by the venture effect in the suction pipe and drainage through gravity. Consequently, the ice cake becomes progressively drier as it grows. The interplay of recrystallization, suction, and gravity facilitates the release of interfacial tension, accelerating the drying process.

An important operational parameter is the size of the ice crystals formed, as this directly affects the drying efficiency. The high-purity ice crystals produced by the HybridICE technology can be obtained without the need for fresh water, making the system both simple and compact. Furthermore,

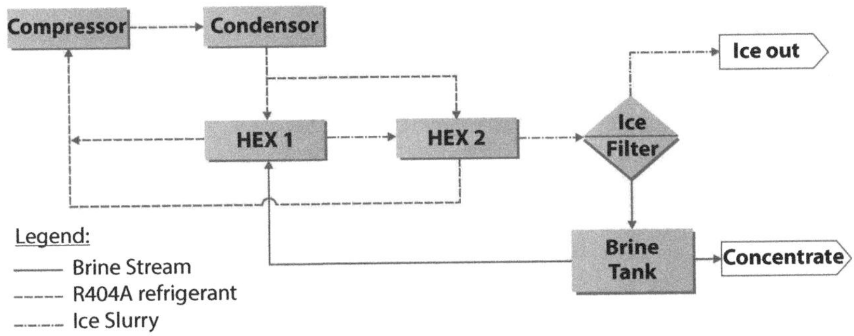

Figure 7.22 Modular block diagram for HybridICE™ desalination plant [40, 41].

the slurry of fine ice crystals generated during the process provides an efficient means of harnessing cooling energy. Figure 7.22 illustrates the block diagram of the HybridICE desalination plant, highlighting its functional components and overall design.

7.2.2.6.2 Ice Filtration Mechanisms

A key challenge in freeze crystallization is the effective separation of ice crystals from the generated ice slurry. This separation process is facilitated by the HybridICE filter, which allows for the transfer of ice slurry, characterized by a known ice fraction, through a centrifugal pump. Upon reaching the larger space within the filter, processes such as recrystallization, flotation, and agglomeration occur at the ice-saline wastewater interface, leading to the growth of an ice column. Factors like impurity concentration and microstructural locations can influence the recrystallization and diffusion processes of the ice [52].

As the ice cake continues to grow, it becomes drier due to suction from the Venturi effect created by the circulating concentrate, whereas gravity drainage enhances the drying process, as modeled by Henry Darcy [53]. The dried ice is subsequently scraped off, with the HybridICE filter serving as a replacement for traditional ice growth tanks. The unique design of this filter allows for the removal of interstitial saline wastewater without the need for washing, aided by larger ice crystals formed through recrystallization and agglomeration. Figure 7.23 illustrates the schematic of the HybridICE slurry filter, which produces pure ice without washing.

Ice slurries containing different ice fractions can be processed using this HybridICE filter that ultimately makes it more applicable for continuous

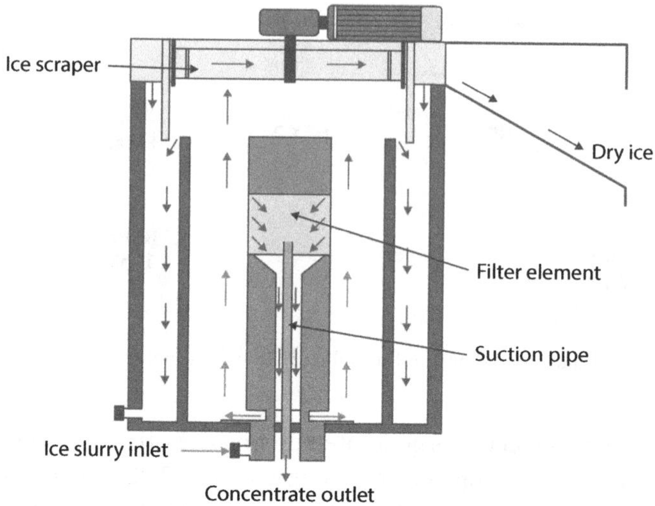

Figure 7.23 Schematic diagram of HybridICE slurry filter [40, 41].

processes where parameters such as feed temperature and concentration change significantly.

Two variations of the HybridICE filter have been developed, differing in filter element design, slurry feeding mode, and separation mechanisms. These include gravitational and hydraulic filters, which continuously form an ice column around the filtering element from its perforations [2]. Maree *et al.* [2] further indicated that the extent of salt removal, ice fraction, and percentage yield can be calculated as follows:

$$\%\ \text{Salt removal} = \frac{\left(C_f - C_i\right)}{C_f} \times 100 \tag{7.17}$$

where C_f is the initial EC of the saline wastewater being fed and C_i is the melted ice EC.

$$\text{Ice fraction} = \frac{m_{ice}}{m_{slurry}} \tag{7.18}$$

where m_{ice} is the mass of the ice and m_{slurry} is the mass of the slurry.

$$\% \text{ yield} = \frac{m_{\text{ice per min}}}{\text{Ice fraction} \times \dot{m}_{slurry}} \times 100 \quad (7.19)$$

where $m_{\text{ice per min}}$ is the mass of ice produced per minute and $\dot{m}_{\text{slurry}}$ is the mass flow rate of the slurry

7.2.2.7 *Heat Pump*

The refrigeration unit that is the heat pump extract heat from saline wastewater and also release it on saline wastewater stream by making use of evaporation or condensation of the refrigerant, respectively. The solute's concentration (freezing point depression) in the solution, coefficient of heat transfer, and temperature difference between the refrigerant and saline wastewater affects the amount of ice that is formed during the process as described by Equation 7.20.

$$Q = \dot{m}_b \dot{C}_p dT + \dot{m}_i H_f \quad (7.20)$$

where Q is the refrigeration unit cooling capacity, $\dot{m}_i$ is the ice mass fraction in the slurry, $\dot{m}_b$ is the saline wastewater mass flow rate, H_f is the ice latent heat of fusion (333 kJ/kg), and dT is the saline wastewater rate of temperature change.

Figure 7.24 describes the heat pump working principle, where evaporation of the refrigerant takes place at lower absolute temperature, T_e, in the evaporator associated with extraction of amount of heat, Q_e, from the source of heat. The existing gaseous refrigerant undergoes compression, and the condensation process in the condenser liberates an amount of heat, Q_c, at higher absolute temperature, T_c. The energy balance can be expressed by Equation 7.21.

$$Q_c = Q_e + W \quad (7.21)$$

where W is the input electrical energy required to drive the compressor. Based on Carnot efficiency, the efficiency of the compressor-driven heat pump can be expressed as a function of coefficient of performance (COP) as defined by Equation 7.22.

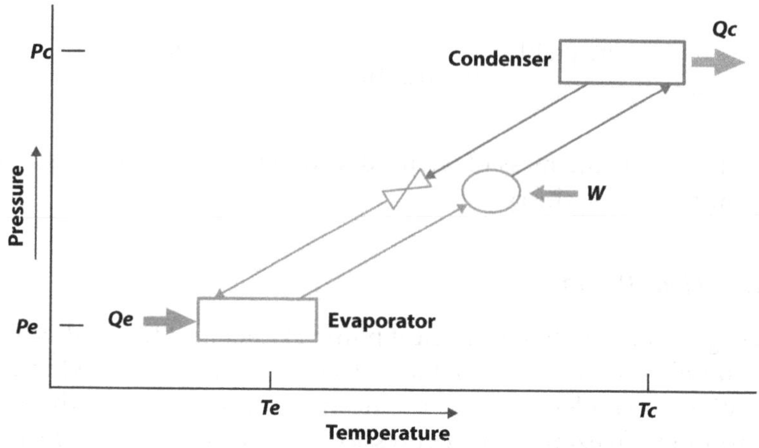

Figure 7.24 Mechanical vapor compression heat pump temperature-pressure plot [40, 41].

$$COP_{TC} = \frac{T_c}{T_c - T_e} \tag{7.22}$$

where COP_{TC} is the highest obtainable performance factor for the refrigeration system, T_c is the higher absolute condensation temperature for gaseous refrigerant, and T_e is the lower absolute evaporative temperature of the refrigerant.

From Equation 7.22, it is seen that the coefficient of performance depends strongly on the temperature difference between heat source and heat sink. Low first-ice points are mostly experienced with saline wastewater of high TDS, and this results in more work being done by the compressor in transporting the heat energy.

7.3 Water Treatment with the Integrated Power Plant

7.3.1 Block Diagram of Power Station

The Kusile Power Plant was designed to generate 4,800 MW (Figure 7.25). The diagram shows the following stages: (i) steam generation by burning coal in air in the burning chamber; (ii) transportation of dry ash (14.2% of the coal) to the Ash Dump; (ii) removal of bottom ash (15%) from the boiler; (iii) electrostatic precipitator (ESP) for removal of solid particles

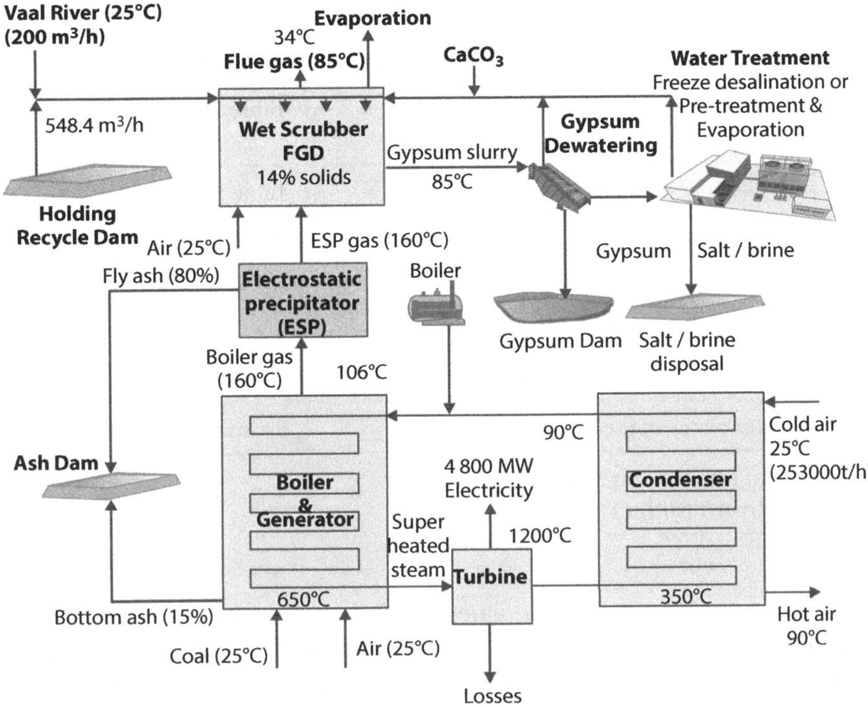

Figure 7.25 Schematic diagram of the Kusile power plant station [9].

from the flue gas scrubber; (iv) cyclones for removal of fly ash before/in/after the ESP; (v) wet scrubber after the ESP for removal of SO_2 as gypsum, where flue gas is contacted with water, $CaCO_3$, and air and where a small portion (5%) fly ash is trapped; (vi) dewatering of the gypsum slurry (15% $CaSO_4{\cdot}2H_2O$); (vii) water treatment plant where wastewater is treated with Na_2CO_3 to allow $CaCO_3$ crystallization, followed by evaporation; (viii) boiler-turbine-condenser-boiler where water/steam is recycled to generate power; and (ix) condenser where saturated steam is condensed by dry cooling.

7.3.2 Water Losses

Water losses at Kusile Power Station are significant, with approximately 681.6 m³/h lost, which is replaced by drawing water from the Osuthu Dam located on the Vaal River. Table 7.3 provides a detailed breakdown of these losses. The IX plant is responsible for producing boiler feed water by treating 156 m³/h of Vaal River water. In the FGD plant scrubber, evaporation

Table 7.3 Water losses at Kusile [9].

Process stages	Flow rate (m^3/h)	Comment
Evaporation on FGD plant	491.6	Table 5.4
Gypsum crystal water	22.4	Table 5.4
Disposal by EnviroServ	6.6	Pers.com., E Ramabina, 2021
FGD intake water	520.6	
Ash conditioning	5.0	Estimate
Boiler water make-up	156.0	Estimate
Total intake from Vaal River/Osothu Dam	681.6	

losses account for 491.6 m^3/h, whereas 22.4 m^3/h of water contains gypsum, with 6.6 m^3/h being disposed of by EnviroServ. Additionally, an estimated 5 m^3/h is used for ash conditioning and dust suppression. The boiler feed water make-up rate is approximately 156 m^3/h, helping to balance the water loss.

Wastewater produced by the FGD plant is managed by disposing of it at a waste disposal site, incurring a cost of R2,000/t. Between 1 January 2019 and 30 April 2019, a total of 9.67 ml of wastewater was generated, which translates to a rate of 3.36 m^3/h for three operational units, or 6.72 m^3/h for 6 units. The total disposal cost for this period reached R9.67 million per month, as shown in Table 7.1. This ongoing water loss and wastewater disposal highlights the significant operational and environmental costs associated with maintaining the plant's water requirements and disposal activities.

7.3.3 Coal Usage and Ash Production

Tau *et al.* (2022) calculated the chemical composition of the FGD wastewater from the chemical composition and load of the coal and the limestone (Table 7.4).

For SO_2 removal, 62.29 t/h of limestone (as $CaCO_3$) was required. The large volume of flue gas results from the 80% nitrogen content in the air.

Table 7.4 Typical South African coal analysis.

Parameter	Value
Proximate Analyses	
Fixed C (%)	56.6
Volatile material (%)	27.1
Moisture content (%)	2.1
Ash (%)	14.2
Ultimate Analyses	
C (%)	67.5
H (%)	4.26
N (%)	1.76
SO_2 as S (%)	1.00
O_2 (by difference) (%)	9.18
Gross calorific value (MJ/kg)	27.37

Ash production was divided between 15% bottom ash and 85% fly ash, with a small amount of fly ash entering the scrubber due to inefficiencies. Ash was transported in a dry form, with minimal water used for dust suppression.

Absorption loads of SO_2, Cl, and B in the FGD scrubber resulting from coal combustion are examined, whereas the dissolution loads of Ca, Mg, Na, and K from limestone reacting with SO_2 in the FGD plant are also explored. This study also estimates the mass of gypsum crystallizing due to limited solubility and the composition of FGD wastewater requiring treatment. The air volume needed to cool condensate at Kusile (a dry power station) and water consumption for cooling condensate in a wet power station are examined. In addition, 39.9 t/h of SO_2 was produced along with 17 643 t/h of flue gas. If the flue gas temperature from the combustion chamber was 160°C and decreased to 85°C after passing through the FGD scrubber, then the flue gas heat dropped by 385.95 MW, calculated as follows: 17,643 × 1,000 / 3,600 × 1.05 × (160 – 85) / 1,000. After the fly ash was removed in the ESP, the gas was passed through the FGD plant to remove SO_2. During scrubbing, 491.6 m^3/h of water evaporated, which corresponds with the flow of 548.4 m^3/h (13,162 m^3/d) from the Holding Recycle Dam to compensate for water losses due to evaporation and the water in the gypsum sludge (54.6 m^3/h).

7.3.4 Quality of FGD Wastewater

The chemical composition of FGD wastewater, taking into account both the mass and chemical makeup of the coal, as well as the flow rate and chemical composition of the intake water, was calculated. These calculated values align closely with the actual chemical composition of the wastewater.

7.4 Treatment Options

Tau *et al.* [9] compared the treatment cost of evaporation with that of freeze crystallization. A spreadsheet was used to calculate the following parameters for the two treatment options: coal usage, water usage, gas flow, chemical usage, and wastewater chemical composition from FGD plant.

7.4.1 Evaporation

The Kusile Power Station made provision for a treatment plant for the wastewater from the FGD plant. The treatment plant includes the the following stages: pre-treatment, evaporation of wastewater, and crystallization [5]. Calcium and magnesium were needed to be removed in the pre-treatment stage to protect the evaporation system from chemical scaling. Lime [$Ca(OH)_2$] and soda ash (Na_2CO_3) with pH adjustment, to between 9 and 10, would be used for water softening. The softened water flows into the evaporation process or brine concentrator where the temperature is increased to about 96°C and further processed in the falling-film type evaporator. Commissioning was planned for December 2021. OLI software simulations by Tau *et al.* [9] showed that Mg^{2+} and Ca^{2+} can be removed by dosing Na_2CO_3. With a dosage of 25 195 mg/L, the TDS was increased from the 41,818 to 42,773 mg/L due to Na^+ addition, precipitation of Mn^{2+} as MnO_2, and partial removal of Mg^{2+} as $Mg(OH)_2$. At a price of R5 000/t Na_2CO_3 cost amounts to R125.98/m^3 (Table 7.5).

When NaOH is used for removal Mg^{2+} and Ca^{2+}, a NaOH dosage of 16,651 mg/L was needed. The TDS was increased from the 41,862 mg/L

Table 7.5 Treatment of FGD effluent with Na_2CO_3 [9].

Na2CO3 (mg/L)	**pH**	**Mn(+2) Aq (mg/L) (Y2)**	**Mg(+2) Aq (mg/L) (Y2)**	**Ca(+2) Aq (mg/L) (Y2)**	**Na(+1) Aq (mg/L) (Y2)**	**TDS (mg/L)**	**R/m³**
0	6.4	5,823	3,340	578	3,257	41,862	0.00
3,176	7.4	4,174	3,339	563	4,634	41,527	15.88
6,342	7.6	2,525	3,337	548	6,011	41,190	31.71
9,500	8.0	878	3,336	533	7,386	40,855	47.50
12,650	9.2	3	3,331	99	8,759	41,186	63.25
15,796	9.2	0	2,951	3	10,136	43,136	78.98
18,935	9.3	0	2,258	3	11,524	43,966	94.67
22,067	9.3	0	1,590	2	12,915	44,910	110.34
25,195	**9.5**	**0**	**989**	**1**	**14,306**	**46,137**	**125.98**
28,323	9.6	0	560	1	15,693	48,039	141.62
31,455	9.7	0	357	1	17,073	50,779	157.27

Temp. = 25°C; Na_2CO_3 = R5,000

to 46,237 mg/L due to Na^+ addition, precipitation of Mn^{2+} as MnO_2, and partial removal of Mg^{2+} as $Mg(OH)_2$. At a price of R9,000/t NaOH, the chemical cost amounted to R149.86/m^3 (Table 7.6).

The electricity cost for evaporation would amount to R4.10 million/month (at an electricity price of R1/kWh) and Na_2CO_3 cost to R7.31 million/month. The total chemical and electricity cost would amount to R11.41 million/month.

The total treatment cost of pre-treatment plus evaporation amounts to R2,077/m^3. This is similar to the cost of waste disposal.

Table 7.6 Treatment of FGD effluent with NaOH [9].

NaOH (mg/L)	pH	Mn(+2) Aq (mg/L) (Y2)	Mg(+2) Aq (mg/L) (Y2)	Ca(+2) Aq (mg/L) (Y2)	Na(+1) Aq (mg/L) (Y2)	TDS (mg/L)	R/m^3
0	6.4	5 823	3,340	578	3,257	21,126	0.00
2,397	8.7	4 181	3,339	562	4,634	20,862	21.57
4,788	8.8	2 531	3,337	547	6,011	20,590	43.09
7,172	9.0	882	3,336	533	7,386	20,319	64.55
9,551	9.2	404	2,818	524	8,759	20,703	85.96
11,923	9.3	306	2,131	515	10,131	21,300	107.31
14,290	9.3	209	1,446	507	11,501	21,896	128.61
16,651	**9.5**	**111**	**762**	**497**	**12,870**	**22,491**	**149.86**
19,006	10.0	12	79	488	14,238	23,085	171.05
21,376	12.6	0	0	599	15,607	24,493	192.38
23,741	12.8	0	0	299	16,977	25,582	213.67

Temp. = 25°C; NaOH = R9.000/t

7.4.2 Freeze Crystallization

Tau *et al.* [9] conducted a study aimed at assessing freeze crystallization as an alternative to evaporation for treatment processes. Freeze crystallization involves several key steps: nucleation, crystal growth, separation, and melting. Crystallization methods include direct contact with a cooling agent, indirect contact, and vacuum operations [3]. These methods result in ice formation on cold surfaces, which then must be removed. ROC Water Technologies designed a system to prevent ice from adhering to these surfaces. Control of the process relies on parameters such as cooling rate, temperature, refrigerant, and flow rate. As heat is withdrawn from brine to reach its freezing point, ice forms while impurities are excluded, with the ice closely resembling pure water [10]. Energy demands for evaporation are seven times higher than for freezing, which explains the growing interest in freeze crystallization [3]. The heat of fusion for ice is 333 kJ/kg, significantly lower than water's heat of evaporation at 2,260 kJ/kg.

7.4.2.1 *Ice Crystals and Concentrate Recovery*

Clean ice was successfully produced in the freeze-crystallization unit and was able to melt under ambient conditions. Ice began to form at a temperature of −1°C, following the processing of feed wastewater through the unit. It took an hour for the cooling process and another hour for the ice to form. Once formed, the ice crystals were left on a filter for 30 min to allow dewatering before removal (Figure 7.26). The initial feed water had a TDS concentration of 4,820 mg/L, whereas the ice produced after 7 h had a TDS of 1 003 mg/L, and the brine had a TDS of 6 392 mg/L (Table 7.7). Over the 7-h operation, the purity of the ice improved, reducing from 4,263 mg/L TDS to 1,003 mg/L. In total, 49.46 kg of ice was produced over a 9-h period (Table 7.7). Continued concentration of the brine will lead to the crystallization of Na2SO4·10H2O, as the solubility of Na_2SO_4 at 0°C is limited to 45 g/L. NaCl will crystallize once it reaches a concentration of 350 g/L if present in the feed water.

Figure 7.26 Ice crystals formation during freeze crystallization.

Table 7.7 Water quality of feed, brine, and melted ice samples [9].

Sample	Sample	Parameter			
Name	**Number**	**pH**	**TDS (mg/L)**	**Electrical conductivity (mS/cm)**	**Amount of ice (kg)**
Feed	1.00	7.17	4,820	7.59	
Concentrate	1.00	8.13	5,585	11.24	
	2.00	8.07	5,591	11.57	
	3.00	7.85	5,958	11.94	
	4.00	7.73	6,115	12.27	
	5.00	7.67	6,296	12.62	
	6.00	7.26	6,392	12.72	
Melted Ice	1.00	7.96	4,263	8.53	
	2.00	7.98	2,984	6.01	14.27
	3.00	8.02	1,573	3.15	
4.00 8.07			1,472	2.96	14.46
	5.00	8.12	1,318	2.66	
	6.00	8.30	1,003	2.02	20.73
Total					49.46

Feed volume = 117 L

Energy utilization is an important consideration in the selection of the treatment method. Table 7.8 shows the brine temperature and actual and theoretical energy utilization. On average, the COP (coefficient of performance) for cooling from 21°C to −1°C was 0.86 and for ice formation 0.31. For a larger system, these COP values will be higher than 2 and 1, respectively, due to less heat losses.

Table 7.9 shows how the chemical composition of FGD wastewater changed as water was removed stepwise through ice formation. $CaSO_4{\cdot}2H_2O$ crystallized out first, up to 76% removal of water. $Na_2SO_4{\cdot}10H_2O$ removal was only observed after 86% removal of water.

Table 7.8 Process control data for freeze crystallization [9].

Time	Temp	Energy	Accumulated energy	Ice recovery	Cooling		COP
					E used (kWh/t) (actual)	E utilized (theory)	
h	°C	kWh	kWh	kg	kWh/t	kWh/t	
0.00	21		0				
0.50	7	2.53	2.53		21.6	16.3	0.75
1.00	2	3.00	5.53		6.0	5.8	0.97
1.50	−1	1.30	6.83		4.0	3.5	0.87
			Average				0.86
					Ice formation (kWh/t)		
2.00	−1		6.83				
3.50	−1	4.28	11.11	14.27	299.9	91.7	0.31
6.00	−1	4.33	15.44	14.46	299.4	91.7	0.31
9.00	−1	6.19	21.63	20.73	298.6	91.7	0.31
			Average		299.3	91.7	0.31
			Total	49.46			

Chiller temp. = −7°C; volume = 117 L

Table 7.10 outlines the costs involved in treating FGD effluent using freeze crystallization. Unlike other methods, no pre-treatment with Na_2CO_3 was necessary, and the primary expense was electricity for the chiller. The electricity cost was calculated at R821,772 per month, based on achieving 86.4% water recovery and 90.0% salt rejection. The TDS of the melted ice came out to 4,840 mg/L, whereas the brine had a TDS of 225,073 mg/L. Additionally, 2.68 g of $Na_2SO_4{\cdot}10H_2O$/kg of water was produced as a solid byproduct. The total cost of the freeze-crystallization process was estimated at R958 388 per month, equating to R198.22/m^3 of water.

Table 7.9 Treatment of FGD effluent without chemical pre-treatment (OLI simulation) [9].

% Water removal	H2O (mg)	pH	S(+6) Aq (mg) (Y2)	TDS (mg)	TDS (mg/L)	CaSO4.2H2O (Gypsum) (mg) (Y2)	Na2SO4.10H2O (Mirabilite) (mg) (Y2)
0.00	1,000,000	6.15	20,432	41,312	41,318	956	0
9.60	904,000	6.12	20,337	41,178	45,560	1,125	0
19.20	808,000	6.08	20,243	41,044	50,810	1,294	0
28.80	712,000	6.04	20,148	40,910	57,478	1,463	0
38.40	616,000	6.00	20,054	40,776	66,225	1,633	0
48.00	520,000	5.94	19,958	40,640	78,203	1,804	0
57.60	424,000	5.87	19,860	40,501	95,603	1,980	0
67.20	328,000	5.77	19,757	40,356	123,185	2,164	0
76.80	232,000	5.62	19,644	40,195	173,583	2,367	0
86.40	**136,000**	**5.34**	**18,714**	**38,828**	**289,738**	**2,597**	**2,684**

Temp. = 0°C

Table 7.10 Treatment cost of freeze crystallization.

Parameter	**Feed**	**Freeze crystallization**		
		Melted ice	**Brine**	**Solids**
Salt rejection (%)		90.0		
Water recovery (%)		86.4		
Flow rate (m^3/h) (Eskom value)	6.63	5.73	0.90	

(*Continued*)

Table 7.10 Treatment cost of freeze crystallization. (*Continued*)

Parameter	Feed	Freeze crystallization		
Chemical composition				
	Mass (mg) (from originally in 1 L)			
H_2O (mg)	1,000,000	864,000	136,000	136,000
Cl^- (mg)	8,000	800	7,200	-
SO_4^{2-} (mg)	20,939	2,094	16,620	2,225
Na^+ (mg)	3,257	326	2,548	383
K^+ (mg) 44		4	40	-
Mg^{2+}	3,340	334	3,006	-
Ca^{2+} (mg)	417	42	35	340
Mn^{2+} (mg)	5,823	582	5,241	-
TDS (mg/kg)	41,821	4,182	34,690	2,949
Cation (meq)	650	65	552	34
Anion (meq)	662	66	549	46
	Concentration (mg/L)			
H_2O (mg)	1,000,000	864,000	136,000	136,000
Cl^- (mg)	8,000	926	52,941	-
SO_4^{2-} (mg)	20,939	2,424	122,206	16,363
Na^+ (mg)	3,257	377	18,735	2,817
K^+ (mg)	44	5	291	-
Mg^{2+} (mg)	3,340	387	22,103	-
Ca^{2+} (mg)	417	48	260	2,503
Mn^{2+} (mg)	5,823	674	38,537	-
TDS (mg/kg)	41,821	4,840	255,073	21,683
Cation (meq)	650	75	4,057	248
Anion (meq)	662	77	4,037	341

(*Continued*)

Table 7.10 Treatment cost of freeze crystallization. (*Continued*)

Parameter	Feed	Freeze crystallization		
Sludge composition				
$CaSO_4 \cdot 2H_2O$ mg/kg H_2O			2,597.42	
$Na_2SO_4 \cdot 10H_2O$ mg/kg H_2O			2,684.01	
Suspended solids mg/kg H_2O				
Electricity cost				
Energy needed for Freeze (m·Hv) (kW)		751		
Electricity cost (R/kWh)		1.50		
Electricity cost for cooling and freezing (R/month)		821,772		
Chemical and electricity cost (R/m³)		169.96		
Capital cost				
Capital cost [R/(ML/d)]		45,000,000		
Capital cost for this flow rate [R/(ML/d)]		10,337,864	Eq. 1	
Capital redemption cost (R/m³)		28.26		
Capital and electricity cost (R/m³)		198.22		
Capital and electricity cost (R/month)		958,388		

Note: Cap cost 1 = Cap cost 2 × (Flow 1/Flow 2)$^{0.8}$
Interest = 10%/a
Period = 120 months
Source: (Tau, *et al.*, 2022)(Tau, *et al.*, 2022)

7.5 Conclusions

The wastewater from Kusile's FGD plant had a high TDS concentration of 50,000 mg/L. This was attributed to the absorption of SO_2, Cl, and boron from the coal, as well as the dissolution of elements like Na, K, Mg, Mn, and Ca from the limestone when it reacted with the acid gases from combustion. A significant portion of $CaSO_4$ in the wastewater precipitated as gypsum due to its low solubility. The disposal cost for wastewater from a 4,800-MW plant to a hazardous waste site reached R9.7 million per month. When treated with lime or Na_2CO_3 followed by evaporation, costs rose to R11.41 million per month. In contrast, freeze crystallization, which does not require chemical pre-treatment, reduced the cost significantly to R0.96 million per month. Freeze crystallization was identified as the most cost-effective method, as it avoids chemical treatment and uses less energy than evaporation—330 kJ/kg for freezing compared to 2,260 kJ/kg for evaporation. Energy consumption for the process was 100 kWh/kg ice, with a feed TDS of 50,000 mg/L, whereas the melted ice had a TDS below 3,000 mg/L.

7.6 Recommendations

Freeze crystallization will provide a cost-effective solution for the treatment of wastewater from a FGP plant and should be demonstrated on-site at Kusile Power Station. The findings of this research work will enable the power industry to refrain from their current practices of brine or wastewater disposal in evaporation ponds, ash slurries, toxic sites, and ash dams as these industrial practices pose risk to both the environment and human life. Thus, the adoption of FGD wastewater treatment using freeze desalination other than disposal will have improved operational cost as evaporation-desalination can be energy consumptive and with freeze crystallization, some of the recovered water can be recycled in the cooling towers, fed to the reverse osmosis, thus producing demineralized water for boiler feed water used during steam generation. Other major components from wastewater such as salts, for example, calcium carbonate ($CaCO_3$), can be recovered using freeze desalination and recycled back in the industrial process feed, for the alkaline slurry generation in the wet FGD process, whereas some can be recovered as sellable products to relevant industries, thus minimizing environmental pollution.

Acknowledgements

The authors thank TIA Seed Fund, F'SAGRI, University of Limpopo, and ROC Water Technologies (PTY) Ltd. for financial support, as well as NIPMO that sponsored the grant for the innovation award of 2021 NSTF-South32 that was received by ROC Water. Dr Kelley Reynolds-Clausen, Ms Akashnie Raghu, Mr Tshitso Tamane, and Mr Evans Ramabina provided valuable information on Power Stations operations. Tshwane University of Technology made analysis service facilities.

References

1. Kramer, A., *Water use at South Africa's Coal Power Stations: a Legal Analysis [Master of Laws: Environmental Law and Governance]*, North-West University, North West, South Africa, 2022.
2. Adeniyi, A., Maree, J.P., Mbaya, R.K.K., Popoola, A.P.I., Mtombeni, T., Zvinowanda, C.M., HybridICE® filter: ice separation in freeze desalination of mine waste waters. *Water Sci. Technol.*, 69, 9, 1820–7, 2014. https://doi.org/10.2166/wst.2014.078.
3. Zikalala, N., Maree, J.P., Zvinowanda, C.M., Akinwekomi, V., Mtombeni, T., Mpenyana-Monyatsi, L., Treatment of sulphate wastewater by freeze desalination. *Desalin. Water Treat.*, 79, 93–102, 2017. https://doi.org/10.5004/dwt.2017.20927.
4. Department of Environmental Affairs, *National Environmental Management: Waste Act, 2008 (Act No. 59 of 2008): Landfill, National Norms and Standards for Disposal of Waste to Landfill*, Government of South Africa, Pretoria, Department of Water and Environmental Affairs (Ed.), 2013.
5. Pillay, L., Moodley, N., Kusile power station flue gas desulphurization wastewater treatment plant, in: *Water Institute of Southern Africa Biennial Conference*, Cape Town, South Africa, WISA (Ed.), 2012.
6. Otter Tail Power, *Otter Tail Power Company Big Stone Generating Station Unit 1 Best Available Retrofit Technology Analysis*, Burns & McDonnel, 2009.
7. Schutte, C., *Value Added Utilisation Possibilities of Coal Combustion Products in South Africa [Master of Engineering]*, North West University, Potchefstroom, South Africa, 2017.
8. Carpenter, A.M., Low water FGD technologies, in: *IEA Clean Coal Centre*, 2012.
9. Tau, A.L., Maree, J.P., Adeniyi, A., Onyango, M.S., Treatment of Flue Gas Desulphurisation Effluent with Freeze Crystallization, in: *35th International Conference on Chemical, Biological and Environmental Engineering*, Johannesburg, South Africa, IICBE (Ed.), pp. 6–7, 2022, https://doi.org/10.17758/IICBE4.C1122223.

10. El Kadi, K. and Janajreh, I., Desalination by freeze crystallization: an overview. *Int. J. Therm. Environ. Eng.*, 15, 2, 103–10, 2017. https://doi.org/10.5383/ijtee.15.02.004.
11. Janajreh, I., Zhang, H., El Kadi, K., Ghaffour, N., Freeze desalination: Current research development and future prospects. *Water Res.*, 229, 119389, 2023. https://doi.org/10.1016/j.watres.2022.119389.
12. Hendrickson, H.M. and Moulton, R.W., *Research and Development of Processes for Desalting Water by Freezing*, United States Department of the Interior, Office of Saline Water, Seattle, Washington, 1956, Report No.: 10 Contract No.: 10.
13. Lewis, A., Nathoo, J., Reddy, S., Randall, D., Zibi, L., Jivanji, R., *Novel Technology for Recovery of Water and Solid Salts from Hypersaline Brines: Eutectic Freeze Crystallization*, Water Research Commission, South Africa, Pretoria, 2008, Report No.: 2.
14. McCloskey, J.P., Karlsson, J.O., Temporally resolved imaging of ice nucleation and growth in highly supercooled water, in: *38th Annual Northeast Bioengineering Conference*, Philadelphia, USA, IEEE, IEEE (Ed.), pp. 195–6, 2012, https://doi.org/10.1109/NEBC.2012.6207030.
15. More, K.S., Maree, J.P., Mahlangu, M., Indirect Freeze Crystallization—An Emerging Technology for Valuable Resource Recovery from Wastewater. *Minerals*, 14, 4, 427, 2024. https://doi.org/10.3390/min14040427.
16. Randall, D.G. and Nathoo, J., Resource recovery by freezing: A thermodynamic comparison between a reverse osmosis brine, seawater and stored urine. *J. Water Process Eng.*, 26, 242–9, 2018. https://doi.org/10.1016/j.jwpe.2018.10.020.
17. Samsuri, S., Amran, N.A., Yahya, N., Jusoh, M., Review on progressive freeze concentration designs. *Chem. Eng. Communic.*, 203, 3, 345–63, 2016. https://doi.org/10.1080/00986445.2014.999050.
18. Zvinowanda, C.M., Zikalala, N.N., Mtombeni, T., Maree, J.P., Louw, W.J., Oosthuizen, F., Process optimization of freeze desalination of brine using HybridICETM pilot plant. *Desalin. Water Treat.*, 52, 34-36, 6315–24, 2014. https://doi.org/10.1080/19443994.2013.822175.
19. Lewis, A.E., Nathoo, J., Thomsen, K., Kramer, H.J., Witkamp, G.J., Reddy, S.T., *et al.*, Design of a Eutectic Freeze Crystallization process for multicomponent waste water stream. *Chem. Eng. Res. Des.*, 88, 9, 1290–6, 2010. https://doi.org/10.1016/j.cherd.2010.01.023.
20. Randall, D.G., Nathoo, J., Lewis, A.E., A case study for treating a reverse osmosis brine using Eutectic Freeze Crystallization—Approaching a zero waste process. *Desalination*, 266, 1, 256–62, 2011. https://doi.org/10.1016/j.desal.2010.08.034.
21. Randall, D.G., Nathoo, J., Lewis, A.E., Seeding for selective salt recovery during eutectic freeze crystallization. *International Mine Water Conference*, Pretoria, South Africa, 19-23 October, IMWA, IMWA (Ed.), pp. 639–46, 2009.

22. Dalmolen, J., *Synthesis and Application of New Chiral Amines in Dutch Resolution: Family Behaviour in Nucleation [Doctor of Philosophy]*, University of Groningen, Netherlands, 2005.
23. Giulietti, M., Seckler, M., Derenzo, S., Ré, M., Cekinski, E., Industrial crystallization and precipitation from solutions: state of the technique. *Braz. J. Chem. Eng.*, 18, 4, 423–40, 2001. https://doi.org/10.1590/S0104-66322001000400007.
24. Chivavava, J., Rodriguez-Pascual, M., Lewis, A.E., Effect of Operating Conditions on Ice Characteristics in Continuous Eutectic Freeze Crystallization. *Chem. Eng. Technol.*, 37, 8, 1314–20, 2014. https://doi.org/10.1002/ceat.201400094.
25. Löffelmann, M. and Mersmann, A., How to measure supersaturation? *Chem. Eng. Sci.*, 57, 20, 4301–10, 2002. https://doi.org/10.1016/S0009-2509(02)00347-0.
26. Himawan, C., Vaessen, R.J.C., Kramer, H.J.M., Seckler, M.M., Witkamp, G.J., Dynamic modeling and simulation of eutectic freeze crystallization. *J. Cryst. Growth*, 237-239, 2257–63, 2002. https://doi.org/10.1016/S0022-0248(01)02240-0.
27. Kubota, N., Doki, N., Yokota, M., Sato, A., Seeding policy in batch cooling crystallization. *Powder Technol.*, 121, 1, 31–8, 2001. https://doi.org/10.1016/S0032-5910(01)00371-0.
28. McGinty, J., Wheatcroft, H., Price, C.J., Sefcik, J., Modelling solution speciation to predict pH and supersaturation for design of batch and continuous organic salt crystallisation processes. *Fluid Phase Equilib.*, 565, 113676, 2023. https://doi.org/10.1016/j.fluid.2022.113676.
29. Richardson, J.F., *Coulson and Richardson's Chemical Engineering*, 1st, Elsevier, Great Britain, 2002.
30. Loï Mi Lung-Somarriba, B., Moscosa-Santillan, M., Porte, C., Delacroix, A., Effect of seeded surface area on crystal size distribution in glycine batch cooling crystallization: a seeding methodology. *J. Cryst. Growth*, 270, 3, 624–32, 2004. https://doi.org/10.1016/j.jcrysgro.2004.07.015.
31. Myerson, A.S. and Ginde, R., Handbook of Industrial Crystallization (Second Edition), in: *Crystals, Crystal Growth, and Nucleation*, A.S. Myerson (Ed.), pp. 33–65, Butterworth-Heinemann, Woburn, 2002, https://doi.org/10.1016/B978-075067012-8/50004-5.
32. Rabizadeh, T., Stawski, T.M., Morgan, D.J., Peacock, C.L., Benning, L.G., The Effects of Inorganic Additives on the Nucleation and Growth Kinetics of Calcium Sulfate Dihydrate Crystals. *Cryst. Growth Des.*, 17, 2, 582–9, 2017. https://doi.org/10.1021/acs.cgd.6b01441.
33. Jones, A.G., *Crystallization Process Systems*, Elsevier, London, 2002.
34. van der Ham, F., Witkamp, G.J., de Graauw, J., van Rosmalen, G.M., Eutectic freeze crystallization simultaneous formation and separation of two solid phases. *J. Cryst. Growth*, 198-199, 744–8, 1999. https://doi.org/10.1016/S0022-0248(98)01003-3.

35. Dhanaraj, G., Byrappa, K., Prasad, V., Dudley, M., *Springer Handbook of Crystal Growth*, p. 1818, Springer, Heidelberg, 2010, https://doi.org/10.1007/978-3-540-74761-1.
36. Mersmann, A., *Crystallization Technology Handbook*, CRC press, New York, 2001.
37. Erlbeck, L., Wössner, D., Kunz, T., Rädle, M., Methner, F.J., Investigation of freeze crystallization and ice pressing in a semi-batch process for the development of a novel single-step desalination plant. *Desalination.*, 448, 76–86, 2018. https://doi.org/10.1016/j.desal.2018.10.002.
38. Akyurt, M., Zaki, G., Habeebullah, B., Freezing phenomena in ice–water systems. *Energy Convers. Manag.*, 43, 14, 1773–89, 2002. https://doi.org/10.1016/S0196-8904(01)00129-7.
39. Mogashane, T., Maree, J., Nyamutswa, N., Vogel, J., Mujuru, M., Mphahlele-Makgwane, M., Evaluation of Forward Osmosis for Treatment of Sodium sulphate Rich Brine, in: *International Conference on Science, Engineering, Technology & Waste Management*, Johannesburg, South Africa, SETWM-20 (Ed.), pp. 69–75, 2020.
40. Mtombeni, T., *Brine Treatment Using Freeze Desalination Technology*, Tshwane University of Technology, Pretoria, 2013.
41. Mtombeni, T., Maree, J.P., Zvinowanda, C.M., Asante, J.K.O., Oosthuizen, F.S., Louw, W.J., Evaluation of the performance of a new freeze desalination technology. *Int. J. Environ. Sci. Technol.*, 10, 3, 545–50, 2013. https://doi.org/10.1007/s13762-013-0182-7.
42. Dai, H., Lu, X., Peng, Y., Yang, Z., Zhu, H., Characteristics of metastable zone in the crystallization process: A case study of sparingly soluble hydroxyapatite. *Desalin. Water Treat.*, 62, 192–9, 2017. https://doi.org/10.5004/dwt.2017.1647.
43. Kadi, K.E., Janajreh, I., High-Fidelity Simulation and Parametric Analysis of Static Indirect Freeze Desalination Process, in: *2018 6th International Renewable and Sustainable Energy Conference (IRSEC)*, Rabat, Morocco, 5-8 December, IEEE, IEEE (Ed.), pp. 1–6, 2018, https://doi.org/10.1109/IRSEC.2018.8702851.
44. Rahman, M.S., Ahmed, M., Chen, X.D., Freezing-melting process and desalination: I. Review of the state-of-the-art. *Sep. Purif. Rev.*, 35, 02, 59–96, 2006. https://doi.org/10.1080/15422110600671734.
45. Jayakody, H., Al-Dadah, R., Mahmoud, S., Cryogenic energy for indirect freeze desalination—numerical and experimental investigation. *Processes.*, 8, 1, 19, 2019. https://doi.org/10.3390/pr8010019.
46. Miyawaki, O., Kato, S., Watabe, K., Yield improvement in progressive freeze-concentration by partial melting of ice. *J. Food Eng.*, 108, 3, 377–82, 2012. https://doi.org/10.1016/j.jfoodeng.2011.09.013.
47. Clayton, R., *Desalination for Water Supply: A Review of Current Knowledge*, Foundation for Water Research, Buckinghamshire, United Kingdom, 2015.

48. Fujioka, R., Wang, L.P., Dodbiba, G., Fujita, T., Application of progressive freeze-concentration for desalination. *Desalin.*, 319, 33–7, 2013. https://doi.org/10.1016/j.desal.2013.04.005.
49. van der Ham, F., Witkamp, G.J., de Graauw, J., van Rosmalen, G.M., Eutectic freeze crystallization: Application to process streams and waste water purification. *Chem. Eng. Process Process Intensif.*, 37, 2, 207–13, 1998. https://doi.org/10.1016/S0255-2701(97)00055-X.
50. Nathoo, J., Jivanji, R., Lewis, A., Freezing your brines off: eutectic freeze crystallization for brine treatment, in: *International Mine Water Association Conference*, Pretoria, South Africa, IMWA, Water Institute of Southern Africa, International Mine Water Association (Ed.), pp. 431–7, 2009.
51. Ahmad, M.M., *Assessment of Freezing Desalination Technologies*, Swansea University, United Kingdom, 2012.
52. Oosthuizen, F.S., *Inventor Method and Apparatus for Filtering a Multi-Phase Fluid*, The German Patent and Trademark Office, Germany, 2008, Germany patent No. DE 10 2008 063 291 A1.
53. Brown, G., Henry Darcy and the making of a law. *Water Resour. Res.*, 38, 7, 11-1–2, 2002. https://doi.org/10.1029/2001WR000727.

8

Sustainabilities and Challenges of Safe Drinking Water Provisions in Low- and Middle-Income Countries

Siphelele Nduli[1,2], Memory Tekere[1], Vhahangwele Masindi[1,2*] and Beauclair Nguegang[3]

[1]Department of Environmental Science, College of Agriculture and Environmental Sciences (CAES), University of South Africa (UNISA), Florida, South Africa
[2]Magalies Water (MW), Scientific Services (SS), Research & Development (R&D) Division, Brits, South Africa
[3]Department of Chemical Sciences, Faculty of Science, University of Johannesburg, Auckland Park, Johannesburg, South Africa

Abstract

To ensure public health safety, drinking water has to be treated to the required standard as stipulated in different risk-based water quality indicators using various water treatment techniques. However, due to deteriorated raw water contaminated by various pollutants including inorganic, organic, and microbial contaminants, water treatment facilities faced a worrying problem of clean and safe water provision. This is exasperated by numerous factors such as aging pipeline materials, high chlorine demand, biofilm, climatic factors, microbial regrowth, and long water supply network. Furthermore, water leakages as part of the non-revenue water compound this problem, leading to low chlorine residual and, ultimately, microbial failures. This will eventually pose health risk to water consumers in the distribution network and further afield, unless chlorine booster stations are installed to supplement residual chlorine and cater for novel and resistant chlorine residuals. Herein, this chapter seeks to give detailed insights into different sources of raw water characteristics, treatment processes, microbial diversity, culturing methods, sequencing methods, different disinfectants used for water treatment and sterilization, and challenges thereof. In addition, the feasibilities of using booster stations to supplement residual

**Corresponding author*: emasinv@unisa.ac.za; masindivhahangwele@gmail.com

Elvis Fosso-Kankeu, Vhahangwele Masindi, Johannes Maree and Bhekie Mamba (eds.) Customized Technologies for Sustainable Management of Industrial Wastewater: A Circular Economy Approach, (311–356) © 2025 Scrivener Publishing LLC

chlorines along with recurring microbial contaminants was also unpacked. Ways to maintain chlorine residuals were also unpacked in this book chapter. Factors that underpin chlorine residuals were also discussed in detail. Future perspective and outlooks were also discussed. Insights from this comprehensive book chapter will ensure consistent supply of drinking water that is safe and suitable for human consumption in different terminal points and avert potential toxicological effects that could potentially emanate from utilizing water that is not fit for defined purposes.

Keywords: Drinking water treatment, novel and emerging aquatic contaminants, microbial contaminants, disinfections, boosters, chlorine residual

8.1 Introduction

Water quality degradation in drinking water distribution systems (DWDSs) may be due to different reasons, for example, the quality of the source water and inefficient treatment processes [1–3]. In situations where water quality deteriorates after leaving the treatment plant, researchers tend to investigate the cause of such deterioration, and microbial community composition has emerged as a potential contributor. This is due to the ability of microorganisms to survive in challenging environments and adapt accordingly in terms of metabolism and function [4, 5]. There is a significant amount of literature on microbial community composition and the influence of various physical, chemical, mechanical, and environmental factors on water quality and microbial community composition in DWDSs [6–13]. The main focus of such research has been on the role that disinfectants, nutrient availability, environmental conditions, and hydraulics play in shaping the DWDS microbiome [7, 14–17]. Determining the microbial community composition of water is essential in understanding the microbial contamination beyond microbial indicators. Such information is essential for public health as it enables the identification of pathogens that may cause harm to consumers. Furthermore, certain microorganisms have the ability to alter the physicochemical properties of water leading to the formation of hazardous chemical substances. Therefore, microbiome studies provide information that can be used to inform comprehensive water quality monitoring regulations to safeguard public health.

8.2 Sources of Surface Water Contamination

Water is a basic need and is required by all living organisms, including humans, as it is a major constituent of the human body (WHO-UNICEF, 2017). Access to clean drinking water remains a challenge for developing countries around the world, including South Africa [18]. The costs of implementing water quality treatment and distribution infrastructure, as well as the skills required for their operation and maintenance, are a well-known bottleneck to water provision [19]. Furthermore, this has been worsened by the increasing contamination of already scarce water resources such as rivers and dams [3, 20–22]. The contaminated surface water increases the costs of water treatment as more inputs are required to produce water that meets set quality standards. Agriculture, human settlements, industrialization, as well as the failure of wastewater treatment entities to treat sewage to acceptable effluent standards are the major contributors to surface water contamination [20, 21, 23].

The various surface water contamination streams exist because of population growth that has led to an increase in industrialization and urbanization [1, 21, 24]. As the world population grows, the demand for housing also increases, thus leading to the formation of new human settlements and overcrowding in already existing settlements [25, 26]. Furthermore, the population growth increases the demand and the consumption of agricultural crop and animal products, thus requiring farmers to scale-up on their production and utilise more land [27–29]. As more land is used for housing and agricultural purposes, the pollution risk to surrounding water bodies also increases. This is due to the fact that increased land-use disturbs the natural soil structure and also reduces the percentage of land occupied by vegetation, thus making soil more vulnerable to erosion and deposition into receiving surface water bodies [27–29].

Erosion of soil into surface water is undesirable as it results in murky water, which limits light penetration and, therefore, affects aquatic life that relies on light for metabolic activities [30, 31]. The soil particles tend to form sediments at receiving water bodies and these sediments serve as hotspots for concentration of pollutants [24, 31]. Furthermore, these soil particles can be mobilized and serve as transport vessels for contaminants to other water bodies and downstream users of surface water such as DWTPs [32]. Depending on the land use, contaminants mobilized to surface water with eroded soil can be of chemical, nutrient, or microbial in nature. For example, land use dominated by commercial or subsistence agricultural activities may be polluted with potentially toxic chemicals (over-application of pesticides) and increased levels of nutrients (nitrogen and phosphorus from the application of fertilizers) or microbial contaminants (from animal defecation and the application of manure) [1, 24].

Pesticide use in agriculture is beneficial to farmers as it controls pests that may have a negative impact on the quantity and the quality of produce [1, 24]. These chemical products, which include fungicides, herbicides, insecticides, nematicides, and rodenticides, are usually applied at different stages of the cropping process and contain both active and inert constituents that might be harmful to the environment [1, 24, 33]. Essentially, contact between these constituents and the environment presents a risk of metabolite formation, which, in some cases, may be more toxic that the original pesticide [1, 24]. The migration of pesticides to surface water may occur as run-off from agricultural lands caused by excessive irrigation or rainfall [24, 28].

Similar to pesticides, the use of fertilizers is of benefit to both commercial and subsistence farmers as it improves the nutrient composition of the soli, leading to better produce [28, 29]. The fertilizers can be obtained and applied as inorganic (for example, nutrient formulations of nitrogen, phosphorus, and potassium), which is commonly used by commercial farmers, or organic (for example, plant compost and animal manure) in the case of subsistence farmers [34, 35]. Coupled with the use of irrigation systems and/or sufficient rainfall, crop farmers benefit from fertilizer use and increase their yield; however, excessive irrigation also increases the risk of surface water contamination by fertilizers through run-offs [27, 28]. The risk arises from the high-water solubility of fertilizer components such as nitrates, thus making it easier for them to dissolve and migrate to receiving water bodies as run-off [28]. This risk is further exacerbated by the over-application of inorganic fertilizers as not all the applied fertilizer is utilized by the crops, leading to an accumulation of nutrients and subsequent mobilization to receiving surface waters after high rainfall or irrigation incidents [28, 29, 35].

Once in surface water, both pesticides and fertilizers have undesirable impacts on the quality and health of aquatic ecosystems. The undesirable impact may include pesticides causing problems to the ecosystem through ingestion and concentrating in tissues of aquatic species such as fish, which are later consumed by humans [1, 24, 33]. Moreover, these pesticides may increase in toxicity due to mixing and accumulation in surface water, as well as through the formation of metabolites after interaction with the environment [33, 36]. The pesticides have been detected in surface water bodies, thus making them potential hazards to humans particularly through ingestion of contaminated food or water [1, 33]. Furthermore, evidence suggests that conventional DWTPs are not 100% efficient in removing all pesticide components in finished water, a problem that is further exacerbated by the potential mixing of pesticides and the formation of metabolites, as reported removal efficiencies are often investigated using known commercial pesticides [1, 33, 36]. Therefore, the reported conventional DWTP pesticide removal efficiencies cannot be used to estimate the removal efficiencies of mixed pesticides and pesticide metabolites that form in the environment.

The fertilizers, on the other hand, may not be as toxic as pesticides; however, their ability to migrate with run-off and concentrate in surface water bodies presents a risk to aquatic life and downstream users [28]. The accumulation of nutrients in surface water, particularly phosphorus and nitrogen, results in anthropogenic eutrophication, which occurs at a more rapid rate than that caused by natural processes such as degradation of rocks [37, 38]. Eutrophication increases the rate of growth of algae and other aquatic plant species, thereby creating aesthetic and organoleptic damage among other effects to water bodies; moreover, excessive growth of algal species will result in higher depletion of oxygen when bacterial plant degradation occurs [37, 38]. The reduced levels of oxygen in water bodies negatively affect the aquatic life of organisms that depend on oxygen for their metabolism, and, as a consequence, ecosystems depending on aquatic life for survival are also impacted [37, 38].

In addition to soil-erosion–related surface water contamination, human settlements and the activities that occur within those settlements pose a notable risk to contamination of surface water bodies with microbial contaminants, nutrients overload, and solid waste contaminants [25, 26]. This is mainly due to the formation of densely populated settlements and the consequent overload on the available waste management services [25, 26]. Furthermore, the emergence and continued increase in the number of informal settlements in areas that were not initially intended for human settlements put more strain to waste management authorities as a majority of these areas do not have waste management services [26, 37]. It has been reported that these settlements generate sewage–related waste on the daily basis, as well as graywater waste and solid litter [37].

Inadequate handling of these waste streams does not only present an ecological or aesthetic problem but also increases the risks of contamination of nearby surface water bodies [26, 37]. The sewage handling infrastructure, where it exists, is subjected to increased strain when settlements populations increase beyond design capacity [3, 26, 39]. As a consequent, blockages and leaks occur on the sewage transportation pipeline, leading to the contamination of the environment and surface water bodies [3, 23]. When untreated sewage leaks into the environment, surface water contamination can occur as a result of heavy rainfall, leading to run-off or contaminants washing off into storm-water drainage systems, and transportation to receiving surface waters [25]. Furthermore, communities in the rural areas and informal settlements, where little to no sewage removal infrastructure exists [18], present a risk to surface water contamination as sewage can easily reach the environment [25, 39].

In addition to sewage waste, graywater disposal and reuse present another surface water contamination stream arising from human settlements [40, 41]. The risk is particularly higher in rural and informal settlements where graywater is disposed into the environment due to the limited availability of sewer systems [18, 39]. Graywater obtained from common household uses such as bathing and cleaning contains contaminants that accumulate in the soil and contaminates groundwater through seepage and surface water through run-offs [39–41]. Due to the limited availability of clean water, communities normally reuse graywater several times before disposal, thus creating a further accumulation of contaminants [41, 42].

The other form of surface water pollution is littering, which has also increased with the growth in population and the emergence of new and densely populated settlements [25, 26]. The communities generate solid waste in the form of household litter such as plastics, paper, tins, disposable nappies, and other general waste generated as part of daily activities [43, 44]. This waste can be transported to nearby surface waters as a result of run-off, storm water drainage, and the close proximity of some settlements to river banks [18]. Once in the water, waste can degrade and form toxic metabolites or morph into smaller particles that can be ingested by aquatic organisms such as fish [45, 46]. Furthermore, solid waste can also serve as a means of transport by which contaminants migrate to other water bodies or to other downstream users of surface water such as agriculture and industry [47, 48]. In unurbanized areas, solid waste is often disposed into the environment as there is little to no waste disposal services, and, therefore, solid waste can reach surface water easily [18, 43]. Albeit solid waste from urbanized areas is often removed by waste handling entities and disposed of in landfills, a possibility of surface water pollution still exists due to leaches and run-offs from these disposal sites, which often contain high concentrations of heavy metals [43, 44].

The quality of surface water is also affected by industrial activities happening around the catchments as industries generate both solid and liquid waste, which may be detrimental to both the environment and human health [49]. Industries producing chemicals, paint, and paper as an example use raw material as input and process these to produce intermediate and final products and, in this process, generate solid waste that may be disposed in landfills and contribute to the accumulation of toxic wastes in the sites if not adequately treated [49, 50]. Industries often use water in their production process and also to clean equipment used during production and, as a result, generate wastewater possibly contaminated with toxic constituents [51]. Prior to disposal to the environment or municipal sewers, this waste must be treated to levels suitable for the environment, but, in some cases, the discharged waste effluent still contains traces of toxic contaminants [20].

The mining industry presents a significant risk to water pollution; although the industry implements water management strategies, inadequately treated effluents and wastewater stored in tailing dams can still accidentally end up in the environment [20, 21, 52]. Mining activities can contribute to surface water pollution when acid water generated from acid rock and acid mine drainage, heavy metal leachate from excavated rocks, and when leaks and leachates from processing chemicals migrate to nearby water bodies [22, 52, 53]. The textile industry is a significant contributor to water pollution as their production value chain relies on substantial water intake and generates large amounts of wastewater. Wastewater generated by this industry is highly toxic and contains constituents with detrimental effects to the environment and human health, for example, persistent dyes, heavy metals, surfactants, as well as substances with carcinogenic and mutagenic effects [51, 54]. The disposal of poorly treated wastewater to the environment or municipal sewers poses a substantial risk to both surface and drinking water pollution as WWTP and DWTP are not designed to remove high levels of such contaminants [20, 54].

Wastewater treatment plants receive influent wastewater from municipal sewer systems and wastewater disposal truckers and are responsible for treating the influent wastewater to effluent standards required for disposal to a natural resource such as a river or stream [20]. To achieve this, conventional sewage treatment processes are followed, but their efficiency has been largely reduced due to overloading of the system, as well as poor maintenance [3, 51]. As a result, poorly treated effluent may reach surface water and, ultimately, water treatment plants, as they use surface water as their raw water source.

8.3 Common Microbial Contaminants of Water and Their Public Health Implications

The lack of decent sanitation and the failure of wastewater treatment plants to treat sewage contaminated water to levels compliant with general limits for disposal of sewage effluents to a natural resource have increased the risk of fecal contamination of water resources in South Africa [3, 20, 23]. The majority of waterborne diseases are caused by pathogenic microorganisms associated with fecal contamination [2]. Furthermore, it is acknowledged that South Africa has an extremely high risk of microbial infection caused by domestic and recreational use of contaminated surface water [55–57]. The microbial contaminants are efficiently reduced at different stages of conventional treatment starting from coagulation to filtration and finally reduced to non-detectable levels during disinfection [58–60].

Due to the large number of microorganisms present in water, indicator organisms such as *Escherichia coli*, fecal coliforms and total coliforms (TCs) are used to assess water quality [61, 62]. *E. coli*, TC bacteria, and heterotrophic plate counts (HPCs) are commonly used indicators in South Africa for determining the bacterial contamination of water, with coliphages predominantly used as indicators for viral contamination due to their ease of quantification [63–65]. SANS 241:2015 and Blue Drop require microbial indicators to be monitored on source, i.e., tap water, and to be treated on a bi-weekly basis [66]. Furthermore, the standard specifies that *E. coli* and fecal coliforms must not be detected in 100 mL of drinking water and that HPC bacteria must not exceed 1,000 colony forming units per 1 mL of sample tested.

E. coli, a species of gram-negative bacteria normally inhabiting the gastrointestinal tract of healthy mammals, is predominantly non-pathogenic with a few infective strains causing severe abdominal cramps, bloody diarrhea, and vomiting [6, 67]. Furthermore, the species includes some Enterohemorrhagic strains producing toxins that can adversely damage the lining of intestines and kidneys of humans, sometimes leading to fatal hemolytic uremic syndrome [6, 67]. Together with TCs, a group of non-pathogenic bacteria closely related bacteria with similar characteristics, *E. coli*, is a good indicator of fecal contamination and the presence of potentially pathogenic bacteria in water due to their ease of cultivation [68–70]. The HPC counts are also generally used by the water sector to assess the microbial quality of water and particularly the effectiveness of disinfection processes [17, 71, 72]. These organisms and members of their group are effectively removed by disinfection of treated water before it is fed into the distribution system.

It is important to note that not only bacteria are responsible for drinking water contamination; however, protozoa are also a culprit and are responsible for drinking water contamination. *Cryptosporidium* and *Giardia* are of major interest to water entities and form part of microbial drinking water quality requirements in South Africa [73]. Although rarely deadly, ingestion of water contaminated by these organisms may lead to Cryptosporidiosis, a disease caused by protozoan parasite *Cryptosporidium* and Giardiasis, a disease caused by *Giardia intestinalis*, *Giardia lamblia*, or *Giardia duodenalis* [73, 74]. Interestingly, both illnesses result in gastrointestinal diseases with symptoms very similar to diarrhea. The removal of these organisms using conventional treatment processes is possible; however, full inactivation requires optimal performance of the entire process from coagulation and filtration to disinfection due to their resistance [59, 73, 75].

The other common biological contaminates of surface water are viruses, and they too require monitoring for ascertaining drinking water quality as members of the enterovirus group have previously been detected in Southern Africa [63, 76, 77]. The conventional water treatment technologies are efficient in removing viral contamination from source water. However, the reliability of this data is questionable due to the difficulty of quantifying viruses in low concentration [78, 79]. Essentially, viruses are normally quantified using cell culture–based methods that require highly specialized facilities and personnel, and, as a result, somatic coliphages are used as indicators of enteric viruses in water, although poor correlation between coliphages and enteric viruses has been reported [79–81].

Despite that most microbial contaminants can be removed during treatment processes, source water remains a major contributor to contamination of water in the distribution network [82–84]. It is important not to ignore contamination incidents caused by repair of pipeline infrastructure and leakages, and persistent microbial contamination cannot be attributed to these events as they normally cause a short-term deterioration in water quality [12]. Secondly, the microbial contaminants introduced during repairs are often corrected using chlorination, which efficiently disinfects planktonic microorganisms, thus reducing its contribution to persistent microbial regrowth in the DWDS [7, 85, 86]. There is a growing consensus among researchers that biofilms inhabiting inner-pipe surfaces of DWDS are the most significant cause of microbial regrowth, and this has subsequently led to an increase in microbial ecology studies focusing on DWDS [11, 12, 16, 83, 87–90].

8.4 Processes for Surface Water Treatment for Drinking Purposes

Surface water treatment has various process involved and are being discussed here below.

8.4.1 Screening

The treatment of surface water to potable water is an important process in the water provision value chain and is routinely performed by water treatment entities in South Africa and across the world [19]. Surface water resources are polluted by both large and small particles requiring screening to remove large materials such as plants, plastics, fish, disposable nappies, metals, or any form of debris to prevent clogging and damage to downstream equipment and piping [91]. As the first unit operation, screening is generally categorized into coarse and fine screens operated either manually or mechanically [92, 93]. The coarse screens (typically 1.5 mm to 6 mm) filter larger materials and typically have openings of 6 mm or larger, whereas finer screens typically remove smaller material, although still visible to the naked eye, with the potential of creating operation and maintenance problems in downstream processes [60, 92]. Screening improves the efficiency of downstream drinking water treatment processes and reduces treatment costs. For instance, screening improves pre-oxidation by removing suspended particles and debris that may shield natural organic matter (NOM) from the oxidant, reduced oxidant dispersion, reduced contact time between the oxidant and water, and increased oxidant demand. Furthermore, the removal of larger particles leaves behind smaller particles that are more likely to coagulate, flocculate, and sediment. Consequently, water clarification is improved, and contaminant shielding is reduced, thereby enhancing disinfection.

8.4.2 Pre-Oxidation

Pre-oxidation is mainly a precursor to the coagulation process aimed at eliminating organic coatings from particulates in solution after the screening process. In the past, chlorine and chloride compounds have been preferred as a pre-oxidation agents but due to its hazardous by-products, its use has widely been discouraged [94, 95]. For example, chlorine reacts with organic matter releasing trihalomethanes (THMs) [96]. Other pre-oxidation agents include hydrogen peroxide, ozone, and permanganate [95, 97]. Permanganate has an enhanced ability to coagulate surface water particularly with high turbidity levels [98, 99]. Potassium ferrate (VI) is equally as popular due to its strong redox potential across the pH range [100]. Pre-oxidation facilitates the reduction of contaminants from surface water including inorganic, organic, and biological. For example, dissolved inorganic contaminants including iron and manganese are oxidized into insoluble forms, which can easily be removed during sedimentation [101].

Furthermore, pre-oxidation facilitates the breakdown of complex organic molecules (e.g., humic acids) to form simpler molecules that can easily be removed by downstream drinking water treatment processes [99, 101]. In a study using ultrasound-ozonation as a pre-oxidant, a removal efficiency of 84% NOM, 46% total organic carbon (TOC), and 33% turbidity was achieved [97]. Additionally, biological contaminants, including Cyanobacteria, are deactivated during the pre-oxidation processes, albeit care needs to be taken to avoid the release of algal organic matter that may promote the formation of disinfection by-products [99, 102].

8.4.3 Coagulation-Flocculation

Coagulation and flocculation are considered the most critical processes in most drinking water treatment plants. These processes facilitate the removal of colloidal and dissolved impurities, which, through various interactions and reactions, shape the biological, physiochemical, and organoleptic characteristics of water [98, 103, 104]. Through gentle mixing, colloidal and other finer dissolved impurities bond flocculate forming larger size particles from sub-microscopic micro-flocs to visible suspended particulates that further undergo chemical collisions forming pin-flocs and eventually macro-flocs as they increase in size [103, 105]. This is aided by inorganic or organic polymers (coagulant aids) added to the medium which help bridge, bind, strengthen, and increase weight, thus improving settling rate and making it easier for suspends to be eliminated either through sedimentation or air flocculation [97, 103, 106].

The coagulation process can be described into three sequential steps: Coagulant addition, particle destabilization, and inter-particle collisions [92, 103, 107]. The first two steps take place in a rapid mixing tank, whereas the latter, a slower process, encourages agglomeration but occurs in the flocculation tank [92, 105]. The coagulation process has also been proven to adequately remove NOM, therefore increasing disinfection efficiency and reducing the formation of harmful disinfection by-products [97, 103]. Addition of coagulant chemicals with charges opposite to that of the suspended negative non-settleable solids prepares suspends for sedimentation and enhances contaminant removal. Specifically, Kuc *et al.* [106] used aluminium chloralhydrate as a coagulant to successfully reduce TOC (88.7%), turbidity (96.2%), color (98.0%), and total suspended solids (97.8%) in river water samples from two regions in Poland. In another study by Setareh *et al.* [97], coagulation using poly aluminium chloride (pre-treated with ultrasound-ozonation) resulted in a 95%, 75%, and 50% removal of NOM, TOC, and turbidity, respectively.

8.4.4 Sedimentation

Sedimentation (clarification) is a process in drinking water treatment required for the removal of suspended sediments and flocs formed during coagulation-flocculation [92, 108]. Several sedimentation techniques are available, including horizontal flow, radial flow, inclined plate, and ballasted floc and floc blanket sedimentation, with choice dependant on the intricacy of the treatment plant [92, 94]. During this process, suspended sediments or flocs settle out to the bottom under the effect of gravity forming a layer known as sludge, whereas persistent suspends are removed as scum in a clarifier [1, 94]. The sludge and scum that are removed contain contaminants emanating from source water and from preceding drinking water treatment processes [94, 109]. For instance, sludge from two DWTPs in Brazil contained elevated levels of contaminants originating from source water (arsenic, barium, sodium, and manganese) and from coagulant dosage (aluminium, chloride, and iron) [108]. Although used in tandem with air flotation occasionally, sedimentation has gradually been replaced by air flotation [104, 110].

8.4.5 Air Flotation

An alternative method to sedimentation is flotation. In this technique, adsorptive bubble separation processes use selective adsorption of impurities at the gas/liquid or gas/solid interfaces that act as transport mediums [94, 110, 111]. The buoyant gas bubbles, either in soluble or insoluble form are carried to the top where a separation reactor removes them from the aqueous system. Various bubble generation techniques exist, either belonging to the broad class of dispersed-air flotation methods in which electro-flotation is often included or dissolved-air flotation (DAF) based on Henry's law [110, 112]. The choice of method depends on level of scientific skill to ensure adequate implementation for optimum results. Furthermore, the improved removal efficiency of turbidity and NOM in highly eutrophic source water makes DAF a better alternative to conventional sedimentation [104, 105, 113]. However, over time, the air flotation shorter processing time has gained popularity replacing the conventional sedimentation process for clarification, although its demand of more careful control has been a discouragement to users [110, 114].

8.4.6 Filtration

Escaped or persistent particulates and floats from previous treatment attempts are removed through filtration. These escaped impurities may cause a level of turbidity, color, microorganisms, and particulates if not removed from treated water [75, 115]. During filtration, partially treated water is channelled through a pore structure/filter consisting of a variety of bed materials usually composed of the following [58, 59, 116, 117]:

(i). a bed of sand (sand filtration) or
(ii). layer of diatomaceous earth (diatomaceous earth filtration) or
(iii). combination of coarse anthracite coal overlaying finer sand (dual- and tri-media filtration).

The removal of suspends occurs by straining through the pores in the filter bed, absorption of the particles to the filter grains, sedimentation of particles while in the media pores, coagulation while passing through pores, and by biological mechanisms such as soil and sand filtration. During this filtration process, a contaminant riddled biofilm aided by the schmutzdecke (top layer of sand and particulate materials, i.e., fine soil particles, plant debris, and algae, free-living on non-pathogenic protozoa), ensures the removal of organic, inorganic, and microbial contaminants [58, 116, 117], although physical processes such as straining are still capable of removing larger microorganisms even in the absence of a biofilm [59].

Filtration is dependent on filter physical characteristics such as grain size, shape, porosity, and bed depth/media grain size ratio [115]. McGivney and Kawamura [118] suggest using *L/de* ratio in designing filter beds where *L* is the depth of the filter bed (mm) and *de* is the effective size of the filter medium. *L/de* ranges between 1,000 and 2,000 for different filter beds with 1,000 for ordinary mono-medium sand and dual media beds, 1,250 for regular tri-media beds (coal, sand, and garnet), 1,250 to 1,500 for coarse deep mono-medium beds in which *de* is 1.2 mm to 1.4 mm, and 1,500 to 2,000 for very coarse mono-medium beds in which *de* is 1.5 mm to 2.0 mm [118]. Although filters can be complex, they still remain efficient in ridding water of suspended physical impurities and biological pathogens particularly *Cryptosporidium*, a common protozoan parasite among other biological floats [119–121].

Micro-chemical contaminants including anionic and cationic constituents such as chlorides, fluorides, and sulfates are more common in ground water than in surface water. However, surface water may be contaminated with ammonia, nitrates, and nitrites emanating from sewage contamination [2, 20, 23]. These contaminants are not efficiently removed by previously discussed conventional water treatment techniques and may, at times, require specialized technologies such as distillation, ion-exchange, reverse osmosis, and the potent sodium hypochlorite agent, which has shown promise, particularly in ammonia removal [122].

8.4.7 Disinfection

All water treatment plants are mandated to be equipped with a disinfection phase especially when surface water is used to produce drinking water [71, 94]. Conventional disinfection methods have been proven to be reliable and efficient over many decades. These methods are generally divided between chemical and physical processes. Firstly, chemical processes make use of disinfecting substances such as (i) chlorine, (ii) chlorine dioxide, (iii) sodium hypochlorite, and (iii) ozone [15, 71].

Chemical processes:

(i) Chlorine and chlorine dioxide

Chlorine is the most universally used disinfectant since late 1940s due to its efficiency in inactivating waterborne pathogens [123]. The application is usually in the form of gaseous chlorine, chloramines, and sodium hypochlorite [71, 92, 123, 124]. Water-dissolved chlorine hydrolyzes to hydrochloric acid (HCL) and hypochlorous acid (HOCL) through Equation 8.1:

$$Cl_2 + H_2O \leftrightarrow H^+ \, Cl^- + HOCL \tag{8.1}$$

Chlorine blocks vital activities through a powerful bacterial action by modifying the chemical structures of enzymes that form the basis of chemical bacterial nutrition. The potency and cost-effectiveness of chlorine disinfection are well documented. However, the ability of chlorine to react with organic matter to form harmful disinfection by-products including THMs and haloacetic acids remains a huge risk to public health [123]. Consequently, chlorine dioxide has emerged as an alternative to chlorine disinfection due to its superior oxidative power, applicability across a wider pH range, and its reduced likelihood to form disinfection by-products [101, 123].

Chlorine dioxide works similarly to chlorine but with a more potent bacterial action [71, 101, 123]. The more potent bacterial action enables disinfection of chlorine tolerant pathogens including *Cryptosporidium* [125]. Chlorine dioxide is an unstable water soluble yellow-green gas with a strong smell, produced by sodium hypochlorite and hydrochloric acid through Equation 8.2:

$$5\ NaClO_2 + 4\ HCl \rightarrow 4\ ClO_2 + 5\ NaCl + 2\ H_2O \tag{8.2}$$

Chlorine dioxide, with its higher oxidative power compared to chlorine, effectively inactivates enzymatic systems and interrupts protein synthesis, thereby inhibiting bacterial growth and survival [71, 125].

(ii) Ozone

Ozone is formed by the dissociation of oxygen molecules in atomic oxygen and is produced by electrolysis, photolytic reactions, and radiochemical reactions induced by electric shocks. It is a highly reactive oxidizing agent that has shown to outperform chlorine as a disinfectant in water treatment plants [71, 95, 126]. Equation 8.3 shows this reaction:

$$O_3 \rightarrow O_2 + O \tag{8.3}$$

Use of electrolysis to produce ozone dates back to the 19th century although the method has since developed and, at present, corona discharge is preferred [127, 128]. Although the method is unpopular due its low current production, complicated production systems, and instability of electrode materials, a simpler approach through the use of "sandwich" configuration, i.e., diamond anode/solid polymer electrolyte (SPE)/cathode sandwich, has made the approach attractive to most users [129, 130]. Ozone is a potent disinfectant capable of inactivating a wide range of microorganisms, including pathogenic bacteria (*Salmonella* and *Campylobacter*), viruses (hepatitis A, norovirus, and rotavirus), fungi (*Aspergillus* and *Candida*), and protozoan parasites (*Cryptosporidium* and *Giardia*) [95, 131, 132]. However, Ozone's lack of a disinfectant residual and its high energy demand make it unfavorable for drinking water treatment plants with long distribution networks [95].

(iii) Sodium hypochlorite

Sodium hypochlorite (NaOCl) is an effective disinfectant used to reduce or eliminate various microbiological contaminants. The agent is formed using salt, water, and electricity and has been used since the 18th century. It effectively eliminates bacterial pathogens, viruses, fungi, and parasites by producing hypochlorous acid (HOCl), which reacts with pathogens in the water through oxidation, thus eliminating them [71]. Equation 8.4 shows this oxidation reaction:

$$NaOCl + H_2O \rightarrow HOCl + NaOH^+ \tag{8.4}$$

Use of this agent is relatively simple and cost-effective and does not require extensive technical expertise. The solution is diluted with water in a mixing tank and introduced into the water by means of a chemical pump at a controlled rate [71]. Its simple operational process makes this approach attractive to most. Sodium hypochlorite has been successfully used to reduce *Cyanobacteria* (*Microcystis*) and bacteriophage MS2 in DWTPs and at points of use [124, 133].

Although the abovementioned chemical process has proven reliable over the decades, as they eliminate microorganisms and protect water against recontamination, their reaction by-products may pose a danger to human health [96].

8.5 Factors Shaping Microbial Community Composition of Water

8.5.1 Raw Water Source, Treatment Regime, and Pipe Material

The understanding of the microbial community composition of any environment is important as it gives an insight into the type of microorganism's present and their relative abundance as well as their interaction with each other and the environment [7, 83, 84, 88, 134]. This is also necessary for water, with the need for understanding microbial ecology further emphasized by the deterioration in surface water quality and water treatment infrastructure. It has been reported in various studies dealing with the value chain of water provision, from source to end consumers, and these studies have revealed notable variability in the microbial community of these ecosystems [7, 12, 83, 84, 88, 134]. The other interesting observation is that results from these studies have indicated correlations with various physicochemical parameters, pipe material, and source water origin as well as disinfection regimes [7, 15, 134].

For example, Eichler *et al.* [134] investigated whether source water had a significant influence on the microbial community structure of tap water received by consumers. Their approach was to use polymerase chain reaction (PCR) and sequencing to analyze samples obtained from two raw water sources and track them through subsequent water treatment processes until the taps of consumers. The findings from their study identified notable differences in community structure of tap water obtained from the two water sources. Furthermore, the samples taken after the flocculation and filtration process showed no significant shift in community structure compared to the original water source. However, notable shifts in community structure were observed after disinfection stage, suggesting that disinfection was the most influential process in shaping the community structure of water received by consumers [134].

In a similar study that compared water produced from two treatment plants with significant differences in design [135], results indicated that both process designs were efficient in reducing microbial abundance and diversity from raw to treated water. Furthermore, their findings indicated significant shifts in microbial community structure after disinfection for both processes. These findings were similar to those reported by Eichler *et al.* [134]; moreover, both studies reported that the shift in community structure favored growth of nitrifying bacteria. In addition, Shaw *et al.* [135] observed microbial regrowth further in the DWDS even though the final treated water had no microbial growth.

The findings from the above studies confirm the effectiveness of disinfection in the reduction of microbial contamination regardless of the water source and differences in treatment process design. This is consistent with literature that has seen that disinfection become an integral part of any water purification process [7, 14, 15, 72, 136]. Most importantly, these findings also suggest that disinfection causes shifts in microbial community structure and favors proliferation of microorganisms with metabolic activities that may interfere with disinfectant residuals, thus promoting regrowth in the DWDS. This has prompted further research investigating the effects of different disinfectants on DWDS microbiomes [7, 14].

To this effect, Waak *et al.* [7] demonstrated that the use of chloramine favored growth of nitrifying bacteria that were not observed for the system with no monochloramine residual. Furthermore, Liu *et al.* [14] reported that a combination of UV-free chlorine disinfection was more effective in restricting planktonic microbial regrowth than using free chlorine alone. However, both disinfection regimes were equally ineffective in eliminating potential pathogens in the biofilm phase Liu *et al.*, 2019 [14]. The encouragement of nitrifying bacteria growth was further confirmed by Zhang *et al.* [15] when comparing differences in microbial community composition caused by using sodium hypochlorite or chlorine dioxide as a disinfectant. The results from their study indicated that both disinfectants promoted growth of *Firmicutes* bacteria [15].

In addition to disinfectants, other factors contributing to microbial regrowth in DWDS have also been investigated. These studies have focused on the availability of growth promoting nutrients and its effect on microbial community structure. In other studies conducted on a stratified reservoir, differences were observed in community composition that the authors attributed to difference in the availability and abundance of key nutrients, dissolved oxygen, and varied temperature at different levels of the reservoir [9]. Similarly, availability of other growth promoting nutrients such as ammonia and nitrogen have also been shown to contribute to microbial regrowth in DWDS [137]. The availability of a carbon source has also been shown to contribute to microbial regrowth [138] together with the availability of sulfate [7].

The DWDSs, as an engineered ecosystem, provide an interesting case for studying microbial community composition due to the many factors that exist within the networks. Previous research has shown that the majority of these factors encourage formation of biofilms in the DWDS, which then becomes the main driver of microbial regrowth and water quality degradation [7, 86, 87, 139]. The occurrence of biofilms in DWDS has been associated with opportunistic pathogens [7, 14, 68, 70, 140, 141], water taste and odor issues [139, 142], nitrification [7, 72, 87, 137], and corrosion [7, 8, 15, 143, 144].

8.5.2 Biofilms as a Contributing Factor to Microbial Community Composition of Water

The DWDS is considered as a harsher environment for microbial survival and proliferation compared to natural environments such as surface water. This is due to a number of biological, chemical, mechanical, and physical stressors that necessitate microbial adaptation for survival in DWDS, with biofilms being a prime example of such adaptation [8, 16, 72]. The survival of microorganisms in a biofilm under DWDS stressors is largely due to the formation of an extracellular polymeric substance (EPS), which serves as a nutrient reservoir and disinfectant inhibitor and aids in the strengthening and inter-linkages that reduce the impacts of sheer stresses and promote microbiome development [16, 145]. The biofilm communities are complex in nature and utilise sophisticated communication systems such as quorum sensing and intracellular second messengers, which occur between similar or different bacterial species, including microorganisms from other kingdoms such as fungi [146–148].

8.5.3 Biofilm Formation and Development

Quorum sensing ensures communication between community members in response to environmental stressors through the production and sensing of autoinducer signaling molecules, leading to a coordinated genetic response [149–152]. Signaling occurs differently in gram-negative and gram-positive bacteria, with the former utilizing the acylated-homoserine lactone system, and the latter using the peptide-based system [149, 150, 152]. Although, different, both systems involve the accumulation of respective autoinducer molecules as cell numbers increase to a required threshold, followed by detection of the signal by corresponding receptors in each cell and then the triggering of a synchronized community-level genetic response [149, 150, 152]. Quorum sensing plays a critical role in DWDS biofilm formation and dispersal as it has been shown for *V. cholerae*, where two autoinducers work in parallel to signal for biofilm formation (autoinducer CAI-1) in low cell-density situations and autoinducer AI-2, which signals for dispersal of cells from the biofilm once cell numbers have the accumulated to a high-density [148].

Intracellular secondary messenger molecule cyclic dimericguanosine monophosphate (ci-di-GMP) also plays a role in biofilm development and dispersal at a genetic level [146, 153]. In response to environmental stressors, microbial cells can increase their ci-di-GMP levels through the synthesis of ci-di-GMP, using two Guanosine triphosphate (GTP) percussor molecules and diguanylate cyclase enzyme activity or decrease levels through the degradation of ci-di-GMP into GMP or phosphoguanylyl guanosine and using the phosphodiesterase enzyme [146, 153]. An increase in ci-di-GMP has been reported to encourage lifestyle switch from planktonic to biofilm state, whereas the inverse leads to dispersal of biofilm bacteria back into planktonic state [146, 153]. Several studies have shown how genetic expressions lead to the formation of various proteins, which control biofilm formation through inhibition of motility appendages and/or secretion of EPS, and others that encourage dispersal by coding for motility [146, 147, 153, 154].

The biofilm formation is a sequential process that begins with the initial attachment of planktonic bacteria to the inner-pipe surface encouraged by hydrophobic properties of bacterial cells. This process is followed by auto-inducers controlled expression of genes coding for permanent attachment through the inactivation of mobility appendages and secretion of an EPS matrix which ensure the maturation into a three-dimensional structure with water channels for nutrient flow [16, 56, 67, 145, 147]. Biofilms can exist as mono- or multi-species; however, previous research indicates that bacteria are more abundant in biofilms due to their ability to easily adapt, as a function of their small size and metabolism, and their ability to produce EPS matrices [4, 7, 68, 70, 89, 155, 156]. Microorganisms from other kingdoms such as algae, archaea, fungi, and protozoa can also be found in DWDS biofilms, although not in dominant numbers as bacteria [82, 155, 157, 158].

8.5.4 Planktonic and Biofilm Bacteria Community Structure

The dominance of bacteria in DWDS manifests itself throughout all stages of biofilm formation, from initial attachment to dispersal. Microbial community studies on DWDS biofilms have revealed that Proteobacteria is the dominant phyla, even under different test conditions [7, 8, 16, 87, 89, 159]. This is a large phylum of gram-negative bacteria, which consists of five classes, of which Alphaproteobacteria, Betaproteobacteria, and Gammaproteobacteriahave been commonly detected in DWDS microbial community studies [7, 8, 16, 87, 89, 159].

Alphaproteobacteria class consists of 10 sub-classes and orders; however, previous research indicates that Hyphomicrobiales [7, 8, 87], Maricaulales [7], Parvularculales [7], Rhodospirillales [87], and Sphingomonadales [8, 159] are found in DWDS. Furthermore, results from these studies suggest that members of the Alphaproteobacteria class are more dominant in DWDS with chlorine as a primary disinfectant and become less abundant in chloraminated systems as well as in the early stages of biofilm formation [7, 8, 159].

The abundance in chlorinated systems may be due to the ability of members of this class such as *Hyphomicrobium* and *Methylobacterium*, capable to metabolize methane and other complex carbon compounds such as THMs [160]. THMs are a common by-product of chlorination, and their metabolism leads to the release of carbon, which is an energy source for these organisms, as well as other members of the biofilm community.

Nitrogen fixation is another possible contribution that members of Alphaproteobacteria may play in biofilm as previous studies have confirmed the presence of *Bradyrhizobium, Hyphomicrobium*, and *Methylobacterium* in DWDS [7, 8, 137]. Although their nitrogen-fixing ability manifests when in symbiotic relationships with plants, the ability to facilitate natural genetic transformation may be a mechanism by which these organisms pass genes involved in the nitrogen cycle to other members of the community [87, 137, 161].

Betaproteobacteria are another class of Proteobacteria that have been detected in microbial community studies of DWDS [4, 8, 16, 87, 156]. Similar to Alphaproteobacteria, this class has been found to be dominant in DWDS systems with different disinfectant regimes [7, 15, 86], including in the early stages of biofilm formation [159]. Several orders exist within this class, with Burkholderiales, Nitrosomonadales, and Rhodocyclales detected in previous DWDS community studies [7, 8, 16, 87, 159]. Nitrosomonadales order (*Nitrosomonas*) was noticeably more abundant in chloraminated samples at all stages of biofilm formation but only become more abundant toward the late stages in a chlorinated system [7, 159].

Nitrosomonas are ammonia-oxidizing bacteria; therefore, their presence in chloraminated systems indicates a role that such organisms play in DWDS biofilm development by reducing the efficacy of chloramine [72, 137]. Interestingly, Betaproteobacteria particularly *Dechloromonas* genus were more abundant on cast-iron pipes irrespective of disinfectant type [8]. This might suggest that members of this genus play a role in pipe-corrosion induced iron release, thus providing a nutrient source for iron bacteria in the biofilm community. The presence of iron-oxidizing bacteria in cast-iron DWDS [4, 8] confirms observations of Waak *et al.* [7] of the dominance of *Gallionella* and *Sideroxydans* in corrosion tubercles.

DWDS microbial community studies have also revealed the presence of a third class of Proteobacteria called Gammaproteobacteria [159]. From this class, order Legionellales, Methylococcales, Pseudomonadales, and Xanthomonadales have been detected at different stages of biofilm formation [7, 8, 16, 89, 159]. Furthermore, results from these studies highlighted the abundance of *Pseudomonas*, a genus in the order Pseudomonadales. The dominance of this genus was seen to be more common at the early stages of biofilm formation [89, 159] and at intermediate flow velocities [16], thus confirming the significant role that species in the *Pseudomonas* genus play in biofilm formation [146, 147, 153]. Apart from *Pseudomonas*, Gammaproteobacteria are known to have genera with potentially pathogenic organisms such as *Escherichia*, *Salmonella*, *Vibrio*, and *Legionella*, which are known contaminants of water, and have been detected in DWDS samples [8, 68, 70, 140].

The other classes of the Proteobacteria phylum have not been reported to occur in DWDS as commonly as the previously discussed. *Deltaproteobacteria* genus *Desulfovibrio* has previously been detected in bulk water and corrosion tubercles and in close association with antibiotic resistant bacteria [7, 141, 162]. Apart from the Proteobacteria phylum, Actinobacteria of the Terrabacteria group are common community members of DWDS microbial communities [7]. Out of this phylum, *Actinomycetia* is mostly abundant in DWDS, with the human pathogenic genus *Mycobacterium* found dominant in both chloraminated [7, 87] and chlorinated [8, 163] systems, irrespective of flow velocity [16].

Nitrospirae is another phylum that is commonly found in DWDS, particularly where chloramine is the primary disinfectant [8, 87]. *Nitrospira*, the largest genus in this phylum, has been detected in previous studies in correlation with the abundance of archaea and Alphaproteobacteria [8, 87]. The presence of *Nitrospira* in chloraminated systems further confirms the correlation between chloroamination and the proliferation of microorganisms capable of ammonia and other nitrogen compound metabolism [72, 87, 137]. Furthermore, the co-occurrence of this genus with known nitrogen fixing *Thaumarchaeota* (Archaea) and *Bradyrhizobium* (Alphaproteobacteria) suggests that the type of disinfectant has a major contribution in shaping microbial community composition [10, 87].

The occurrence of multi-species biofilms in DWDS suggests that interdependency exists between drinking water microorganisms, ensuring collective survival in inherently harsh conditions. Various species of biofilms consisting of archaea and bacteria have been reported in studies discussed previously; although those including fungi do exist, they are not often the subject of investigation as the former [89, 159]. This has contributed to the current limited understanding of the role fungi play in multi-species biofilms. Siqueira *et al.* [158] investigated the occurrence of filamentous fungi in a full-scale DWDS and confirmed their presence together with that of algae and bacteria; however, they were not able to explain the role those filamentous fungi played. Their role was explained to a certain extent by Douterelo *et al.* [159], who found the presence of fungal genera such as *Acremonium* and *Neocomospora*, capable of utilizing chloride and of producing metabolites that interfere with disinfectants, particularly in the early stages of biofilm formation, thus suggesting that fungi play a role in disinfectant decay during biofilm formation where bacteria are more vulnerable. The occurrence of fungi in DWDS, particularly filamentous fungi, has been reported by other researchers [5, 164, 165]; however, there appears to be no definitive answer to the role fungi play in multi-species biofilm formation and subsequent development. In a study conducted by Rosales *et al.* [89], fungal genera normally associated with skin infections were detected, but the role they played in the community was not clear. However, both Douterelo *et al.* [159] and Rosales *et al.* [89] found less diversity in fungal populations in the community compared to bacteria, irrespective of biofilm age or phosphate concentration.

8.6 Sampling for Microbial Community Composition Studies

8.6.1 Bulk Water-Phase Sampling

Sampling is the first step of any microbial diversity study and a critical step of the process that is often taken lightly. Several guidelines have been documented for sampling water from different sources, and these are often customized for each country; for example, SANS 5667 series is used in South Africa as a guideline for compliance monitoring companies to sample and analyze samples collected (SABS, 1991). These techniques are applied when sampling water for the purposes of regulatory monitoring using culture-dependent techniques.

Although these sampling guidelines were primarily developed for culture-dependent techniques, they have been adopted and used for culture-independent analysis as well with slight modifications. For example, most molecular microbiology water studies follow these sampling guidelines with modifications to the volumes of sample collected based on the expected quantity of genetic material in that environment and the choice of method for microbial community analysis [7, 8, 87, 159, 166]. As such, the volume of sample collected and processed for studies using shotgun metagenomic sequencing may be significantly higher than 16S rRNA gene sequencing, due to initial DNA quantity requirements [89, 167].

8.6.2 Biofilm-Phase Sampling

In recent times, the focus in microbial ecology studies has been to collect samples from not only the bulk water phase but also the biofilm phase. This is based on the growing understanding that there is more diversity in this phase, and that answers to what exactly happens in the DWDS obtained from studying naturally forming biofilms [7]. Secondly, there is a growing consensus that modeling systems such as reactors [145] do not adequately mimic real DWDS and, therefore, provide limited information. However, as desired as studying natural forming biofilms may be, sampling of biofilms in real-life DWDS remains a major challenge because the majority of DWDS pipelines are buried underground.

Arguably, the best representative biofilm sample would be obtained by cutting out a section of a natural DWDS pipeline and scrape of the biofilms that are on the inner-pipe surfaces. This approach was used by Kelly *et al.* [168] to study the microbial diversity and abundance of biofilms obtained from a real DWDS over an 18-month period. In their study, they found that diversity was relatively low over the sampling period as biofilms were dominated by only four taxa, although their abundance differed over time. The abundancy was attributed to temperature changes as well as changes in concentrations of parameters such as ammonia, nitrates, monochloramine, total chlorine, alkalinity, and hardness. Most important, the highest diversity and abundance was observed from samples with the highest concentration of nitrates, thus suggesting a stronger correlation than the other conditions. Other studies [68, 87, 169] have used this sampling technique to study biofilms in different DWDS to good effect; however, the difficult and labor-intensive processes required to cut and remove pipes as well as potential water contamination make this technique unsuitable for most systems.

An alternative sampling methodology which does not include cutting the pipe for biofilm sampling are devices such as coupons which are commonly used to sample and study biofilms over a selected period of time without having to dig and cut sections of the pipeline [89, 159, 170]. These devices are inserted into an existing DWDS and used to study biofilm dynamics *in situ* or are removed for further analysis using both culture-dependent and non-dependent techniques. These devices have been used to study biofilms in both simulated and real-life DWDS [89, 159, 170] to good effect, although questions remain on how such devices disrupt normal hydraulic conditions.

Perhaps an alternative and less resource requiring method for biofilm sampling is flushing. Changes in hydraulic conditions have been reported as one of the major causes of biofilm detachment and flushing allows for detachment of biofilms for purposes of microbial ecology studies [145]. Flushing as a biofilm sampling technique has been used successfully [171, 172] and results have demonstrated its ability to mobilise pipe-wall biofilms, however, optimization is required to achieve full detachment based on pipe material and size. Flushing to mobilise and sample biofilms from an operational DWDS has been used successfully to determine bacterial and fungal community composition of collected biofilm samples [5, 164]. Abundance and diversity of the community was consistent with that obtained from chlorinated DWDS studies sampled using pipe-cuts or coupons [7, 86, 167]. Although the flushing and pipe-cut sampling technique can produce comparable results, the pipe-cut technique remains the preferred options as it produces more representative biofilm samples. Flushing induces sheer stresses that disrupt the biofilm and cause loosely attached biofilm members to detach and resuspend back to the water phase [173]. Therefore, biofilm samples sampled using the flushing technique may not include all the taxa that was part of the biofilm community prior to flushing.

8.7 Approaches to Studying Microbial Community Composition

8.7.1 Culture-Dependent Methods

Traditionally, culture-dependent techniques have been used to study microbial community composition of various matrices including water. The cost of performing these analyses, especially those of indicator organisms, is relatively cheap and requires less sophisticated equipment and moderately skilled personnel. However, the major drawback associated with these methods are well-documented and their inability to enumerate nonculturable microorganisms, as well as their non-specificity makes then an unpopular choice for those intending to study microbial ecology [17, 68, 80, 135, 166].

However, traditional culture–dependent methods are still widely used for compliance monitoring purposes and regulatory authorities in most countries prescribe their use [66, 174]. For example, SANS 241:2015 prescribes *E. coli* and TCs to be monitored using either membrane filtration (MF) methods or methods based on the multiple tube fermentation (MTF) technique. In the past, these methods required longer incubation periods, followed by further incubation for confirmation of presumptive positive colonies [69, 175, 176]. However, the emergence of chromogenic medium has led to reductions in incubation times and complete elimination of confirmatory steps, leading to less than 24-h turnaround times. Both MF and MTF techniques use chromogenic media containing a particular substrate that can only be cleaved by a target organism, leading to color development whether in colonies for MF or in tubes for MTF methods [175, 176].

Monitoring of HPC bacteria is also used by water purification companies to measure efficiency of disinfection of their treatment processes [66]. This is done using non-selective media, which supports the growth of all heterotrophic bacteria and fungi that utilize carbon as an energy source [69]. These methods are very simple to perform and can be done using either pour-plate, spread-plate, or spiral-plate techniques and incubated at temperatures ranging from 22°C to 37°C for 2 to 7 days [69, 177]. After the required incubation period, colonies that grow within and on the surface of the culture medium are counted and compared against predetermined criteria to conclude whether water is safe for human consumption.

E. coli, TC, and HPC are useful to monitor microbial quality of water for bacteria and to a limited extent fungus; however, they are not suitable indicators for viral contamination [69]. Cell culture remains the gold standard for viral water quality assessments but is very difficult to perform, which require sophisticated equipment and highly skilled personnel [69]. For this reason, somatic coliphages are widely accepted by regulatory authorities as a suitable indicator, albeit some contestation remains [65, 80, 81, 178]. This method can be performed by following either the single-layer or double-layer agar techniques with both methods resulting in the formation of zones of clearance (plaques), indicating the presence of a bacteriophage capable of infecting the host culture lawn. As this is an indicator method, it is often necessary to confirm any positive sample using cell culture–based methods, and this might explain why molecular based methods have been usually preferred in microbial ecology studies than this method.

Although culture-dependent technique limitations are well documented, their continued use in modern times indicates their importance, especially for compliance monitoring purposes. Furthermore, one cannot disregard the contribution that these techniques have made in the development and improvement of molecular-based techniques.

8.7.2 Culture-Independent Methods

The development in culture-independent techniques, coupled with the desire to understand microbial communities at a genetic level, has added to the growing popularity of molecular based approaches. These molecular techniques have been used in ecology studies and provided useful information that would not have been obtained using culture-based methods. For example, 16S rRNA gene–based sequencing technology has been successfully used to identify distinct community structure differences between microorganisms found in surface water and sediment samples [179], to identify 42 phyla and 689 genera in a stratified reservoir [9], as well as to study the phylogenetic diversity of microorganisms from bulk water and biofilm samples from different DWDSs [7, 87, 180].

8.7.2.1 *Polymerase Chain Reaction*

Polymerase chain reaction (PCR) techniques have become a common tool for detection and relative quantification of microorganisms in various sample matrices including water [7, 68, 166]. The relative simplicity of these techniques makes them an ideal tool for diagnostic laboratories with moderate levels of molecular biology skills. In summary, this approach involves the extraction of DNA/RNA either from a cultivated colony or directly from the sample, followed by amplification of target segments of these nucleic acids using complimentary oligonucleotides (primers) and visualization by gel electrophoresis–based methods [7, 68, 166]. Alternatively, visualization of amplicons can be done in real time using fluorescent-labeled probes that are released every time a new strand is synthesized [7, 8]. The amount of fluorescent signal emitted is then used to determine the initial quantity of target DNA/RNA present in the sample.

Both versions of these techniques have been used to study microbial diversity in DWDS, but their use has decreased, as evidenced by the limited number of recent publications, due to the emergence of new technologies. In the past, PCR was used in combination with fingerprinting techniques such as denaturing gradient gel electrophoresis to determine microbial diversity in biofilms obtained from different pipe surfaces and to predict the occurrence of pathogenic organisms using non-fecal indicators [181, 182]. The recent use of this technique is often as a preliminary step prior to sequencing or in the quantitative detection of microorganisms and genes associated with specific traits such as disinfectant decay and nitrification [7, 72, 87]. The PCR approaches are useful in microbial diversity studies but do not provide enough information for microbial community studies that often seek to understand interactions between different microorganisms, and their environment as well as the functionality of these community members in that environment [89, 167]. To counteract these shortcomings, amplicons obtained from PCR reactions need to be analyzed further using post-PCR techniques such as 16S rRNA gene, ITS region, 18S rDNA, and whole-metagenome sequencing [7, 89, 159].

8.7.2.2 *Amplicon Metagenomic Sequencing*

Metagenomic sequencing can be target based depending on the type of information required. Commonly, assays are designed to target genes such as the 16S rRNA for prokaryotes, 18S for eukaryotes, and ITS regions for fungal classification [7, 159]. The 16S rRNA gene sequencing involves the sequencing of amplified hypervariable regions of the bacterial 16S rRNA gene that contains phylogenetic information that can be used to understand microbial community composition to species level in an environment [4, 87]. The technology offers a rapid and relatively cheap way of studying communities due to its targeted nature and the availability of primers targeting different regions. In the previous studies, this technology has been used to evaluate the efficiency of different water treatment options due to its ability to identify and quantify the abundance of bacteria within a given community. For example, Waak *et al.* [7] used this technique to study the differences in bacterial communities from DWDS systems of two WTPs with different disinfection treatments.

The 16S rRNA sequencing has been used in other similar studies comparing the effectiveness of UV disinfection followed by chlorination and chlorination on its own [14] and the effects of chlorine dioxide in biofilm community composition and pipe corrosion [15]. The results from these studies showed differences in bacterial community composition at the genus level for the different treatment practices, thus confirming the usefulness of this technique in studying bacterial community diversity. However, results from the above studies all indicated that 16S rRNA sequencing alone is useful in identifying community members but does not provide information on what functions those members perform in their community as well as their interaction with each other and the environment.

Using gene-targeted quantitative polymerase chain reaction (QPCR) in combination with 16S rRNA, Shaw *et al.* [135] were able to observe regrowth of bacteria in the distribution network of two systems and subsequently attributed it to the presence of nitrifying bacteria and bacteria with genes coding for resistance to the type of disinfectant used. These findings were consistent with those reported by Waak *et al.* [7] who, in their study, used both QPCR and 16SrRNA sequencing and successfully demonstrated that 16S rRNA sequencing in combination with 16S rRNA-targeted QPCR was able to explain the interaction between bacterial communities and the environment and the effect that those interactions had on end user water quality. However, DWDSs are complex environments, and developing QPCR assays targeting each gene of interest would be a long and laborious exercise.

8.7.2.3 Whole-Metagenome DNA Sequencing

Depending on the level of detail required, sequencing techniques that cover the entire genome of microbial community members present in an environmental sample may be necessary. Shotgun sequencing is increasingly becoming an ideal solution for this problem as it allows for sequencing of entire genomes in very short periods of time. This is achieved by fragmenting genomic DNA into smaller parts that are then sequenced individually and then later combined to form the initial template, thus allowing for the whole genome to be sequenced and not just regions of interest [5, 89, 158, 167].

This approach has been used to sequence the entire genome of biofilm and bulk water communities obtained from a DWDS with results from the study confirming that this technique goes beyond taxonomic level as is the case with 16S rRNA gene sequencing [158, 167]. Brumfield *et al.* [183] used this approach to study the microbiome of four different types of potable water and found presence of microorganisms from each of the kingdoms; however, specific trait analysis such as the presence of genes coding for antimicrobial resistance revealed inconclusive results as quantities detected were below the confidence threshold levels. The inconclusive nature of their study could perhaps be attributed to low quantities of DNA available in selected samples, for example, tap water, which has an expected lower microbial diversity and abundance as compared to DWDS.

8.7.3 Developments in Sequencing Platforms

8.7.3.1 Sanger and Clone-Based Sequencing

The Sanger and clone-based sequencing offers an extensively accurate and detailed microbial classification information from samples than PCR-based methods and the more commonly used approach [159, 184]. Similar to PCR, this technique also requires nucleic acid extraction from samples, followed by amplification using primers specific to a target region or gene. Subsequently, vectors are used to build clone libraries and then clones are selected and sequenced to yield nucleotide sequences that are then compared to existing databases to obtain phylogenetic information [166, 184]. Although clone-based sequencing has been used for various microbial diversity studies to great success, the selection of clones is a labor-intensive process, and the sequencing of each colony makes it expensive and time-consuming [184, 185].

8.7.3.2 Next-Generation Sequencing

The developments in the field of genomics have made sequencing quicker and more affordable than traditional Sanger sequencing. These advances, referred to as next-generation sequencing (NGS), have made it possible to reduce the time required to sequence a genome with the use of parallel systems and complex bioinformatics applications [88, 89]. There are different NGS platforms and techniques used in microbial ecology studies, for example, 454 pyrosequencing, which uses sequencing by synthesis principle characterized by the detection of light emitted whenever a new nucleotide is incorporated and pyrophosphate is released [16, 186].

Pyrosequencing has been used in microbial ecology studies of DWDSs communities to study microbial diversity of biofilms under different flow velocities, different pipe materials, and from four different phases of a DWDS [16, 171, 186–188]. In all these studies, this technique was able to identify and classify bacteria to at least the genus level, thus making it a suitable alternative to Sanger sequencing.

A second platform for NGS is the Illumina system, which like pyrosequencing uses sequencing by synthesis, but, in this case, each of the four nucleotide bases has a fluorescently labeled terminator that competitively binds to a complementary base on the template strand [7, 8]. After binding, light excitation is applied, and the fluorescence emitted for each base has a characteristic wavelength and color that is captured by an imaging system [7, 8]. The use of this platform in DWDS microbial communities studies is widespread and has achieved success in studies aiming to investigate the presence of nitrifying bacteria and their potential contribution to disinfectant decay, diversity of biofilms at different stages of formation, and the effect of pipe-material and disinfectant regime [4, 7, 8, 87, 159]. As is the case with pyrosequencing, Illumina sequencing provides an alternative to Sanger sequencing due to its resolution to at least genus level. Furthermore, this platform is comparatively cheaper and has a higher throughput than the former two, albeit with a lower accuracy due its shorter read length [4, 87].

When using sequencing-based techniques for microbial community studies, it is important to ensure a wider coverage of species present in a sample by not subjecting the analysis to similar limitations discussed for culture-dependent methods. For example, earlier studies used nucleic acids extracted from colonies grown on culture media and thus could only study the small percentage of microorganisms that is culturable [68, 166]. This is where metagenomic-based approaches are of particular importance as they ensure a wider coverage by analyzing the total DNA extracted directly from an environmental sample and, therefore, eliminating biases linked to culturing of microorganisms [167]. Once metagenomic DNA is isolated, a suitable NGS technique can be selected on the basis of whether the aim is to study the entire genome or to target genes of particular interest.

8.7.3.3 *Third-Generation Sequencing*

The previously discussed NGS platforms are characterized by short-read lengths, which, although useful, the inherent systematic errors associated with these methods do not allow for production of high-accuracy complete genome sequences [189]. This can be overcome by third-generation sequencing platforms that utilize single-molecule real-time sequencing technology [189, 190]. Such technology allows for generation of longer reads, which are crucial in the sequencing of complex genomes and in linking a gene with its carrier species and also in identifying flanking genes coding in similar pathways [190].

Pacific Biosciences is a dominant player in this technology, particularly for laboratory-scale application, whereas Oxford Nanopore has the edge on portable platforms and on-site applications [189, 190]. Initially, this technology was characterized by high-error rates, thus limiting its applications; however, this has been overcome by the introduction of circular consensus sequencing, which significantly reduces the error rate of the consensus sequence, by generation of random errors rather than systematic errors [190, 191]. The random errors minimize the occurrence of overlapping errors as the number of sequences increase, thus allowing for complete high-accuracy full genomes to be sequenced [191].

8.8 Conclusion

This book chapter successfully summarized the sustainabilities and challenges of safe drinking water provisions in low- and middle-income countries. Factors such as aging water supply network, pipeline materials, high chlorine demand, biofilm, climatic factors, microbial regrowth, and long water supply network were evaluated and reported. Furthermore, water leakages as part of the non-revenue water were identified as part of the leading problem toward low chlorine residual and microbial failures, thus posing health risk to water consumers in the distribution network and further afield. Chlorine booster stations could be installed to supplement residual chlorine and cater for novel and resistant chlorine residuals, hence sterilizing the distribution network. Characteristics of raw water, treatment processes, different disinfectants used for water treatment and sterilization, and challenges thereof were successfully reported. Avenues to maintain the chlorine residuals were also explored, unpacked, and documented. Factors that underpin chlorine residuals were also elucidated. In essence, this chapter demonstrated the feasibility of ensuring consistent supply of safe water for human consumption and avert potential eco-toxicological effects that could derive from consuming unsafe drinking water.

Acknowledgments

The authors of this manuscript would like to express their sincere gratitude to the University of South Africa and Magalies Water for extending their facilities in support of the objectives of this project.

References

1. Elfikrie, N., Ho, Y.B., Zaidon, S.Z., Juahir, H., Tan, E.S.S., Occurrence of pesticides in surface water, pesticides removal efficiency in drinking water treatment plant and potential health risk to consumers in Tengi River Basin, Malaysia. *Sci. Total Environ.*, 712, 136540, 2020.
2. Verlicchi, P. and Grillini, V., Surface water and groundwater quality in South Africa and mozambique—Analysis of the Most critical pollutants for drinking purposes and challenges in water treatment selection. *Water*, 12, 305, 2020.
3. Mnguni, E.S., Pollution Of The Vaal River System In South Africa: A Case Study. *Waste Manage. Environ. Impact XI*, 257, 17, 2022.
4. Kimbell, L.K., Lamartina, E.L., Kappell, A.D., Huo, J., Wang, Y., Newton, R.J., Mcnamara, P.J., Cast iron drinking water pipe biofilms support diverse microbial communities containing antibiotic resistance genes, metal resistance genes, and class 1 integrons. *Environ. Sci. Water Res. Technol.*, 7, 584–598, 2021.
5. Douterelo, I., Dutilh, B.E., Arkhipova, K., Calero, C., Husband, S., diversity, M., ecological networks and functional traits associated to materials used in drinking water distribution systems. *Water Res.*, 173, 115586, 2020.
6. Ibekwe, A.M. and Murinda, S.E., Linking microbial community composition in treated wastewater with water quality in distribution systems and subsequent health effects. *Microorganisms*, 7, 660, 2019.
7. Waak, M.B., Hozalski, R.M., Hallé, C., Lapara, T.M., Comparison of the microbiomes of two drinking water distribution systems—with and without residual chloramine disinfection. *Microbiome*, 7, 1–14, 2019.
8. Li, W., Tan, Q., Zhou, W., Chen, J., Li, Y., Wang, F., Zhang, J., Impact of substrate material and chlorine/chloramine on the composition and function of a young biofilm microbial community as revealed by high-throughput 16S rRNA sequencing. *Chemosphere*, 242, 125310, 2020.
9. Shilei, Z., Yue, S., Tinglin, H., Ya, C., Xiao, Y., Zizhen, Z., Yang, L., Zaixing, L., Jiansheng, C., Xiao, L., Reservoir water stratification and mixing affects

microbial community structure and functional community composition in a stratified drinking reservoir. *J. Environ. Manage.*, 267, 110456, 2020.
10. Kitajima, M., Cruz, M.C., Williams, R.B., Wuertz, S., Whittle, A.J., Microbial abundance and community composition in biofilms on in-pipe sensors in a drinking water distribution system. *Sci. Total Environ.*, 766, 142314, 2021.
11. Siponen, S., Jayaprakash, B., Hokajärvi, A.-M., Gomez-Alvarez, V., Inkinen, J., Ryzhikov, I., Räsänen, P., Ikonen, J., Pursiainen, A., Kauppinen, A., Composition of active bacterial communities and presence of opportunistic pathogens in disinfected and non-disinfected drinking water distribution systems in Finland. *Water Res.*, 248, 120858, 2024.
12. Abkar, L., Moghaddam, H.S., Fowler, S.J., Microbial ecology of drinking water from source to tap. *Sci. Total Environ.*, 908, 168077, 2024.
13. Wolf-Baca, M. and Siedlecka, A., Community composition and antibiotic resistance of tap water bacteria retained on filtration membranes. *Diversity*, 15, 427, 2023.
14. Liu, L., Xing, X., Hu, C., Wang, H., Lyu, L., Effect of sequential UV/free chlorine disinfection on opportunistic pathogens and microbial community structure in simulated drinking water distribution systems. *Chemosphere*, 219, 971–980, 2019.
15. Zhang, H., Tian, Y., Kang, M., Chen, C., Song, Y., Li, H., Effects of chlorination/chlorine dioxide disinfection on biofilm bacterial community and corrosion process in a reclaimed water distribution system. *Chemosphere*, 215, 62–73, 2019.
16. Liu, Y., Shan, R., Chen, G., Liu, L., Linking flow velocity-regulated EPS production with early-stage biofilm formation in drinking water distribution systems. *Water Supply*, 20, 1253–1263, 2020.
17. Speksnijder, R., *Spatial and Temporal Relation Between Drinking Water Temperature and Indicators for Microbial Water Quality in the Drinking Water Distribution System of Amsterdam*, pp. 1–5, Ferrara, Italy, 2021.
18. Statistics South Africa, *General Household Survey*, S.S. Africa (Ed.), Petroria, South Africa, 2022.
19. Desye, B., Belete, B., Asfaw Gebrezgi, Z., Terefe Reda, T., Efficiency of treatment plant and drinking water quality assessment from source to household, gondar city, Northwest Ethiopia. *J. Environ. Public Health*, 2021, 1–8, 2021.
20. Iloms, E., Ololade, O.O., Ogola, H.J., Selvarajan, R., Investigating industrial effluent impact on municipal wastewater treatment plant in Vaal, South Africa. *Int. J. Environ. Res. Public Health*, 17, 1096, 2020.
21. Pacheco, F.A.L., do Valle Junior, R.F., de Melo, M.M.A.P., Pissarra, T.C.T., de Souza Rolim, G., de Melo, M.C., Valera, C.A., Moura, J.P., Fernandes, L.F.S., Geochemistry and contamination of sediments and water in rivers affected by the rupture of tailings dams (Brumadinho, Brazil). *Appl. Geochem.*, 152, 105644, 2023.
22. Millán-Becerro, R., Cánovas, C.R., Macías, F., Roetting, T.S., Siddorn, L., Stanley, P., Nieto, J.M., Passive remediation of mine waters from Parys

Mountain (Wales): Laboratory column experiments. *J. Clean. Prod.*, 425, 138872, 2023.

23. Singh, K.P., *Physicochemical Analysis Of Greywater And Its Impact On Surface Water Quality*, Shri Jagdishprasad Jhabarmal Tiberewala University, 2020. PhD thesis.
24. de Souza, R.M., Seibert, D., Quesada, H.B., de Jesus Bassetti, F., Fagundes-Klen, M.R., Bergamasco, R., Occurrence, impacts and general aspects of pesticides in surface water: a review. *Process Saf. Environ. Prot.*, 135, 22–37, 2020.
25. Chetty, S. and Pillay, L., Assessing the influence of human activities on river health: a case for two South African rivers with differing pollutant sources. *Environ. Monit. Assess.*, 191, 1–11, 2019.
26. Mutyambizi, C., Mokhele, T., Ndinda, C., Hongoro, C., Access to and satisfaction with basic services in informal settlements: results from a baseline assessment survey. *Int. J. Environ. Res. Public Health*, 17, 4400, 2020.
27. Wu, G.L., Liu, Y.F., Cui, Z., Liu, Y., Shi, Z.H., Yin, R., Kardol, P., Trade-off between vegetation type, soil erosion control and surface water in global semi-arid regions: A meta-analysis. *J. Appl. Ecol.*, 57, 875–885, 2020.
28. Opoku-Kwanowaa, Y., Furaha, R.K., Yan, L., Wei, D., Effects of planting field on groundwater and surface water pollution in China. *CLEAN–Soil Air Water*, 48, 1900452, 2020.
29. Gani, M.A., Sajib, A.M., Siddik, M.A., Moniruzzaman, M., Assessing the impact of land use and land cover on river water quality using water quality index and remote sensing techniques. *Environ. Monit. Assess.*, 195, 449, 2023.
30. Rodrigues, J.N., Ortega, J.C., Petsch, D.K., Padial, A.A., Moi, D.A., Figueiredo, B.R., A meta-analytical review of turbidity effects on fish mobility. *Rev. Fish Biol. Fish.*, 33, 1113–1127, 2023.
31. Allibhai, I., Zanghi, C., How, M.J., Ioannou, C.C., Increased water temperature and turbidity act independently to alter social behavior in guppies (Poecilia reticulata). *Ecol. Evol.*, 13, e9958, 2023.
32. Lin, L., Li, C., Yang, W., Zhao, L., Liu, M., Li, Q., Crittenden, J.C., Spatial variations and periodic changes in heavy metals in surface water and sediments of the Three Gorges Reservoir, China. *Chemosphere*, 240, 124837, 2020.
33. Syafrudin, M., Kristanti, R.A., Yuniarto, A., Hadibarata, T., Rhee, J., Al-Onazi, W.A., Algarni, T.S., Almarri, A.H., Al-Mohaimeed, A.M., Pesticides in drinking water—a review. *Int. J. Environ. Res. Public Health*, 18, 468, 2021.
34. Adam, R.I., David, S., Cairns, J.E., Olsen, M., A review of the literature on gender and chemical fertiliser use in maize production in sub-Saharan Africa. *J. Agric. Rural Dev. Tropics Subtropics (JARTS)*, 122, 91–102, 2021.
35. Xiang, T., Malik, T.H., Nielsen, K., The impact of population pressure on global fertiliser use intensity, 1970–2011: an analysis of policy-induced mediation. *Technol. Forecast. Soc. Change*, 152, 119895, 2020.
36. Klarich Wong, K.L., Webb, D.T., Nagorzanski, M.R., Kolpin, D.W., Hladik, M.L., Cwiertny, D.M., LeFevre, G.H., Chlorinated byproducts of

neonicotinoids and their metabolites: an unrecognized human exposure potential? *Environ. Sci. Technol. Lett.*, 6, 98–105, 2019.
37. Koekemoer, L., van Vuuren, S.J., Levanets, A., The influence of land use-impacted tributaries on water quality and phytoplankton in the Mooi River, North West Province, South Africa. *Bothalia-African Biodivers. Conserv.*, 51, 1–22, 2021.
38. Adams, J., Taljaard, S., Van Niekerk, L., Lemley, D., Nutrient enrichment as a threat to the ecological resilience and health of South African microtidal estuaries. *Afr. J. Aquat. Sci.*, 45, 23–40, 2020.
39. Becciu, G. and Mambretti, S., Water and Wastewater Management in Informal Settlements: A Case Study, in: *Environmental Performance and Social Inclusion in Informal Settlements*, pp. 155–175, Springer, Switzerland, 2020.
40. Turner, R.D., Warne, M.S.J., Dawes, L.A., Thompson, K., Will, G.D., Greywater irrigation as a source of organic micro-pollutants to shallow groundwater and nearby surface water. *Sci. Total Environ.*, 669, 570–578, 2019.
41. Putri, R., Hadisoebroto, R., Hendrawan, D., Analysis of pollutant load due to greywater from riverbanks settlement on Ciliwung River segment 2, in: *Journal of Physics: Conference Series*, IOP Publishing, p. 022099, 2019.
42. Magaba, S.L., Sustainable disposal and potential reuse of greywater produced at households in informal settlements: a case study of Monwabisi Park, Cape Town, South Africa, in: *Cape Peninsula University of Technology*, 2020.
43. Gohain, S.B. and Bordoloi, S., Impact of municipal solid waste disposal on the surface water and sediment of adjoining wetland Deepor Beel in Guwahati, Assam, India. *Environ. Monit. Assess.*, 193, 1–19, 2021.
44. Shaylinda, N., Amani, N., Mohd-Salleh, S., Shahar, M., Investigation of heavy metals pollution in Piyungan Landfill underground and surface water, in: *IOP Conference Series: Earth and Environmental Science*, IOP Publishing, p. 012080, 2020.
45. Wang, W., Ge, J., Yu, X., Bioavailability and toxicity of microplastics to fish species: A review. *Ecotoxicol. Environ. Saf.*, 189, 109913, 2020.
46. Zhang, C., Wang, J., Zhou, A., Ye, Q., Feng, Y., Wang, Z., Wang, S., Xu, G., Zou, J., Species-specific effect of microplastics on fish embryos and observation of toxicity kinetics in larvae. *J. Hazard. Mater.*, 403, 123948, 2021.
47. Bowley, J., Baker-Austin, C., Porter, A., Hartnell, R., Lewis, C., Oceanic hitchhikers–assessing pathogen risks from marine microplastic. *Trends Microbiol.*, 27, 107–116, 2020.
48. Capriotti, M., Cocci, P., Bracchetti, L., Cottone, E., Scandiffio, R., Caprioli, G., Sagratini, G., Mosconi, G., Bovolin, P., Palermo, F.A., Microplastics and their associated organic pollutants from the coastal waters of the central Adriatic Sea (Italy): Investigation of adipogenic effects *in vitro*. *Chemosphere*, 263, 128090, 2021.

49. Manoj, S., RamyaPriya, R., Elango, L., Long-term exposure to chromium contaminated waters and the associated human health risk in a highly contaminated industrialised region. *Environ. Sci. Pollut. Res.*, 28, 4276–4288, 2021.
50. Rahman, M.M., Alam, K., Velayutham, E., Is industrial pollution detrimental to public health? Evidence from the world's most industrialised countries. *BMC Public Health*, 21, 1–11, 2021.
51. Ramzan, A., Aiman, V., Intisar, A., Afzal, A., Hussain, T., Abid, M.A., Hussain, N., Microbial remediation of emerging pollutants from wastewater, in: *Advances in Chemical Pollution, Environmental Management and Protection*, pp. 207–226, Elsevier, Netherland, 2023.
52. E.d.C. Richard, G.C.D., Bechtold, J.P., de Aguiar Duarte Jr., H., Maioli, B.G., de Freitas, A.H.A., Warner, K.E., Figueiredo, L.H.M., Water and sediment quality in the coastal zone around the mouth of Doce River after the Fundão tailings dam failure. *Integr. Environ. Assess. Manage.*, 16, 643–654, 2020.
53. Luo, C., Routh, J., Dario, M., Sarkar, S., Wei, L., Luo, D., Liu, Y., Distribution and mobilization of heavy metals at an acid mine drainage affected region in South China, a post-remediation study. *Sci. Total Environ.*, 724, 138122, 2020.
54. Kishor, R., Purchase, D., Saratale, G.D., Saratale, R.G., Ferreira, L.F.R., Bilal, M., Chandra, R., Bharagava, R.N., Ecotoxicological and health concerns of persistent coloring pollutants of textile industry wastewater and treatment approaches for environmental safety. *J. Environ. Chem. Eng.*, 9, 105012, 2021.
55. Upfold, N.S., Luke, G.A., Knox, C., Occurrence of human enteric viruses in water sources and shellfish: A focus on Africa. *Food Environ. Virol.*, 13, 1–31, 2021.
56. Kabir, M.N., Aras, S., Wadood, S., Chowdhury, S., Fouladkhah, A.C., Fate and biofilm formation of wild-type and pressure-stressed pathogens of public health concern in surface water and on abiotic surfaces. *Microorganisms*, 8, 408, 2020.
57. Sekgobela, J.M., Murei, A., Khabo-Mmekoa, C.M., Momba, M.N.B., Identification of fecal contamination sources of groundwater in rural areas of Vhembe District Municipality, Limpopo Province, South Africa. *Water Environ. Res.*, 95, e10965, 2023.
58. Zhao, X., Yang, Y., Feng, K., Wang, X., Liu, B., Xie, G., Xing, D., Self-regulating microbiome networks ensure functional resilience of biofilms in sand biofilters during manganese load fluctuations. *Water Res.*, 188, 116473, 2021.
59. Maurya, A., Singh, M.K., Kumar, S., Biofiltration technique for removal of waterborne pathogens, in: *Waterborne Pathogens*, pp. 123–141, Elsevier, Netherland, 2020.
60. Hell, F., Shaheen, M., Liebminger, L.A., Water, D., River Water Purification Systems, Case Study El Raswa/Egypt, in: *Handbook of Water and Used Water Purification*, pp. 425–436, Springer, Singapore, 2024.

61. Mahmud, Z.H., Islam, M.S., Imran, K.M., Hakim, S.A.I., Worth, M., Ahmed, A., Hossan, S., Haider, M., Islam, M.R., Hossain, F., Occurrence of Escherichia coli and faecal coliforms in drinking water at source and household point-of-use in Rohingya camps, Bangladesh. *Gut Pathogens*, 11, 1–11, 2019.
62. Abubakar, M.M. and Abubakar, I.K., Physicochemical and Microbiological Assessment of Drinking Water in Kari Estate Bauchi Nigeria. *J. Biotechnol. Res.*, 6, 12–17, 2020.
63. Potgieter, N., Karambwe, S., Mudau, L.S., Barnard, T., Traore, A., Human enteric pathogens in eight Rivers used as rural household drinking water sources in the northern region of South Africa. *Int. J. Environ. Res. Public Health*, 17, 2079, 2020.
64. Tau, P., Anyasi, R., Mearns, K., Evaluating the Pollution of the Apies River in Pretoria South Africa, in: *E3S Web of Conferences*, EDP Sciences, p. 01004, 2021.
65. Toribio-Avedillo, D., Martín-Díaz, J., Jofre, J., Blanch, A.R., Muniesa, M., New approach for the simultaneous detection of somatic coliphages and F-specific RNA coliphages as indicators of fecal pollution. *Sci. Total Environ.*, 655, 263–272, 2019.
66. SABS, Drinking Water, Part 1: Microbiological, Physical, Chemical, Aesthetic and Chemical Determinants, in: *SABS Standards Division*, Pretoria, South Africa, 2015.
67. Mohanty, S. and Paul, S., Biofilm formation, problems, and diseases: Methods for film eradication, a nanostructured material based approach, in: *Microbial Ecology of Wastewater Treatment Plants*, pp. 469–492, Elsevier, Netherland, 2021.
68. Fu, Y., Peng, H., Liu, J., Nguyen, T.H., Hashmi, M.Z., Shen, C., Occurrence and quantification of culturable and viable but non-culturable (VBNC) pathogens in biofilm on different pipes from a metropolitan drinking water distribution system. *Sci. Total Environ.*, 764, 142851, 2021.
69. Baird, R.B., Standard Methods for the Examination of Water and Wastewater, in: *23rd, Water Environment Federation*, American Public Health Association, American, 2017.
70. Hemdan, B.A., El-Taweel, G.E., Goswami, P., Pant, D., Sevda, S., The role of biofilm in the development and dissemination of ubiquitous pathogens in drinking water distribution systems: an overview of surveillance, outbreaks, and prevention. *World J. Microbiol. Biotechnol.*, 37, 1–18, 2021.
71. Collivignarelli, M.C., Abbà, A., Benigna, I., Sorlini, S., Torretta, V., Overview of the main disinfection processes for wastewater and drinking water treatment plants. *Sustainability*, 10, 86, 2018.
72. Shi, Y., Babatunde, A., Bockelmann-Evans, B., Li, Q., Zhang, L., On-going nitrification in chloraminated drinking water distribution system (DWDS) is conditioned by hydraulics and disinfection strategies. *J. Environ. Sci.*, 96, 151–162, 2020.

73. Adeyemo, F.E., Singh, G., Reddy, P., Bux, F., Stenström, T.A., Efficiency of chlorine and UV in the inactivation of Cryptosporidium and Giardia in wastewater. *PLoS One*, 14, e0216040, 2019.
74. Betancourt, W., *Cryptosporidium spp*, Michigan State University, Michigan, USA, 2019.
75. Sarma, J., Filtration and chemical treatment of waterborne pathogens, in: *Waterborne Pathogens*, pp. 105–122, Elsevier, Netherland, 2020.
76. Tubatsi, G. and Kebaabetswe, L.P., Detection of enteric viruses from wastewater and river water in Botswana. *Food Environ. Virol.*, 14, 157–169, 2022.
77. Khabo-Mmekoa, C.M., Genthe, B., Momba, M.N.B., Enteric pathogens risk factors associated with household drinking water: a case study in Ugu District Kwa-Zulu Natal Province, South Africa. *Int. J. Environ. Res. Public Health*, 19, 4431, 2022.
78. Farkas, K., Mannion, F., Hillary, L.S., Malham, S.K., Walker, D.I., Emerging technologies for the rapid detection of enteric viruses in the aquatic environment. *Curr. Opin. Environ. Sci. Health*, 16, 1–6, 2020.
79. Bofill-Mas, S. and Rusiñol, M., Recent trends on methods for the concentration of viruses from water samples. *Curr. Opin. Environ. Sci. Health*, 16, 7–13, 2020.
80. Nasser, A. and Nitzan, Y., Coliphages as Indicators for the Microbial Quality of Treated Wastewater Effluents. *Food Environ. Virol.*, 13, 170–178, 2021.
81. Truchado, P., Garre, A., Gil, M.I., Simón-Andreu, P.J., Sánchez, G., Allende, A., Monitoring of human enteric virus and coliphages throughout water reuse system of wastewater treatment plants to irrigation endpoint of leafy greens. *Sci. Total Environ.*, 782, 146837, 2021.
82. Douterelo, I., Sharpe, R.L., Husband, S., Fish, K.E., Boxall, J.B., Understanding microbial ecology to improve management of drinking water distribution systems. *Wiley Interdiscip. Rev.: Water*, 6, e01325, 2019.
83. Jing, Z., Lu, Z., Mao, T., Cao, W., Wang, W., Ke, Y., Zhao, Z., Wang, X., Sun, W., Microbial composition and diversity of drinking water: A full scale spatial-temporal investigation of a city in northern China. *Sci. Total Environ.*, 776, 145986, 2021.
84. Favere, J., Barbosa, R.G., Sleutels, T., Verstraete, W., De Gusseme, B., Boon, N., Safeguarding the microbial water quality from source to tap. *NPJ Clean Water*, 4, 1–6, 2021.
85. Kennedy, L.C., Miller, S.E., Kantor, R.S., Nelson, K.L., Effect of disinfectant residual, pH, and temperature on microbial abundance in disinfected drinking water distribution systems. *Environ. Sci. Water Res. Technol.*, 7, 78–92, 2021.
86. Fish, K.E., Reeves-Mclaren, N., Husband, S., Boxall, J., Unchartered waters: the unintended impacts of residual chlorine on water quality and biofilms. *NPJ Biofilms Microbiomes*, 6, 1–12, 2020.

87. Cruz, M.C., Woo, Y., Flemming, H.-C., Wuertz, S., Nitrifying niche differentiation in biofilms from full-scale chloraminated drinking water distribution system. *Water Res.*, 176, 115738, 2020.
88. Li, W., Li, Y., Wang, F., Chen, S., Chen, J., Flow cytometric assessment of the chlorine/chloramine efficacy of particle-associated bacteria in drinking water. *Environ. Technol.*, 43, 1–17, 2021.
89. Rosales, E., Del Olmo, G., Calero Preciado, C., Douterelo, I., Phosphate Dosing in Drinking Water Distribution Systems Promotes Changes in Biofilm Structure and Functional Genetic Diversity. *Front. Microbiol.*, 11, 3226, 2020.
90. Rosenqvist, T., Danielsson, M., Schleich, C., Ahlinder, J., Brindefalk, B., Pullerits, K., Dacklin, I., Salomonsson, E.N., Sundell, D., Forsman, M., Succession of bacterial biofilm communities following removal of chloramine from a full-scale drinking water distribution system. *NPJ Clean Water*, 6, 41, 2023.
91. Karapanagioti, H., Water Management, Treatment and Environmental Impact, *Encyclopedia of Food and Health*, pp. 453–457, 2016.
92. García-Ávila, F., Avilés-Anazco, A., Sánchez-Cordero, E., Valdiviezo-Gonzáles, L., Ordonez, M.D.T., The challenge of improving the efficiency of drinking water treatment systems in rural areas facing changes in the raw water quality. *S. Afr. J. Chem. Eng.*, 37, 141–149, 2021.
93. WHO, *Water Safety Plan Manual: Step-by-Step Risk Management for Drinking-Water Suppliers*, World Health Organization, Geneva, 2023.
94. Moran, S., *An Applied Guide to Water and Effluent Treatment Plant Design*, Butterworth-Heinemann, Netherlands, 2018.
95. Gorito, A.M., Pesqueira, J.F., Moreira, N.F., Ribeiro, A.R., Pereira, M.F.R., Nunes, O.C., Almeida, C.M.R., Silva, A.M., Ozone-based water treatment (O3, O3/UV, O3/H2O2) for removal of organic micropollutants, bacteria inactivation and regrowth prevention. *J. Environ. Chem. Eng.*, 9, 105315, 2021.
96. Li, R.A., Mcdonald, J.A., Sathasivan, A., Khan, S.J., A multivariate Bayesian network analysis of water quality factors influencing trihalomethanes formation in drinking water distribution systems. *Water Res.*, 190, 116712, 2021.
97. Setareh, P., Khezri, S.M., Hossaini, H., Pirsaheb, M., Coupling effect of ozone/ultrasound with coagulation for improving NOM and turbidity removal from surface water. *J. Water Process Eng.*, 37, 101340, 2020.
98. Cui, H., Huang, X., Yu, Z., Chen, P., Cao, X., Application progress of enhanced coagulation in water treatment. *RSC Adv.*, 10, 20231–20244, 2020.
99. Chen, J., Ling, J., Sun, B., Wang, J., Zhou, B., Guan, X., Sun, Y., Trace organic contaminants abatement by permanganate/bisulfite pretreatment coupled with conventional water treatment processes: Lab-and pilot-scale tests. *J. Hazard. Mater.*, 401, 123380, 2021.

100. Zhang, H., Zheng, L., Li, Z., Pi, K., Deng, Y., One-step Ferrate (VI) treatment as a core process for alternative drinking water treatment. *Chemosphere*, 242, 125134, 2020.
101. Gan, W., Ge, Y., Zhong, Y., Yang, X., The reactions of chlorine dioxide with inorganic and organic compounds in water treatment: kinetics and mechanisms. *Environ. Sci. Water Res. Technol.*, 6, 2287–2312, 2020.
102. Ghernaout, D. and Elboughdiri, N., Eliminating Cyanobacteria and Controlling Algal Organic Matter—Short Notes. *Open Access Lib. J.*, 7, 1–17, 2020.
103. Abiola, O.N., Polymers for coagulation and flocculation in water treatment, in: *Polymeric Materials for Clean Water*, pp. 77–92, Springer, Switzerland, 2019.
104. Wang, L.K., Wang, M.-H.S., Shammas, N.K., Hahn, H.H., Physicochemical treatment consisting of chemical coagulation, precipitation, sedimentation, and flotation, in: *Integrated Natural Resources Research*, pp. 265–397, 2021.
105. Wang, L.K. and Wang, M.-H.S., Laboratory Procedures and Apparatus for Chemical Coagulation/Precipitation, Sedimentation, Dissolved Air Flotation, Dispersed Air Flotation, and Filtration, in: *Environmental Science, Technology, Engineering, and Mathematics (STEM)*, pp. 1–51, 2022.
106. Kuc, J., Thomas, M., Grochowalska, I., Kulczyk, R., Mikosz, G., Mrózek, F., Janik, D., Korta, J., Cwynar, K., Determination and removal of selected pharmaceuticals and total organic carbon from surface water by aluminum chlorohydrate coagulant. *Molecules*, 27, 5740, 2022.
107. Liu, Z., Wei, H., Li, A., Yang, H., Enhanced coagulation of low-turbidity micro-polluted surface water: properties and optimization. *J. Environ. Manage.*, 233, 739–747, 2019.
108. Fiore, F.A., Rodgher, S., Ito, C.Y.K., V.S.d.S. Bardini, L.M.G., Klinsky, Quality of surface water and generation of sludge at water treatment plants. *Rev. Ambiente Água*, 15, e2565, 2020.
109. Nayeri, D. and Mousavi, S.A., A comprehensive review on the coagulant recovery and reuse from drinking water treatment sludge. *J. Environ. Manage.*, 319, 115649, 2022.
110. Mwaisowa, F., Hoko, Z., Mapenzauswa, C.F., Feasibility of dissolved air flotation for drinking water treatment for Harare. *Water Pract. Technol.*, 18, 102–118, 2023.
111. Oh, H.J., Design and application of a high-rate dissolved air floatation process in a drinking water treatment plant: A field study for turbidity and algae removal. *J. Korean Soc. Water Wastewater*, 36, 391–402, 2022.
112. Kyzas, G.Z. and Matis, K.A., Flotation in water and wastewater treatment. *Processes*, 6, 116, 2018.
113. Gonzales-Galvis *et al.*, Comparison of Dissolved Air Flotation and Ballasted Sedimentation for the Treatment of an Algal Impacted Water, *Desalin. Water Treat.*, 245, 45–53, 2022.

114. Zhang, R., Sun, Z., Cui, B., Ren, P., Du, Z., Chai, X., Lu, J., Improvement on algae and turbidity removal in an integrated flotation and sedimentation unit using side flow-inclined plate settlers: evidence from a full-scale field experiment. *J. Water Supply Res. Technol.—AQUA*, 67, 84–98, 2018.
115. Cescon, A. and Jiang, J.-Q., Filtration Process and Alternative Filter Media Material in Water Treatment. *Water*, 12, 3377, 2020.
116. Chen, L., Zhai, Y., Van Der Mark, E., Liu, G., Van Der Meer, W., Medema, G., Microbial community assembly and metabolic function in top layers of slow sand filters for drinking water production. *J. Clean. Prod.*, 294, 126342, 2021.
117. Chan, S., Pullerits, K., Riechelmann, J., Persson, K.M., Rådström, P., Paul, C.J., Monitoring biofilm function in new and matured full-scale slow sand filters using flow cytometric histogram image comparison (CHIC). *Water Res.*, 138, 27–36, 2018.
118. McGivney, W. and Kawamura, S., *Cost Estimating Manual for Water Treatment Facilities*, Wiley Online Library, Switzerland, 2008.
119. Betancourt, W.Q. and Rose, J.B., Drinking water treatment processes for removal of Cryptosporidium and Giardia. *Vet. Parasitology*, 126, 219–234, 2004.
120. Andreoli, F.C. and Sabogal-Paz, L.P., Coagulation, flocculation, dissolved air flotation and filtration in the removal of Giardia spp. and Cryptosporidium spp. from water supply. *Environ. Technol.*, 40, 654–663, 2019.
121. Lanrewaju, A.A., Enitan-Folami, A.M., Sabiu, S., Swalaha, F.M., A review on disinfection methods for inactivation of waterborne viruses. *Front. Microbiol.*, 13, 991856, 2022.
122. Sholichah, H., Application of sodium hypochlorite (NaOCl) for ammonia removal in wastewater effluent of PT. HS. *J. Environ. Eng. Waste Manage.*, 5, 156–169, 2020.
123. Han, J., Zhang, X., Li, W., Jiang, J., Low chlorine impurity might be beneficial in chlorine dioxide disinfection. *Water Res.*, 188, 116520, 2021.
124. Tang, A., Bi, X., Li, X., Li, F., Liao, X., Zou, J., Sun, W., Yuan, B., The inactivation of bacteriophage MS2 by sodium hypochlorite in the presence of particles. *Chemosphere*, 266, 129191, 2021.
125. Gelete, G., Gokcekus, H., Ozsahin, D.U., Uzun, B., Gichamo, T., Evaluating disinfection techniques of water treatment. *Desalin. Water Treat.*, 177, 408–415, 2020.
126. Ding, W., Jin, W., Cao, S., Zhou, X., Wang, C., Jiang, Q., Huang, H., Tu, R., Han, S.-F., Wang, Q., Ozone disinfection of chlorine-resistant bacteria in drinking water. *Water Res.*, 160, 339–349, 2019.
127. Zhu, Y., Chen, C., Shi, J., Shangguan, W., A novel simulation method for predicting ozone generation in corona discharge region. *Chem. Eng. Sci.*, 227, 115910, 2020.
128. Wattieaux, G., Ferrer, V., Sarrette, J.-P., Merbahi, N., Eichwald, O., Experimental mapping of the ozone distribution in a pulsed positive corona

discharge to estimate the efficiency of ozone production. *Phys. Scr.*, 97, 125608, 2022.

129. Kraft, A., Electrochemical water disinfection: a short review. *Platinum Met. Rev.*, 52, 177–185, 2008.
130. Bakheet, B., Islam, M.A., Beardall, J., Zhang, X., McCarthy, D., Effective electrochemical inactivation of Microcystis aeruginosa and degradation of microcystins *via* a novel solid polymer electrolyte sandwich. *Chem. Eng. J.*, 350, 616–626, 2018.
131. Lou, Y., Kasler, D.R., Hawkins, Z.L., Li, Z., Sannito, D., Fritz, R.D., Yousef, A.E., Inactivation kinetics of selected pathogenic and non-pathogenic bacteria by aqueous ozone to validate minimum usage in purified water. *Front. Microbiol.*, 14, 1258381, 2024.
132. He, Z., Fan, X., Jin, W., Gao, S., Yan, B., Chen, C., Ding, W., Yin, S., Zhou, X., Liu, H., Chlorine-resistant bacteria in drinking water: Generation, identification and inactivation using ozone-based technologies. *J. Water Process Eng.*, 53, 103772, 2023.
133. Mokoena, M.M., Mudau, L.S., Mokgobu, M.I., Mukhola, M.S., The use of sodium hypochlorite at point-of-use to remove microcystins from water containers. *Toxins*, 13, 207, 2021.
134. Eichler, S., Christen, R., Höltje, C., Westphal, P., Bötel, J., Brettar, I., Mehling, A., Höfle, M.G., Composition and dynamics of bacterial communities of a drinking water supply system as assessed by RNA-and DNA-based 16S rRNA gene fingerprinting. *Appl. Environ. Microbiol.*, 72, 1858–1872, 2006.
135. Shaw, J.L., Monis, P., Weyrich, L.S., Sawade, E., Drikas, M., Cooper, A.J., Using amplicon sequencing to characterize and monitor bacterial diversity in drinking water distribution systems. *Appl. Environ. Microbiol.*, 81, 6463–6473, 2015.
136. Léziart, T., Dutheil De La Rochere, P.-M., Cheswick, R., Jarvis, P., Nocker, A., Effect of turbidity on water disinfection by chlorination with the emphasis on humic acids and chalk. *Environ. Technol.*, 40, 1734–1743, 2019.
137. Potgieter, S.C., Dai, Z., Venter, S.N., Sigudu, M., Pinto, A.J., Microbial Nitrogen Metabolism in Chloraminated Drinking Water Reservoirs. *mSphere*, 5, 1–19, 2020.
138. Bruno, A., Sandionigi, A., Bernasconi, M., Panio, A., Labra, M., Casiraghi, M., Changes in the drinking water microbiome: effects of water treatments along the flow of two drinking water treatment plants in a urbanized area, Milan (Italy). *Front. Microbiol.*, 9, 2557, 2018.
139. Gholami-Borujeni, F., Relationship Between Environment Temperature, TOC, Turbidity, and HPC on Free Residual Chlorine in Municipal Drinking Water Distribution Networks (WDNs): A Case Study (Abadan, Iran), in: *Proceedings of the 4th International Conference on Advances in Civil and Ecological Engineering Research: ACEER2022*, Springer, pp. 305–314, 2022.

140. Perrin, Y., Bouchon, D., Héchard, Y., Moulin, L., Spatio-temporal survey of opportunistic premise plumbing pathogens in the Paris drinking water distribution system. *Int. J. Hyg. Environ. Health*, 222, 687–694, 2019.
141. Chen, J., Li, W., Zhang, J., Qi, W., Li, Y., Chen, S., Zhou, W., Prevalence of antibiotic resistance genes in drinking water and biofilms: The correlation with the microbial community and opportunistic pathogens. *Chemosphere*, 259, 127483, 2020.
142. Abhijith, G.R. and Ostfeld, A., Modeling the Formation and Propagation of 2, 4, 6-trichloroanisole, a Dominant Taste and Odor Compound, in Water Distribution Systems. *Water*, 13, 638, 2021.
143. Kimbell, L.K., Wang, Y., Mcnamara, P.J., The impact of metal pipe materials, corrosion products, and corrosion inhibitors on antibiotic resistance in drinking water distribution systems. *Appl. Microbiol. Biotechnol.*, 104, 7673–7688, 2020.
144. Hu, Y., Dong, D., Wan, K., Chen, C., Yu, X., Lin, H., Potential shift of bacterial community structure and corrosion-related bacteria in drinking water distribution pipeline driven by water source switching. *Front. Environ. Sci. Eng.*, 15, 1–10, 2021.
145. Fish, K., Osborn, A., Boxall, J., Biofilm structures (EPS and bacterial communities) in drinking water distribution systems are conditioned by hydraulics and influence discolouration. *Sci. Total Environ.*, 593, 571–580, 2017.
146. Andersen, J.B., Kragh, K.N., Hultqvist, L.D., Rybtke, M., Nilsson, M., Jakobsen, T.H., Givskov, M., Tolker-Nielsen, T., Induction of native c-di-GMP phosphodiesterases leads to dispersal of Pseudomonas aeruginosa biofilms. *Antimicrob. Agents Chemother.*, 65, 1–13, 2021.
147. Park, S., Dingemans, J., Gowett, M., Sauer, K., Glucose-6-Phosphate Acts as an Extracellular Signal of SagS To Modulate Pseudomonas aeruginosa c-di-GMP Levels, Attachment, and Biofilm Formation. *Msphere*, 6, 1–17, 2021.
148. Bridges, A.A. and Bassler, B.L., The intragenus and interspecies quorum-sensing autoinducers exert distinct control over Vibrio cholerae biofilm formation and dispersal. *PLoS Biol.*, 17, e3000429, 2019.
149. Kumar, R.V., Singh, R.P., Mishra, P., Endophytes as emphatic communication barriers of quorum sensing in Gram-positive and Gram-negative bacteria—a review. *Environ. Sustainability*, 2, 455–468, 2019.
150. Prabhu, M., Naik, M., Manerikar, V., Quorum Sensing-Controlled Gene Expression Systems in Gram-Positive and Gram-Negative Bacteria, in: *Implication of Quorum Sensing and Biofilm Formation in Medicine, Agriculture and Food Industry*, pp. 11–20, Springer, Singapore, 2019.
151. Afonso, T.B., Simões, L.C., Lima, N., Effect of quorum sensing and quenching molecules on inter-kingdom biofilm formation by Penicillium expansum and bacteria. *Biofouling*, 36, 965–976, 2020.
152. Vanetti, M.C.D., Carneiro, D.G., De Almeida, F.A., Vargas, E.L.G., Lima, E.M.F., Pinto, U.M., Autoinducer-1 Quorum Sensing Communication

Mechanism in Gram-Negative Bacteria, in: *Trends in Quorum Sensing and Quorum Quenching*, pp. 9–29, 2020.

153. Cai, Y.-M., Hutchin, A., Craddock, J., Walsh, M.A., Webb, J.S., Tews, I., Differential impact on motility and biofilm dispersal of closely related phosphodiesterases in Pseudomonas aeruginosa. *Sci. Rep.*, 10, 1–12, 1–18, 2020.
154. Cherny, K.E. and Sauer, K., Pseudomonas aeruginosa requires the DNA-specific endonuclease EndA to degrade extracellular genomic DNA to disperse from the biofilm. *J. Bacteriol.*, 201, 1–18, 2019.
155. Afonso, T.B., Simões, L.C., Lima, N., Occurrence of filamentous fungi in drinking water: their role on fungal-bacterial biofilm formation. *Res. Microbiol.*, 172, 103791, 2020.
156. Ahmad, J.I., Dignum, M., Liu, G., Medema, G., Van Der Hoek, J.P., Changes in biofilm composition and microbial water quality in drinking water distribution systems by temperature increase induced through thermal energy recovery. *Environ. Res.*, 194, 110648, 2021.
157. Inkinen, J., Jayaprakash, B., Siponen, S., Hokajärvi, A.-M., Pursiainen, A., Ikonen, J., Ryzhikov, I., Täubel, M., Kauppinen, A., Paananen, J., Active eukaryotes in drinking water distribution systems of ground and surface waterworks. *Microbiome*, 7, 1–17, 2019.
158. Siqueira, V.M., Oliveira, H., Santos, C., Paterson, R., Gusmão, N., Lima, N., Biofilms from a Brazilian water distribution system include filamentous fungi. *Can. J. Microbiol.*, 59, 183–188, 2013.
159. Douterelo, I., Fish, K., Boxall, J., Succession of bacterial and fungal communities within biofilms of a chlorinated drinking water distribution system. *Water Res.*, 141, 74–85, 2018.
160. Tsagkari, E. and Sloan, W.T., Impact of Methylobacterium in the drinking water microbiome on removal of trihalomethanes. *Int. Biodeterior. Biodegrad.*, 141, 10–16, 2019.
161. Abe, K., Nomura, N., Suzuki, S., Biofilms: hot spots of horizontal gene transfer (HGT) in aquatic environments, with a focus on a new HGT mechanism. *FEMS Microbiol. Ecol.*, 96, fiaa031, 2020.
162. Gomez-Smith, C.K., Lapara, T.M., Hozalski, R.M., Sulfate reducing bacteria and mycobacteria dominate the biofilm communities in a chloraminated drinking water distribution system. *Environ. Sci. Technol.*, 49, 8432–8440, 2015.
163. Waak, M.B., Lapara, T.M., Halle, C., Hozalski, R.M., Nontuberculous mycobacteria in two drinking water distribution systems and the role of residual disinfection. *Environ. Sci. Technol.*, 53, 8563–8573, 2019.
164. Del Olmo, G., Husband, S., Briones, C.S., Soriano, A., Preciado, C.C., Macian, J., Douterelo, I., The microbial ecology of a Mediterranean chlorinated drinking water distribution systems in the city of Valencia (Spain). *Sci. Total Environ.*, 754, 142016, 2021.

165. Afonso, T.B., Simões, L.C., Lima, N., *In vitro* assessment of inter-kingdom biofilm formation by bacteria and filamentous fungi isolated from a drinking water distribution system. *Biofouling*, 35, 1041–1054, 2019.
166. Douterelo, I., Boxall, J.B., Deines, P., Sekar, R., Fish, K.E., Biggs, C.A., Methodological approaches for studying the microbial ecology of drinking water distribution systems. *Water Res.*, 65, 134–156, 2014.
167. Douterelo, I., Calero-Preciado, C., Soria-Carrasco, V., Boxall, J.B., Whole metagenome sequencing of chlorinated drinking water distribution systems. *Environ. Sci. Water Res. Technol.*, 4, 2080–2091, 2018.
168. Kelly, J.J., Minalt, N., Culotti, A., Pryor, M., Packman, A., Temporal variations in the abundance and composition of biofilm communities colonizing drinking water distribution pipes. *PLoS One*, 9, e98542, 2014.
169. Liu, G., Bakker, G., Li, S., Vreeburg, J., Verberk, J., Medema, G., Liu, W., Van Dijk, J., Pyrosequencing reveals bacterial communities in unchlorinated drinking water distribution system: an integral study of bulk water, suspended solids, loose deposits, and pipe wall biofilm. *Environ. Sci. Technol.*, 48, 5467–5476, 2014.
170. Douterelo, I., Sharpe, R., Boxall, J., Bacterial community dynamics during the early stages of biofilm formation in a chlorinated experimental drinking water distribution system: implications for drinking water discolouration. *J. Appl. Microbiol.*, 117, 286–301, 2014.
171. Liu, J.-Q., Luo, Z.-F., Liu, K., Zhang, Y.-F., Peng, H.-X., Hu, B.-L., Ren, H.-X., Zhou, X.-Y., He, X.-F., Ye, P., Effect of flushing on the detachment of biofilms attached to the walls of metal pipes in water distribution systems. *J. Zhejiang University-SCIENCE A*, 18, 313–328, 2017.
172. Fish, K.E., Husband, S.P., Boxall, J.B., The impact of chlorine concentration on the discolouration response of biofilms in drinking water distribution systems, in: *WDSA/CCWI Joint Conference Proceedings*, 2018.
173. Calero Preciado, C., Soria-Carrasco, V., Boxall, J., Douterelo, I., Climate change and management of biofilms within drinking water distribution systems. *Front. Environ. Sci.*, 10, 962514, 2022.
174. WHO, Guidelines for drinking-water quality, in: *WHO chronicle*, pp. 104–108, 2011.
175. T. ISO, *9308-1-Water Quality-Determination and Counting of Escherichia coli and coliform bacteria-Part 1: Membrane Filtration Method for low Bacterial Ground Water (ISO 9308-1: 2014)*, Turkish Standards Institute, Ankara, 2014.
176. Pitkänen, T., *Water Quality-Enumeration of Escheria coli and coliform bacteria-Part 2: Most Probable Number Method*, Springer, Switzerland, 2012.
177. ISO EN 6222, Water quality—Enumeration of culturable micro-organism—colony count by inoculation in a nutrient agar culture medium, in: *International Organization for Standardization*, 1999, (EN ISO Standard 6222–1999).
178. Lin, J. and Ganesh, A., Water quality indicators: bacteria, coliphages, enteric viruses. *Int. J. Environ. Health Res.*, 23, 484–506, 2013.

179. Wang, L., Zhang, J., Li, H., Yang, H., Peng, C., Peng, Z., Lu, L., Shift in the microbial community composition of surface water and sediment along an urban river. *Sci. Total Environ.*, 627, 600–612, 2018.
180. Potgieter, S., Pinto, A., Sigudu, M., Du Preez, H., Ncube, E., Venter, S., Long-term spatial and temporal microbial community dynamics in a large-scale drinking water distribution system with multiple disinfectant regimes. *Water Res.*, 139, 406–419, 2018.
181. Yu, J., Kim, D., Lee, T., Microbial diversity in biofilms on water distribution pipes of different materials. *Water Sci. Technol.*, 61, 163–171, 2010.
182. Lu, J., Struewing, I., Vereen, E., Kirby, A., Levy, K., Moe, C., Ashbolt, N., Molecular detection of Legionella spp. and their associations with Mycobacterium spp., Pseudomonas aeruginosa and amoeba hosts in a drinking water distribution system. *J. Appl. Microbiol.*, 120, 509–521, 2016.
183. Brumfield, K.D., Hasan, N.A., Leddy, M.B., Cotruvo, J.A., Rashed, S.M., Colwell, R.R., Huq, A., A comparative analysis of drinking water employing metagenomics. *PLoS One*, 15, e0231210, 2020.
184. Waak, M., *Investigation of the Microbiomes in Two Full-Scale Drinking Water Distribution Systems*, PhD thesis, University of Minnesota, California, 2018.
185. Kormas, K.A., Neofitou, C., Pachiadaki, M., Koufostathi, E., Changes of the bacterial assemblages throughout an urban drinking water distribution system. *Environ. Monit. Assess.*, 165, 27–38, 2010.
186. El-Chakhtoura, J., Saikaly, P.E., Van Loosdrecht, M., Vrouwenvelder, J.S., Impact of distribution and network flushing on the drinking water microbiome. *Front. Microbiol.*, 9, 2205, 2018.
187. Ren, H., Wang, W., Liu, Y., Liu, S., Lou, L., Cheng, D., He, X., Zhou, X., Qiu, S., Fu, L., Pyrosequencing analysis of bacterial communities in biofilms from different pipe materials in a city drinking water distribution system of East China. *Appl. Microbiol. Biotechnol.*, 99, 10713–10724, 2015.
188. Zhang, C., Li, C., Zheng, X., Zhao, J., He, G., Zhang, T., Effect of pipe materials on chlorine decay, trihalomethanes formation, and bacterial communities in pilot-scale water distribution systems. *Int. J. Environ. Sci. Technol.*, 14, 85–94, 2017.
189. Garner, E., Davis, B.C., Milligan, E., Blair, M.F., Keenum, I., Maile-Moskowitz, A., Pan, J., Gnegy, M., Liguori, K., Gupta, S., Next generation sequencing approaches to evaluate water and wastewater quality. *Water Res.*, 194, 116907, 2021.
190. Ardui, S., Ameur, A., Vermeesch, J.R., Hestand, M.S., Single molecule real-time (SMRT) sequencing comes of age: applications and utilities for medical diagnostics. *Nucleic Acids Res.*, 46, 2159–2168, 2018.
191. Wenger, A.M., Peluso, P., Rowell, W.J., Chang, P.-C., Hall, R.J., Concepcion, G.T., Ebler, J., Fungtammasan, A., Kolesnikov, A., Olson, N.D., Accurate circular consensus long-read sequencing improves variant detection and assembly of a human genome. *Nat. Biotechnol.*, 37, 1155–1162, 2019.

Part III

OPTIMIZATION OF THE RECOVERY OF VALUABLE BY-PRODUCTS FROM THE TREATMENT PROCESS (CIRCULAR ECONOMY)

9

Avenues for the Recovery and Synthesis of Valuable Minerals from Municipal Wastewater and Their Valorization

Collen Nepfumbada[1], Tavenga Nikita[1], Beauclair Nguegang[2] and Vhahangwele Masindi[3]*

[1]Department of Chemistry, School of Mathematics and Natural Sciences, University of Venda, Thohoyandou, South Africa
[2]Department of Chemical Sciences, Faculty of Science, University of Johannesburg, Johannesburg, South Africa
[3]Department of Environmental Sciences, College of Agriculture and Environmental Sciences (CAES), University of South Africa (UNISA), Florida, South Africa

Abstract

Recently, municipal wastewater (MWW) treatment plants have received significant attention owing to the high concentration level of pollutants present, which negatively affect the environment and human health. Regulatory structures expect MWW effluents to be treated before being released into the environment to conform to specific limits and reduce its potential ecological impacts. For this reason, different strategies have been established, conducted, and applied for the treatment of MWW and obtain multitude of ecological benefits, hence encouraging the concept of circular economy, sustainability, and beneficiation of MWW. Nowadays, there are many different approaches to integrating MWW treatment systems that combine chemical and physical treatment or any other hybrid strategy. These approaches have received a lot of attention and are described in this article. This initiative allows for the recovery of valuable minerals from the treatment of MWW. This critical and comprehensive review concentrate on the recovery of valuable minerals from MWWs, valorization, beneficiation, technologies, and the mechanisms involved during the recovery process. In addition, the economic prospects of such processes are assessed. Possible future directions for the

**Corresponding author*: emasinv@unisa.ac.za; masindivhahangwele@gmail.com

Elvis Fosso-Kankeu, Vhahangwele Masindi, Johannes Maree and Bhekie Mamba (eds.) Customized Technologies for Sustainable Management of Industrial Wastewater: A Circular Economy Approach, (359–388) © 2025 Scrivener Publishing LLC

recovery of mineral from wastewater are also recommended with the pursuit of highlighting the concept of circular economy and sustainable development. The principal objective of this review is to clarify latest advancement and challenges in terms of MWW treatment, beneficiation routes, and valorization. Different technologies that have been used to recover minerals from MWW and their limitations will thoroughly be unpacked. This review will also underline the challenges hampering water treatment. Insights toward potential research avenues and enlightenment of future research perspectives will also be discussed in detail. This study will present possible, careful, and practical avenues for valorization and beneficiation of MWW via recovering of minerals while insights into prospects and areas of interest will also be enlightened and explained herein.

***Keywords*:** Synthesis and recovery, valuable minerals, municipal wastewater, valorizations, circular economy, drinking water reclamation

9.1 Introduction

The demand for natural resources, such as water, renewable energy, and minerals, has considerably increased because of the rapid growth in the global population. As a result, soil amendment productions are currently facing several challenges as 80%–90% of phosphate demand in the universe is used for food production [1–3], although the residual available phosphate rock is expected to be completely depleted in 100–300 years to come [4]. But the depletion of natural sources of valuable minerals and fresh water is of emerging concerns [5, 6] and gradually lead to the development of strict regulation restrictive on waste discharges from wastewater treatment plants (WWTPs) [7]. In addition, the role of phosphorus in fertilizer production can never be replaced by any other materials. For instance, the industrial Haber–Bosch process, which converted atmospheric nitrogen into NH_3-N, is continually required to produce ammonia fertilizer. However, the amount of NH_3-N produced by human activity exceeds that will be transformed by natural processes, disrupting the nature of the nitrogen cycle, which leaves the environment exposed to severe risk [8]. Most importantly, the industrial Haber–Bosch process utilizes a significant amount of energy during ammonia production, which account for roughly 2% of annual global energy supply being utilized and attendant greenhouse gas emissions [9, 10]. Furthermore, throughout the nitrification-denitrification process, a substantial amount of chemical input is needed to support as the electron donor. The nitrate by-product from the nitrification-denitrification process lacks significant marketplace and usually needs to be purified further before it can be released [11]. Therefore, recovering and reusing of water and raw materials have gained essential interest to obtain sustainable fresh water and certain valuable raw materials to meet their demand.

Now, considering the risk that pollutants from municipal wastewater treatment plants (MWWTPs) has to the environment, human health, and water quality, new strategies for their recovery and removal is crucial. However, during the treatment process in MWWTPs, a significant amount of sludge is produced, which seriously endanger the ecosystem [12, 13]. Fortunately, MWWTP treatment techniques are evolving in a sense that makes the wastewater treatment industry more sustainable owing to the growing market price of the elements contained in MWWTPs that can be recovered [14, 15]. Therefore, the emphasis is on wastewater being regarded as renewable resource and one that is becoming more and more valued to factors such as economic, environmental, and social concerns [16]. For this reason, MWWTPs are being used as natural resource recovery facilities rather than waste streams that aim at discharging treated wastewater into waterways because they comprise resources that can be recovered using various processes into reusable energy, water, fertilizers, and other valuable by-products [17–21] Furthermore, this shift from the removal of waste products to the recovery of valuable minerals provides wastewater sectors with broad understanding of the circular economy. The most common problem in MWWTPs is the presence of abundant nutrients such as ammonia, nitrate, and phosphate, which have the potential to create significant environmental threat to water bodies, if not handled properly. The excessive nutrient enrichment in biota could also lead to the rapid growth of aquatic plants [22, 23]. Recovering energy and nutrients from MWWTPs, such as phosphorus and ammonia, displays potential for the circular economy and deserves special attention due to the increase in the inflexibility of effluent discharge guidelines and questions associated with minable phosphorus resource for food safety [16]. Utilizing the biogas produced from wastewater organic content can reduce the amount of energy needed to treat waste [24]. Alternatively, recovery of nutrients from MWWTPs can lessen the upkeep cost of wastewater treatment facilities and, thus, prevent environmental impacts. The recovered nutrients from municipal wastewater (MWW) can also be used to improve soil quality or as fertilizers for agricultural purposes [25].

Although various academics have conducted substantial research on methods to recover valuable resources from wastewater in recent years, there is currently a lack of widespread implementation of these technologies in MWWTPs, which can be justified mainly by technical and non-technical reasons.

Hence, the tools and methods for recovering resources from MWWTPs must be environmentally friendly and economically attractive and must have the potential to recycle water, energy, nutrients, and other valuable minerals from wastewater effectively. Ammonium ion is valuable because it prevails in anaerobic reactor effluents and can be beneficial for soil amendments. On the other hand, phosphorus recovery as a struvite is important for the prevention of eutrophication in the water system and for easing economic reliance on phosphate rocks [25]. Thus, precipitation of phosphorus in the form of struvite from wastewater streams has increasingly gained more attention from many researchers because of its possible use in fertilizing, given that phosphate rock, a non-renewable resource, is expected to run out in the upcoming years [26].

To remove nitrogen and phosphorus from WWTPs, a variety of physical, biological, and chemical technologies have been used. These technologies include membrane processes, adsorption, chemical precipitation, membrane processes, electrochemical processes, ion exchange processes, biological processes, physical and physico-chemical processes, and conventional activated sludge [27–31]. Most of these methods are challenging as they generate low-purity products and are cost-effective or operationally complex. The most used method in WWTPs treatment is the conventional activated sludge process, through which aerobic microbes metabolize the organic substrate present in water under constant oxygen supply and crystallization. Albeit the conventional activated sludge process succeeds in regulated effluent quality standards, it is regarded as unsustainable due to its limitations, such as low potential for resource recovery, cost-effectiveness, high energy demand, and large environmental footprint [18]. However, crystallization is a widely used method with a moderately high resource recovery and considerable economic effectiveness as it produces beneficial products with limited environmental challenges [32, 33]. These properties reveal the potential for crystallization process for the recovery of nitrogen and phosphorus from wastewater. Therefore, the potential of calcium- and magnesium-based material on the recovery of nitrogen and phosphorus from WWTPs, including sustainable mechanisms and regeneration methods of the utilized calcium- and magnesium-based material, has been investigated.

The recovery of precious minerals in MWWTPs and the related challenges have been the subject of critical concern corresponding to the number of review articles published worldwide [34, 35]. To the best of our knowledge, few of those published reviews are centered on the beneficiation and valorization of MWW effluent for circular economy. Moreover, most of those reviews lack information on the environmental benefits associated with the recovered minerals [36, 37]. Therefore, this study demonstrates the milestones that have been reached by different researchers toward environmental examination of nutrients in wastewater. Additionally, the summary of the methods used by different studies to evaluate nutrients in the aquatic environments is provided. Furthermore, the emphasis on the role of calcium- and magnesium-based material in mineral recovery from WWTPs and their final disposal is largely explained. This paper's main goals are to (1) investigate the mechanisms and effectiveness of calcium- and magnesium-based material for resource recovery; (2) discuss the methods for resource recovery during water treatment; (3) demonstrate possible reuse of recovered resources; (4) increase interests of resource recovery from wastewater using calcium- and magnesium-based material; (5) put forward the prospects of future research studies in calcium- and magnesium-based resource recovery; and (6) the economic and technical feasibility of nitrogen and phosphorus recovery through crystallization process and evaluate the recovered products.

9.2 Streams and Their Composition

9.2.1 Influent and Effluent Composition

MWWTPs are very important structures to safeguard public health and ecosystem. The amount of waste entering MWWTPs is continually increasing due to population growth and development [38]. These wastes are generated from anthropogenic activities such as bathing, washing, and cooking. Treating MWWs is important to keep the environment safe and to sustain the wastewater discharge standard quality established by Environmental Protection Agency (EPA) and other regulatory bodies. Therefore, insufficient waste removal during wastewater treatment leads to the frequent identification of organic compounds such as personal care products, pharmaceuticals, disinfection by-products, and their metabolites [39–42], along with toxic heavy metals that can be disposed into water bodies and, consequently, pose severe threat to the natural ecosystem [43].

9.2.2 Dewatering Stream

The conventional WWTPs face great challenges on water produced in dewatering of digested sewage sludge [44]. Reports suggest that the phosphorus concentration in WWTPs influent is significantly lower than that of the sludge or the solution that separates from the sludge during the dewatering process [19]. However, concentrated phosphorus is present in the filtrate from the dewatering process, which is often returned to the traditional wastewater treatment procedures. Because of this, phosphorus is currently circulated between sludge management and conventional wastewater treatment systems, which results in ineffective phosphorus removal from MWWTPs.

9.2.3 Sludge

Treating MWW resolves one challenge but, at the same time, creates another problem, which is excessive production of sludge that requires precise disposal and controlling mechanism [14]. Wastewater biosolid, also known as sludge from the biological treatment process, is a by-product of both water and wastewater. Sludge, also known as slurry or semi-solid waste, is essentially a by-product of sewage handling processes, which needs to undergo further treatment procedures before being disposed of or applied to land [45, 46]. Sewage sludge is attained during wastewater treatment in sewage treatment sectors. Consequently, the generated sludge comprises two fundamental forms: primary sludge and secondary sludge, which, in the event of an activated sludge process, is also referred to as activated sludge and contains a variety of organic and inorganic compounds [47, 48]. However, among all these pollutants, heavy metals are very toxic to animals, plants, and human health and limit the application of sewage sludge for agricultural purposes [49]. There are number of alternate methods for sludge disposal, including energy recoveries, adsorbent preparations, landfills, incineration, and agricultural field applications [50].

9.3 Ecological Impacts of Municipal Wastewater

9.3.1 Effects on Agriculture

The use of wastewater for agriculture has drawn more attention as a viable and effective waste management technique [51]. Due to the global shortage of freshwater and increased stress on irrigation water, utilization of wastewater for agricultural purposes has gained essential interest [52]. However, the ecological and toxicological effects of effluents from MWW on soil properties and plants are not being considered [53]. Several contaminants, such as heavy metals, resistant pathogens, and emerging pollutants (EPs), enter soils because of reusing wastewater during irrigation for agricultural practice [54]. Toxic materials are deposited by the soils after usage and are frequently connected to the ingestion process. Consequently, this causes damage and disruption to the metabolic processes and functional roles that soil organisms play [55, 56]. These soil organisms enhance soil fertility as they provide different functions, such as improving organic matter and enhancing soil structure, which is valuable for plants. Thus, the biological fertility of the soil is most likely affected by pollutants that constitute the MWW [57, 58]. Conversely, MWW effluents comprise nutrients that can act as fertilizer for crop production; however, occasionally, these nutrients become more accessible than the plant could need, which could result in a threat related to excessive vegetative growth, delayed or uneven maturity, or decreased quality [59]. For instance, some toxic heavy metals are mostly detected in organs from the tomatoes grown within the effluent discharge area of the MWWTPs [60].

9.3.2 Effects on Living Organisms

Despite the lowering of water quality, water pollution can disrupt aquatic life in the water bodies. Due to the existence of several chemical components, this may likely be toxic to aquatic organisms even at very low quantities [61]. Thus, it is well recognized that the major source of toxic contaminants entering aquatic habitats is through the discharge of effluent from MWWTPs. Most pollutants found in WWTPs are not entirely removed by WWTPs, which leaves these substances in surface waters. This increases concerns regarding the potential for wastewater discharge to impact aquatic organisms [62]. Specifically, organisms that are mainly impacted by the wastewater effluents are those of fish habitat [63]. According to the study performed by [64] and [65], there was clear indication of neuroendocrine and reproductive defects in humans and animals with the detection of such contaminants. The detrimental effects, such as changes in metabolism, development, growth, and reproduction of organisms such as fish, algae, and amphibians in relation to the exposure to endocrine disruptors, were documented [66, 67]. For instance, Jasinska *et al.* [68] and Bahamonde *et al.* [69] have indicated that sewage waste could negatively influence the characteristics of fish physiology, such as reproduction, general condition, and liver function. Although the excessive nutrient load from WWTPs has been associated with increased growth and condition of fish [70], exposure to WWTP effluent compounds, such as organophosphorus with hormonal activity, may have been linked with increasing prevalence of different metabolic disorders, such as obesity and type 2 diabetes [71, 72]. Furthermore, when different metals from diverse industrial operations are discharged into soil and water, their high solubility in WWTPs turns them into extremely harmful and toxic pollutants. The physiological effects of these metals are principally caused by differences in their relative oxidation states, which also determine their toxicity. Thus, when living organisms are exposed to these metals, they merge with enzymes, proteins, and DNA molecules, forming highly stable bio-toxic compounds, changing their correct functioning, and stopping them from the bioreactions. Furthermore, most of these metals are extremely toxic and may cause carcinogenic, mutagenic, and genotoxic effects [73].

9.3.3 Effects on the Economy

MWWTPs are essential facilities for protecting the environment and public health. However, they rank among the top state-owned users of resources and energy across the globe, particularly in developing countries. Most developed country towns are anticipated to have the most energy-intensive public operations in the form of MWWTPs. About 3%–4% of the entire electric expenditure has been associated with water cycle, including MWWTPs [74]. Utilization of waste from MWWTPs for energy generation can benefit the economy, as this does not require an expensive process.

9.3.4 Effects on River Health

WWTPs have become a serious emerging problem associated with water pollution in most developing countries [75] due to the discharge of waste by different sectors, such as industrial, agricultural, municipal, and sewage effluents into waterways, which then results in the deterioration of water quality threatening the diversity of life [76]. Receiving streams become vulnerable to receive contaminants from anthropogenic sources, including organic and inorganic compounds [77]. Consequently, the discharge of nutrients, such as phosphorus, nitrogen, and ammonia in water bodies through point sources such as WWTPs, contributes to inadequate environmental conditions, leading to eutrophication and subsequent occurrences of harmful algal blooms as well as the decrease in aquatic biodiversity [78]. Furthermore, WWTP effluents are also recognized as significant sources for EPs, such as pharmaceuticals, biocides, endocrine disruptors, and personal care products, which can also pose a serious threat to water quality in receiving bodies [79].

9.4 Municipal Wastewater Treatment Methods

9.4.1 Biological Treatment Method

Biological methods for wastewater treatment have been employed for over 100 years and continue to maintain their standard within the environmental protection infrastructure [80]. These methods primarily involved two stages, where, in the initial stage, wastewater is treated to remove organic compounds and nutrients, and the latter stage involves sludge disposal of waste-activated sludge [81]. When compared to chemical treatment approaches, the biological treatment method utilizes low energy and chemicals, resulting in a reasonable removal efficiency [82], which makes it the most economic and environmentally friendly method for MWW treatment [83, 84]. Activated sludge, aerobic, and anaerobic biological treatment technologies are the three primary categories [85]. The most common biological technique is activated sludge, especially for treating MWW [86–88]. Wastewater containing a high load of organic substances can be processed with activated sludge to produce biogas, which is inexpensive to produce and simple to develop [89]. Biological treatment process based on aerobic and anaerobic technologies have been continuously designed to treat wastewater and are currently employed to reduce the contaminants discharge levels into the environment [90].

9.4.2 Chemical Treatment Method

Although biological processes are the most widespread for treating wastewater, chemical processes may also be useful in situations where biological methods are unable to produce acceptable outcomes [61]. Chemical processes for wastewater treatment consist of different chemical reactions that convert contaminants into less toxic compounds or more environmentally friendly compounds [91]. For this reason, chemical processes use many chemical agents, such as hydrogen peroxide (H_2O_2), ozone (O_3), chlorine (Cl_2), and transition metal–based catalysts, to treat wastewater. Moreover, chemical processes are identified as conventional and advanced oxidation processes [92]. Furthermore, chemical precipitation, coagulation/flocculation, chemical oxidation, adsorption, electrochemical processes, and ion-exchange are the nature of processes involved in chemical treatment methods [93].

9.4.3 Integrated or Multi-Staged Approach

To date, chemical and biological treatment procedures have been assessed, so far, to remove contaminants from the MWW matrix. However, these methods lack the ability to completely remove pollutants from MWW [94–96]. Consequently, integrated treatment approach has been developed by utilizing the distinctive removal potential for various treatment procedures [97]. Besides the hybrid systems, an integrated approach can be used to treat MWW, allowing valuable minerals to be recovered at different stages. This method has been effectively employed to treat MWW, removing all pollutants at different stages. It exploits its advantage by using separate mechanisms that control the removal of pollutants from MWWTPs. Precipitation, adsorption, and filtration work together in this technique to remove pollutants from MWWTPs and recover precious minerals at different phases. Figure 9.1 demonstrates two approaches: the dissipative treatment and the recovery of raw materials. Waste products produced during the treatment phase are not employed in the dissipative treatment process, and, after the treatment is finished, the effluents are released into the environment. On the other hand, energy can be obtained from the by-products created during resource recovery.

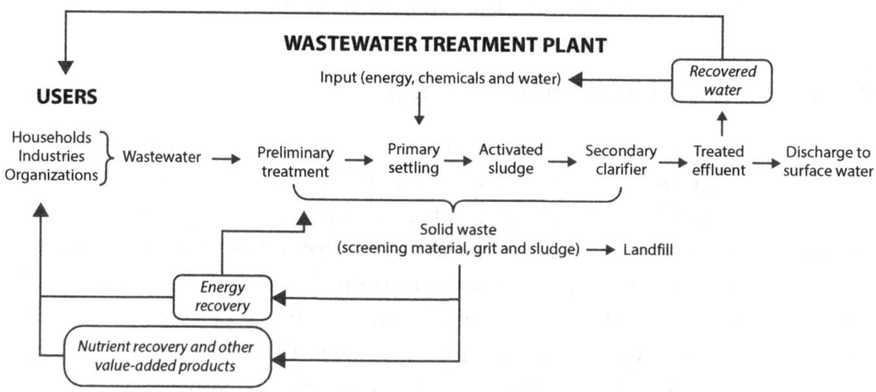

Figure 9.1 A typical multi-stage process occurred in MWWTPs where tiny arrows specify dissipative treatment, and wide arrows and italic font highlight resource recovery integrated in wastewater treatment.

9.5 Water Treatment Challenges

MWWTPs play a substantial role in preserving the health of animals, humans, and the environment by reducing pollutant discharges into bodies of water. However, MWWTPs are becoming problematic due to inefficient pollutant removal. Various recalcitrant substances are presented into the environment, and the traditional treatment methods show poor removal efficiency to many of these contaminants. For instance, wastewater treatment structures fail to effectively remove EPs, and, as such, the development of efficient, cost-effective treatment technique that can be utilized to a wide range of concentration in wastewater is essential [98, 99]. On the other hand, water treatment process can be affected by seasonal changes such as water temperatures and light.

9.6 Beneficiation and Valorization of Products

Beneficiation and valorization of minerals from MWWTPs is a promising aspect toward a circular economy. For this reason, the process of sustainable development will be achieved, which, as a result, the environmental footprint of wastewater is reduced. Thus, society will begin to distinguish MWWTPs as a resource, not a waste stream sector.

9.6.1 Biogas Production

It is well acknowledged that the acceleration of population growth has increased the rate at which wastewater is generated, hence increasing the demand for energy. More than 30 million tons of sewage sludge are generated per annum in developed countries alone [100]. Biogas production from sewage sludge is believed to be an economic solution and environmentally friendly to sustainable development [101]. Because a lot of energy is required to treat waste in WWTPs, biogas production at the WWTPs for internal electricity is believed to be the most suitable and renewable notion for the development of plant's energy independence and the solution to free up some energy for other important uses. The energy in this case is produced by anaerobic digestion (AD) using waste materials [102]. Anaerobic digestion is a term used to describe a natural microbiological process in which bacteria use methane gas as a basic energy source to convert organic materials into biogas energy when there is no oxygen present. The AD can be placed at agricultural biogas plants or WWTPs, respectively, to produce biogas energy from crops like rye, triticale, wheat, grass, or maize, and to stabilize sewage sludge [103]. Furthermore, biogas produced over stabilization of sewage sludge happens to be a by-product that becomes a source of energy when used in a collective heat and power unit to generate on-site heat and electricity [104]. Most of biogas products constitute about 40%–60% and 35%–55% of CH_4 and CO_2, respectively, together with trace levels of hydrogen sulfide (H_2S), moisture, and other toxins. The process of AD involves four stages: hydrolysis, acidogenesis, acetogenesis, and methanogenesis [105]. Methanogenesis stage is directed by a specific microbial group of archaea that are extremely sensitive to changes in temperature, pH, and C/N ratio and sustaining the optimal conditions for AD process, which is a key for enhancing biogas yield [106]. A common challenge typically faced during biogas production is low biogas yield obtained. Furthermore, many research studies using single-stage AD revealed minimal CH_4 levels in biogas [107, 108]. The WWTP sludge valorization by AD process allows the energy, which was available, to be rehabilitated with little cost and no pollution [101].

9.6.2 Fertilizer Production

Traditional wastewater management strategies are based on the use of end-of-pipe methodologies to eradicate pollutants in wastewater before being discharged. Although this approach had an impact toward human health safety for years, pollutant removal and disposal methods require high energy and material investment and overshadow the importance of nutrients in wastewater. To date, MWW has gained significant interest in recovering nutrients such as nitrogen and phosphorus because their quantity is quite considerable [109]. The recovery of nutrients from MWW may deliver magnificent benefits to the environment such as easing the problem of fertilizer production and reducing the occurrence of eutrophication in water bodies together with the problem associated with phosphorus rock being depleted [110]. Several nutrient recovery techniques have been established and applied in experimental or industrial scale, including chemical precipitation and crystallization processes where phosphorus precipitates as struvite ($MgNH_4PO_4{\cdot}6H_2O$) or HAP [hydroxyapatite, $Ca_5(PO_4)_3OH$]. These recovered by-products may be used as soil amendments for agricultural purposes [111].

9.6.3 Biochar Synthesis

The production of sludge during wastewater treatment has received serious interest from several researchers due to its toxicity and safe disposal to the environment. The inadequate disposal of sludge produced during wastewater treatment as management strategy could pose a severe threat to water quality and the environment. Therefore, synthesis of biochar from sewage sludge has gained great interest from many researchers due to its efficiency in enhancing soil quality [112–114] and decreasing heavy metal consumption [115]. The synthesis of biochar sewage sludge is achieved by carbonized thermochemical transformation of biomass in the oxygen-limited environment. The biomass comprises lignin, cellulose, and minerals, with carbon, oxygen, hydrogen, nitrogen, and sulfur as the main elements [116]. Sewage sludge biochar is highly porous with great surface area for the adsorption of heavy metals [58].

9.6.4 Clean Water Reclamation

Water is considered as the most significant resource on earth that living organisms rely on for survival. Currently, the need for clean water supply rises significantly with the rapid population growth and active industrial improvement. The quality of water supplies is constantly declining due to the significant rise in industrialization and urbanization, and, as a result, around 70% of the global population faces challenges in accessing clean water because approximately 2% of fresh water is available for human consumption [117]. The continuous contamination of freshwater with various organic and inorganic pollutants is a major challenge in water supply chain [118]. The main contaminants that harm the aquatic ecosystem are heavy metals, dyes, and fertilizers, among other organic and inorganic pollutants found in WWTPs. A reliable quality water supply is essential for preserving ecosystem and ensuring the safety of drinking water. Therefore, removing waste from WWTPs before discharging it into water bodies should be the focus of study in order to maintain environmental stability and achieve a more sustainable future [119]. Furthermore, the reclamation and reuse of water reduce not only the undesirable impacts of wastewater on the environment but also provide an alternative to withdrawal from natural water resources, forming a closed-loop water supply chain [120]. Investigations are still being conducted on several methods to recover drinking water. However, filtration process has been identified as a special technique for reclaiming drinking water without threatening the biodiversity compared to precipitation and adsorption processes that pose secondary pollution and are selective in their effectiveness. Precipitation method may reclaim water appropriate for discharge or irrigation, whereas adsorption process cannot reclaim water. Due to the increasing content of dissolved toxic metals and water hardness, filtration system faces challenges of fouling, scaling, and damage because of different pH and chemicals in water [121, 122]. In light of that, researchers, engineers, and the scientific community have placed their focus on the practice of distinct pre-treatment tools and water reclamation. For instance, the filtration process behaviors and fouling mechanisms utilizing polyacrylamide flocculation were investigated for application in water treatment streams [123].

9.6.5 Acid Mine Drainage Treatment

The search for alternative strategies for waste stream management is essential. The co-treatment of MWW and acid mine drainage (AMD) using activated sludge as an advanced method for treating AMD was investigated [124]. This method takes advantage of the MWWs alkalinity as well as activated sludge's adsorptive properties from MWW particulates and biomass from activated sludge for metal removal from AMD-polluted waters and acidity neutralization. The ability of these materials to remediate AMD was studied in batch method using high-strength synthetic AMD (pH 2.8, Al at 120–200 mg L^{-1}, Mn at 18–30 mg L^{-1}, Zn at 36–60 mg L^{-1}, Cu at 18–30 mg L^{-1}, and Fe at 324–540 mg L^{-1}). In another study, the domestic wastewater sludge was employed as a carbon source in microbial bioreactors for a passive AMD biotic treatment to create anaerobic conditions for the removal of sulfate and chemical oxygen demand (COD) as well as pH neutralization [125]. For the 1:1 ratio AMD treatment, a synthetic solution that mimicked domestic wastewater sludge was used. In bioreactors, the removal of COD and sulfate was investigated at various incubation temperatures with or without a biofilm, and, after 26 days, the removal efficiencies of COD and sulfate were 60.8% and 96%, respectively, and then a plateau was attained.

9.7 Valorization Challenges and Sustainability

Changing from a direct to circular economy has an impact on how the sustainability of products can be evaluated [126]. The energy recovery from WWTPs depends on its organic matter [127]. Another significant issue is the need for low operational temperatures in anaerobic digesters because many microorganisms thrive in environments between 15°C and 35°C. Variations in these temperatures cause the organisms to operate inefficiently, which lowers the overall kinetics process and reduces the generation of component energy like H_2 and CH_4 [128, 129], which then reduces the performance of the AD to produce enough energy from the wastewater. Furthermore, WWTPs are primarily comprised of different organic compounds and nitrogenous wastes. Thus, most of the biodegradable materials in wastewater are converted to methane gas, and it is highly likely that nitrogenous wastes will partially nitrate to produce nitrous gas, a potent greenhouse gas. This could lower the pH and oxygen levels in the digester, which is important for various microorganisms to survive. These limitations inhibit some technologies from wide application. Contrary, there are also challenges associated with the recovery of nutrients from WWTPs as they precipitate as struvite. The formation of struvite may block pipes, valves, centrifuge bowls, and pumps, decreasing the flow rate capacity and subsequent equipment failure. Struvite precipitation from nutrient recovery adds another problem due to toxic metal ion presence and EPs in wastewater, which significantly reduces struvite purity and reliability for agricultural application [130–132]. Even though struvite crystallization has been thoroughly studied in laboratories [22, 26], its full-scale application remained limited, and only a few WWTPs in Germany, Denmark, United Kingdom, and Netherlands have yet to put this technology into practice [133]. Another problem involves the chemical reagent cost necessary for adjusting pH and magnesium improvement [114].

9.8 Conclusions

As was previously mentioned, MWWTPs can be thought of as sustainable resources capable of producing vital minerals like biogas, struvite, and calcium phosphate. Reclamation of safe drinking water for human consumption is made possible by MWWTPs. The activated sludge biochar, which can be used as an adsorbent, can be created during the MWW treatment procedure. Because of the revenue that is recovered, synthesized, and reclaimed, the treatment process operating expenses will be decreased, enabling this structure to become self-sustaining. Precipitation has shown to be the most successful method for recovering minerals from MWW, whereas filtration techniques are useful for recovering water. The recovery of minerals and the valuation of wastewater present several challenges, such as ongoing sludge generation, cost related to recovering minerals, low recovery factor, and sustainability. Because adsorbents quickly become saturated and require unselective regeneration, adsorption techniques have been acknowledged as being unsuitable for beneficiation and minerals recovery. Future research should concentrate on the fractional and sequential recovery of important minerals from MWWTPs through various methods. Additionally, a hybrid technique should be used in the future to recover minerals from MWWTPs, as it has proven to be the most effective method when compared to other approaches.

9.9 Future Outlook and Research Avenues

The main goals of wastewater treatment technology research, which has gained a lot of attention recently, are to deliver the circular economy concept, value minerals, and beneficiate MWW streams. Still, given the significance of these fields, future studies ought to concentrate on the following:

- Upscaling laboratory research on circular economy and bridging the lab-piloting interface that will reduce the shortcomings related to technology failure during upscaling: Additionally, this will guarantee large-scale bulk material manufacturing and avoid misconceptions from laboratory scale.
- Emphasis on increasing the purity of minerals recovered from MWWTPs is essential and requires advanced technologies that focus on enhancing the mineral production quality and avoid the production of low-grade minerals that will have a negative impact to the end users.
- Utilizing accountable, practical, and environmentally friendly methods to upscale the freeze desalination process from pilot scale to implementation while demonstrating a zero-liquid-discharge process on a pilot scale: By doing this, the difficulties in scaling up technology and the release of harmful pollutants into the environment will be eliminated.
- Incorporating techno-economic and feasibility analysis on technologies that focus on wastewater treatment and mineral recovery that have economic value.
- To minimize ecological consequences and promote ecological conservation, life cycle assessment of the technologies and processes must be pursued. This will highlight the environmental footprints of the suggested technologies and their possibilities for leveraging environmental advantages.
- Ultimately, doing a thorough examination of the market demand for chemicals generated from MWW and optimizing efforts toward compounds with high market demand.

This will remove unwanted outputs with little to no end use and strengthen the circular economy's financial sustainability in wastewater treatment.

Acknowledgments

The authors of this book chapter would like to express their sincere gratitude to the University of Venda and the University of South Africa for the resources extended toward the completion of this project.

References

1. Wendling, L.A., Blomberg, P., Sarlin, T., Priha, O., Arnold, M., Phosphorus sorption and recovery using mineral-based materials: Sorption mechanisms and potential phytoavailability. *Appl. Geochemistry.*, 37, 157–169, 2013. https://doi.org/10.1016/j.apgeochem.2013.07.016.
2. Peng, L., Dai, H., Wu, Y., Peng, Y., Lu, X., A comprehensive review of phosphorus recovery from wastewater by crystallization processes. *Chemosphere*, 197, 768–781, 2018. https://doi.org/10.1016/j.chemosphere.2018.01.098.
3. Hallas, J.F., Mackowiak, C.L., Wilkie, A.C., Harris, W.G., Struvite phosphorus recovery from aerobically digested municipal wastewater. *Sustain.*, 11, 1–12, 2019. https://doi.org/10.3390/su11020376.
4. Leyva-Díaz, J.C., Monteoliva-García, A., Martín-Pascual, J., Munio, M.M., García-Mesa, J.J., Poyatos, J.M., Moving bed biofilm reactor as an alternative wastewater treatment process for nutrient removal and recovery in the circular economy model. *Bioresour. Technol.*, 299, 122631, 2020. https://doi.org/10.1016/j.biortech.2019.122631.
5. Pramanik, B.K., Thangavadivel, K., Shu, L., Jegatheesan, V., A critical review of membrane crystallization for the purification of water and recovery of minerals. *Rev. Environ. Sci. Bio/Technol.*, 15, 411–439, 2016. https://doi.org/10.1007/s11157-016-9403-0.
6. Puyol, D., Batstone, D.J., Hülsen, T., Astals, S., Peces, M., Krömer, J.O., Resource recovery from wastewater by biological technologies: Opportunities, challenges, and prospects. *Front. Microbiol.*, 7, 1–23, 2017. https://doi.org/10.3389/fmicb.2016.02106.
7. Arola, K., Van der Bruggen, B., Mänttäri, M., Kallioinen, M., Treatment options for nanofiltration and reverse osmosis concentrates from municipal wastewater treatment: A review. *Crit. Rev. Environ. Sci. Technol.*, 49, 2049–2116, 2019. https://doi.org/10.1080/10643389.2019.1594519.
8. Yan, T., Ye, Y., Ma, H., Zhang, Y., Guo, W., Du, B., Wei, Q., Wei, D., Ngo, H.H., A critical review on membrane hybrid system for nutrient recovery from

wastewater. *Chem. Eng. J.*, 348, 143–156, 2018. https://doi.org/10.1016/j.cej.2018.04.166.

9. Nancharaiah, Y.V., Venkata Mohan, S., Lens, P.N.L., Recent advances in nutrient removal and recovery in biological and bioelectrochemical systems. *Bioresour. Technol.*, 215, 173–185, 2016. https://doi.org/10.1016/j.biortech.2016.03.129.
10. Perera, M.K., Englehardt, J.D., Dvorak, A.C., Technologies for Recovering Nutrients from Wastewater: A Critical Review. *Environ. Eng. Sci.*, 36, 511–529, 2019. https://doi.org/10.1089/ees.2018.0436.
11. Ye, Y., Ngo, H.H., Guo, W., Liu, Y., Chang, S.W., Nguyen, D.D., Liang, H., Wang, J., A critical review on ammonium recovery from wastewater for sustainable wastewater management. *Bioresour. Technol.*, 268, 749–758, 2018. https://doi.org/10.1016/j.biortech.2018.07.111.
12. Kelessidis, A. and Stasinakis, A.S., Comparative study of the methods used for treatment and final disposal of sewage sludge in European countries. *Waste Manag.*, 32, 1186–1195, 2012. https://doi.org/10.1016/j.wasman.2012.01.012.
13. Gherghel, A., Teodosiu, C., De Gisi, S., A review on wastewater sludge valorisation and its challenges in the context of circular economy. *J. Clean. Prod.*, 228, 244–263, 2019. https://doi.org/10.1016/j.jclepro.2019.04.240.
14. Masrura, S.U., Dissanayake, P., Sun, Y., Ok, Y.S., Tsang, D.C.W., Khan, E., Sustainable use of biochar for resource recovery and pharmaceutical removal from human urine: A critical review. *Crit. Rev. Environ. Sci. Technol.*, 51, 3016–3048, 2021. https://doi.org/10.1080/10643389.2020.1818497.
15. Song, D., Yang, R., Long, F., Zhu, A., Applications of magnetic nanoparticles in surface-enhanced Raman scattering (SERS) detection of environmental pollutants. *J. Environ. Sci. (China)*, 80, 14–34, 2019. https://doi.org/10.1016/j.jes.2018.07.004.
16. Ansari, A.J., Hai, F.I., Price, W.E., Drewes, J.E., Nghiem, L.D., Forward osmosis as a platform for resource recovery from municipal wastewater - A critical assessment of the literature. *J. Memb. Sci.*, 529, 195–206, 2017. https://doi.org/10.1016/j.memsci.2017.01.054.
17. Mohammadi, R., Ramasamy, D.L., Sillanpää, M., Enhancement of nitrate removal and recovery from municipal wastewater through single- and multi-batch electrodialysis: Process optimisation and energy consumption. *Desalination*, 498, 114726, 2021. https://doi.org/10.1016/j.desal.2020.114726.
18. Kehrein, P., Van Loosdrecht, M., Osseweijer, P., Posada, J., Exploring resource recovery potentials for the aerobic granular sludge process by mass and energy balances-energy, biopolymer and phosphorous recovery from municipal wastewater. *Environ. Sci. Water Res. Technol.*, 6, 2164–2179, 2020. https://doi.org/10.1039/d0ew00310g.
19. Papa, M., Foladori, P., Guglielmi, L., Bertanza, G., How far are we from closing the loop of sewage resource recovery? A real picture of municipal wastewater treatment plants in Italy. *J. Environ. Manage.*, 198, 9–15, 2017. https://doi.org/10.1016/j.jenvman.2017.04.061.

20. Huang, X., Guida, S., Jefferson, B., Soares, A., Economic evaluation of ion-exchange processes for nutrient removal and recovery from municipal wastewater. *npj Clean Water*, 3, 2020. https://doi.org/10.1038/s41545-020-0054-x.
21. Qadir, M., Drechsel, P., Jiménez Cisneros, B., Kim, Y., Pramanik, A., Mehta, P., Olaniyan, O., Global and regional potential of wastewater as a water, nutrient and energy source. *Nat. Resour. Forum*, 44, 40–51, 2020. https://doi.org/10.1111/1477-8947.12187.
22. Mavhungu, A., Masindi, V., Foteinis, S., Mbaya, R., Tekere, M., Kortidis, I., Chatzisymeon, E., Advocating circular economy in wastewater treatment: Struvite formation and drinking water reclamation from real municipal effluents. *J. Environ. Chem. Eng.*, 8, 103957, 2020. https://doi.org/10.1016/j.jece.2020.103957.
23. You, X., Valderrama, C., Soldatov, V., Cortina, J.L., Phosphate recovery from treated municipal wastewater using hybrid anion exchangers containing hydrated ferric oxide nanoparticles. *J. Chem. Technol. Biotechnol.*, 93, 358–364, 2018. https://doi.org/10.1002/jctb.5361.
24. Batstone, D.J., Hülsen, T., Mehta, C.M., Keller, J., Platforms for energy and nutrient recovery from domestic wastewater: A review. *Chemosphere*, 140, 2–11, 2015. https://doi.org/10.1016/j.chemosphere.2014.10.021.
25. Akyol, Ç., Foglia, A., Ozbayram, E.G., Frison, N., Katsou, E., Eusebi, A.L., Fatone, F., Validated innovative approaches for energy-efficient resource recovery and re-use from municipal wastewater: From anaerobic treatment systems to a biorefinery concept. *Crit. Rev. Environ. Sci. Technol.*, 50, 869–902, 2020. https://doi.org/10.1080/10643389.2019.1634456.
26. Mavhungu, A., Mbaya, R., Masindi, V., Foteinis, S., Muedi, K.L., Kortidis, I., Chatzisymeon, E., Wastewater treatment valorisation by simultaneously removing and recovering phosphate and ammonia from municipal effluents using a mechano-thermo activated magnesite technology. *J. Environ. Manage.*, 250, 109493, 2019. https://doi.org/10.1016/j.jenvman.2019.109493.
27. Solon, K., Volcke, E.I.P., Spérandio, M., Van Loosdrecht, M.C.M., Resource recovery and wastewater treatment modelling. *Environ. Sci. Water Res. Technol.*, 5, 631–642, 2019. https://doi.org/10.1039/c8ew00765a.
28. Perera, M.K., Englehardt, J.D., Cohn, J.L., Dauer, E.A., Shukla, D., Electrohydromodulation for phosphate recovery from wastewater. *Sep. Purif. Technol.*, 247, 116909, 2020. https://doi.org/10.1016/j.seppur.2020.116909.
29. Kasina, M., Wendorff-Belon, M., Kowalski, P.R., Michalik, M., Characterization of incineration residues from wastewater treatment plant in Polish city: a future waste based source of valuable elements? *J. Mater. Cycles Waste Manag.*, 21, 885–896, 2019. https://doi.org/10.1007/s10163-019-00845-1.
30. You, X., Valderrama, C., Cortina, J.L., Nutrients recovery from treated secondary mainstream in an urban wastewater treatment plant: A financial assessment case study. *Sci. Total Environ.*, 656, 902–909, 2019. https://doi.org/10.1016/j.scitotenv.2018.11.420.

31. Chae, S., Murugesan, B., Kim, H., Duvvuru, D.K., Lee, T., Choi, Y.H., Baek, M.H., Nadagouda, M.N., Advanced Phosphorus Recovery from Municipal Wastewater using Anoxic/Aerobic Membrane Bioreactors and Magnesium Carbonate-Based Pellets. *ACS ES T Water*, 1, 1657–1664, 2021. https://doi.org/10.1021/acsestwater.0c00300.
32. Tarayre, C., De Clercq, L., Charlier, R., Michels, E., Meers, E., Camargo-Valero, M., Delvigne, F., New perspectives for the design of sustainable bioprocesses for phosphorus recovery from waste. *Bioresour. Technol.*, 206, 264–274, 2016. https://doi.org/10.1016/j.biortech.2016.01.091.
33. Yan, D., Li, Y., Huo, J., Chen, R., Dai, L., Wang, S., Defect Chemistry of Nonprecious-Metal Electrocatalysts for Oxygen Reactions. *Adv. Mater.*, 29, 2017. https://doi.org/10.1002/adma.201606459.
34. Chrispim, M.C., Scholz, M., Nolasco, M.A., Phosphorus recovery from municipal wastewater treatment: Critical review of challenges and opportunities for developing countries. *J. Environ. Manage.*, 248, 109268, 2019. https://doi.org/10.1016/j.jenvman.2019.109268.
35. Egle, L., Rechberger, H., Krampe, J., Zessner, M., Phosphorus recovery from municipal wastewater: An integrated comparative technological, environmental and economic assessment of P recovery technologies. *Sci. Total Environ.*, 571, 522–542, 2016. https://doi.org/10.1016/j.scitotenv.2016.07.019.
36. Quist-Jensen, C.A., Jørgensen, M.K., Christensen, M.L., Treated seawater as a magnesium source for phosphorous recovery from wastewater—A feasibility and cost analysis. *Membr. (Basel)*, 6, 2016. https://doi.org/10.3390/membranes6040054.
37. Malila, R., Lehtoranta, S., Viskari, E.L., The role of source separation in nutrient recovery - Comparison of alternative wastewater treatment systems. *J. Clean. Prod.*, 219, 350–358, 2019. https://doi.org/10.1016/j.jclepro.2019.02.024.
38. Gillis, P.L., McInnis, R., Salerno, J., de Solla, S.R., Servos, M.R., Leonard, E.M., Municipal wastewater treatment plant effluent-induced effects on freshwater mussel populations and the role of mussel refugia in recolonizing an extirpated reach. *Environ. Pollut.*, 225, 460–468, 2017. https://doi.org/10.1016/j.envpol.2017.03.010.
39. Schaider, L.A., Rodgers, K.M., Rudel, R.A., Review of Organic Wastewater Compound Concentrations and Removal in Onsite Wastewater Treatment Systems. *Environ. Sci. Technol.*, 51, 7304–7317, 2017. https://doi.org/10.1021/acs.est.6b04778.
40. Wang, W. and Kannan, K., Fate of Parabens and Their Metabolites in Two Wastewater Treatment Plants in New York State, United States. *Environ. Sci. Technol.*, 50, 1174–1181, 2016. https://doi.org/10.1021/acs.est.5b05516.
41. Isidori, M., Lavorgna, M., Russo, C., Kundi, M., Žegura, B., Novak, M., Filipič, M., Mišík, M., Knasmueller, S., de Alda, M.L., Barceló, D., Žonja, B., Česen, M., Ščančar, J., Kosjek, T., Heath, E., Chemical and toxicological characterisation of anticancer drugs in hospital and municipal wastewaters

from Slovenia and Spain. *Environ. Pollut.*, 219, 275–287, 2016. https://doi.org/10.1016/j.envpol.2016.10.039.

42. Arlos, M.J., Bragg, L.M., Parker, W.J., Servos, M.R., Distribution of selected antiandrogens and pharmaceuticals in a highly impacted watershed. *Water Res.*, 72, 40–50, 2015. https://doi.org/10.1016/j.watres.2014.11.008.
43. Vimala, R.T.V., Lija Escaline, J., Murugan, K., Sivaramakrishnan, S., An overview of organic matters in municipal wastewater: Removal via self-assembly flocculating mechanism and the molecular level characterization. *J. Environ. Manage.*, 266, 110572, 2020. https://doi.org/10.1016/j.jenvman.2020.110572.
44. Podstawczyk, D., Witek-Krowiak, A., Dawiec-Liśniewska, A., Chrobot, P., Skrzypczak, D., Removal of ammonium and orthophosphates from reject water generated during dewatering of digested sewage sludge in municipal wastewater treatment plant using adsorption and membrane contactor system. *J. Clean. Prod.*, 161, 277–287, 2017. https://doi.org/10.1016/j.jclepro.2017.05.105.
45. Demirbas, A., Taylan, O., Kaya, D., Biogas production from municipal sewage sludge (MSS). *Energy Sources Part A Recover. Util. Environ. Eff.*, 38, 3027–3033, 2016. https://doi.org/10.1080/15567036.2015.1124944.
46. Demirbas, A., Edris, G., Alalayah, W.M., Sludge production from municipal wastewater treatment in sewage treatment plant. *Energy Sources Part A Recover. Util. Environ. Eff.*, 39, 999–1006, 2017. https://doi.org/10.1080/15567036.2017.1283551.
47. Bharathiraja, B., Yogendran, D., Ranjith Kumar, R., Chakravarthy, M., Palani, S., Biofuels from sewage sludge- A review. *Int. J. Chem. Tech. Res.*, 6, 4417–4427, 2014.
48. Chen, G., Ekama, G.A., van Loosdrecht, M.C.M., Brdjanovic, D., Biological wastewater treatment. *Biol. Wastewater Treat.*, 1–840, Springer Nature, Switzerland, 2020. https://doi.org/10.1201/b18368-4.
49. Fijalkowski, K., Rorat, A., Grobelak, A., Kacprzak, M.J., The presence of contaminations in sewage sludge – The current situation. *J. Environ. Manage.*, 203, 1126–1136, 2017. https://doi.org/10.1016/j.jenvman.2017.05.068.
50. Faria, W.M., de Figueiredo, C.C., Coser, T.R., Vale, A.T., Schneider, B.G., Is sewage sludge biochar capable of replacing inorganic fertilizers for corn production? Evidence from a two-year field experiment. *Arch. Agron. Soil Sci.*, 64, 505–519, 2018. https://doi.org/10.1080/03650340.2017.1360488.
51. Natasha, M., Khalid, S., Murtaza, B., Anwar, H., Shah, A.H., Sardar, A., Shabbir, Z., Niazi, N.K., A critical analysis of wastewater use in agriculture and associated health risks in Pakistan. *Environ. Geochem. Health*, 45, 5599–5618, 2023. https://doi.org/10.1007/s10653-020-00702-3.
52. Gashaye, D., Wastewater-irrigated urban vegetable farming in Ethiopia: A review on their potential contamination and health effects. *Cogent Food Agric.*, 6, 1–17, 2020. https://doi.org/10.1080/23311932.2020.1772629.
53. Mkhinini, M., Boughattas, I., Alphonse, V., Livet, A., Bousserrhine, N., Banni, M., Effect of treated wastewater irrigation in East Central region

of Tunisia (Monastir governorate) on the biochemical and transcriptomic response of earthworms Eisenia andrei. *Sci. Total Environ.*, 647, 1245–1255, 2019. https://doi.org/10.1016/j.scitotenv.2018.07.449.
54. Ibekwe, A.M., Gonzalez-Rubio, A., Suarez, D.L., Impact of treated wastewater for irrigation on soil microbial communities. *Sci. Total Environ.*, 622–623, 1603–1610, 2018. https://doi.org/10.1016/j.scitotenv.2017.10.039.
55. Vasseur, P. and Bonnard, M., Ecogenotoxicology in earthworms: A review. *Curr. Zool.*, 60, 255–272, 2014. https://doi.org/10.1093/czoolo/60.2.255.
56. Pelosi, C., Barot, S., Capowiez, Y., Hedde, M., Vandenbulcke, F., Pesticides and earthworms. A review. *Agron. Sustain. Dev.*, 34, 199–228, 2014. https://doi.org/10.1007/s13593-013-0151-z.
57. Liang, J., Xia, X., Yuan, L., Zhang, W., Lin, K., Zhou, B., Hu, S., The reproductive responses of earthworms (Eisenia fetida) exposed to nanoscale zero-valent iron (nZVI) in the presence of decabromodiphenyl ether (BDE209). *Environ. Pollut.*, 237, 784–791, 2018. https://doi.org/10.1016/j.envpol.2017.10.130.
58. Li, Z., Ma, Z., van der Kuijp, T.J., Yuan, Z., Huang, L., A review of soil heavy metal pollution from mines in China: Pollution and health risk assessment. *Sci. Total Environ.*, 468–469, 843–853, 2014. https://doi.org/10.1016/j.scitotenv.2013.08.090.
59. Pedrero, F., Kalavrouziotis, I., Alarcón, J.J., Koukoulakis, P., Asano, T., Use of treated municipal wastewater in irrigated agriculture-Review of some practices in Spain and Greece. *Agric. Water Manag.*, 97, 1233–1241, 2010. https://doi.org/10.1016/j.agwat.2010.03.003.
60. Arslan Topal, E.I., Topal, M., Öbek, E., Assessment of heavy metal accumulations and health risk potentials in tomatoes grown in the discharge area of a municipal wastewater treatment plant. *Int. J. Environ. Health Res.*, 32, 93–405, 2022. https://doi.org/10.1080/09603123.2020.1762071.
61. Ullah, A., Hussain, S., Wasim, A., Jahanzaib, M., Development of a decision support system for the selection of wastewater treatment technologies. *Sci. Total Environ.*, 731, 139158, 2020. https://doi.org/10.1016/j.scitotenv.2020.139158.
62. Melvin, S.D., Short-term exposure to municipal wastewater influences energy, growth, and swimming performance in juvenile Empire Gudgeons (Hypseleotris compressa). *Aquat. Toxicol.*, 170, 271–278, 2016. https://doi.org/10.1016/j.aquatox.2015.06.003.
63. Simmons, D.B.D., Miller, J., Clarence, S., McCallum, E.S., Balshine, S., Chandramouli, B., Cosgrove, J., Sherry, J.P., Altered expression of metabolites and proteins in wild and caged fish exposed to wastewater effluents *in situ*. *Sci. Rep.*, 7, 1–14, 2017. https://doi.org/10.1038/s41598-017-12473-6.
64. Helander, B., Axelsson, J., Borg, H., Holm, K., Bignert, A., Ingestion of lead from ammunition and lead concentrations in white-tailed sea eagles

(Haliaeetus albicilla) in Sweden. *Sci. Total Environ.*, 407, 5555–5563, 2009. https://doi.org/10.1016/j.scitotenv.2009.07.027.
65. Dickerson, S.M. and Gore, A.C., Estrogenic environmental endocrine-disrupting chemical effects on reproductive neuroendocrine function and dysfunction across the life cycle. *Rev. Endocr. Metab. Disord.*, 8, 143–159, 2007. https://doi.org/10.1007/s11154-007-9048-y.
66. Jackson, J. and Sutton, R., Sources of endocrine-disrupting chemicals in urban wastewater, Oakland, CA. *Sci. Total Environ.*, 405, 153–160, 2008. https://doi.org/10.1016/j.scitotenv.2008.06.033.
67. Pal, A., Gin, K.Y.H., Lin, A.Y.C., Reinhard, M., Impacts of emerging organic contaminants on freshwater resources: Review of recent occurrences, sources, fate and effects. *Sci. Total Environ.*, 408, 6062–6069, 2010. https://doi.org/10.1016/j.scitotenv.2010.09.026.
68. Jasinska, E.J., Goss, G.G., Gillis, P.L., Van Der Kraak, G.J., Matsumoto, J., de Souza Machado, A.A., Giacomin, M., Moon, T.W., Massarsky, A., Gagné, F., Servos, M.R., Wilson, J., Sultana, T., Metcalfe, C.D., Assessment of biomarkers for contaminants of emerging concern on aquatic organisms downstream of a municipal wastewater discharge. *Sci. Total Environ.*, 530–531, 140–153, 2015. https://doi.org/10.1016/j.scitotenv.2015.05.080.
69. Bahamonde, P.A., Fuzzen, M.L., Bennett, C.J., Tetreault, G.R., McMaster, M.E., Servos, M.R., Martyniuk, C.J., Munkittrick, K.R., Whole organism responses and intersex severity in rainbow darter (Etheostoma caeruleum) following exposures to municipal wastewater in the Grand River basin, ON, Canada. *Part A Aquat. Toxicol.*, 159, 290–301, 2015. https://doi.org/10.1016/j.aquatox.2014.11.023.
70. Fuzzen, M.L.M., Bennett, C.J., Tetreault, G.R., McMaster, M.E., Servos, M.R., Severe intersex is predictive of poor fertilization success in populations of rainbow darter (Etheostoma caeruleum). *Aquat. Toxicol.*, 160, 106–116, 2015. https://doi.org/10.1016/j.aquatox.2015.01.009.
71. Lyche, J.L., Nourizadeh-Lillabadi, R., Karlsson, C., Stavik, B., Berg, V., Skåre, J.U., Alestrøm, P., Ropstad, E., Natural mixtures of POPs affected body weight gain and induced transcription of genes involved in weight regulation and insulin signaling. *Aquat. Toxicol.*, 102, 197–204, 2011. https://doi.org/10.1016/j.aquatox.2011.01.017.
72. Miralles-Cuevas, S., Oller, I., Agüera, A., Llorca, M., Sánchez Pérez, J.A., Malato, S., Combination of nanofiltration and ozonation for the remediation of real municipal wastewater effluents: Acute and chronic toxicity assessment. *J. Hazard. Mater.*, 323, 442–451, 2017. https://doi.org/10.1016/j.jhazmat.2016.03.013.
73. Manzoor, M.M., *Environmental Biotechnology: For Sustainable Future*, Springer, Switzerland, 2020, https://doi.org/10.1007/978-3-030-40333-1_14.
74. Daneshgar, S., Callegari, A., Capodaglio, A.G., Vaccari, D., The potential phosphorus crisis: Resource conservation and possible escape technologies: A review. *Resources*, 7, 1–22, 2018. https://doi.org/10.3390/resources7020037.

75. Haque, N., Exploratory analysis of fines for water pollution in Bangladesh. *Water Resour. Ind.*, 18, 1–8, 2017. https://doi.org/10.1016/j.wri.2017.05.001.
76. Paul, D., Research on heavy metal pollution of river Ganga: A review. *Ann. Agrar. Sci.*, 15, 278–286, 2017. https://doi.org/10.1016/j.aasci.2017.04.001.
77. Galib, S.M., Mohsin, A.B.M., Parvez, M.T., Lucas, M.C., Chaki, N., Arnob, S.S., Hossain, M.I., Islam, M.N., Municipal wastewater can result in a dramatic decline in freshwater fishes: A lesson from a developing country. *Knowl. Manag. Aquat. Ecosyst.*, 419, 1–12, 2018. https://doi.org/10.1051/kmae/2018025.
78. Qi, M., Yang, Y., Zhang, X., Zhang, X., Wang, M., Zhang, W., Lu, X., Tong, Y., Pollution reduction and operating cost analysis of municipal wastewater treatment in China and implication for future wastewater management. *J. Clean. Prod.*, 253, 120003, 2020. https://doi.org/10.1016/j.jclepro.2020.120003.
79. Harth, F.U.R., Arras, C., Brettschneider, D.J., Misovic, A., Oehlmann, J., Schulte-Oehlmann, U., Oetken, M., Small but with big impact? Ecotoxicological effects of a municipal wastewater effluent on a small creek. *J. Environ. Sci. Heal. - Part A Toxic/Hazard. Subst. Environ. Eng.*, 53, 1149–1160, 2018. https://doi.org/10.1080/10934529.2018.1530328.
80. Xiao, Y., De Araujo, C., Sze, C.C., Stuckey, D.C., Toxicity measurement in biological wastewater treatment processes: A review. *J. Hazard. Mater.*, 286, 15–29, 2015. https://doi.org/10.1016/j.jhazmat.2014.12.033.
81. Zhang, Z., Zhang, M., Yang Wu, X., Chang, Z., Lee, Y.I., Huy, B.T., Sakthivel, K., Fu Liu, J., Bin Jiang, G., Upconversion fluorescence resonance energy transfer-a novel approach for sensitive detection of fluoroquinolones in water samples. *Microchem. J.*, 124, 181–187, 2016. https://doi.org/10.1016/j.microc.2015.08.024.
82. Quijano, G., Arcila, J.S., Buitrón, G., Microalgal-bacterial aggregates: Applications and perspectives for wastewater treatment. *Biotechnol. Adv.*, 35, 772–781, 2017. https://doi.org/10.1016/j.biotechadv.2017.07.003.
83. Wei, W., Wang, Q., Zhang, L., Laloo, A., Duan, H., Batstone, D.J., Yuan, Z., Free nitrous acid pre-treatment of waste activated sludge enhances volatile solids destruction and improves sludge dewaterability in continuous anaerobic digestion. *Water Res.*, 130, 13–19, 2018. https://doi.org/10.1016/j.watres.2017.11.050.
84. Liu, Y., Ngo, H.H., Guo, W., Peng, L., Wang, D., Ni, B., The roles of free ammonia (FA) in biological wastewater treatment processes: A review. *Environ. Int.*, 123, 10–19, 2019. https://doi.org/10.1016/j.envint.2018.11.039.
85. Tang, J., Zhang, C., Shi, X., Sun, J., Cunningham, J.A., Municipal wastewater treatment plants coupled with electrochemical, biological and bio-electrochemical technologies: Opportunities and challenge toward energy self-sufficiency. *J. Environ. Manage.*, 234, 396–403, 2019. https://doi.org/10.1016/j.jenvman.2018.12.097.

86. Gude, V.G., Wastewater treatment in microbial fuel cells - An overview. *J. Clean. Prod.*, 122, 287–307, 2016. https://doi.org/10.1016/j.jclepro.2016.02.022.
87. Selvaratnam, T., Henkanatte-Gedera, S.M., Muppaneni, T., Nirmalakhandan, N., Deng, S., Lammers, P.J., Maximizing recovery of energy and nutrients from urban wastewaters. *Energy*, 104, 16–23, 2016. https://doi.org/10.1016/j.energy.2016.03.102.
88. Singhal, N. and Perez-Garcia, O., Degrading organic micropollutants: The next challenge in the evolution of biological wastewater treatment processes. *Front. Environ. Sci.*, 4, 1–5, 2016. https://doi.org/10.3389/fenvs.2016.00036.
89. Dos Santos, I.F.S., Barros, R.M., Tiago Filho, G.L., Electricity generation from biogas of anaerobic wastewater treatment plants in Brazil: An assessment of feasibility and potential. *J. Clean. Prod.*, 126, 504–514, 2016. https://doi.org/10.1016/j.jclepro.2016.03.072.
90. Mesquita, D.P., Quintelas, C., Amaral, A.L., Ferreira, E.C., Monitoring biological wastewater treatment processes: recent advances in spectroscopy applications. *Rev. Environ. Sci. Bio/Technol.*, 16, 395–424, 2017. https://doi.org/10.1007/s11157-017-9439-9.
91. Saleh, I.A., Zouari, N., Al-Ghouti, M.A., Removal of pesticides from water and wastewater: Chemical, physical and biological treatment approaches. *Environ. Technol. Innov.*, 19, 101026, 2020. https://doi.org/10.1016/j.eti.2020.101026.
92. Salimi, M., Esrafili, A., Gholami, M., Jonidi Jafari, A., Rezaei Kalantary, R., Farzadkia, M., Kermani, M., Sobhi, H.R., Contaminants of emerging concern: a review of new approach in AOP technologies. *Environ. Monit. Assess.*, 189, 1–22, 2017. https://doi.org/10.1007/s10661-017-6097-x.
93. Meghwal, K., Agrawal, R., Kumawat, S., Jangid, N.K., Ameta, C., *Chemical and biological treatment of dyes*, Springer, Singapore, 2019, https://doi.org/10.4018/978-1-7998-0311-9.ch009.
94. Mohapatra, D.P. and Kirpalani, D.M., Advancement in treatment of wastewater: Fate of emerging contaminants. *Can. J. Chem. Eng.*, 97, 2621–2631, 2019. https://doi.org/10.1002/cjce.23533.
95. Donkadokula, N.Y., Kola, A.K., Naz, I., Saroj, D., A review on advanced physico-chemical and biological textile dye wastewater treatment techniques. *Rev. Environ. Sci. Bio/Technol.*, 19, 543–560, 2020. https://doi.org/10.1007/s11157-020-09543-z.
96. Grandclément, C., Seyssiecq, I., Piram, A., Wong-Wah-Chung, P., Vanot, G., Tiliacos, N., Roche, N., Doumenq, P., From the conventional biological wastewater treatment to hybrid processes, the evaluation of organic micropollutant removal: A review. *Water Res.*, 111, 297–317, 2017. https://doi.org/10.1016/j.watres.2017.01.005.
97. Dhangar, K. and Kumar, M., Tricks and tracks in removal of emerging contaminants from the wastewater through hybrid treatment systems: A

review. *Sci. Total Environ.*, 738, 140320, 2020. https://doi.org/10.1016/j.scitotenv.2020.140320.

98. Lim, T.T., Yap, P.S., Srinivasan, M., Fane, A.G., TiO2/AC composites for synergistic adsorption-photocatalysis processes: Present challenges and further developments for water treatment and reclamation. *Crit. Rev. Environ. Sci. Technol.*, 41, 1173–1230, 2011. https://doi.org/10.1080/10643380903488664.
99. Sutherland, D.L. and Ralph, P.J., Microalgal bioremediation of emerging contaminants - Opportunities and challenges. *Water Res.*, 164, 114921, 2019. https://doi.org/10.1016/j.watres.2019.114921.
100. Jupp, A.R., Beijer, S., Narain, G.C., Schipper, W., Slootweg, J.C., Phosphorus recovery and recycling-closing the loop. *Chem. Soc. Rev.*, 50, 87–101, 2021. https://doi.org/10.1039/d0cs01150a.
101. Kalloum, S., Bouabdessalem, H., Touzi, A., Iddou, A., Ouali, M.S., Biogas production from the sludge of the municipal wastewater treatment plant of Adrar city (southwest of Algeria). *Biomass Bioenergy*, 35, 2554–2560, 2011. https://doi.org/10.1016/j.biombioe.2011.02.012.
102. Kumar, S.S., Kumar, V., Kumar, R., Malyan, S.K., Bishnoi, N.R., Ferrous sulfate as an *in-situ* anodic coagulant for enhanced bioelectricity generation and COD removal from landfill leachate. *Energy*, 176, 570–581, 2019. https://doi.org/10.1016/j.energy.2019.04.014.
103. Bunzel, K., Kattwinkel, M., Schauf, M., Thrän, D., Energy crops and pesticide contamination: Lessons learnt from the development of energy crop cultivation in Germany. *Biomass Bioenergy*, 70, 416–428, 2014. https://doi.org/10.1016/j.biombioe.2014.08.016.
104. Kosse, P., Lübken, M., Wichern, M., Urban lignocellulosic biomass can significantly contribute to energy production in municipal wastewater treatment plants - A GIS-based approach for a metropolitan area. *Biomass Bioenergy*, 81, 568–573, 2015. https://doi.org/10.1016/j.biombioe.2015.08.013.
105. Ghosh, P., Kumar, M., Kapoor, R., Kumar, S.S., Singh, L., Vijay, V., Vijay, V.K., Kumar, V., Thakur, I.S., Enhanced biogas production from municipal solid waste via co-digestion with sewage sludge and metabolic pathway analysis. *Bioresour. Technol.*, 296, 122275, 2020. https://doi.org/10.1016/j.biortech.2019.122275.
106. Mao, C., Feng, Y., Wang, X., Ren, G., Review on research achievements of biogas from anaerobic digestion. *Renew. Sustain. Energy Rev.*, 45, 540–555, 2015. https://doi.org/10.1016/j.rser.2015.02.032.
107. Bona, D., Papurello, D., Flaim, G., Cerasino, L., Biasioli, F., Silvestri, S., Management of Digestate and Exhausts from Solid Oxide Fuel Cells Produced in the Dry Anaerobic Digestion Pilot Plant: Microalgae Cultivation Approach. *Waste Biomass Valorization*, 11, 6499–6514, 2020. https://doi.org/10.1007/s12649-019-00931-3.
108. Lanzini, A., Ferrero, D., Papurello, D., Santarelli, M., Reporting Degradation from Different Fuel Contaminants in Ni-anode SOFCs. *Fuel Cells*, 17, 423–433, 2017. https://doi.org/10.1002/fuce.201600184.

109. Theregowda, R.B., González-Mejía, A.M., Ma, X., Garland, J., Nutrient Recovery from Municipal Wastewater for Sustainable Food Production Systems: An Alternative to Traditional Fertilizers. *Environ. Eng. Sci.*, 36, 833–842, 2019. https://doi.org/10.1089/ees.2019.0053.
110. Ye, Y., Ngo, H.H., Guo, W., Liu, Y., Li, J., Liu, Y., Zhang, X., Jia, H., Insight into chemical phosphate recovery from municipal wastewater. *Sci. Total Environ.*, 576, 159–171, 2017. https://doi.org/10.1016/j.scitotenv.2016.10.078.
111. Zhou, K., Barjenbruch, M., Kabbe, C., Inial, G., Remy, C., Phosphorus recovery from municipal and fertilizer wastewater: China's potential and perspective. *J. Environ. Sci. (China)*, 52, 151–159, 2017. https://doi.org/10.1016/j.jes.2016.04.010.
112. Singh, S., Kumar, V., Dhanjal, D.S., Datta, S., Bhatia, D., Dhiman, J., Samuel, J., Prasad, R., Singh, J., A sustainable paradigm of sewage sludge biochar: Valorization, opportunities, challenges and future prospects. *J. Clean. Prod.*, 269, 122259, 2020. https://doi.org/10.1016/j.jclepro.2020.122259.
113. Kahiluoto, H., Kuisma, M., Ketoja, E., Salo, T., Heikkinen, J., Phosphorus in manure and sewage sludge more recyclable than in soluble inorganic fertilizer. *Environ. Sci. Technol.*, 49, 2115–2122, 2015. https://doi.org/10.1021/es503387y.
114. Martí, N., Pastor, L., Bouzas, A., Ferrer, J., Seco, A., Phosphorus recovery by struvite crystallization in WWTPs: Influence of the sludge treatment line operation. *Water Res.*, 44, 2371–2379, 2010. https://doi.org/10.1016/j.watres.2009.12.043.
115. Khanmohammadi, Z., Afyuni, M., Mosaddeghi, M.R., Effect of sewage sludge and its biochar on chemical properties of two calcareous soils and maize shoot yield. *Arch. Agron. Soil Sci.*, 63, 198–212, 2017. https://doi.org/10.1080/03650340.2016.1210787.
116. Chen, D., Mei, J., Li, H., Li, Y., Lu, M., Ma, T., Ma, Z., Combined pretreatment with torrefaction and washing using torrefaction liquid products to yield upgraded biomass and pyrolysis products. *Bioresour. Technol.*, 228, 62–68, 2017. https://doi.org/10.1016/j.biortech.2016.12.088.
117. Jayaswal, K., Sahu, V., Gurjar, B.R., Pollution, W., Human Health and Remediation, Energy. *Environ. Sustain.*, 65, 11–27, 2018. https://doi.org/10.1007/978-981-10-7551-3_2.
118. Anjum, M., Miandad, R., Waqas, M., Gehany, F., Barakat, M.A., Remediation of wastewater using various nano-materials. *Arab. J. Chem.*, 12, 4897–4919, 2019. https://doi.org/10.1016/j.arabjc.2016.10.004.
119. Mon, M., Bruno, R., Ferrando-Soria, J., Armentano, D., Pardo, E., Metal-organic framework technologies for water remediation: Towards a sustainable ecosystem. *J. Mater. Chem. A*, 6, 4912–4947, 2018. https://doi.org/10.1039/c8ta00264a.
120. Rezaei, N., Sierra-Altamiranda, A., Diaz-Elsayed, N., Charkhgard, H., Zhang, Q., A multi-objective optimization model for decision support in water

reclamation system planning. *J. Clean. Prod.*, 240, 118227, 2019. https://doi.org/10.1016/j.jclepro.2019.118227.
121. Nleya, Y., Simate, G.S., Ndlovu, S., Sustainability assessment of the recovery and utilisation of acid from acid mine drainage. *J. Clean. Prod.*, 113, 17–27, 2016. https://doi.org/10.1016/j.jclepro.2015.11.005.
122. Masindi, V., Madzivire, G., Tekere, M., Reclamation of water and the synthesis of gypsum and limestone from acid mine drainage treatment process using a combination of pre-treated magnesite nanosheets, lime, and CO2 bubbling. *Water Resour. Ind.*, 20, 1–14, 2018. https://doi.org/10.1016/j.wri.2018.07.001.
123. Long, Y., You, X., Chen, Y., Hong, H., Liao, B.Q., Lin, H., Filtration behaviors and fouling mechanisms of ultrafiltration process with polyacrylamide flocculation for water treatment. *Sci. Total Environ.*, 703, 135540, 2020. https://doi.org/10.1016/j.scitotenv.2019.135540.
124. Hughes, T.A. and Gray, N.F., Co-treatment of acid mine drainage with municipal wastewater: Performance evaluation. *Environ. Sci. Pollut. Res.*, 20, 7863–7877, 2013. https://doi.org/10.1007/s11356-012-1303-4.
125. Van Den Berg, M.F., Botes, M., Slabbert, E., Cloete, T.E., Evaluating sulphate removal and identifying the bacterial community present in acid mine drainage treated with synthetic domestic wastewater sludge. *Water SA*, 42, 475–482, 2016. https://doi.org/10.4314/wsa.v42i3.13.
126. Sfez, S., De Meester, S., Vlaeminck, S.E., Dewulf, J., Improving the resource footprint evaluation of products recovered from wastewater: A discussion on appropriate allocation in the context of circular economy. *Resour. Conserv. Recycl.*, 148, 132–144, 2019. https://doi.org/10.1016/j.resconrec.2019.03.029.
127. Liu, Y.J., Gu, J., Liu, Y., Energy self-sufficient biological municipal wastewater reclamation: Present status, challenges and solutions forward. *Bioresour. Technol.*, 269, 513–519, 2018. https://doi.org/10.1016/j.biortech.2018.08.104.
128. Malaeb, L., Katuri, K.P., Logan, B.E., Maab, H., Nunes, S.P., Saikaly, P.E., A hybrid microbial fuel cell membrane bioreactor with a conductive ultrafiltration membrane biocathode for wastewater treatment. *Environ. Sci. Technol.*, 47, 11821–11828, 2013. https://doi.org/10.1021/es4030113.
129. Bhatia, S.K. and Yang, Y.H., Microbial production of volatile fatty acids: current status and future perspectives. *Rev. Environ. Sci. Bio/Technol.*, 16, 327–345, 2017. https://doi.org/10.1007/s11157-017-9431-4.
130. Rouff, A.A. and Juarez, K.M., Zinc interaction with struvite during and after mineral formation. *Environ. Sci. Technol.*, 48, 6342–6349, 2014. https://doi.org/10.1021/es500188t.
131. Pizzol, M., Smart, J.C.R., Thomsen, M., External costs of cadmium emissions to soil: A drawback of phosphorus fertilizers. *J. Clean. Prod.*, 84, 475–483, 2014. https://doi.org/10.1016/j.jclepro.2013.12.080.

132. Xie, M., Shon, H.K., Gray, S.R., Elimelech, M., Membrane-based processes for wastewater nutrient recovery: Technology, challenges, and future direction. *Water Res.*, 89, 210–221, 2016. https://doi.org/10.1016/j.watres.2015.11.045.
133. Desmidt, E., Ghyselbrecht, K., Zhang, Y., Pinoy, L., Van Der Bruggen, B., Verstraete, W., Rabaey, K., Meesschaert, B., Global phosphorus scarcity and full-scale P-recovery techniques: A review. *Crit. Rev. Environ. Sci. Technol.*, 45, 336–384, 2015. https://doi.org/10.1080/10643389.2013.866531.

10

Recovery, Valorization, and Beneficiation of Valuable Minerals From Natural Acid Mine Drainage and Their Respective Application in Wastewater Treatment

Khathutshelo Lilith Muedi[1], **Vhahangwele Masindi**[2] **and Hendrik Gideon Brink**[1*]

[1]*Department of Chemical Engineering, Faculty of Engineering, Built Environment and Information Technology, University of Pretoria, Private Bag X20, Hatfield, Pretoria, South Africa*

[2]*Department of Environmental Sciences, School of Agriculture and Environmental Sciences, University of South Africa (UNISA), Florida, South Africa*

Abstract

Acid mine drainage (AMD) is a significant environmental concern globally, with diverse mine effluents posing considerable ecological and economic challenges. This chapter provides an in-depth exploration of AMD mechanisms, diverse effluent compositions, and the ramifications on human health, aquatic ecosystems, water quality, economy and infrastructure. Treatment technologies, including passive, active, and integrated remediation methods, are reviewed. Additionally, the chapter investigates the opportunities for exploiting valuable minerals from AMD, including recovery, valorization, and beneficiation processes. These minerals find applications in industrial processes and environmental remediation, offering economic potential. Furthermore, the chapter examines heavy metal contaminants in wastewater, such as arsenic and chromium, detailing their impacts and removal technologies. Paramount water contaminants like Congo red dye, ammonia, nitrates, phosphates, and sulfates are also discussed, along with removal technologies. Adsorption phenomena are elucidated, encompassing categorizations of mechanisms, kinetics, types of adsorbents, and their applications and advantages.

**Corresponding author*: deon.brink@up.ac.za

Elvis Fosso-Kankeu, Vhahangwele Masindi, Johannes Maree and Bhekie Mamba (eds.) Customized Technologies for Sustainable Management of Industrial Wastewater: A Circular Economy Approach, (389–466) © 2025 Scrivener Publishing LLC

This comprehensive review contributes to understanding the complexities of AMD treatment and its applications in wastewater treatment strategies.

Keywords: Industrial acid mine drainage, mineral recovery, selective precipitation, oxyanions, nutrients, adsorption, municipal wastewater, industrial effluents

List of Abbreviations

Al	Aluminum
AMD	Acid mine drainage
AOB	Ammonia-oxidizing bacteria
As(III)	Arsenite
As(V)	Arsenate
ATP	Adenosine triphosphate
Cd	Cadmium
CO_2	Carbon dioxide
Cr(III)	Trivalent chromium
Cr(VI)	Hexavalent chromium
CSIR	Council for Scientific and Industrial Research
Cu	Copper
Fe	Iron
MCL	Maximum contaminant level
Mn	Manganese
MOFs	Metal-organic frameworks
MRI	Magnetic resonance imaging
MS	Membrane separation
NOB	Nitrite-oxidizing bacteria
pH	Potential hydrogen
REEs	Rare earth elements
RO	Reverse osmosis
ROS	Reactive oxygen species
SRB	Sulfate-reducing bacteria
S	Sulfur
WRC	Water Research Commission
XRF	X-ray fluorescence
Zn	Zinc

10.1 Introduction

Acid mine drainage (AMD) is a significant environmental issue arising from mining activities, characterized by the release of acidic water containing various pollutants. This book chapter comprehensively explores the mechanisms behind AMD formation, the elemental composition of different mine effluents, associated challenges, and the diverse treatment technologies employed. Furthermore, it investigates the potential for exploiting valuable minerals recovered from AMD and their application in wastewater treatment.

10.2 Acid Mine Drainage and Its Mechanisms

Acid mine drainage (AMD) is a multifaceted environmental issue arising from the oxidation of sulfide minerals, particularly pyrite (FeS_2), upon exposure to oxygen and water. This oxidation process is catalyzed by microbial activity and leads to the production of sulfuric acid (H_2SO_4) [1]. Mine drainage remains a focal point of concern as one of the foremost environmental challenges confronting the global mining industry. This is primarily attributed to the deleterious effects that they inflict upon the environment, characterized by heightened concentrations of heavy metals, sulfate, trace metals, and pH alterations. Despite these challenges, the mining sector persists in its expansion to meet the increasing demand for minerals, thereby driving the growth and development of the global economy [2].

Mining operations, encompassing both open-cast and underground activities, entail the extraction of substantial volumes of rock to access desired mineral deposits. Throughout this mineral extraction process, whether from hard rock, such as gold, or soft rock, such as coal, exposure to air and water is inevitable [3]. Water presence during mining operations may manifest as surface runoff or groundwater, utilized for various purposes including cooling, dust suppression, and lubrication. Research by Madzivire *et al.* [4] highlights that mine water, originating from tailings, can exhibit a range of pH levels, spanning from acidic to neutral or alkaline, contingent upon the geological context of the mining site. Table 10.1 delineates sulfide-bearing materials that significantly contribute to AMD formation depending on the geological area, along with their respective chemical compositions.

Table 10.1 Sulfide-bearing minerals that significantly contribute to AMD formation depending on the geological area, and their chemical composition [5].

Mineral ore	Chemical formula	Element composition
Pyrite	FeS_2	Iron (Fe), Sulfur (S)
Covellite	CuS	Copper (Cu), Sulfur (S)
Chalcopyrite	$CuFeS_2$	Copper (Cu), Iron (Fe), Sulfur (S)
Arsenopyrite	FeAsS	Iron (Fe), Arsenic (As), Sulfur (S)
Cinnabar	HgS	Mercury (Hg), Sulfur (S)
Marcasite	FeS_2	Iron (Fe), Sulfur (S)
Pyrrhotite	Fe_xS_x	Iron (Fe), Sulfur (S)
Chalcocite	Cu_2S	Copper (Cu), Sulfur (S)
Molybdenite	MoS_2	Molybdenum (Mo), Sulfur (S)
Millerite	NiS	Nickel (Ni), Sulfur (S)
Galena	PbS	Lead (Pb), Sulfur (S)
Sphalerite	ZnS	Zinc (Zn), Sulfur (S)
Bornite	Cu_5FeS_4	Copper (Cu), Iron (Fe), Sulfur (S)

The acidification resulting from sulfide oxidation triggers reactions with minerals in the surrounding rock, leading to the liberation of metals and metalloids. This phenomenon liberates a spectrum of metal ions, comprising iron (Fe), aluminum (Al), and manganese (Mn), along with hazardous heavy metals like cadmium (Cd), zinc (Zn), and copper (Cu) [6].

Equation 10.1 depicts the sulfide oxidation reaction of pyrite (FeS_2) in the presence of water, which is fundamental in AMD formation:

$$FeS_{2(s)} + 7O_{2(g)} + 2H_2O_{(l)} \rightarrow 2Fe^{2+}{}_{(aq)} + 4SO_4{}^{2-}{}_{(aq)} + 4H^+{}_{(aq)} \tag{10.1}$$

This reaction highlights the conversion of pyrite (FeS_2) to ferrous ions (Fe^{2+}) and sulfate ions ($SO_4{}^{2-}$) in the presence of oxygen and water, resulting in acidification.

Equations (10.2) to (10.4) represent the dissolution of various metal-containing minerals under acidic conditions:

$$\textit{Fe-containing mineral} + 2H^+ \rightarrow Fe^{2+} + \textit{other products} \quad (10.2)$$

$$\textit{Al-containing mineral} + 3H^+ \rightarrow Al^{3+} + \textit{other products} \quad (10.3)$$

$$\textit{Mn-containing mineral} + 2H^+ \rightarrow Mn^{2+} + \textit{other products} \quad (10.4)$$

These equations describe the dissolution of minerals containing iron, aluminum, and manganese, leading to the release of respective metal ions.

Similarly, Equations (10.5), to (10.7) illustrate the release of heavy metals such as cadmium (Cd), zinc (Zn), and copper (Cu) from their respective mineral sources:

$$\textit{Cd-containing mineral} + 2H^+ \rightarrow Cd^{2+} + \textit{other products} \quad (10.5)$$

$$\textit{Zn-containing mineral} + 2H^+ \rightarrow Zn^{2+} + \textit{other products} \quad (10.6)$$

$$\textit{Cu-containing mineral} + 2H^+ \rightarrow Cu^{2+} + \textit{other products} \quad (10.7)$$

These equations delineate the acid-driven release of heavy metal ions into the surrounding environment, contributing to the environmental hazards associated with AMD.

Recent research has shed light on the microbial-mediated processes driving sulfide oxidation in the formation of AMD. Microbial taxa such as *Thiobacillus* spp. and *Acidithiobacillus* spp. are key players in catalyzing sulfide oxidation reactions. These microorganisms utilize reduced sulfur compounds, such as sulfides, elemental sulfur, and thiosulfate, as energy sources, with the concomitant production of sulfuric acid as a metabolic by-product [1].

Moreover, genomic studies have unveiled the genetic mechanisms underlying the metabolic capabilities of these microorganisms. For instance, genomic analysis of *Acidithiobacillus ferrooxidans* has revealed the presence of gene clusters encoding enzymes involved in the oxidation of ferrous iron and sulfur compounds, elucidating its role in AMD formation [7]. Additionally, meta-transcriptomic studies have provided insights into the transcriptional activity of microbial communities in AMD environments, highlighting the upregulation of genes associated with sulfur oxidation pathways under acidic conditions [8].

Recent studies have delved into the mechanisms and consequences of metal release induced by AMD. Advanced spectroscopic techniques, such as X-ray fluorescence (XRF) mapping, have provided insights into the speciation and spatial distribution of metal ions in AMD-affected environments [9, 10]. These studies have highlighted the intricate interplay between mineralogical composition, redox conditions, and microbial activity in influencing metal mobilization and attenuation processes.

Moreover, recent research has emphasized the importance of microbial-metal interactions in modulating metal mobility and bioavailability in AMD-affected environments [11]. Microbial communities associated with AMD often possess metal resistance mechanisms, including metal ion efflux pumps and metal-binding proteins, which influence the fate and transport of metals in aquatic ecosystems. The acidic characteristic of AMD environments facilitate the dissolution and mobilization of metals, thereby augmenting their bioavailability and potential toxicity to aquatic biota [12]. This underscores the urgent need for comprehensive research efforts to elucidate the mechanisms governing metal behavior in AMD-affected systems, in order to develop targeted interventions for environmental remediation and conservation.

Additionally, isotopic tracing techniques, including stable isotopes of metals and metalloids, have been utilized to trace the sources and pathways of metal contamination in AMD-impacted ecosystems. Isotopic signatures of metals such as iron, sulfur, and lead have been instrumental in discerning between natural background sources and anthropogenic inputs of metals into aquatic systems affected by AMD [13].

The acidic pH levels resulting from AMD contribute significantly to the dissolution and mobilization of metals, thereby amplifying their bioavailability and potential adverse effects on aquatic organisms [12]. Under such acidic conditions, metals like iron, aluminum, and manganese are released from mineral sources and become more soluble in water, facilitating their uptake by biota and leading to potential ecological disruptions [14].

This phenomenon highlights the critical role of pH in regulating metal speciation and toxicity in AMD-affected environments. As pH decreases, metal solubility generally increases, exacerbating the risks posed to aquatic ecosystems [15]. Therefore, comprehensive investigations into the dynamic interactions between pH, metal speciation, and biological responses are essential for devising effective management strategies to mitigate AMD-related pollution and safeguard freshwater habitats. Figure 10.1 shows the formation of AMD and pathways thereof.

In essence, AMD is intricately linked to sulfide oxidation processes catalyzed by microbial activity, resulting in the production of sulfuric acid and subsequent leaching of metals from the surrounding geological materials, as illustrated in Figure 10.1. This cascade of chemical reactions ultimately leads to the release of acidic, metal-laden water into the environment, posing significant ecological and human health risks. This nexus between microbial activity, geochemical transformations, and metal mobilization underscores the complexity of AMD contamination and emphasizes the importance of interdisciplinary research approaches in unraveling its mechanisms and devising effective management interventions [1, 17].

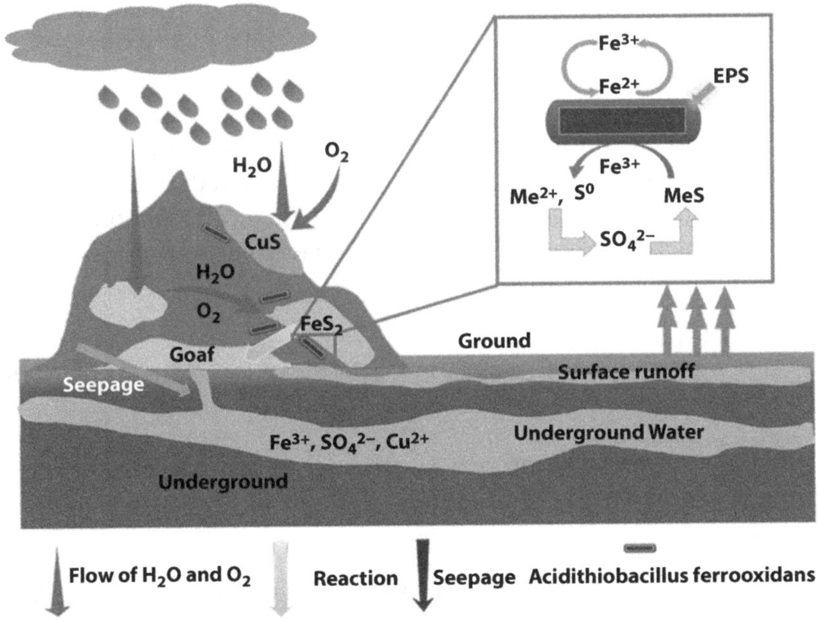

Figure 10.1 Sources, pathway formation, and mechanisms of AMD (modified from [16]).

10.3 Diverse Mine Effluents and Their Elemental Composition

In the realm of mining and its environmental repercussions, understanding the diverse nature of mine effluents and their elemental composition is paramount. Mine effluents encompass a wide range of discharge types, each with unique chemical characteristics and potential environmental impacts. This section delves into the varied nature of mine effluents, exploring their elemental compositions, and the implications for environmental management and remediation strategies. Through a comprehensive examination of acidic, neutral, and basic mine effluents, along with their associated elemental profiles, we aim to shed light on the multifaceted challenges posed by mining activities and the importance of proactive approaches to mitigate their environmental footprint.

10.3.1 Acid Mine Effluents

Acidic mine effluents, a result of sulfide mineral oxidation, pose significant environmental hazards due to their high acidity and elevated metal concentrations, including Fe, Al, Mn, Cd, Zn, and Cu. These effluents are a consequence of the exposure of sulfide minerals, like pyrite (FeS_2), to oxygen and water during mining activities or natural weathering processes [18]. Brown and Burgos [19] elucidated the reactive transport modeling of AMD, emphasizing the complex interactions between minerals, water, and microbial activity in generating and transporting acid drainage.

The formation of AMD involves a series of chemical reactions. Initially, sulfide minerals undergo oxidation, releasing ferrous iron and sulfate ions, as depicted by Equation 10.8:

$$4FeS_2 + 14O_2 + 4H_2O \rightarrow 4Fe^{2+} + 8SO_4^{2-} + 8H^+ \tag{10.8}$$

Subsequently, ferrous iron reacts with oxygen and water to form ferric hydroxide precipitates:

$$2Fe^{2+} + \frac{3}{2}O_2 + 3H_2O \rightarrow 2Fe(OH)_3 \downarrow \tag{10.9}$$

The hydrolysis of ferric hydroxide further contributes to acidity, as depicted by Equation 10.10:

$$Fe(OH)_3 + 3H^+ \rightarrow Fe^{3+} + 3H_2O \tag{10.10}$$

However, the presence of carbonate minerals can reduce acid generation through the neutralisation of protons, as depicted by Equation 10.11:

$$2H^+ + CO_3^{2-} \rightarrow H_2O + CO_2 \tag{10.11}$$

Countries with extensive mining operations, such as the United States, particularly in regions like Appalachia, experience significant issues related to AMD [20]. Similarly, in South Africa, the Witwatersrand Basin, renowned for its gold mining history, faces substantial challenges associated with acid mine effluents [21]. These examples underscore the global prevalence of acid mine effluents and the urgent need for effective mitigation strategies to address their environmental consequences.

10.3.2 Neutral Mine Effluents

Neutral mine effluents, characterized by a pH close to neutral, present a unique environmental challenge despite their seemingly benign acidity levels. While these effluents may not exhibit the extreme acidity of their acidic counterparts, they can still harbor significant levels of metals, notably iron and aluminum, which can exert adverse effects on aquatic ecosystems. The composition of neutral mine effluents is often influenced by the depletion or reduced abundance of sulfide minerals in the mining area, resulting in decreased acid generation compared to acidic mine drainage scenarios [22]. However, the presence of metals in these effluents remains a concern, as they can accumulate in water bodies and disrupt ecological balance.

Recent studies underscore the importance of monitoring and mitigating the impacts of neutral mine effluents on aquatic environments [23]. In countries like South Africa, where mining activities are prevalent, neutral mine effluents can be found in regions such as the Bushveld Igneous Complex, where sulfide mineral occurrences may be less pronounced [24].

An illustration of an equation depicting the formation and constituent components of neutral mine effluents is provided in Equation 10.12, where, during the dissolution process, aluminum can be released into the solution.

$$Al_2Si_2O_5(OH)_4 + 6H^+ \rightarrow 2Al^{3+} + 5H_2O + 2SiO_2 \quad (10.12)$$

This reaction illustrates the release of aluminum into solution, contributing to the elemental composition of neutral mine effluents.

10.3.3 Basic Mine Effluents

Basic mine effluents, distinguished by their elevated pH levels attributed to alkaline minerals like calcite ($CaCO_3$) or dolomite [$CaMg(CO_3)_2$], present notable environmental considerations despite being less prevalent compared to acidic counterparts. These effluents, while less common, can still pose significant challenges to environmental management, particularly in regions where carbonate deposits abound or where mining activities introduce carbonate minerals into the ecosystem [25]. Azzie [26] additionally elucidated the fundamental characteristics of mine drainage originating from a coal mine situated in Vryheid, South Africa.

The dissolution of carbonate minerals plays a pivotal role in the elevation of pH levels observed in basic mine effluents. For example, the dissolution of calcite can be described by Equation 10.13:

$$CaCO_3 + H_2CO_3 \rightarrow Ca^{2+} + 2HCO_{3^-} \quad (10.13)$$

Similarly, dolomite dissolution results in the release of calcium and magnesium ions, accompanied by the formation of bicarbonate ions, as illustrated in Equation 10.14:

$$CaMg(CO_3)_2 + 2H^+ \rightarrow Ca^{2+} + Mg^{2+} + 2HCO_{3^-} \quad (10.14)$$

These chemical reactions highlight the buffering capacity of carbonate minerals, which neutralize acidity and contribute to the alkalinity observed in basic mine effluents.

10.3.4 Saline Mine Effluents

Oceanographers delineate salinity within the framework of seawater characteristics, influenced by dilution processes, and with emphasis on either conductivity or chlorinity metrics. Existing literature highlights a prevailing notion of low sodium concentrations in mine water; however, investigations reveal a notable prevalence of high sodium chloride content in coastal mine locations [27]. Geological formations in certain regions feature sodium-calcium-chloride–rich rocks, further contributing to the salinity of mine drainage. Particularly, underground coal mines discharge drainage characterized by elevated levels of alkali and halogen elements, consequently imparting alkaline properties to the mine water. Saline mine drainage exhibits pronounced levels of calcium and magnesium ions, alongside relatively low metal concentrations albeit moderate levels of iron and sulfate may still be present [27].

10.4 Ramifications of Acid Mine Drainage

In the realm of mining's environmental impacts, few issues loom as large as AMD. From its inception to its far-reaching consequences, the ramifications of AMD extend beyond the confines of mining sites, affecting entire ecosystems and communities downstream. This section delves into the multifaceted repercussions of AMD, exploring its origins, ecological effects, and socio-economic implications. By examining the chemical processes underlying AMD formation, its impact on water quality, aquatic life, and human health, as well as the economic burdens that it imposes on affected regions, we aim to illuminate the urgent need for comprehensive mitigation strategies. Understanding the ramifications of AMD is essential for stakeholders, policymakers, and communities alike as they navigate the complexities of sustainable resource extraction and environmental stewardship.

10.4.1 Impacts on Human Health

The impact of AMD on human health is multifaceted and profound, encompassing a spectrum of adverse effects stemming from exposure to acidic waters and elevated concentrations of heavy metals. A study by CSIR [28] underscores the diverse health issues associated with AMD contamination, including skin irritation, respiratory ailments, and gastrointestinal disorders. The acidic nature of AMD can irritate the skin upon contact and exacerbate existing dermatological conditions, leading to discomfort and potential complications. Inhalation of acidic aerosols or vapours generated from AMD-contaminated water bodies can irritate the respiratory tract, causing symptoms such as coughing, wheezing, and shortness of breath. Furthermore, ingestion of water contaminated with toxic metals, including lead, arsenic, and mercury, poses grave long-term health risks to affected communities [28]. These heavy metals accumulate in the body over time, leading to systemic toxicity and increasing the risk of neurological disorders, cardiovascular diseases, and various cancers. Communities residing downstream from AMD-affected areas are particularly vulnerable to these health impacts, as they rely on contaminated water sources for drinking, bathing, and agricultural purposes [29].

10.4.2 Ecological Ramifications on Aquatic Flora and Fauna

The ecological ramifications of AMD on aquatic flora and fauna are profound and far-reaching, with consequences that extend beyond individual species to entire ecosystems. A study conducted underscores the detrimental effects of AMD on aquatic biodiversity, driven by the acidic conditions and elevated concentrations of heavy metals in affected water bodies. These adverse conditions disrupt the delicate balance of aquatic ecosystems, leading to population declines and even extinctions of vulnerable species [30]. Fish populations, in particular, are highly susceptible to the toxic effects of AMD-contaminated waters. Some studies highlight how AMD-induced toxicity impairs fish reproduction, growth, and survival rates, posing a significant threat to aquatic biodiversity [31]. Moreover, the loss of keystone species and disruption of ecological functions can have cascading effects throughout entire ecosystems, exacerbating the ecological consequences of AMD contamination.

10.4.3 Phytotoxicity Effects

The phytotoxic effects of AMD on terrestrial vegetation are a critical aspect of its environmental impact, with implications for ecosystem health and biodiversity. Studies have shown that elevated concentrations of heavy metals in AMD-contaminated soils can inhibit plant growth, impairing crop yields and compromising soil fertility [32]. The toxic effects of metals such as lead, arsenic, and cadmium on plant physiology disrupt essential metabolic processes, including photosynthesis and nutrient uptake, leading to stunted growth and reduced biomass production [33]. Moreover, the accumulation of heavy metals in plant tissues poses risks to human health through the consumption of contaminated agricultural products, highlighting the interconnectedness of environmental and human health concerns [34].

10.4.4 Impacts on Water Quality

The degradation of water quality due to AMD is a critical environmental concern, with significant implications for both aquatic ecosystems and human welfare. AMD contamination profoundly impairs the ecological integrity of affected water bodies, rendering them inhospitable to aquatic life and unsuitable for various human uses [35]. Kibria [32] highlighted the adverse effects of AMD on water quality, characterized by low pH levels and elevated concentrations of heavy metals. These conditions create hostile environments for aquatic organisms, leading to population declines and biodiversity loss. Moreover, AMD-contaminated waters pose risks to human health, as they are unsuitable for drinking water supply and recreational activities. The transport of acidic waters and metal pollutants downstream can have far-reaching effects, impacting entire river systems and exacerbating the environmental footprint of AMD contamination [16].

10.4.5 Economic Impacts

The economic ramifications stemming from AMD are considerable, encompassing both direct and indirect costs that reverberate throughout affected communities and industries. Remediation efforts aimed at mitigating the impacts of AMD contamination entail substantial financial investments, encompassing the deployment of remediation technologies and the rehabilitation of degraded ecosystems [36]. These endeavors often entail significant expenditures, as evidenced by the extensive infrastructure required to treat and manage AMD-affected waters effectively [30]. Furthermore, the deleterious effects of AMD extend beyond the direct costs of remediation, encompassing broader economic losses arising from the impairment of ecosystem services. The decline in water quality resulting from AMD contamination undermines critical ecosystem functions, such as fisheries productivity, recreational tourism, and reliable water supply, thereby eroding the economic foundations of affected regions [37].

Recent studies have underscored the multifaceted nature of AMD's economic impacts, shedding light on previously overlooked dimensions of its influence. For example, Munyai *et al.* [38] elucidated the role of AMD-induced land degradation in exacerbating soil erosion and reducing agricultural productivity, thereby amplifying the economic strain on rural communities reliant on farming as a livelihood. Moreover, advancements in economic modeling techniques, as demonstrated by Pell [39], have facilitated more accurate assessments of the long-term economic costs associated with AMD contamination, providing policymakers with valuable insights into the potential benefits of proactive intervention and remediation measures. Additionally, the socio-economic implications of AMD contamination extend beyond immediate financial costs, encompassing broader societal impacts such as the loss of cultural heritage and the erosion of community cohesion.

10.4.6 Impacts on Infrastructure

The impact of AMD on infrastructure extends beyond environmental degradation, encompassing tangible damage to critical systems such as roads, bridges, and pipelines. The corrosive nature of acidic waters and metal-laden sediments associated with AMD poses a significant threat to the structural integrity of infrastructure elements, leading to accelerated deterioration and heightened maintenance requirements [40]. The consequences of AMD-induced infrastructure degradation are manifold, encompassing both immediate safety concerns and long-term economic burdens. For instance, the weakening of bridge supports and road foundations due to corrosion can compromise public safety and disrupt transportation networks, impeding access to essential services and hindering socio-economic development in affected regions.

Recent research has shed light on the complex interactions between AMD contamination and infrastructure resilience, highlighting the need for proactive mitigation measures to mitigate adverse impacts. For example, studies by Ekolu *et al.* [40] have underscored the role of innovative materials and construction techniques in enhancing the durability of infrastructure exposed to AMD-affected environments, offering promising avenues for reducing maintenance costs and enhancing long-term resilience. Moreover, the indirect impacts of AMD-induced infrastructure degradation on local economies cannot be overlooked. The disruption of transportation networks and utilities can impede commercial activities, hamper tourism, and undermine the viability of businesses reliant on reliable infrastructure services, thereby exacerbating the economic strain on affected communities.

10.5 Treatment Technologies of Acid Mine Drainage

Treatment technologies for AMD encompass a spectrum of approaches, ranging from passive to active remediation methods, with emerging holistic strategies aimed at addressing the multifaceted nature of AMD contamination. In South Africa, where AMD poses significant environmental and socio-economic challenges, several successful treatment technologies have been developed and implemented.

10.5.1 Passive Remediation Technologies

10.5.1.1 Natural and Constructed Wetlands

Passive remediation technologies represent a sustainable approach to mitigating the impacts of AMD by harnessing natural processes to treat contaminated water. In South Africa, where AMD poses significant environmental challenges, the successful implementation of passive remediation technologies has been instrumental in restoring water quality and ecosystem health. Natural wetlands serve as exemplars of passive remediation, capitalizing on the intricate interactions among vegetation, soil, and microorganisms to facilitate the removal of contaminants from AMD-affected waters [41]. By leveraging the natural filtration and retention capabilities of wetland ecosystems, these sites effectively sequester metals and neutralize acidity, thereby improving water quality downstream [42].

Constructed wetlands, engineered systems designed to emulate the functions of natural wetlands, offer a versatile and cost-effective solution for AMD remediation [43]. By incorporating engineered features such as substrate layers, emergent vegetation, and controlled hydraulic regimes, these systems enhance the natural processes of sedimentation, adsorption, and microbial degradation, leading to effective pollutant removal [44]. Recent advancements in passive remediation technologies have further expanded their applicability and efficacy in addressing AMD contamination. For example, research by Thi-Kim-Quyen Vo *et al.* [45] has explored the integration of floating treatment wetlands into existing passive treatment systems, demonstrating enhanced removal efficiencies for metals and acidity. Additionally, a study by Gupta *et al.* [46] has investigated the role of biochar amendments in augmenting the performance of constructed wetlands, highlighting the potential for a sustainable enhancement of treatment capacities.

10.5.1.2 Microbial Degradation

Microbial degradation remediates AMD by leveraging the metabolic activities of microorganisms to transform and detoxify contaminants present in AMD-affected waters. Microbial sulfate reduction, a key process facilitated by sulfate-reducing bacteria (SRB), plays a pivotal role in AMD remediation by promoting the precipitation of metal sulfides and the neutralization of acidity [47]. By harnessing the metabolic capabilities of SRB, indigenous microbial communities within AMD-affected environments can effectively mitigate the release of toxic metals and restore ecological balance [48].

Bioremediation, which encompasses a suite of microbial-based techniques for pollutant removal, offers versatile and cost-effective solutions for AMD treatment in South Africa. Advancements in molecular microbiology techniques, such as next-generation sequencing and metagenomics, have enabled comprehensive characterization of microbial communities inhabiting AMD-impacted environments [49]. By elucidating the functional diversity and metabolic potential of indigenous microorganisms, these molecular approaches provide valuable insights into the mechanisms driving microbial degradation processes in AMD treatment systems, facilitating the design of targeted remediation strategies. In addition to *in situ* microbial degradation, *ex situ* bioremediation technologies offer opportunities for the treatment of AMD-affected waters in controlled laboratory settings [50].

10.5.1.3 Sedimentation

Sedimentation is a fundamental process in the treatment of AMD, involving the settling of suspended solids and metal precipitates from contaminated water. Passive sedimentation ponds represent a widely utilized treatment method for AMD, leveraging gravity-driven settling to remove particulate matter and metal precipitates from contaminated water streams [51]. By facilitating the removal of sediment-bound contaminants, these treatment facilities play a crucial role in improving water quality and ecosystem health downstream.

Innovative sedimentation technologies, such as high-rate settling processes and engineered sedimentation basins, offer enhanced efficiencies for AMD remediation in South Africa. Tarpagkou and Pantokratoras [52] have explored the use of inclined plate settlers and lamella clarifiers to accelerate the sedimentation of fine particles and metal precipitates, thereby optimizing treatment performance and reducing footprint requirements. These advanced sedimentation systems offer scalable solutions for the treatment of large volumes of AMD-affected water, making them well-suited for application in mining-impacted regions.

10.5.2 Active Remediation Technologies

Active remediation technologies involve the direct treatment of AMD through physical, chemical, or biological processes. In South Africa, where AMD contamination is prevalent, the development and implementation of active remediation technologies have played a crucial role in mitigating the environmental and socio-economic impacts of mining-related pollution [53].

One successful active remediation technology utilized in South Africa is lime neutralization, which involves the addition of lime (calcium hydroxide) to AMD-contaminated waters to raise pH levels and facilitate the precipitation of metals as hydroxides or carbonates [36]. Studies conducted at AMD-impacted sites, such as the Witwatersrand Goldfields, have demonstrated the efficacy of lime neutralization in reducing metal concentrations and acidity in treated effluents. By leveraging readily available and cost-effective reagents, lime neutralization offers a practical solution for AMD treatment in South Africa. Furthermore, advancements in chemical precipitation technologies have expanded the treatment options for AMD remediation in South Africa. Innovative approaches such as iron(III) hydroxide precipitation, utilizing ferric iron salts to facilitate the removal of metals through precipitation and co-precipitation mechanisms, have shown promising results in laboratory and field-scale studies [54, 55]. By optimizing process parameters and treatment conditions, iron(III) hydroxide precipitation offers an effective and environmentally sustainable method for AMD treatment in South Africa. Biological treatment technologies, such as microbial sulfate reduction and bioremediation, represent another active remediation approach for AMD treatment. Microbial sulfate reduction, facilitated by SRB, promotes the precipitation of metal sulfides and the neutralization of acidity in AMD-affected waters. Research conducted by Luptakova and Kusnierova [56] has demonstrated the role of indigenous microbial communities in mediating AMD remediation processes, highlighting the potential of biological treatment strategies for sustainable pollution control.

Moreover, electrochemical treatment technologies offer innovative solutions for AMD remediation in South Africa and globally. Electrocoagulation, which involves the application of electric current to induce the coagulation and precipitation of contaminants, has shown promise for the removal of metals and acidity from AMD-affected waters [57]. By harnessing electrochemical processes, these technologies offer scalable and energy-efficient alternatives for AMD treatment in South Africa.

10.5.3 Integrated Remediation Technologies

Integrating active and passive technologies for AMD treatment presents a holistic approach to remediation, leveraging both engineered interventions and natural processes to effectively mitigate environmental impacts. These integrated systems capitalize on the strengths of both active and passive methods, achieving sustainable and efficient treatment outcomes. Among the successful technologies, integrated anaerobic/aerobic bioreactors stand out, where anaerobic bioreactors facilitate sulfate reduction and metal precipitation, complemented by aerobic bioreactors promoting further metal oxidation and removal [1]. Additionally, hybrid constructed wetlands integrate surface and subsurface flow wetlands, exhibiting substantial reductions in metal concentrations and acidity [58].

Sequential treatment systems combine multiple stages, including chemical precipitation and passive filtration, effectively removing heavy metals and neutralizing acidity in AMD-contaminated water [59]. Electrocoagulation-constructed wetlands hybrid systems enhance metal removal efficiency through electrocoagulation [60], whereas solar-powered hybrid systems, as explored by Chidiac *et al.* [61], offer sustainable solutions by integrating renewable energy sources with active and passive treatment components. These integrated systems provide robust and sustainable solutions for AMD treatment, addressing short-term needs and long-term environmental considerations.

Furthermore, the integration of sedimentation with other treatment processes, such as neutralization and flocculation, enhances overall treatment efficiencies and pollutant removal capacities. Studies by Simate and Ndlovu [18] have investigated the synergistic effects of combined treatment strategies, demonstrating improved removal rates for metals and acidity compared to standalone sedimentation processes.

10.6 Acid Mine Drainage Exploitation Opportunities

Acid mine drainage presents a complex environmental challenge resulting from the oxidation of sulfide minerals in mining activities. However, amidst this challenge lies a spectrum of opportunities for environmental remediation and resource recovery. This section delves into three key avenues for exploiting AMD, namely, extraction of valuable minerals, valorization of oxides and other derivatives, and beneficiation of AMD derivatives. Each avenue offers promising prospects for mitigating environmental pollution while harnessing valuable resources for economic gain and sustainable development.

10.6.1 Reclamation of Valuable Minerals From Acidic Mine Effluents

Reclamation of valuable minerals from AMD offers a dual benefit of environmental remediation and economic viability. Recent studies have highlighted the presence of various valuable minerals in AMD, with rare earth elements (REEs) being particularly noteworthy due to their importance in high-tech industries such as electronics and renewable energy [54]. Selective extraction processes have been developed to recover REEs from AMD, thereby reducing environmental contamination and lessening reliance on traditional mining sources [62].

Furthermore, AMD is known to contain significant concentrations of other valuable metals such as copper, zinc, and nickel. Advanced extraction techniques, including ion exchange, solvent extraction, and precipitation methods, have been explored to recover these metals from AMD effluents [63]. These recovered metals can then undergo further processing and find applications in various industrial sectors, thus contributing to resource sustainability and circular economy initiatives.

10.6.1.1 Recovery Technologies for Valuable Minerals From AMD

a) Ion exchange

Ion exchange is one of the cornerstone techniques in the repertoire for recovering valuable metals from acid mine effluents. This method operates on the fundamental principle of selective ion exchange between metal ions present in the AMD solution and ions affixed to a solid phase, typically comprising a resin or zeolite matrix [64]. As the AMD solution traverses through an ion exchange column packed with the resin or zeolite material, metal ions like copper, zinc, and nickel within the solution undergo selective capture by the exchange sites immobilized on the solid phase.

The efficacy of ion exchange in metal recovery from AMD resides in the precision of the exchange mechanism, allowing for the targeted removal of desired metals while preserving the integrity of other ions within the solution. This selectivity is meticulously orchestrated through the careful selection of the resin or zeolite material and fine-tuning of operational parameters such as pH and flow rate [64]. Consequently, ion exchange furnishes a meticulous and efficient avenue for concentrating valuable metals from AMD effluents. Subsequent to the adsorption of metal ions onto the solid phase, their retrieval is facilitated through either elution or regeneration of the resin or zeolite material [65].

Elution entails the passage of a suitable eluent through the ion exchange column, liberating the captured metal ions from the exchange sites and thereby releasing them into a separate solution for subsequent processing [66]. Alternatively, regeneration of the resin or zeolite material can be achieved by treating it with a regenerating agent, such as acid or base, which displaces the captured metal ions and restores the exchange sites to their original state, poised for reutilization [67].

The versatility and efficiency embodied by ion exchange render it a method of choice for metal recovery from wastewater across various industrial and environmental contexts. Its ability to selectively sequester specific metals facilitates the generation of high-purity metal concentrates amenable to downstream processing and utilization [67].

b) Solvent extraction

Solvent extraction is one of the techniques used for the selective separation and recovery of valuable metals from acidic mine effluents [68]. This method harnesses the differential solubility of metal ions in organic solvents versus aqueous phases to achieve targeted extraction. By manipulating factors such as pH and the composition of solvent and aqueous phases, specific metal ions can be preferentially transferred into the organic phase, thereby facilitating their isolation and subsequent recovery [69]. Solvent extraction exploits the distinct chemical properties of various metal ions present in AMD effluents. By carefully adjusting parameters such as pH and temperature, researchers can fine-tune the extraction process to selectively target desired metals, including but not limited to copper, zinc, nickel, and REEs [70].

In solvent extraction, the aqueous phase containing the metal ions of interest is brought into contact with an organic solvent possessing a high affinity for these ions. This initiates a transfer of metal ions from the aqueous phase to the organic phase, forming complexes or chelates with the solvent molecules [71]. Subsequent separation of the metal-laden organic phase from the aqueous phase allows for the isolation of the targeted metal ions, typically accomplished through techniques such as centrifugation or decantation [71].

Post-separation, the metal-laden organic phase undergoes further processing to facilitate metal recovery. This entails stripping or back-extraction steps, where metal ions are selectively released from the organic phase into a fresh aqueous solution or electrolyte. Following this, precipitation or electrochemical methods are employed to concentrate and purify the recovered metals, rendering them suitable for downstream applications.

c) Precipitation

Precipitation involves the addition of reagents to AMD solutions, thereby inducing the formation of insoluble metal compounds. These precipitates can then be easily separated from the solution through processes such as filtration or sedimentation. Commonly employed precipitating agents include lime (calcium hydroxide) and sodium hydroxide, which serve to elevate the pH of the AMD solution, thus facilitating the precipitation of metal hydroxides or sulfides [72]. Precipitation methods rely on their ability to selectively isolate target metals in solid form, effectively concentrating them for subsequent processing. Following precipitation, the resultant metal precipitates can undergo further treatment to extract and refine the valuable metals they contain. Techniques such as leaching or smelting may be employed to dissolve the metal compounds within the precipitates, allowing for the recovery of high-purity metal concentrates [73].

d) Biological treatment

Biological treatment leverages the metabolic prowess of microorganisms to facilitate metal reclamation. This approach capitalizes on certain bacteria and fungi to bioaccumulate metals from AMD solutions, concentrating them within their biomass. Once enriched, the metal-laden biomass can be harvested and subjected to processing techniques aimed at recovering the valuable metals that they contain [30].

Certain microbial species possess specialized metal uptake mechanisms, allowing them to absorb and accumulate metals from their surrounding environment. Through metabolic processes such as biosorption and bioaccumulation, these microorganisms sequester metals within their cellular structures, thereby concentrating them for subsequent extraction and recovery [74].

10.6.2 Valorization of Recovered Minerals From Acid Mine Drainage

Recent scholarly endeavors have concentrated on the valorization of valuable minerals salvaged from AMD. The envisioned commercial exploitation of AMD remediation by-products represents a pivotal stride toward realizing economically sustainable technologies for AMD mitigation. Predominantly recovered minerals from AMD, notably iron and aluminum hydroxides, emerge as focal points due to their prevalence in AMD discharges and associated environmental hazards. Other frequently retrieved minerals include manganese and gypsum, all of which present a diverse array of industrial and environmental uses, thus rendering their recovery a financially viable avenue for AMD remediation.

10.6.2.1 Industrial Applications

The industrial sector is renowned for consuming large volumes of water of varying quality. The type of water used depends on the process and the quality of product required. For many processes, high grade or potable water is not necessary [75]. It has been found that water of a quality lower than that required for potable use is sufficient for cooling purposes, with leachate being a prime example. High grade water usage in these situations is an unnecessary cost.

In many cases, industry is being pressured to substitute potable water and further reduce waste by using wastewater for lower grade applications. These practices are in line with the world trend to sustainable industry [76]. The use of treated AMD water will offset high water usage costs by industry and provide a cheaper, alternative water source.

10.6.2.2 Environmental Remediation

Several studies have shown that the environmental remediation using AMD recovered minerals will be thoroughly investigated with a case-by-case study of affected water sources, which may have prospects of applications in mineral recovery technique design [77]. An estimation of AMD-affected water sources is in the scope of 10,000–20,000 throughout the world [78, 79]. A potentially large benefit may be in the prevention of further damaging environments in remote or developing areas, which are in proximity to modern mining operations.

With the application of a two-stage neutralization technique and one that is optimized for selectivity, the result is a water source that meets drinking water standards and a range of high-quality industrial waters. This may enable recovery systems to be installed at contaminated water sources, allowing on-site treatment and prevention of further environmental damage [80].

The mineral recovery process from AMD proposed allows for a large percentage of contaminants to be removed from the affected water source. The recovery of iron oxides can be utilized as an effective sorbent for the treatment of contaminated aqueous systems [81]. Heavy metal contaminants within the source water are removed by dissolving a neutralizing agent. This, in turn, promotes hydrolysis of these metals, forming settleable and recoverable metal hydroxide precipitates. The neutralizing agent is an alkali such as lime or caustic soda and may be associated with the pH increase of the water [82]. The generation of gypsum neutralises acidity which is one of the problematic characteristics of AMD. A conceptual process of heavy metals removal in the form of sludge from AMD by neutralization is shown in Figure 10.2.

When natural AMD seeps into the environment, it causes environmental pollution and may have adverse effects on human, animal, and plant life. Mostly, the only action taken is to divert the pathway of AMD to prevent clean water from being contaminated by constructing expensive

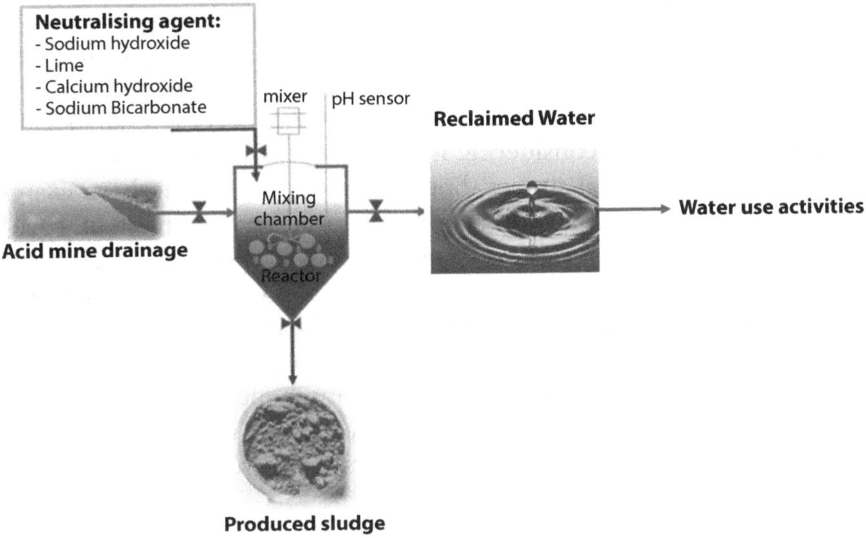

Figure 10.2 Heavy metals removal in the form of sludge from AMD by neutralization [Modified from (Masindi *et al.*, 2022)].

passive remediation systems [83]. The removal of contaminants from these sources is a minor occurrence. The result is that remediation of affected environments is a costly and seldom fully effective process.

10.6.2.3 Economic Potential

A report by NEDLAC [84] estimated the value of the downstream iron products at R5.5 billion and the aluminum product at R20 billion. This is a gold mine opportunity of which there are few others existing in South Africa today. Minerals worth of millions can potentially be recovered from mine water. This is economically significant for the reality that mine water reclamation can be a financially self-adequate industry [85]. This denotes that the South African government and its taxpayers could alleviate themselves of the fiscal responsibility for mine drainage treatment in the long term. The money made from the sale of the minerals reclaimed from mine water could be injected back into the mining industry to financially promote further mine water reclamation and treatment research.

Additionally, the implementation of eco-taxes can be an attractive solution. Eco-taxes are market-based instruments that impose taxes on polluters for the detrimental environmental impact caused by their business activities [86]. By placing taxes on environmental damage caused by AMD, the relative cost of AMD treatment would be higher compared to metal recovery. Consequently, this would incentivize mining companies to improve their pollution control measures and prioritize the recovery of metals.

The economic value of metal recovery from AMD is especially significant for South Africa, considering the depletion of resources in the gold mining industry. Several studies, both historical and recent, have examined the economic feasibility of metal recovery and have consistently generated positive results [54]. Notably, metals recovery provides a significant economic avenue which would help offset the cost of sludge disposal that has detrimental impacts on the environment. The recovery of value in sulfidic minerals can be evaluated in terms of the potential revenue from sulfate and water minimization and from the linear increase in value from metal recovery. By comparing the payback processes of mineral recovery and AMD remediation, the potential benefits can be realized.

10.6.2.4 Iron Oxides

The valorization of oxides and other derivatives found in AMD presents a compelling opportunity for both environmental remediation and resource utilization. Among these derivatives, iron oxide stands out prominently, owing to its abundance in AMD resulting from the oxidative weathering of iron sulfide minerals [87]. Iron oxide nanoparticles synthesized from AMD have garnered significant attention due to their unique properties and diverse applications across various fields. These nanoparticles have shown promise in wastewater treatment, catalysis, and environmental remediation [88].

Iron oxide nanoparticles derived from AMD exhibit exceptional adsorption capabilities, owing to their large surface area and surface reactivity. This makes them highly effective in removing pollutants from aqueous solutions, including heavy metals like lead, cadmium, and arsenic (Bhateria and Singh, 2019), as well as organic compounds like dyes [89]. The utilization of AMD-derived iron oxide nanoparticles in water treatment technologies offers a dual benefit by not only mitigating AMD pollution but also providing a sustainable solution for addressing water contamination issues in various industrial and environmental settings [90].

Iron oxides encompass a diverse array of compounds, each with unique structural and chemical properties that contribute to their varied applications in fields ranging from materials science to environmental remediation. The most common forms of iron oxides include hematite (α-Fe_2O_3), magnetite (Fe_3O_4), and goethite (α-FeOOH), each exhibiting distinct crystal structures and physicochemical characteristics [91].

- **Hematite**

Hematite, characterized by its distinctive red color and rhombohedral crystal structure, occupies a prominent position among iron oxides due to its prevalence in natural settings. Beyond its aesthetic allure, hematite possesses remarkable thermal stability, making it indispensable in various applications such as paints, coatings, and ceramics, where its vibrant red pigment is highly valued [91, 92]. However, the utility of hematite transcends its traditional roles, as advancements in nanotechnology have unveiled novel applications for hematite nanoparticles across diverse fields [93].

In recent years, hematite nanoparticles have attracted considerable attention for their exceptional properties, particularly in biomedical imaging, drug delivery, and environmental remediation endeavors. These nanoparticles exhibit outstanding biocompatibility, rendering them well-suited for various biomedical applications [94]. Moreover, the inherent magnetic properties of hematite nanoparticles render them invaluable in targeted drug delivery systems, where external magnetic fields guide and localize drug carriers to specific anatomical sites within the body. Additionally, hematite nanoparticles function as contrast agents in magnetic resonance imaging (MRI), augmenting the visualization of biological tissues and enhancing diagnostic precision [95].

In the realm of environmental remediation, hematite nanoparticles offer a versatile solution for tackling pollution challenges. Their elevated surface area and reactivity render them effective adsorbents for diverse contaminants, including heavy metals and organic pollutants, in aqueous environments [96].

By immobilizing contaminants onto their surfaces, hematite nanoparticles facilitate their extraction from water systems, thereby mitigating environmental hazards and reinstating water quality. Furthermore, the magnetic responsiveness of hematite nanoparticles streamlines their separation from aqueous solutions, simplifying recovery and reuse processes in wastewater treatment applications [97].

- **Magnetite**

Magnetite (Fe_3O_4), characterized by its distinctive black hue and cubic crystal lattice, stands out in the realm of materials science owing to its exceptional magnetic attributes, which originate from its mixed-valence configuration [98]. At the nanoscale, magnetite nanoparticles demonstrate superparamagnetic behavior, a phenomenon crucial for various applications such as magnetic separation, hyperthermia therapy, and environmental remediation [99].

The superparamagnetic nature of magnetite nanoparticles facilitates their manipulation and separation in aqueous media, offering significant advantages in wastewater treatment and pollutant elimination processes. This feature underscores their suitability for addressing environmental challenges, including the removal of heavy metals and organic pollutants from contaminated water sources [100]. Furthermore, the unique magnetic properties of magnetite nanoparticles enable their utilization in targeted drug delivery systems, where controlled release mechanisms are desirable [101].

- **Goethite**

Goethite [α-FeO(OH)], distinguished by its characteristic yellow-brown hue and orthorhombic crystalline structure, occupies a prominent position in both geological formations and environmental matrices. Primarily occurring as a secondary mineral resulting from the weathering processes acting upon primary iron-bearing minerals, goethite manifests remarkable adsorption properties attributable to its elevated surface area and surface reactivity [102]. This inherent trait renders goethite an efficacious agent for the sequestration of contaminants from aqueous solutions. The utilization of goethite nanoparticles in environmental remediation endeavors has garnered considerable attention, particularly in addressing challenges posed by heavy metals, organic pollutants, and emerging contaminants present in wastewater and contaminated soils [103].

Through mechanisms such as surface complexation and ion exchange, goethite nanoparticles exhibit a propensity for adsorbing a diverse array of pollutants, thereby facilitating their removal from aqueous matrices [103]. Moreover, the adaptability of goethite/iron-based remediation strategies extends beyond conventional pollutants, encompassing emerging contaminants such as pharmaceuticals and personal care products (PPCPs) [104]. This underscores the versatility and applicability of goethite nanoparticles in addressing contemporary environmental challenges.

10.6.2.5 Aluminum Oxides

Aluminum oxides, notably alumina (Al_2O_3), which are commonly found in AMD precipitates, demonstrate exceptional adsorption properties, particularly toward heavy metals and metalloids [105]. The unique surface characteristics of alumina, including its high surface area and abundance of active sites, contribute to its effectiveness in sequestering contaminants from AMD-affected waters [106].

Studies have shown that alumina-based materials can efficiently remove pollutants through mechanisms such as ion exchange and surface complexation [107]. During ion exchange, metal ions in the AMD solution are exchanged with ions on the surface of alumina particles, leading to the immobilization of contaminants [108]. Surface complexation involves the formation of chemical bonds between pollutant species and functional groups present on the alumina surface, further enhancing adsorption [109].

Furthermore, the versatility of alumina-based materials allows for their tailored application in AMD remediation strategies. By adjusting parameters such as particle size, surface modification, and porosity, the adsorption capacity and selectivity of alumina can be optimized to target specific contaminants present in aqueous systems [110]. This versatility underscores the potential of alumina as a cost-effective and sustainable solution for mitigating the environmental impacts of AMD contamination.

Recent advancements in material synthesis techniques, such as the development of nanostructured alumina materials, have further enhanced the adsorption performance of alumina-based adsorbents [111]. These advancements hold promise for improving the efficiency of AMD remediation processes and expanding the applicability of alumina-based materials in addressing emerging environmental challenges.

10.6.2.6 Other Derivatives

The alkaline properties exhibited by calcium oxides and magnesium oxides found within AMD residues constitute a fundamental mechanism for neutralizing acidity and facilitating the precipitation of metal ions through both pH adjustment and chemical precipitation reactions [112]. When introduced into AMD-contaminated waters, these alkaline earth oxides engage in reactions with acidic constituents, notably sulfate ions, triggering chemical transformations that result in the formation of insoluble precipitates [36]. Recent research has underscored the efficacy of calcium oxides and magnesium oxides in elevating pH levels and mitigating acidity within AMD solutions. This pH modulation fosters conditions conducive to the precipitation of metal ions, thereby diminishing their solubility and bioavailability in water matrices [113, 114]. Consequently, the overall acidity of the water is diminished, contributing to the amelioration of environmental impacts associated with AMD contamination.

Furthermore, the formation of insoluble metal precipitates catalyzes their removal from aqueous solutions, thereby mitigating the risk of metal leaching and subsequent environmental harm. This process of chemical precipitation effectively immobilizes metals within the solid phase, consequently reducing their mobility and potential ecological toxicity [115]. Leveraging the alkaline properties of calcium oxides and magnesium oxides within AMD residues not only aids in AMD remediation but also presents opportunities for resource recovery and utilization. These oxides can be retrieved and repurposed for various applications, including soil conditioning, agricultural nutrient supplementation, and as raw materials in industrial processes [1]. This approach not only addresses the environmental challenges posed by AMD but also promotes sustainable resource management practices.

10.6.3 Beneficiation of Acid Mine Drainage Derivatives

The beneficiation of AMD derivatives encompasses a multifaceted approach aimed at mitigating environmental contamination and harnessing valuable resources [54]. Although AMD is characterized by its acidic nature and elevated concentrations of heavy metals and metalloids that pose significant ecological risks, its derivatives, through beneficiation processes, can be transformed into materials with potential environmental and industrial applications. These metal oxides exhibit alkaline properties that can be leveraged to neutralize acidity and precipitate metal ions through pH adjustment and chemical precipitation mechanisms, as well as soil conditioning, agricultural nutrient supplementation, and industrial processes [54, 116].

The industrial utilization of minerals extracted from AMD exhibits extensive diversity across various sectors. In the realm of building and construction, iron hydroxides serve as viable pigments in a spectrum of hues, including red, brown, and yellow, within concrete, ceramic, and paint formulations [117, 118]. This transformative process, known as calcination, entails the thermal treatment of iron hydroxide to yield iron oxide, renowned for its robust tinting capabilities and steadfast color retention, while also boasting eco-friendly attributes [119]. Moreover, this pigment finds application in a plethora of plastic compositions [120].

Furthermore, distinct variants of iron oxide, such as magnetite, alongside aluminum hydroxides, demonstrate multifaceted utility, notably in arsenic mitigation within certain water sources. Leveraging iron oxides emerges as a premier approach for arsenic abatement, distinguished by its diminished sludge output, high metal removal efficiency, and uncomplicated, safe implementation. Additionally, the cost-effectiveness of arsenic removal using iron oxides renders it an attractive solution for small-scale community water systems in developing nations [9]. Beyond this, iron and aluminum hydroxides serve as catalysts in chemical synthesis and fillers in rubber manufacturing, among other industrial applications [121].

Aligned with the thematic focus of our investigation, recent scholarly endeavors have concentrated on the valorization of valuable minerals salvaged from AMD. The envisioned commercial exploitation of AMD remediation by-products represents a pivotal stride toward realizing economically sustainable technologies for AMD mitigation. Predominantly recovered minerals from AMD, notably iron and aluminum hydroxides, emerge as focal points due to their prevalence in AMD discharges and associated environmental hazards [122]. Other frequently retrieved minerals include manganese and gypsum, all of which present a diverse array of industrial and environmental uses, thus rendering their recovery a financially viable avenue for AMD remediation [123].

Overall, the beneficiation of AMD derivatives offers a holistic approach to addressing environmental contamination while simultaneously recovering valuable resources. By implementing effective beneficiation strategies, environmental scientists and engineers can mitigate the environmental impacts of AMD and contribute to the sustainable management of AMD-affected sites [90].

10.6.4 Comparison of Acid Mine Drainage Technologies in South Africa

The growing population in South Africa, along with increasing pressure on water treatment infrastructure and rising water contamination has driven the continuous development of wastewater treatment methods [124]. This includes technologies specifically designed to address AMD issues, which are exacerbated by the country's extensive mining activities. Many of these technologies produce sludge by-products that can be processed into raw materials or valuable secondary commodities, promoting resource recovery and sustainable management [125, 126]. Table 10.2 presents an overview of selected AMD treatment technologies developed and deployed for mineral recovery and pollutant removal.

Understanding the intricate interplay of factors governing metal release, transport, and transformation in environments impacted by AMD is essential for devising successful management strategies aimed at mitigating metal pollution and preserving aquatic ecosystems. The integration of diverse disciplines such as geochemistry, microbiology, and environmental engineering is imperative for tackling the multifaceted challenges presented by AMD contamination.

Table 10.2 AMD treatment technologies developed and deployed for mineral recovery and pollutant removal.

Treatment technology	Treatment cost	Potential profit	Pollutant removal technique	Recovered pollutants/ minerals	Advantages	Disadvantages	References
Neutralization	Moderate	Low	Chemical precipitation	Metal sludge	- Effective pollutant removal - Widely applicable	- Moderate to high chemical usage - Produces metal-laden sludge	Weinberg, 2022
Membrane Separation (NF)	Moderate	Low	Membrane filtration	Multivalent ions and dissolved compounds	- Low power consumption - High efficiency - Ease of operation	- Fouling at high water recovery rates	Zhao *et al.*, 2023; Ang *et al.*, 2023
Reverse Osmosis (RO)	High	High	Membrane filtration	High-quality water	- Produces high-quality water - Minimizes discharge	- High capital and operating costs	Masindi, 2017
ABC Method	High	Low	Chemical precipitation	Metal hydroxides and gypsum	- Potential for reuse opportunities - Effective pollutant removal	- High waste sludge production - High capital and operating costs	Mulopo *et al.*, 2012; Motaung *et al.*, 2008; Swanepoel *et al.*, 2012

(Continued)

Table 10.2 AMD treatment technologies developed and deployed for mineral recovery and pollutant removal. (*Continued*)

Treatment technology	Treatment cost	Potential profit	Pollutant removal technique	Recovered pollutants/ minerals	Advantages	Disadvantages	References
MASRO Freeze Crystallization	High	Moderate	Chemical precipitation	Metal hydroxides, gypsum, and calcium carbonate	- Efficient metal recovery - Improves water quality	- High initial costs - Additional processing costs	Masindi *et al.*, 2018; Masindi *et al.*, 2019
SAVMIN Process	High	Moderate	Chemical precipitation	Metal hydroxides, gypsum, and calcite	- Produces high-quality by-products - Effective pollutant removal	- Significant waste sludge production	Petterson, 2018; van Rooyen and van Staden, 2020
SPARRO Process	High	Variable	Membrane desalination	Water	- Treats AMD with variable recoveries - Membrane fouling challenge	- Membrane fouling - High capital cost	Simate *et al.*, 2014
GYP-CIX Process	Moderate	Low	Ion exchange	Calcium and sulfate precipitates	- Inexpensive chemicals - Efficient water recoveries	- Generates significant sludge - Expensive disposal method	Robertson *et al.*, 1993

(*Continued*)

Table 10.2 AMD treatment technologies developed and deployed for mineral recovery and pollutant removal. (*Continued*)

Treatment technology	Treatment cost	Potential profit	Pollutant removal technique	Recovered pollutants/ minerals	Advantages	Disadvantages	References
THIOPAQ	High	Low	Biotechnological	Metal sulfides	- Biotechnological approach - Effective pollutant removal	- Expensive chemicals - Reduced attractiveness due to cost	Sullivan *et al.*, 1997; Dhir, 2018
Rhodes BioSURE Process	Moderate	Low	Biological treatment	Acid removal	- Uses waste materials - Low landfill loads	- Requires sufficient stock of waste materials - Limited reagents	Rose, 2013; Rose *et al.*, 2004; Corbett, 2001
HiPRO Process	High	Moderate	Membrane filtration	Water, calcium sulfate, and metal sulfates	- Good water recovery - Potential for by-product sales	- Challenges in waste sludge disposal - By-product variability	Hutton *et al.*, 2009
Electrochemical Treatment	High	Moderate	Electrochemical	Metal precipitates and sulfuric acid	- Virtually no chemical usage - Lower sludge volumes	- High initial capital costs - Membrane fouling	Brewster, 2020

10.7 Heavy Metal Contaminants in Wastewater

Heavy metals, characterized by their high atomic weights and densities exceeding 5 g/cm^3, are intrinsic components of the Earth's crust. These elements occur naturally and can be released into the environment through various geological processes and human activities [127, 128]. In wastewater, heavy metals are prevalent contaminants due to industrial discharges, mining activities, urban runoff, and agricultural practices (Aziz *et al.*, 2023). Among the most common heavy metals found in wastewater are lead (Pb), cadmium (Cd), mercury (Hg), chromium (Cr), arsenic (As), and nickel (Ni), although others such as zinc (Zn), copper (Cu), and manganese (Mn) are also frequently encountered [29].

Persistent and bio-accumulative, heavy metals pose significant risks to human health and ecosystems due to their persistence in the environment and ability to accumulate in living organisms (Mitra *et al.*, 2022). Heavy metals pose serious health risks to humans, even at low concentrations. Chronic exposure to heavy metals through contaminated water, food, or air can lead to various health problems, including neurological disorders, kidney damage, respiratory illnesses, cardiovascular diseases, and cancer [29]. Children and pregnant women are particularly vulnerable to the adverse effects of heavy metal exposure, which can hinder cognitive development and fetal growth [129].

Aquatic organisms are highly sensitive to heavy metal contamination in water bodies. Heavy metals can accumulate in aquatic organisms through waterborne exposure or dietary intake, leading to physiological disturbances, reproductive impairments, and population declines [130]. Fish, invertebrates, and algae are particularly susceptible to the toxic effects of heavy metals, which can disrupt their metabolism, growth, and reproductive success [131]. Heavy metals can also adversely affect plant growth and development, primarily through the inhibition of photosynthesis, disruption of nutrient uptake, and oxidative stress [132]. Plants exposed to elevated levels of heavy metals may exhibit symptoms of toxicity, such as chlorosis, stunted growth, and reduced yields. Heavy metal contamination in soil and irrigation water can compromise agricultural productivity and food safety, posing risks to human health through dietary exposure [133].

Various treatment technologies are available for the removal of heavy metals from contaminated water sources, aiming to mitigate environmental pollution and safeguard human health [134]. Adsorption processes, which involve the use of adsorbent materials, such as activated carbon, zeolites, and clay minerals, has been used to capture heavy metal ions from aqueous solutions through surface interactions. Adsorption is an effective and widely used method for heavy metal removal, offering high removal efficiencies and versatile application in diverse water treatment scenarios [135]. On the other hand, chemical precipitation methods involve the addition of precipitating agents, such as lime, calcium hydroxide, or sodium hydroxide, to wastewater to induce the formation of insoluble metal hydroxide or sulfide precipitates [136]. Precipitation processes are commonly employed for the treatment of industrial effluents and municipal wastewater to remove dissolved heavy metals and comply with regulatory discharge limits [137]. Ion exchange processes utilize ion exchange resins or chelating agents to selectively remove heavy metal ions from aqueous solutions by exchanging them with less harmful ions immobilized on the exchange medium. Ion exchange is effective for treating low to moderate concentrations of heavy metals in water and is often used in combination with other treatment technologies for enhanced removal efficiency [108].

10.7.1 Arsenic

Arsenic, a naturally occurring element ubiquitous in the Earth's crust, manifests in diverse chemical forms, predominantly in environmental settings. Among these forms, inorganic arsenic compounds predominate, with arsenite [As(III)] and arsenate [As(V)] being the most prevalent [138]. These compounds exhibit varying oxidation states and toxicity levels, influencing their environmental behavior and health impacts. Additionally, certain organic arsenic compounds, including arsenobetaine and arsenocholine, are present in seafood and marine organisms, albeit with lower toxicity profiles compared to their inorganic counterparts [139]. These distinctions in chemical speciation highlight the complex nature of arsenic in environmental matrices and underscore the importance of understanding its diverse forms for assessing risks and implementing effective mitigation strategies.

10.7.1.1 Impacts of Arsenic on Human Health

Arsenic contamination in drinking water represents a significant public health concern, with profound implications for human health and well-being. Chronic exposure to elevated levels of arsenic has been associated with a myriad of health problems, ranging from acute toxic effects to long-term chronic conditions [140]. One of the most visible and early manifestations of chronic arsenic exposure is the development of skin lesions. Prolonged ingestion of arsenic-contaminated water can lead to the formation of characteristic skin lesions, including hyperpigmentation, hyperkeratosis, and the development of skin cancers such as basal cell carcinoma and squamous cell carcinoma [141].

Epidemiological studies have linked chronic arsenic exposure to an increased risk of cardiovascular diseases, including hypertension, atherosclerosis, and ischemic heart disease. Arsenic-induced oxidative stress and inflammation contribute to endothelial dysfunction, vascular remodeling, and thrombotic events, predisposing individuals to cardiovascular morbidity and mortality [142]. Growing evidence suggests that prenatal and early-life exposure to arsenic may adversely affect neurodevelopment in children, leading to cognitive impairments, learning disabilities, and behavioral disorders. Arsenic-induced neurotoxicity is mediated through multiple mechanisms, including disruption of synaptic function, neuronal apoptosis, and dysregulation of neurotransmitter systems [143].

Arsenic is classified as a Group 1 carcinogen by the International Agency for Research on Cancer (IARC), indicating sufficient evidence for its carcinogenicity in humans [138]. Chronic arsenic exposure via drinking water has been strongly associated with the development of various cancers, including the skin, lung, bladder, and kidney cancer. Mechanistic studies suggest that arsenic promotes tumorigenesis through genotoxic, epigenetic, and oxidative stress-mediated pathways [144]. In regions where arsenic contamination is endemic, populations are at risk of developing arsenicosis, a chronic condition characterized by systemic symptoms, including skin lesions, peripheral neuropathy, gastrointestinal disturbances, and cardiovascular complications. Arsenicosis represents a significant public health burden in affected areas, necessitating comprehensive interventions for mitigation and management [139].

10.7.1.2 Impacts of Arsenic on Aquatic Organisms

Arsenic contamination exerts detrimental effects on aquatic organisms, compromising their growth and reproductive success. Chronic exposure to arsenic can impede the growth rates of fish, amphibians, and invertebrates, leading to diminished population abundance and biomass [145]. Furthermore, arsenic exposure has been associated with reproductive abnormalities, including decreased fecundity, impaired larval development, and elevated embryo mortality, thereby affecting the long-term viability of affected populations [146]. Aquatic organisms exposed to arsenic may exhibit alterations in behavior and physiological functions, such as impaired locomotor activity, altered feeding behavior, and disrupted osmoregulation [147]. Arsenic-induced neurotoxicity can impair sensory perception and motor coordination in fish and other aquatic species, compromising their ability to forage, evade predators, and navigate their environment [148].

Accumulation of arsenic in fish and shellfish can pose risks to human health through dietary exposure. Bioaccumulation and biomagnification of arsenic in aquatic food webs can result in elevated concentrations of arsenic in commercially important fish species consumed by humans [149]. Consumption of arsenic-contaminated seafood can contribute to chronic arsenic exposure in human populations, increasing the risk of adverse health effects, including cancer and cardiovascular diseases [150]. Arsenic contamination in aquatic habitats can disrupt food chains and ecosystem dynamics, leading to cascading effects on community structure and ecosystem function. Arsenic-induced mortality or sublethal effects in key species can alter trophic interactions, predator-prey dynamics, and energy flow within aquatic ecosystems [151]. Moreover, changes in community composition and species abundance resulting from arsenic contamination can impair ecosystem resilience and stability, increasing susceptibility to further environmental stressors [152].

10.7.1.3 Impacts of Arsenic on Plants

Arsenic toxicity poses significant challenges to plant growth and development. Exposure to elevated levels of arsenic can inhibit root elongation, reduce shoot biomass, and impair overall growth performance in various plant species [153]. Arsenic-induced phytotoxicity disrupts cellular processes, including cell division, cell expansion, and nutrient uptake, leading to stunted growth and decreased productivity [153]. Arsenic exposure interferes with the photosynthetic machinery of plants, disrupting light harvesting, electron transport, and carbon assimilation processes [154]. Arsenic-induced inhibition of photosynthesis results in reduced chlorophyll content, impaired stomatal conductance, and diminished CO_2 fixation rates, compromising plant productivity and yield potential [155].

Arsenic toxicity disrupts nutrient uptake and metabolism in plants, leading to imbalances in essential mineral nutrients and impaired physiological functions [133, 156]. Arsenic interferes with the uptake and translocation of essential nutrients such as phosphorus, iron, and sulfur, resulting in nutrient deficiencies and metabolic disturbances [133]. Moreover, arsenic exposure alters the activity of key enzymes involved in nutrient metabolism, compromising plant nutrient utilization efficiency and metabolic homeostasis [157].

Arsenic accumulation in crops grown on arsenic-contaminated soil poses risks to food safety and human health through dietary exposure. Arsenic uptake by edible plant parts, such as grains, vegetables, and fruits, can result in the contamination of the food supply chain, leading to potential health risks for consumers [158]. Chronic dietary exposure to arsenic-contaminated food products may increase the risk of arsenic-related health effects, including cancer and cardiovascular diseases [159]. Arsenic uptake by plants contributes to the transfer of arsenic contamination from soil to the food chain, exacerbating human exposure to this toxic element [153]. Bioaccumulation and biomagnification of arsenic in terrestrial and aquatic food webs can result in elevated arsenic concentrations in animal-derived food products, further increasing human health risks associated with dietary exposure [160].

10.7.1.4 Technologies for Arsenic Removal From Aqueous Systems

Various treatment technologies have been developed to remove arsenic from contaminated water sources, aiming to mitigate human health risks and environmental impacts. Adsorption processes involve the use of adsorbent materials, such as activated carbon, iron oxide–based materials, and various adsorbent minerals, to remove arsenic from water through surface interactions [161].

Adsorption is an effective and widely employed method for arsenic removal due to its simplicity, efficiency, and versatility in treating diverse water sources. Additionally, recent studies have explored novel adsorbents for arsenic removal [9, 162]. For instance, Das *et al.* [163] investigated the use of graphene oxide–based materials for efficient arsenic removal from water. They demonstrated that functionalized graphene oxide exhibited high adsorption capacity for arsenic ions, offering a promising solution for arsenic remediation in contaminated water sources.

Chemical precipitation methods involve the addition of precipitating agents, such as iron or aluminum salts, to arsenic-contaminated water to induce the formation of insoluble arsenic-containing precipitates [164]. Precipitation processes are commonly used for large-scale arsenic removal in water treatment plants and are effective in reducing arsenic concentrations to meet regulatory standards [165].

Recent advancements in technologies for arsenic removal have focused on improving process efficiency and reducing environmental impacts. For example, Oehmen *et al.* [166] investigated the use of a hybrid ion exchange membrane–coagulation process for arsenic removal from drinking water. They found that combining using this system enhanced arsenic removal efficiency and removed arsenic from drinking water to levels below the MCL of 10 ppb.

Ion exchange processes involve the exchange of arsenic ions in water with ions immobilized on an exchange medium, such as ion exchange resins or zeolites. Ion exchange is effective for arsenic removal, particularly in groundwater treatment applications, and can achieve high removal efficiencies with proper system design and operation [167].

10.7.2 Chromium

Chromium is a naturally occurring transition metal found in the Earth's crust. It exists in various oxidation states, with the most prevalent forms being trivalent chromium [Cr(III)] and hexavalent chromium [Cr(VI)]. Trivalent chromium is an essential micronutrient for humans, playing a crucial role in glucose metabolism, lipid metabolism, and insulin signalling pathways [168]. It is also involved in the metabolism of proteins and nucleic acids, contributing to overall cellular function and homeostasis. Trivalent chromium is commonly found in dietary sources such as meat, whole grains, fruits, and vegetables [169].

In contrast, hexavalent chromium is highly toxic and carcinogenic, posing significant risks to human health and the environment. Exposure to hexavalent chromium compounds can occur through inhalation, ingestion, or dermal contact with contaminated air, water, or soil (Sharma *et al.*, 2022). Hexavalent chromium compounds are extensively used in various industrial processes, including electroplating, metal finishing, and leather tanning, leading to widespread environmental contamination [170].

10.7.2.1 Impacts of Chromium on Human Health

Chromium exposure, especially to hexavalent chromium, can have severe detrimental effects on human health. Inhalation, ingestion, or dermal contact with hexavalent chromium compounds can lead to a range of health issues, affecting various organ systems in the body. Inhalation of hexavalent chromium compounds, particularly chromium(VI) particles or fumes, can cause respiratory problems such as irritation of the nose, throat, and lungs. Prolonged exposure to airborne chromium particles may result in chronic respiratory conditions, including bronchitis and asthma [171]. Ingestion of hexavalent chromium-contaminated water or food can lead to gastrointestinal problems such as nausea, vomiting, abdominal pain, and diarrhea. Chromium compounds can irritate the lining of the stomach and intestines, causing inflammation and digestive disturbances [172].

Direct contact with hexavalent chromium compounds can cause skin irritation, allergic reactions, and dermatitis. Prolonged or repeated exposure to chromium compounds through skin contact can result in skin ulceration, eczema, and allergic dermatitis [168]. Prolonged exposure to high levels of hexavalent chromium has been strongly associated with an increased risk of various cancers, particularly lung cancer, nasal and sinus cancers, and gastrointestinal cancers. Chromium compounds can penetrate deep into the lungs upon inhalation, causing DNA damage, oxidative stress, and mutations in lung cells, leading to the development of cancerous tumors [173].

10.7.2.2 Impacts of Chromium on Aquatic Organisms

Chromium contamination in aquatic ecosystems can have detrimental effects on various organisms, including fish, invertebrates, and algae. Exposure to elevated levels of chromium can impair the growth and reproduction of these aquatic species, leading to population declines and ecosystem disruption [174]. Studies have shown that chronic exposure to chromium can interfere with reproductive processes in fish and invertebrates, resulting in decreased fertility and offspring survival rates [174]. Chromium accumulation in aquatic organisms can also disrupt essential physiological processes, such as ion regulation and enzyme activity. Chromium ions can interfere with ion transport mechanisms in fish and invertebrates, disrupting osmoregulation and electrolyte balance. Additionally, chromium exposure can inhibit enzyme systems involved in metabolism and detoxification, impairing cellular function and compromising organismal health [175].

Chromium toxicity can lead to impaired health and reduced viability in aquatic organisms. Chronic exposure to elevated levels of chromium can cause oxidative stress, DNA damage, and tissue injury in fish, invertebrates, and algae. These physiological disturbances can weaken immune responses, increase susceptibility to diseases, and reduce overall fitness and survival rates in affected populations [176].

The adverse effects of chromium contamination on aquatic organisms can have cascading effects on ecosystem dynamics and functioning. Population declines or extinctions of key species can disrupt food webs and ecological interactions, leading to alterations in nutrient cycling, energy flow, and community structure. Chromium-induced ecosystem disruptions can have far-reaching consequences for aquatic habitats and the services they provide to human societies [177].

10.7.2.3 Impacts of Chromium on Plants

Chromium toxicity poses significant risks to plant health and can adversely affect various physiological processes, ultimately impacting growth, development, and productivity [41]. Chromium exposure can inhibit seed germination in plants, leading to reduced seedling emergence and establishment [41]. The presence of chromium in the germination medium interferes with metabolic processes essential for seed germination, such as water uptake and enzyme activation, thereby impeding the germination process [178]. Chromium contamination inhibits root elongation in plants, limiting the exploration of soil for water and nutrients [119]. Chromium ions can penetrate root tissues and disrupt cellular structures, leading to impaired root growth and reduced root biomass [171]. This inhibition of root elongation hampers the plant's ability to establish a robust root system, thereby compromising nutrient and water uptake efficiency [119].

Chromium accumulation in plant tissues disrupts nutrient uptake and metabolism, leading to nutrient imbalances and physiological disorders [179]. Chromium ions interfere with the absorption of essential nutrients such as nitrogen, phosphorus, potassium, and micronutrients, resulting in nutrient deficiencies and imbalances [180]. This disruption of nutrient uptake and metabolism impairs various biochemical processes vital for plant growth and development, ultimately compromising plant health and productivity. Chromium-induced oxidative stress is a common consequence of chromium toxicity in plants [119, 179]. Chromium ions generate reactive oxygen species (ROS) within plant cells, leading to oxidative damage to cellular components such as lipids, proteins, and nucleic acids [181]. This oxidative damage disrupts cellular homeostasis and impairs cellular functions, exacerbating the adverse effects of chromium toxicity on plant growth and vitality [182].

Chromium-contaminated soils pose risks to food safety and human health through the uptake of chromium by edible crops. Plants grown in chromium-contaminated soils can accumulate chromium in their tissues, leading to potential chromium exposure through the consumption of contaminated food crops [183]. Chronic dietary exposure to chromium-contaminated food can have adverse health effects on humans, including gastrointestinal disorders, carcinogenicity, and genotoxicity [172].

10.7.2.4 *Technologies for Chromium Removal From Aqueous Systems*

Various treatment technologies have been developed to remove chromium from contaminated water sources and soil, aiming to mitigate environmental impacts and human health risks. Chemical precipitation methods involve the addition of precipitating agents, such as lime or iron salts, to chromium-contaminated water to induce the formation of insoluble chromium-containing precipitates [184].

Precipitation processes can effectively reduce chromium concentrations to meet regulatory standards and are commonly used in industrial wastewater treatment [185]. For instance, a study conducted by Gheju and Balcu [186] showcased the effectiveness of minimizing scrap iron and then precipitating the resulting cations with NaOH to completely eliminate chromium from Cr(VI) aqueous solutions.

Adsorption processes utilize adsorbent materials, such as activated carbon, zeolites, or iron oxide–based materials, to remove chromium from water through surface interactions. Adsorption is an efficient and versatile method for chromium removal, offering high removal efficiencies and the ability to treat diverse water sources [89]. For example, Alvarez *et al.* [168] investigated the use of natural zeolite in the removal of chromium (IV) removal from tannery effluents which demonstrated effectiveness in reducing chromium concentrations.

Ion exchange processes involve the exchange of chromium ions in water with ions immobilized on an exchange medium, such as ion exchange resins or zeolites. Ion exchange can achieve high removal efficiencies for both trivalent and hexavalent chromium and is commonly used in both industrial and municipal water treatment applications. A study by Rengaraj *et al.* [187] evaluated the performance of ion exchange resins in the removal of chromium(III) from water and wastewater, highlighting their potential for efficient chromium removal.

Membrane filtration technologies, including reverse osmosis and nanofiltration, can also be used for chromium removal from water. These technologies rely on semi-permeable membranes to selectively remove chromium ions from water, allowing for the production of high-quality treated water [188, 189]. Research by Zolfaghari and Kargar [190] investigated the use of nanofiltration and microfiltration for chromium removal from water, which demonstrated the decrease of chromium concentrations in the treated water.

10.8 Other Paramount Water Contaminants

10.8.1 Congo Red Dye

Congo red, chemically identified as disodium 3,3′-bis(4-dimethylamino-phenyl)-4,4′-bis(phenylamino)-biphenyl-4,4′-disulfonate, represents a synthetic diazo dye celebrated for its remarkable red coloration. This compound has garnered considerable attention across various industries, particularly in textiles, paper manufacturing, and cosmetics, owing to its vivid and enduring hue [191, 192]. The chemical structure of Congo red comprises intricate molecular arrangements, with its distinctive red color attributed to its diazo chromophore. This chromophore exhibits pronounced light absorption in the visible spectrum, resulting in the characteristic red appearance of Congo red solutions [191].

Textile industries extensively employ Congo red in dyeing processes to impart vibrant red shades to fabrics [193]. Similarly, paper manufacturing facilities utilize this dye for coloration purposes, enhancing the visual appeal of paper products [194]. Furthermore, Congo red finds application in the cosmetics industry, where it contributes to the formulation of cosmetic products, including lipsticks and hair dyes, adding vibrancy to the final products [195].

10.8.1.1 Impacts of Congo Red Dye on Human Health

Exposure to Congo red dye has been associated with several adverse effects on human health. Studies have shown that skin contact with Congo red dye can lead to skin irritation and dermatitis [196]. Furthermore, inhalation of dye particles can result in respiratory irritation and allergic reactions [195]. There is also evidence suggesting that Congo red dye may have potential carcinogenic effects. Ingestion of food or water contaminated with Congo red dye may lead to gastrointestinal problems and organ damage [192].

10.8.1.2 Impacts of Congo Red Dye on Aquatic Organisms

Congo red dye has been found to have adverse effects on aquatic organisms. Research indicates that exposure to the dye can lead to decreased oxygen levels in water, negatively affecting fish and other aquatic life. Moreover, it can disrupt the reproductive and developmental processes of aquatic organisms, leading to reduced population viability [197].

10.8.1.3 *Impacts of Congo Red Dye on Plants*

Congo red dye can have detrimental effects on plant growth and development. Studies have shown that the dye, when present in soil or water, can inhibit seed germination [58]. Additionally, it may lead to reduced photosynthesis and chlorophyll degradation in plant leaves, ultimately affecting plant health. High concentrations of Congo red dye can also result in root damage and hinder nutrient absorption [192].

10.8.1.4 *Technologies for Removal of Congo Red Dye From Aqueous Systems*

The removal of Congo red dye from water requires a holistic approach, combining various scientific methods. Adsorption, using materials like activated carbon, clay minerals, and agricultural waste, captures dye molecules on their surfaces. Membrane filtration separates dyes based on size and charge, while coagulation-flocculation destabilizes dye particles for easy removal [198].

Biological treatments utilize microorganisms to degrade or absorb dye pollutants [199]. Advanced oxidation processes generate reactive species to break down dye molecules. Ion exchange resins selectively remove dyes by exchanging ions. Electrochemical methods induce chemical transformations for dye removal. Integrating these approaches offers effective solutions to combat dye pollution and preserve water quality [200].

10.8.2 Ammonia

Ammonia occurs naturally in the environment and is produced through both biotic and abiotic processes. Biological processes such as decomposition of organic matter, animal waste, and nitrogen-fixing bacteria contribute to ammonia production [201]. Additionally, industrial activities such as fertilizer production, livestock farming, and chemical manufacturing are significant anthropogenic sources of ammonia emissions [201].

Ammonia has a wide range of industrial applications, with its primary use being as a nitrogen fertilizer in agriculture. It is also utilized in the production of various chemicals, including ammonium nitrate, urea, and nitric acid. Furthermore, ammonia serves as a refrigerant in industrial cooling systems and as a cleaning agent in household and industrial settings [201].

10.8.2.1 Impacts of Ammonia on Human Health

Ammonia is a colorless gas with a pungent odor, commonly encountered in industrial, agricultural, and household settings. Inhalation of ammonia vapor can lead to irritation of the respiratory tract due to its alkaline nature. Upon exposure, ammonia reacts with moisture in the mucous membranes of the respiratory system, forming ammonium hydroxide, which damages epithelial cells and triggers an inflammatory response. Acute symptoms of ammonia inhalation include coughing, wheezing, and throat irritation. Prolonged or high-level exposure can result in more severe respiratory effects such as pulmonary edema and bronchiolar necrosis, which may be life-threatening [202].

In addition to its respiratory effects, ammonia can also cause dermal and ocular irritation upon direct contact with skin and eyes. The alkaline nature of ammonia disrupts the lipid barrier of the skin, leading to irritation, burns, and dermatitis. Eye exposure to ammonia vapor or liquid can cause immediate pain, redness, and corneal damage. Chronic exposure to low levels of ammonia may result in respiratory sensitization, increasing the risk of asthma and other respiratory disorders [202].

10.8.2.2 Impacts of Ammonia on Aquatic Organisms

Both forms of ammonia, un-ionized ammonia (NH_3) and ionized ammonium (NH_4^+), are toxic to aquatic organisms, albeit through different mechanisms. Un-ionized ammonia, being a weak base, can diffuse across biological membranes and disrupt cellular pH balance, leading to cellular dysfunction and metabolic impairment. Ionized ammonium, on the other hand, can compete with essential cations such as potassium and magnesium, interfering with ion regulation and osmoregulation in aquatic organisms [203].

Fish and other aquatic organisms exposed to elevated concentrations of ammonia may exhibit physiological stress responses, including increased ventilation rate, reduced feeding activity, and altered swimming behavior. Prolonged exposure to sublethal concentrations of ammonia can impair growth, reproduction, and immune function in aquatic species, ultimately affecting population dynamics and ecosystem health [204].

10.8.2.3 Impacts of Ammonia on Plants

In terrestrial ecosystems, ammonia serves as a vital source of nitrogen for plant growth and development. However, excessive concentrations of ammonia in soil can have detrimental effects on plant health and productivity.

Ammonia toxicity in plants typically manifests as leaf chlorosis, necrosis, and stunted growth, resembling symptoms of nitrogen deficiency [205].

The toxic effects of ammonia on plants are attributed to several mechanisms, including disruption of cellular pH homeostasis, inhibition of nitrate assimilation, and interference with root uptake of essential nutrients such as potassium and magnesium. Moreover, ammonia-induced alkalization of soil can alter microbial activity and nutrient cycling processes, further exacerbating nutrient imbalances in plant-soil systems [206].

Certain plant species exhibit varying degrees of tolerance to ammonia toxicity, depending on factors such as species-specific metabolic pathways, root morphology, and symbiotic relationships with nitrogen-fixing microorganisms. Nonetheless, sustained exposure to high levels of ammonia can compromise plant health and reduce crop yields, posing challenges for agricultural productivity and food security [206].

10.8.2.4 Technologies for Removal of Ammonia From Aqueous Systems

Ammonia contamination in aqueous systems, including wastewater, industrial effluents, and agricultural runoff, poses significant environmental and public health concerns. Biological treatment processes harness the metabolic activity of microorganisms to convert ammonia into harmless nitrogen gas through nitrification and denitrification reactions. Nitrification involves the oxidation of ammonia to nitrite (NO_2^-) and subsequently to nitrate (NO^{3-}) by ammonia-oxidizing bacteria (AOB) and nitrite-oxidizing bacteria (NOB), respectively. Denitrification completes the nitrogen cycle by reducing nitrate to nitrogen gas (N_2) under anoxic conditions using denitrifying bacteria [207]. Chemical precipitation methods involve the addition of precipitating agents such as lime (calcium hydroxide), ferric chloride, or aluminum sulfate to aqueous solutions containing ammonia. These precipitants react with ammonia to form insoluble compounds such as struvite ($NH_4MgPO_4 \cdot 6H_2O$) or metal hydroxides, which can be subsequently separated by sedimentation or filtration [208].

Ion exchange involves the removal of ammonia ions from aqueous solutions by exchanging them with other ions immobilized on a solid resin matrix. Cation exchange resins, typically containing sulfonic acid functional groups, selectively bind ammonium ions (NH_4^+) while releasing counter ions such as sodium (Na^+) or hydrogen (H^+). Regeneration of ion exchange resins is achieved by eluting the retained ammonia ions using a concentrated salt solution [209]. Adsorption technologies utilize adsorbent materials such as activated carbon, zeolites, or modified clays to capture ammonia molecules from aqueous solutions through physical or chemical interactions. Surface adsorption mechanisms may include electrostatic attraction, hydrogen bonding, or van der Waals forces. Adsorption capacity and efficiency are influenced by factors such as adsorbent surface area, pore structure, pH, and ammonia concentration [210].

10.8.3 Nitrates

Nitrates (NO_3^-) are nitrogen-oxygen compounds essential to various biological processes and commonly found in aqueous environments. Their chemical structure consists of one nitrogen atom bonded to three oxygen atoms in a trigonal planar arrangement, with a single negative charge. The ubiquity of nitrates in water bodies arises from natural processes such as the decomposition of organic matter and the application of nitrogen-based fertilizers in agriculture [211].

The sources of nitrates in water are diverse and encompass both natural and anthropogenic origins. Agricultural activities, including the use of nitrogen-based fertilizers, represent a major source of nitrate contamination in water bodies. Runoff from agricultural fields can transport excess nitrates into nearby waterways, contributing to elevated nitrate levels [211]. Additionally, animal excreta from livestock farming operations can introduce nitrates into water bodies through surface runoff and leaching into groundwater. Furthermore, discharge from wastewater treatment facilities and the leaching process from septic systems and landfills can also contribute to nitrate pollution in aquatic environments [211].

10.8.3.1 Impacts of Nitrates on Human Health

The presence of nitrates in water can have significant implications for human health, aquatic organisms, and terrestrial ecosystems. High levels of nitrates in drinking water have been associated with adverse health effects, particularly in vulnerable populations such as infants and pregnant women. Nitrate ingestion can lead to methemoglobinemia, commonly known as "blue baby syndrome," a condition characterized by reduced oxygen-carrying capacity of the blood, which can be life-threatening if left untreated [212]. Moreover, nitrates can undergo biological conversion to nitrites in the human body, which can further react with amines to form nitrosamines, compounds with carcinogenic properties [212].

10.8.3.2 Impacts of Nitrates on Aquatic Organisms

In aquatic ecosystems, elevated nitrate levels can fuel excessive algal growth, leading to algal blooms and subsequent oxygen depletion in water bodies. The depletion of dissolved oxygen can have detrimental effects on aquatic organisms, such as fish and invertebrates, by causing hypoxia or even anoxia, which can result in fish kills and disruptions to the food web [213]. Additionally, nitrates can leach into groundwater and contaminate drinking water sources, posing risks to human health and ecosystem integrity [214].

10.8.3.3 Impacts of Nitrates on Plants

In terrestrial ecosystems, excessive nitrate levels can lead to soil acidification and nutrient imbalances, affecting plant growth and productivity. Nitrate leaching from agricultural fields can result in the loss of soil fertility and contribute to nitrate pollution in groundwater, further exacerbating the environmental impacts of nitrate contamination. Mitigating nitrate pollution requires comprehensive management strategies, including the implementation of best management practices in agriculture, the promotion of sustainable wastewater treatment technologies, and the development of policies to regulate nitrate levels in water bodies [215].

Elevated nitrate levels in aquatic ecosystems can lead to eutrophication, promoting the growth of algae and aquatic plants. This can result in reduced oxygen levels, harming fish and other aquatic organisms [213]. Nitrate can also contribute to the formation of harmful algal blooms, which can produce toxins detrimental to aquatic life and human health [215].

10.8.3.4 Technologies for Removal of Nitrates From Aqueous Systems

Various technologies exist for the removal of nitrates from water. Biological denitrification processes utilize microorganisms to convert nitrates into harmless nitrogen gas, which is then released into the atmosphere [211]. Ion exchange involves the removal of nitrates from water by exchanging them with other ions present in an exchange resin. Membrane filtration techniques, such as reverse osmosis and nanofiltration, can remove nitrates from water by selectively rejecting nitrate ions while allowing clean water to pass through [216].

10.8.4 Phosphates

Phosphate ions (PO_4^{3-}) are essential chemical entities consisting of phosphorus and oxygen atoms, fundamental to various biological processes, including cellular energy transfer and ATP synthesis [217]. These ions are commonly encountered in aquatic systems, primarily originating from natural weathering of rocks and minerals, agricultural activities such as fertilizer application, industrial effluents, and domestic sewage [218].

10.8.4.1 Impacts of Phosphate on Human Health

Phosphates are essential nutrients for humans, playing a vital role in various physiological processes, primarily in bone and teeth formation. However, excessive phosphate intake, often through processed foods and soft drinks containing phosphoric acid, has been associated with health issues. High phosphate intake is linked to an increased risk of kidney disease [219]. Elevated phosphate levels in the blood, a condition known as hyperphosphatemia, can be seen in individuals with kidney disease and can lead to complications such as vascular calcification [220].

10.8.4.2 Impacts of Phosphate on Aquatic Organisms

Excess phosphate in aquatic environments, often from agricultural runoff and wastewater discharges, can lead to eutrophication [221]. This can result in decreased oxygen levels in the water, harming aquatic organisms. Fish kills and disruptions in the food chain are common consequences (Kies, 2024). Elevated phosphate levels can promote the growth of harmful cyanobacteria, leading to the production of toxins harmful to aquatic life and humans [213].

10.8.4.3 Impacts of Phosphate on Plants

Phosphates are essential for plant growth and are commonly used as fertilizers. However, excessive use of phosphate fertilizers can lead to soil and water pollution. Runoff from agricultural fields with high phosphate content can result in eutrophication of water bodies, causing harmful algal blooms and negatively affecting aquatic ecosystems [222]. In some cases, excessive phosphate in soil can lead to nutrient imbalances and reduced plant growth due to the inhibition of the uptake of other essential nutrients [223].

10.8.4.4 *Technologies for Phosphate Removal From Aqueous Systems*

Phosphate removal from aqueous systems encompasses several techniques. Phosphate readily forms insoluble compounds with Al and Fe salts, making chemical precipitation possible for phosphate removal from poor quality water [224]. On the other hand, metal hydroxides like aluminum, iron, calcium, and magnesium hydroxides have been used for batch phosphate removal. Adsorption rates are highest between pH 5 and 7, with ferri and aluminum chemistries showing quicker precipitation [225, 226]. Soluble orthophosphate competes with metal hydroxide surface sites for maximum adsorption.

Biological Treatment integrates advanced technology with biological processes for low effluent phosphate concentrations. Phosphorus accumulating organisms (PAO) remove phosphate via luxury uptake during alternating anaerobic and aerobic stages. This process involves intracellular energy reserves buildup, conversion of volatile fatty acids to polyhydroxyalkanoates and exopolymeric substances, and storage of excess phosphate as inorganic polyphosphate [227].

Other phosphate removal technique involves high-capacity ion exchange which selectively removes phosphate ions by exchanging them with ions on the resin. While economically viable for treating low flows of water to ultra-low phosphate levels, conventional ion exchange may not be suitable for large volumes due to high resin requirements [228]. Microfiltration and ultrafiltration remove particulate matter and some phosphorus, but membrane fouling poses challenges [229]. Nanofiltration and reverse osmosis achieve ultra-low phosphorus effluents but are associated with membrane fouling and increased energy demand, making them high-cost options for phosphate removal from waste streams [230].

10.8.5 Sulfates

Sulfate ions (SO_4^{2-}) are chemically composed of one sulfur atom bonded to four oxygen atoms, forming a tetrahedral structure with a negative two charge. This structure is crucial to their chemical and biological roles, including participation in cellular respiration, protein synthesis and sulfur metabolism [231, 232]. Sulfate ions also play a significant role in environmental systems, particularly in the biogeochemical cycling of sulfur and influencing aquatic ecosystems. The sulfate ions are widespread in natural water bodies, originating from various sources. Geological weathering of sulfur-containing rocks and minerals, volcanic emissions which release sulfur dioxide that oxidises into sulfate aerosols, and atmospheric deposition from both natural and human activities contribute to their presence [50]. Additionally, agricultural practices, such as the use sulfate-containing fertilizers and animal waste runoff, further elevate sulfate levels in aquatic environments [233].

10.8.5.1 Impacts of Sulfates on Human Health

Elevated levels of sulfate ions in water bodies can exert significant impacts on human health, aquatic organisms, and terrestrial plants, highlighting the multifaceted consequences of sulfate pollution. One of the primary concerns associated with high sulfate concentrations in drinking water is their potential adverse effects on human health [50]. Research conducted by Backer (2000) indicates that elevated sulfate levels have been linked to various health issues, including gastrointestinal irritation, laxative effects, and dehydration. These effects are particularly concerning in regions where sulfate contamination of drinking water sources is prevalent, posing risks to public health and necessitating measures to mitigate sulfate exposure [234].

10.8.5.2 Impacts of Sulfates on Aquatic Organisms

Sulfate pollution can disrupt the delicate balance of aquatic ecosystems, leading to cascading effects on the health and behavior of aquatic organisms. Studies have demonstrated that sulfate ions can interfere with the osmoregulation mechanisms of freshwater organisms, such as fish and invertebrates, resulting in physiological stress and compromised reproductive success [235]. The disruption of osmoregulation processes can have far-reaching consequences for aquatic biodiversity and ecosystem functioning, underscoring the importance of managing sulfate pollution in aquatic environments [236].

10.8.5.3 Impacts of Sulfates on Plants

In addition to their impacts on aquatic organisms, elevated sulfate levels can also influence terrestrial ecosystems by inhibiting plant growth and development. Research has shown that excessive sulfate concentrations in soil and water can disrupt nutrient uptake and metabolic processes in plants, thereby impeding their growth and productivity. This interference with plant physiology can have detrimental effects on agricultural productivity, ecosystem dynamics, and overall terrestrial biodiversity, highlighting the need for effective strategies to mitigate sulfate pollution in both aquatic and terrestrial environments [237].

10.8.5.4 *Technologies for Sulfate Removal From Aqueous Systems*

To mitigate the adverse effects of sulfate pollution in water bodies, various technologies have been developed for sulfate removal. Chemical precipitation methods involve the addition of chemicals to water, which react with sulfate ions to form insoluble precipitates that can be physically separated from the water [238]. Ion exchange processes utilize ion exchange resins to selectively remove sulfate ions from water by exchanging them with other ions present in the resin [239]. Additionally, membrane-based technologies such as reverse osmosis employ semi-permeable membranes to effectively remove sulfate ions from water by applying pressure to force water molecules through the membrane, while retaining sulfate ions and other dissolved contaminants [240].

10.9 Adsorption Phenomena

Adsorption is a fundamental surface phenomenon wherein molecules or ions from a fluid phase are attracted and adhere to the surface of a solid or liquid phase [241]. This interaction occurs due to various forces, including van der Waals forces, electrostatic interactions, and chemical bonding, depending on the nature of the adsorbate and adsorbent [242].

Adsorption involves the establishment of an equilibrium between the concentration of molecules or ions in the bulk phase (the fluid) and the concentration on the surface of the adsorbent material [241]. The adsorbent surface provides sites for the adsorbate to accumulate, either through physical adsorption, where weak van der Waals forces dominate, or chemical adsorption, where stronger chemical bonds are formed between the adsorbate and surface functional groups of the adsorbent.

Adsorption processes are influenced by factors such as surface area, pore size distribution, surface chemistry, temperature, and pressure [243]. These factors dictate the extent and rate of adsorption and determine the overall efficiency of the process. Understanding adsorption mechanisms is essential for the design and optimization of various applications, including water purification, gas separation, catalysis, and environmental remediation. Recent research in adsorption science has focused on the development of novel adsorbent materials with enhanced properties, such as higher surface area, improved selectivity, and increased stability, to address emerging challenges in pollutant removal, resource recovery, and sustainable manufacturing processes [244, 245].

10.9.1 Categorization of Adsorption Mechanisms

Adsorption can be considered as a process of concentration. In any given system involving an adsorbate, adsorbent, and solution, an equilibrium distribution of the adsorbate between the adsorbent and the solution will be established [241]. Furthermore, modifications in the physical or chemical state of the adsorbate can also impact the quantity of adsorbate present on the adsorbent. This alteration may lead to additional surface reactions accompanying the adsorption process (Williams, 2019). Adsorption mechanisms are classified into different types based on the nature of the interactions between adsorbate molecules and adsorbent surfaces, and they are physisorption and chemisorption.

10.9.1.1 Physical Adsorption (Physisorption)

Physical adsorption, also known as physisorption, involves the attachment of molecules or atoms to the surface of a solid through weak intermolecular forces. This type of adsorption occurs primarily due to van der Waals forces, which arise from interactions between the adsorbate and the adsorbent. The adsorbate molecules are drawn to the adsorbent surface because of the temporary or induced dipoles. Physical adsorption is typically reversible and occurs at relatively low temperatures and pressures [246]. The van der Waals forces responsible for physical adsorption include dispersion forces, dipole-dipole interactions, and hydrogen bonding. These forces are non-specific and operate over short distances, allowing adsorbate molecules to be easily desorbed from the surface under mild conditions. Physical adsorption is commonly observed in gas-solid and liquid-solid systems and plays a significant role in processes such as gas adsorption in porous materials, solvent adsorption on solid surfaces, and the formation of monolayers on surfaces [247]. Figure 10.3 depicts the physisorption mechanism in wastewater treatment.

Figure 10.4 illustrates the intermolecular forces between molecules of the adsorbent and adsorbate.

10.9.1.2 Chemical Adsorption (Chemisorption)

Chemical adsorption, also known as chemisorption, involves the formation of stronger chemical bonds between the adsorbate molecules and the adsorbent surface. These bonds can include covalent bonds, coordinate bonds, or ion-exchange interactions, depending on the nature of the adsorbate and surface functional groups [248]. Chemical adsorption often leads

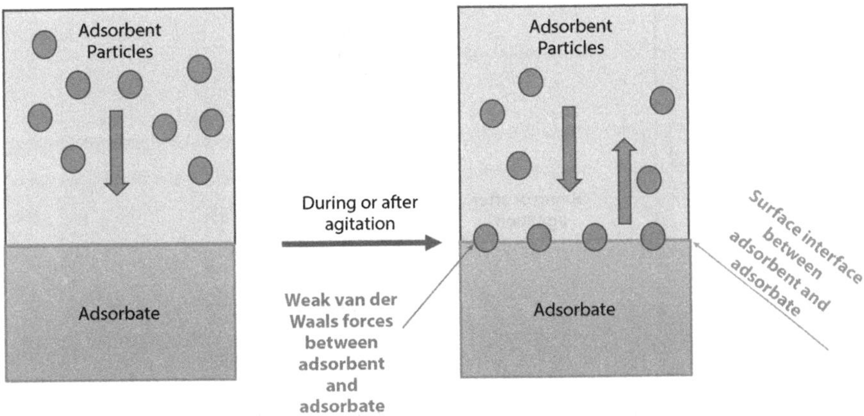

Figure 10.3 Physisorption mechanism in wastewater treatment.

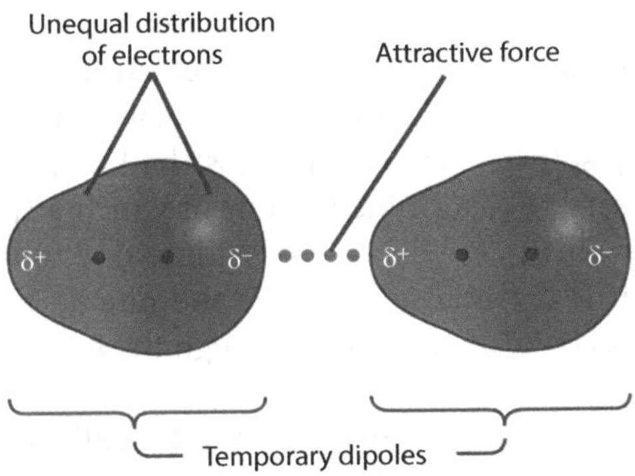

Figure 10.4 Intermolecular forces (van der Waals forces) between molecules of the adsorbent and adsorbate.

to irreversible binding of adsorbate molecules to the adsorbent surface due to the formation of stable chemical bonds. Unlike physical adsorption, which is driven by weak intermolecular forces, chemisorption requires activation energy to overcome the energy barrier associated with bond formation [249]. Chemical adsorption is prevalent in catalytic processes, where reactant molecules are adsorbed onto catalytic surfaces, facilitating chemical reactions. It also plays a crucial role in surface modification, heterogeneous catalysis, and gas-solid reactions. Figure 10.5 illustrates the chemisorption mechanism in wastewater treatment.

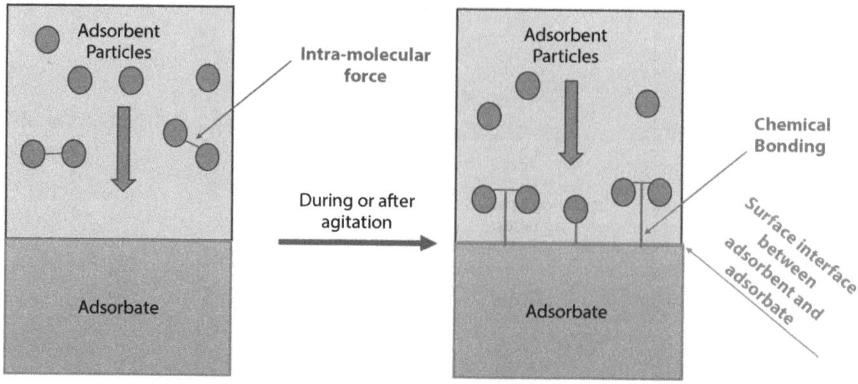

Figure 10.5 Chemisorption mechanism in wastewater treatment.

10.9.2 Adsorption Kinetics

The concept of adsorption kinetics pertains to the speed at which adsorption occurs and the associated mass transfer processes. It is commonly characterized by the rate at which adsorbate accumulates on the adsorbent and the corresponding isotherm [250]. To assess the kinetics of an adsorptive system, it is necessary to determine the rate at which a specific adsorption state is reached, as this rate may vary over time due to surface saturation [251]. Therefore, it is essential to measure the rate as a function of time. Experimental data is typically analyzed through the following steps:

Step 1: Solute molecules transport from the bulk of the solution to the external surface of the adsorbent.

Step 2: These molecules diffuse from the adsorbent surface to the site of adsorption. For porous solids, this involves the movement of molecules through the particle pores.

Step 3: The molecules become adsorbed at the site on the surface.

The rate of this process is influenced by the specific interactions between the solute and adsorbent molecules. Consequently, toxic organic compounds often exhibit high values of adsorption coefficient (Koc), indicating that adsorption is an effective method for removing organics from water [252].

10.9.3 Types of Adsorbents and Their Properties

10.9.3.1 Activated Carbon

Activated carbon is a highly porous form of carbon characterized by its high surface area and extensive network of pores. It is produced by heating carbonaceous materials, such as wood, coal, or coconut shells, to high temperatures in the absence of oxygen, followed by chemical or physical activation processes [253]. Activated carbon exhibits exceptional adsorption capacity and selectivity for a wide range of organic and inorganic contaminants in water and air. Its porous structure provides numerous active sites for adsorption, making it suitable for applications such as water purification, air filtration, and gas storage.

10.9.3.2 Zeolites

Zeolites are crystalline aluminosilicate minerals with a well-defined porous structure composed of interconnected channels and cages. They possess ion-exchange properties and molecular sieving capabilities, allowing for selective adsorption of molecules based on size, shape, and charge. Due to their unique structure, zeolites are utilized in various applications, including water softening, catalysis, gas separation, and odor control. Their ability to selectively adsorb ions and molecules makes them valuable in industries such as wastewater treatment, petrochemical refining, and environmental remediation [254].

10.9.3.3 Silica Gel

Silica gel is a porous form of silicon dioxide (SiO_2) with a high surface area and excellent adsorption properties. It is produced by the gelation of sodium silicate followed by dehydration, resulting in a porous, granular material. Silica gel is commonly used as a desiccant to control humidity and moisture levels in various applications, including packaging, electronics, and pharmaceuticals. Its high surface area and affinity for water molecules make it effective in adsorbing moisture and volatile organic compounds from air and gas streams [255].

10.9.3.4 Metal-Organic Frameworks

Metal-organic frameworks (MOFs) are a class of porous materials composed of metal ions or clusters coordinated with organic ligands to form extended networks with well-defined pore structures. MOFs exhibit exceptionally high surface areas and tunable pore sizes, making them promising adsorbents for gas storage, separation, and catalysis. MOFs offer advantages such as high adsorption capacities, structural diversity, and tailorable properties, allowing for the design of adsorbents with specific functionalities for various applications [256].

10.9.3.5 Iron-Based Adsorbents

Iron-based adsorbents are materials that incorporate iron species and demonstrate adsorption capabilities. These adsorbents are often prepared by impregnating porous substrates like activated carbon or silica with iron oxides or hydroxides, forming composites with enhanced adsorption properties [257]. Iron oxides, such as iron(III) oxide (Fe_2O3) and iron(II, III) oxide (Fe_3O_4), and iron hydroxides, such as ferric hydroxide [$Fe(OH)_3$] or ferrihydrite, are commonly utilized in the synthesis of iron-based adsorbents. These iron species possess high surface areas and surface reactivity, making them effective in the adsorption of various contaminants from water [258].

Iron-based adsorbents have shown promise in the removal of contaminants such as arsenic, fluoride, heavy metals, and organic pollutants from water sources. The high affinity of iron oxides for these species allows for efficient removal through adsorption processes. In particular, iron-based adsorbents are widely studied for arsenic removal due to their ability to form stable complexes with arsenic species, leading to their effective sequestration from water. The presence of iron oxides on the adsorbent surface facilitates the adsorption of arsenic by chemisorption and surface complexation mechanisms [259].

Iron-based adsorbents have also been investigated for fluoride removal from water, where iron oxides and hydroxides interact with fluoride ions through surface complexation or ion exchange processes. The formation of insoluble iron fluoride complexes contributes to fluoride removal from aqueous solutions. Furthermore, iron-based adsorbents demonstrate efficacy in the removal of heavy metals, such as lead, cadmium, and copper, through mechanisms including ion exchange, surface complexation, and precipitation. The presence of iron species on the adsorbent surface facilitates the formation of metal hydroxide or metal oxide precipitates, leading to the removal of heavy metals from water matrices [258].

10.9.3.6 Aluminum-Based Adsorbents

Aluminum-based adsorbents encompass materials containing aluminum species capable of adsorbing a variety of substances. These adsorbents can be synthesized from alumina (Al_2O_3) or aluminum-containing compounds and possess adsorption properties suitable for diverse applications, including water treatment, gas purification, and catalysis [260]. Aluminum-based adsorbents are commonly prepared by the precipitation or impregnation of aluminum species onto porous substrates or support materials. The adsorbent structure can vary depending on the synthesis method, with materials ranging from porous solids to amorphous gels, each offering unique adsorption properties [261].

Aluminum-based adsorbents find applications in various environmental remediation processes, owing to their surface chemistry and ion-exchange properties. These adsorbents have demonstrated effectiveness in the removal of pollutants such as phosphates, fluoride, heavy metals, and organic compounds from aqueous solutions [262]. In water treatment, aluminum-based adsorbents are widely employed for phosphate removal due to their ability to form insoluble aluminum phosphate complexes. The adsorption process involves the precipitation of aluminum hydroxides or oxides in the presence of phosphate ions, leading to the formation of stable complexes that are removed from the solution [263].

Aluminum-based adsorbents are also effective in fluoride removal applications, where aluminum species interact with fluoride ions through surface complexation or ion exchange mechanisms. The formation of insoluble aluminum fluoride complexes facilitates fluoride removal from water matrices [264]. Furthermore, aluminum-based adsorbents exhibit adsorption capabilities toward organic pollutants present in water, including dyes, pesticides, and pharmaceuticals. The porous structure and surface chemistry of these adsorbents allow for interactions with organic molecules, leading to their adsorption and removal from aqueous solutions.

10.9.4 Applications of Adsorption

Adsorption is a surface phenomenon essential in numerous industrial and environmental processes, including water purification, gas separation, and catalysis. It involves the accumulation of molecules or ions from a fluid phase onto the surface of a solid or liquid phase. In water purification, adsorption is utilized to remove contaminants such as heavy metals, organic pollutants, and dyes from wastewater. Adsorbent materials like activated carbon, zeolites, and MOFs are employed to adsorb these pollutants onto their surfaces, effectively purifying water for various applications [265].

Adsorption finds diverse applications in various fields, including wastewater treatment. It has proved to be an effective method for removing pollutants from wastewater, ensuring cleaner water for human consumption. Adsorption technology plays a crucial role in reducing the levels of heavy metals, organic compounds, and other harmful substances present in wastewater. One prominent application of adsorption in wastewater treatment involves the use of activated carbon as an adsorbent. Activated carbon possesses a high adsorption capacity, making it ideal for removing organic compounds, dyes, and other pollutants from water.

Numerous studies have been conducted to evaluate its efficiency in wastewater treatment. For instance, a study by Chang *et al.* [266] demonstrated the successful removal of pharmaceutical compounds from wastewater using activated carbon adsorption. The results highlighted the potential of this adsorption process as an effective treatment approach. Additionally, inorganic adsorbents are utilized in wastewater treatment for the removal of heavy metals. Materials such as zeolites, clay minerals, and activated alumina have shown remarkable adsorption capabilities toward heavy metal ions. In a study by Sayed [267], the researcher investigated the efficiency of zeolite-based adsorbents in the removal of lead and cadmium ions from wastewater. The results showed that zeolite-based adsorbents exhibited excellent adsorption capacity and selectivity for and cadmium and lead removal. Moreover, adsorption techniques are employed in the purification of drinking water. Through the use of specific adsorbents, such as activated carbon or ion exchange resins, contaminants like pesticides, organic pollutants, and chlorine can be effectively eliminated from drinking water sources.

A study by Ebrahimzadeh *et al.* [268] investigated the removal of organic micropollutants from drinking water using activated carbon. The research demonstrated that activated carbon adsorption effectively reduced the concentrations of micropollutants, ensuring safer drinking water. In conclusion, the applications of adsorption, particularly in wastewater treatment, are abundant. Studies have provided evidence of its efficiency in removing various pollutants, such as organic compounds, heavy metals, and micropollutants. The use of adsorption technology in this context contributes to the development of sustainable and effective wastewater treatment processes, ensuring the protection of human health and the environment.

10.9.5 Advantages of Adsorption

The utilization of adsorption in the treatment of wastewater offers several advantages over other physico-chemical methods. This method exhibits a broader range of applicability, allowing for the removal of low pollutant concentrations, and is cost-effective compared to alternative treatments [269]. The applicability of adsorption is demonstrated in various scenarios, such as the removal of individual metals from wastewater, the decolorization of effluent, and the elimination of non-biodegradable organic matter from industrial wastewater [270]. Furthermore, the efficacy of adsorption has been enhanced with the development of newer activated carbon materials. These novel carbon materials possess twice the adsorptive capacity of traditional activated carbon, ensuring efficient pollutant removal [271].

One notable operational advantage of adsorption is the implementation of the packed bed system, which facilitates continuous operation. Once the carbon becomes fully saturated with pollutants, it can be easily removed from the column and sent for regeneration. The regenerated carbon can then be recycled and the contaminants extracted [272]. Apart from its operational benefits, adsorption is also a manageable and efficient treatment system for toxic and hazardous wastewater. Industries dealing with hazardous waste prefer a treatment method that is both efficient and effective, while also allowing for on-site handling due to high costs associated with off-site treatment [273].

Adsorbents often possess high surface areas and porosities, providing ample active sites for the adsorption of target substances. This high surface area-to-volume ratio enhances the adsorption capacity of adsorbents, allowing them to effectively capture and retain a large quantity of contaminants from the surrounding medium [274]. Many adsorbents exhibit selective adsorption properties, enabling them to preferentially adsorb specific molecules or ions while leaving others unaffected. This selectivity allows for the precise removal of target contaminants from complex mixtures, enhancing the efficiency and effectiveness of adsorption processes [134].

In summary, adsorption in wastewater treatment offers various advantages. Its wide applicability allows for the removal of pollutants at low concentrations, whereas its cost-effectiveness makes it a favorable option compared to other methods. The packed bed system enables continuous operation, and the development of newer activated carbon materials enhances adsorptive capacity. Adsorption is also a manageable and efficient system for toxic and hazardous wastewater, with high-pressure liquid and supercritical adsorption offering increased effectiveness for industrial waste treatment.

10.10 Conclusions

In conclusion, the exploration of AMD and its treatment technologies presented in this chapter underscores the multifaceted nature of this environmental challenge and the importance of addressing its ramifications comprehensively. AMD, with its diverse effluent compositions, poses significant risks to human health, aquatic ecosystems, water quality, and infrastructure integrity. However, advancements in passive, active, and integrated remediation technologies offer promising avenues for mitigating these impacts.

Moreover, the exploitation of valuable minerals from AMD presents an intriguing opportunity for both economic gain and environmental remediation. Through recovery, valorization, and beneficiation processes, minerals recovered from AMD can find application in various industrial processes and contribute to the restoration of contaminated sites.

Furthermore, the chapter underscores the critical importance of addressing heavy metal contaminants and other paramount water pollutants in wastewater treatment strategies. Understanding the impacts of contaminants like arsenic, chromium, Congo red dye, ammonia, nitrates, phosphates, and sulfates on human health, aquatic organisms, and plants is crucial for designing effective removal technologies.

Adsorption emerges as a promising method for removing contaminants from aqueous systems, with diverse adsorbents, offering unique advantages. The elucidation of adsorption mechanisms, kinetics, and the properties of different adsorbents provides valuable insights for the development of efficient wastewater treatment systems.

Summarily, this chapter contributes to advancing the understanding of AMD treatment and its broader applications in wastewater treatment, emphasizing the importance of interdisciplinary approaches in addressing complex environmental challenges. Continued research and innovation in this field are essential for developing sustainable solutions to mitigate the impacts of AMD and other contaminants on the environment and human health.

Acknowledgments

Our profound gratitude and thanks go to the National Research Foundation (NRF) South Africa for funding this work through the Postgraduate Scholarship **(Grant ID No: MND200517522430)**. The authors also convey appreciation to the University of Pretoria and the Council for Scientific and Industrial Research for the characterization of the samples.

References

1. Jiao, Y., *et al.*, A review of acid mine drainage: Formation mechanism, treatment technology, typical engineering cases and resource utilization. *Process Saf. Environ. Prot.*, 170, 1240–1260, 2023.
2. INAP. Global Acid Rock Drainage Guide. *The International Network for Acid Prevention*, Canada, 2012.
3. Brusseau, M.L. and Artiola, J.F., *Chapter 12 - Chemical Contaminants, Environmental and Pollution Science (Third Edition)*, pp. 175–190, Academic Press, United Kingdom, 2019.
4. Madzivire, G., *et al.*, Fate of sulphate removed during the treatment of circumneutral mine water and acid mine drainage with coal fly ash: Modelling and experimental approach. *Miner. Eng.*, 24, 1467–1477, 2011.
5. Masindi, V., Mkhonza, P., Tekere, M., Sources of Heavy Metals Pollution, 2021.
6. Rodriguez, M., *et al.*, Remediation of acid mine drainage. *Environ. Chem. Lett.*, 17, 1529–1538, 2019.
7. Ullrich, S., *et al.*, Genome Analysis of the Biotechnologically Relevant Acidophilic Iron Oxidising Strain JA12 Indicates Phylogenetic and Metabolic Diversity within the Novel Genus "Ferrovum". *PLoS One*, 11, 1, 2016.
8. Chen, L.-X., *et al.*, Comparative metagenomic and metatranscriptomic analyses of microbial communities in acid mine drainage. *ISME J.*, 9, 7, 1579–1592, 2015.
9. Muedi, K.L., *et al.*, Effective removal of arsenate from wastewater using aluminium enriched ferric oxide-hydroxide recovered from authentic acid mine drainage. *J. Hazard. Mater.*, 414, 125491, 2021.
10. Masindi, V., *et al.*, Passive remediation of acid mine drainage using cryptocrystalline magnesite: A batch experimental and geochemical modelling approach. *Water SA*, 41, 5, 2015.
11. Roane, T.M., *et al.*, R.M. Maier, I.L. Pepper, C.P. Gerba (Eds.), pp. 421–441 Academic Press, San Diego, 2009.
12. Anekwe, I.M.S. and Isa, Y.M., Bioremediation of acid mine drainage – Review. *Alexandria Eng. J.*, 65, 1047–1075, 2023.
13. Borrok, D.M., *et al.*, Application of iron and zinc isotopes to track the sources and mechanisms of metal loading in a mountain watershed. *Appl. Geochem.*, 24, 7, 1270–1277, 2009.
14. Olivarez, M., *et al.*, Mobility of Heavy Metals in Aquatic Environments Impacted by Ancient Mining-Waste. Water Quality - Factors and Impacts, *IntechOpen*, 2021.

15. Butler, B.A., Effect of pH, ionic strength, dissolved organic carbon, time, and particle size on metals release from mine drainage impacted streambed sediments. *Water Res.*, 43, 5, 1392–402, 2009.
16. Zhang, T., *et al.*, A review: The formation, prevention, and remediation of acid mine drainage. *Environ. Sci. Pollut. Res.*, 30, 52, 111871–111890, 2023.
17. Yuan, J., *et al.*, Resource Utilization of Acid Mine Drainage (AMD): A Review. *Water*, 14, 15, 2385, 2022.
18. Simate, G.S. and Ndlovu, S., Acid mine drainage: Challenges and opportunities. *J. Environ. Chem. Eng.*, 2, 3, 1785–1803, 2014.
19. Brown, J. and Burgos, W., Evaluation of Fe(II) oxidation at an acid mine drainage site using laboratory-scale reactors. *EGU General Assembly Conference Abstracts*, p. 1104, 2010.
20. Schmidt, A., The effects of coal mining on health in Appalachia : global context and social justice implications, 2014.
21. Malatse, M. and Ndlovu, S., The viability of using the Witwatersrand gold mine tailings for brickmaking. *J. South. Afr. Inst. Min. Metall.*, 115, 321–327, 2015.
22. Prepilková, V., *et al.*, Possibilities of remediation of neutral mine drainage - Removal and recovery of potentially hazardous elements. *Soil Water Res.*, 17, 1–17, 2022.
23. Kengni, B., *Strengthening Decision-Making Processes To Promote Water Sustainability In The South African Mining Context: The Role Of Good Environmental Governance*, University of Cape Town, South Africa, 2019.
24. Schouwstra, R., Kinloch, E.D., Lee, C.A., A Short Geological Review of the Bushveld Complex. Platinum Metals Review, 2000.
25. Fourie, P.J., *The Description Of Physico-Chemical Processes In Coal Mine Spoils And Associated Production Of Acid Mine Drainage*, University of Free State, South Africa, 2007.
26. Azzie, B.A.-M., Coal mine waters in South Africa : their geochemistry, quality and classification. Thesis, Department of Geological Sciences, University of Cape Town, 2002.
27. Nordstrom, D.K., Blowes, D.W., Ptacek, C.J., Hydrogeochemistry and microbiology of mine drainage: An update. *Appl. Geochem.*, 57, 3–16, 2015.
28. CSIR, Characterising the risk of human exposure and health impacts from acid mine drainage in South Africa Deliverable for Output VII – "Final Project Report, 2013.
29. Jaishankar, M., *et al.*, Toxicity, mechanism and health effects of some heavy metals. *Interdiscip. Toxicol.*, 7, 2, 60–72, 2014.
30. Rambabu, K., *et al.*, Biological remediation of acid mine drainage: Review of past trends and current outlook. *Environ. Sci. Ecotechnol.*, 2, 100024, 2020.
31. Varshney, S., *et al.*, Ecotoxicological assessment of Cu-rich acid mine drainage of Sulitjelma mine using zebrafish larvae as an animal model. *Ecotoxicol. Environ. Saf.*, 269, 115796, 2024.

32. Kibria, G., Effects of acid mine drainage (AMD) on water quality, aquatic biodiversity, plant crops and human health), 2014.
33. Balali-Mood, M., *et al.*, Toxic Mechanisms of Five Heavy Metals: Mercury, Lead, Chromium, Cadmium, and Arsenic. *Front. Pharmacol.*, 12, 643972, 2021.
34. Rai, P.K., *et al.*, Heavy metals in food crops: Health risks, fate, mechanisms, and management. *Environ. Int.*, 125, 365–385, 2019.
35. Gonah, T., Impact of Acid Mine Drainage on Water Resources in South Africa, pp. 41–65, 2016.
36. Baloyi, J., *et al.*, Recent Progress on Acid Mine Drainage Technological Trends in South Africa: Prevention, Treatment, and Resource Recovery. *Water*, 15, 19, 3453, 2023.
37. Humphries, M., *et al.*, Attenuation of pollution arising from acid mine drainage by a natural wetland on the Witwatersrand. *S. Afr. J. Sci.*, 113, 1–2, 1–9, 2017.
38. Munyai, R., Ogola, H., Modise, D., Microbial Community Diversity Dynamics in Acid Mine Drainage and Acid Mine Drainage-Polluted Soils: Implication on Mining Water Irrigation Agricultural Sustainability. *Front. Sustain. Food Syst.*, 5, Switzerland, 2021.
39. Pell, R.E., Evaluating remote economic benefits of watershed-scale acid mine drainage restoration, in: *Graduate Theses, Dissertations, and Problem Reports*, p. 11625, 2022.
40. Ekolu, S.O., *et al.*, Disintegration of concrete construction induced by acid mine drainage attack. *J. S. Afr. Inst. Civ. Eng.*, 58, 1, 34–42, 2016.
41. Dean, A., *et al.*, Natural Wetlands Are Efficient at Providing Long-Term Metal Remediation of Freshwater Systems Polluted by Acid Mine Drainage. *Environ. Sci. Technol.*, 47, 21, 2013.
42. Dordio, A., Carvalho, A., Pinto, A., Wetlands: Water "living filters"?, 2008.
43. Pat-Espadas, A., *et al.*, Review of Constructed Wetlands for Acid Mine Drainage Treatment. *Water*, 10, 1–25, 2018.
44. Rahman, M.E., *et al.*, Design, Operation and Optimization of Constructed Wetland for Removal of Pollutant. *Int. J. Environ. Res. Public Health*, 17, 8339, 2020.
45. Vo, T.-K-Q., *et al.*, Pilot and full scale applications of floating treatment wetlands for treating diffuse pollution. *Sci. Total Environ.*, 899, 165595, Elsevier, Netherlands, 2023.
46. Gupta, P., Ann, T.-., Lee, S.-M., Use of biochar to enhance constructed wetland performance in wastewater reclamation. *Environ. Eng. Res.*, 21, 1, 36–44, 2015.
47. Ayangbenro, A.S., Olanrewaju, O.S., Babalola, O.O., Sulfate-Reducing Bacteria as an Effective Tool for Sustainable Acid Mine Bioremediation. *Front. Microbiol.*, 9, 1986, 2018.

48. Akimbekov, N.S., *et al.*, Biotechnology of Microorganisms from Coal Environments: From Environmental Remediation to Energy Production. *Biol. (Basel)*, 11, 9, 1306, 2022.
49. Mishel, P.F., *et al.*, *Chapter 1 - Bioremediation: A green technology for environmental cleanup, in Metagenomics to Bioremediation*, V. Kumar and (Eds.), pp. 3–21, Academic Press, United Kingdom, 2023.
50. Bagheri Novair, S., *et al.*, The role of sulphate-reducing bacteria (SRB) in bioremediation of sulphate-rich wastewater: Focus on the source of electron donors. *Process Saf. Environ. Prot.*, 184, 190–207, 2024.
51. Treatment Solutions, A.O.S., *What Is Sedimentation in Water Treatment?*, 2018.
52. Tarpagkou, R. and Pantokratoras, A., The influence of lamellar settler in sedimentation tanks for potable water treatment — A computational fluid dynamic study. *Powder Technol.*, 268, 139–149, 2014.
53. Aguilar-Garrido, A., *et al.*, Remediation potential of mining, agro-industrial, and urban wastes against acid mine drainage. *Sci. Rep.*, 13, 1, 12120, 2023.
54. Masindi, V., *et al.*, Challenges and avenues for acid mine drainage treatment, beneficiation, and valorisation in circular economy: A review. *Ecol. Eng.*, 183, 106740, 2022.
55. Mahmood, M., *et al.*, Removal of Heavy Metals Using Chemicals Precipitation. *Eng. Technol. J.*, 29, 3, 2011.
56. Luptakova, A. and Kusnierova, M., Bioremediation of acid mine drainage contaminated by SRB. *Hydrometallurgy*, 77, 97–102, 2005.
57. Ighalo, J.O., *et al.*, A review of treatment technologies for the mitigation of the toxic environmental effects of acid mine drainage (AMD). *Process Saf. Environ. Prot.*, 157, 37–58, 2022.
58. Kaur, J., *et al.*, Reactive Black-5, Congo Red and Methyl Orange: Chemical Degradation of Azo-Dyes by Agrobacterium. *Water*, 15, 9, 1664, 2023.
59. Wang, Y., *et al.*, A novel method based on membrane distillation for treating acid mine drainage: Recovery of water and utilization of iron. *Chemosphere*, 279, 2021.
60. Nguyen, T., Soda, S., Horiuchi, K., Removal of Heavy Metals from Acid Mine Drainage with Lab-Scale Constructed Wetlands Filled with Oyster Shells. *Water*, 14, 3325, 2022.
61. Chidiac, C., *et al.*, Passive Treatments for Mine Drainage, in: *A Guide for Early Researchers*, 2024.
62. Song, G., *et al.*, Extraction of selected rare earth elements from anthracite acid mine drainage using supercritical CO2 via coagulation and complexation. *J. Rare Earths*, 39, 1, 83–89, 2021.
63. Mwewa, B., *et al.*, Recovery of rare earth elements from acid mine drainage: A review of the extraction methods. *J. Environ. Chem. Eng.*, 10, 3, 107704, 2022.
64. Gaikwad, D.R., Ion exchange system design for removal of heavy metals from acid mine drainage wastewater. *Acta Montan. Slovaca Ročník*, 15, 298–304, 2010.

65. Nleya, Y., *Removal of toxic metals and recovery of acid from acid mine drainage using acid retardation and adsorption processes*, University of Witwatersrand, South Africa, 2016.
66. Yasser, M.M. and Rania, E.M., *Ion Exchange Chromatography - An Overview, in Column Chromatography*, F.M. Dean and B.M. Barbara, Editors. 2013, IntechOpen: Rijeka. p. Ch. 1.
67. Liu, Z., *et al.*, Segmented regeneration of ion exchange resins used for natural organic matter removal. *Sep. Purif. Technol.*, 303, 122271, 2022.
68. Kul, M. and Oskay, K.O., Separation and recovery of valuable metals from real mix electroplating wastewater by solvent extraction. *Hydrometallurgy*, 155, 153–160, 2015.
69. Ola, P.D. and Matsumoto, M., Metal Extraction with Ionic Liquids-Based Aqueous Two-Phase System, in: *Recent Advances in Ionic Liquids*, 2018.
70. Devi, N. and Nayak, B., Liquid-liquid extraction and separation of copper(II) and nickel(II) using LIX®984N. *J. S. Afr. Inst. Min. Metall.*, 114, 937–943, 2014.
71. Lommelen, R., *Mechanism Of Solvent Extraction Of Metal Ions By Basic Extractants*, Arenberg Doctoral School. Faculty Of Science, KU Leuven, Belgium, 2021.
72. Lavinia, L. and Laura, C., *Heavy Metals Removal from Water and Wastewater, in Heavy Metals*, A.A. Basim (Ed.), p. Ch. 21, IntechOpen, Rijeka, 2023.
73. Coetzee, R., Characterization of precipitate formed during the removal of iron and precious metals from sulphate leach solutions. *J. South. Afr. Inst. Min. Metall.*, 117, 771–778, 2017.
74. Ojuederie, O.B. and Babalola, O.O., Microbial and Plant-Assisted Bioremediation of Heavy Metal Polluted Environments: A Review. *Int. J. Environ. Res. Public Health*, 14, 12, 1504, 2017.
75. Dutta, D., Arya, S., Kumar, S., Industrial wastewater treatment: Current trends, bottlenecks, and best practices. *Chemosphere*, 285, 131245, 2021.
76. Silva Rodríguez de San Miguel, J.A., Wastewater Treatment and Reuse for Sustainable Water Resources Management: A Systematic Literature Review. *Sustainability*, 15, 10940, 2023.
77. Feris, L. and Kotzé, L., The Regulation of Acid Mine Drainage in South Africa: Law and Governance Perspectives. *Potchefstroom Electron. Law Journal/Potchefstroomse Elektroniese Regsblad*, 17, 2104, 2015.
78. Kumar, A., Singh, U., Pradhan, B., Enhancing subsurface contamination assessment via ensemble prediction of ground electrical property: A Colorado AMD-impacted wetland case study. *J. Environ. Manage.*, 351, 2024, 119943, 2024.
79. Daraz, U., *et al.*, Remediation technologies for acid mine drainage: Recent trends and future perspectives. *Chemosphere*, 311, Pt 2, 137089, 2023.
80. Masindi, V., Osman, M., Shingwenyana, R., Valorization of acid mine drainage (AMD): A simplified approach to reclaim drinking water and synthesize valuable minerals - Pilot study. *J. Environ. Chem. Eng.*, 7, 3, 2019.

81. Yang, J.-S., *et al.*, Adsorption of Arsenic from Aqueous Solutions by Iron Oxide Coated Sand Fabricated with Acid Mine Drainage. *Sep. Sci. Technol.*, 50, 2, 267–275, 2015.
82. Skousen, J., Overview of Acid Mine Drainage Treatment with Chemicals, in: *Acid Mine Drainage, Rock Drainage, and Acid Sulfate Soils: Causes, Assessment, Prediction, Prevention, and Remediation*, pp. 325–337, 2014.
83. Luís, A.T., *et al.*, Extremely Acidic Eukaryotic (Micro) Organisms: Life in Acid Mine Drainage Polluted Environments-Mini-Review. *Int. J. Environ. Res. Public Health*, 19, 1, 376, 2021.
84. NEDLAC, Fund For Research Into Industrial Development, Growth And Equity (FRIDGE). National Economic Development and Labour Council, South Africa, 2006.
85. Vitti, C. and Arnold, B.J., The Reprocessing and Revalorization of Critical Minerals in Mine Tailings. *Min. Metall. Explor.*, 39, 1, 49–54, 2022.
86. Wright, S. and Mallia, C., The Potential of Eco-Taxes as Instruments for Sustainability. An Analysis of the Critical Design Elements. *J. Transdiscipl. Environ. Stud.*, 2, 2, 2003.
87. Dempsey, B.A., *et al.*, Enhanced iron oxidation to improve AMD treatment. *Geosyst. Eng.*, 15, 3, 203–209, 2012.
88. Kimpiab, E., *et al.*, Synthesis of Stabilized Iron Nanoparticles from Acid Mine Drainage and Rooibos Tea for Application as a Fenton-like Catalyst. *ACS Omega*, 7, 28, 24423–24431, 2022.
89. Muedi, K.L., *et al.*, Effective Adsorption of Congo Red from Aqueous Solution Using Fe/Al Di-Metal Nanostructured Composite Synthesised from Fe(III) and Al(III) Recovered from Real Acid Mine Drainage. *Nanomater. (Basel)*, 12, 5, 776, 2022.
90. Mapukata, S., *et al.*, Acid Mine Drainage Treatment and Control: Remediation Methodologies, Mineral Beneficiation and Water Reclamation Strategies, 2024.
91. Huang, J., *et al.*, Fe(II) Redox Chemistry in the Environment. *Chem. Rev.*, 121, 13, 8161–8233, 2021.
92. Onizuka, T., Fukuda, M., Iwasaki, T., Effects of Coexisting Anions on the Formation of Hematite Nanoparticles in a Hydrothermal Process with Urea Hydrolysis and the Congo Red Dye Adsorption Properties. *Powders*, 2, 2, 338–352, 2023.
93. Rufus, D., Sreeju, N., Philip, D., Synthesis of biogenic hematite (α-Fe2O3) nanoparticles for antibacterial and nanofluid applications. *RSC Adv.*, 6, 94206–94217, 2016.
94. Ponnaiah, S. and Rajan, J., A Review on Iron Oxide Nanoparticles and Their Biomedical Applications. *J. Supercond. Novel Magn.*, 31, 3397–3413, 2018.
95. Pusta, A., *et al.*, Recent Advances in the Development of Drug Delivery Applications of Magnetic Nanomaterials. *Pharmaceutics*, 15, 7, 1872, 2023.
96. Tsedenbal, B., *et al.*, Removal of Heavy Metals from Wastewater using α-Fe2O3 Nanocrystals. *Korean J. Mater. Res.*, 30, 447–452, 2020.

97. Kefeni, K., Msagati, T.A.M., Mamba, B., Acid mine drainage: Prevention, treatment options, and resource recovery: A review. *J. Clean. Prod.*, 151, 475–493, 2017.
98. Blaney, L., Magnetite (Fe3O4)Properties, synthesis and applications. *Lehigh Rev.*, 15, 33–81, 2007.
99. Kirkpatrick, K.M., *et al.*, Quantifying superparamagnetic signatures in nanoparticle magnetite: a generalized approach for physically meaningful statistics and synthesis diagnostics. *Chem. Sci.*, 14, 27, 7589–7594, 2023.
100. Dutta, S., Dixit, R., Sharma, R., Remediation of heavy metals with nanomaterials, pp. 97–138, 2022.
101. Liu, J.F., *et al.*, Use of magnetic fields and nanoparticles to trigger drug release and improve tumor targeting. *WIREs Nanomed. Nanobiotechnol.*, 11, 6, e1571, 2019.
102. Geology Science, in: *Oxides Minerals: Goethite*, 2023.
103. Yi, S., *et al.*, Influencing factors and environmental effects of interactions between goethite and organic matter: A critical review. *Front. Environ. Sci.*, 10, 2022.
104. Wydro, U., *et al.*, A Review on Pharmaceuticals and Personal Care Products Residues in the Aquatic Environment and Possibilities for Their Remediation. *Sustainability*, 16, 1, 2023.
105. Congreve, R., Quezada, C., Varaprasad, K., Aluminum Oxide Nanoparticles: Properties and Applications Overview, pp. 265–288, 2024.
106. Busca, G., The surface of transitional aluminas: A critical review. *Catal. Today*, 226, 2–13, 2014.
107. Mahesh, R., *et al.*, Removal of pollutants from wastewater using alumina based nanomaterials: A review. *Korean J. Chem. Eng.*, 40, 2035–2045, 2023.
108. Qasem, N.A.A., Mohammed, R.H., Lawal, D.U., Removal of heavy metal ions from wastewater: a comprehensive and critical review. *npj Clean Water*, 4, 1, 36, 2021.
109. Bullen, J.C., Landa-Cansigno, O., Weiss, D.J., Teaching Adsorption Chemistry by Constructing Surface Complexation Models (SCM) in PHREEQC. *J. Chem. Educ.*, 101, 5, 1914–1924, 2024.
110. Giles, D.E., *et al.*, Iron and aluminium based adsorption strategies for removing arsenic from water. *J. Environ. Manage.*, 92, 12, 3011–22, 2011.
111. Mahinroosta, M. and Moattari, R.M., Alumina-based nanoadsorbents for wastewater treatment, pp. 205–239, 2023.
112. Roulia, M., *et al.*, Lignite fly ash utilization for acid mine drainage neutralization and clean-up. *Cleaner Mater.*, 6, 100142, 2022.
113. Sulaiman, A., Othman, A., Ibrahim, I., The use of magnesium oxide in acid mine drainage treatment. *Mater. Today Proc.*, 5, 21566–21573, 2018.
114. Liu, F., *et al.*, Effect of calcium oxide on the efficiency of ferrous ion oxidation and total iron precipitation during ferrous ion oxidation in simulated acid mine drainage treatment with inoculation of Acidithiobacillus ferrooxidans. *Water Sci. Technol.*, 73, 6, 1442–1453, 2015.

115. Chen, Q., *et al.*, Comparison of heavy metal removals from aqueous solutions by chemical precipitation and characteristics of precipitates. *J. Water Process Eng.*, 26, 289–300, 2018.
116. Akinwekomi, V., *et al.*, Beneficiation of acid mine drainage (AMD): A viable option for the synthesis of goethite, hematite, magnetite, and gypsum – Gearing towards a circular economy concept. *Miner. Eng.*, 148, 106204, 2020.
117. Bersch, J.D., *et al.*, TiO2-Based Mortars for Rendering Building Envelopes: A Review of the Surface Finishing for Sustainability. *Sustainability*, 15, 24, 16920, 2023.
118. Marcello, R.R., *et al.*, Inorganic pigments made from the recycling of coal mine drainage treatment sludge. *J. Environ. Manage.*, 88, 4, 1280–1284, 2008.
119. Abdullah, J.A.A., *et al.*, Effect of Calcination Temperature and Time on the Synthesis of Iron Oxide Nanoparticles: Green vs. Chemical Method. *Materials*, 16, 1798, 2023.
120. Pfaff, G., Iron oxide pigments. *Phys. Sci. Rev.*, 6, 10, 2021.
121. Mallick, P.K., 2.09 – Particulate and Short Fiber Reinforced Polymer Composites. *Compr. Compos. Mater.*, 2, 291–331, 2000.
122. Florence, K., *et al.*, Iron-mineral accretion from acid mine drainage and its application in passive treatment. *Environ. Technol.*, 37, 1428 – 1440, 2016.
123. Chimanlal, I., *et al.*, Resource recovery from acid mine drainage in membrane distillation crystallization. *Front. Membr. Sci. Technol.*, 2, 2023.
124. Mema, V., Impact of poorly maintained wastewater sewage treatment plants - lessons from South Africa : wastewater management. *ReSource*, 12, 3, 60–65, 2010.
125. Puyol, D., *et al.*, Resource Recovery from Wastewater by Biological Technologies: Opportunities, Challenges, and Prospects. *Front. Microbiol.*, 7, 2106, 2016.
126. WRC, Energy Recovery From Wastewater Sludge – A Review Of Appropriate Emerging And Established Technologies For The South African Industry. *Water Res. Commission*, WRC Report No. TT 752/18, 2018.
127. Tchounwou, P.B., *et al.*, Heavy metal toxicity and the environment. *Exp. Suppl.*, 101, 133–64, 2012.
128. Ali, H., Khan, E., Ilahi, I., Environmental Chemistry and Ecotoxicology of Hazardous Heavy Metals: Environmental Persistence, Toxicity, and Bioaccumulation. *J. Chem.*, 2019, 1, 6730305, 2019.
129. Zinia, S.S., *et al.*, Effects of heavy metal exposure during pregnancy on birth outcomes. *Sci. Rep.*, 13, 1, 18990, 2023.
130. Emon, F., *et al.*, Bioaccumulation and Bioremediation of Heavy Metals in Fishes-A Review. *Toxics*, 11, 510, 2023.
131. Taslima, K., *et al.*, Impacts of heavy metals on early development, growth and reproduction of fish - A review. *Toxicol. Rep.*, 9, 858–868, 2022.
132. Souri, Z., *et al.*, Heavy Metals and Photosynthesis: Recent Developments, pp. 107–134, 2019.

133. Vasilachi, I.C., Stoleru, V., Gavrilescu, M., Analysis of Heavy Metal Impacts on Cereal Crop Growth and Development in Contaminated Soils. *Agriculture*, 13, 1983, 2023.
134. Zaimee, M.Z.A., Sarjadi, M.S., Rahman, M.L., Heavy Metals Removal from Water by Efficient Adsorbents. *Water*, 13, 19, 2659, 2021.
135. Ince, M. and Kaplan Ince, O., An Overview of Adsorption Technique for Heavy Metal Removal from Water/Wastewater: A Critical Review. *Int. J. Pure App. Sci.*, 3, 2, 10–19, 2017.
136. Zhang, Y. and Duan, X., Chemical precipitation of heavy metals from wastewater by using the synthetical magnesium hydroxy carbonate. *Water Sci. Technol.*, 81, 6, 1130–1136, 2020.
137. Vidu, R., *et al.*, Removal of Heavy Metals from Wastewaters: A Challenge from Current Treatment Methods to Nanotechnology Applications. *Toxics*, 8, 4, 101, 2020.
138. IARC, Working Group on the Evaluation of Carcinogenic Risks to Humans. Arsenic, metals, fibres, and dusts. *IARC Monogr. Eval. Carcinog. Risks Hum.*, 100, Pt C, 11–465, 2012, PMID: 23189751; PMCID: PMC4781271.
139. Hughes, M.F., *et al.*, Arsenic exposure and toxicology: a historical perspective. *Toxicol. Sci.*, 123, 2, 305–32, 2011.
140. National Research Council Subcommittee on Arsenic in Drinking, W, in: *Arsenic in Drinking Water*, National Academies Press (US, Washington DC), 1999, Copyright 1999 by the National Academy of Sciences. All rights reserved.
141. Tang, G.T., Elakis, J., Scardamaglia, L., Cutaneous manifestations and treatment of arsenic toxicity: A systematic review. *Skin Health Dis.*, 3, 4, e231, 2023.
142. States, J.C., *et al.*, Arsenic and cardiovascular disease. *Toxicol. Sci. Off. J. Soc. Toxicol.*, 1072, 312–23, 2009.
143. Tyler, C.R. and Allan, A.M., The Effects of Arsenic Exposure on Neurological and Cognitive Dysfunction in Human and Rodent Studies: A Review. *Curr. Environ. Health Rep.*, 1, 2, 132–147, 2014.
144. Kaur, R., *et al.*, Understanding arsenic toxicity: Implications for environmental exposure and human health. *J. Hazard. Mater. Lett.*, 5, 100090, 2024.
145. Eisler, R., Arsenic Hazards to Fish, Wildlife, and Invertebrates: A Synoptic Review, in: *Contaminant Hazard Reviews*, United States Department of the Interior, Fish and Wildlife Service, 1988.
146. Rachamalla, M., *et al.*, Reproductive and Developmental Effects of Sex-Specific Chronic Exposure to Dietary Arsenic in Zebrafish (Danio rerio). *Toxics*, 12, 4, 302, 2024.
147. Vellinger, C., *et al.*, Behavioural and physiological responses of Gammarus pulex exposed to cadmium and arsenate at three temperatures: individual and combined effects. *PLoS One*, 7, 6, e39153, 2012.
148. Thakur, M., *et al.*, Molecular Mechanism of Arsenic-Induced Neurotoxicity including Neuronal Dysfunctions. *Int. J. Mol. Sci.*, 22, 18, 10077, 2021.

149. Nyarko, E., *et al.*, Potential human health risks associated with ingestion of heavy metals through fish consumption in the Gulf of Guinea. *Toxicol. Rep.*, 10, 117–123, 2023.
150. Mandal, P., An insight of environmental contamination of arsenic on animal health. *Emerging Contam.*, 3, 17–22, 2017.
151. Baudrot, V., *et al.*, Effects of contaminants and trophic cascade regulation on food chain stability: Application to cadmium soil pollution on small mammals – Raptor systems. *Ecol. Modell.*, 382, 33–42, 2018.
152. Rapport, D.J., Regier, H.A., Hutchinson, T.C., Ecosystem Behavior Under Stress. *Am. Nat.*, 125, 5, 617–640, 1985.
153. Bali, A.S. and Sidhu, G.P.S., Arsenic acquisition, toxicity and tolerance in plants - From physiology to remediation: A review. *Chemosphere*, 283, 131050, 2021.
154. Aggarwal, A., *et al.*, Metal Toxicity and Photosynthesis, pp. 16: 229–236, 2011.
155. Liang, Y., *et al.*, Regulatory effects of silicon nanoparticles on the growth and photosynthesis of cotton seedlings under salt and low-temperature dual stress. *BMC Plant Biol.*, 23, 1, 504, 2023.
156. Chandrakar, V., Pandey, N., Keshavkant, S., Hasanuzzaman, M., Nahar, K., Fujita, M., *Plant Responses to Arsenic Toxicity: Morphology and Physiology, in Mechanisms of Arsenic Toxicity and Tolerance in Plants*, pp. 27–48, Springer Singapore, Singapore, 2018.
157. Jha, A.B. and Dubey, R.S., Arsenic exposure alters activity behaviour of key nitrogen assimilatory enzymes in growing rice plants. *Plant Growth Regul.*, 43, 3, 259–268, 2004.
158. Khan, M.I., *et al.*, Arsenic Exposure through Dietary Intake and Associated Health Hazards in the Middle East. *Nutrients*, 14, 10, 2136, 2022.
159. Rahaman, M.S., *et al.*, Environmental arsenic exposure and its contribution to human diseases, toxicity mechanism and management. *Environ. Pollut.*, 289, 117940, 2021.
160. Chandel, M., *et al.*, Poison in the water: Arsenic's silent assault on fish health. *J. Appl. Toxicol.*, 44, 9, 1282–1301, 2024.
161. Xanthopoulou, M. and Katsoyiannis, I.A., Enhanced Adsorption of Chromate and Arsenate Ions from Contaminated Water with Emphasis on Polyethylenimine Modified Materials: A Review. *Separations*, 10, 8, 2023.
162. Mojiri, A., *et al.*, Adsorption methods for arsenic removal in water bodies: a critical evaluation of effectiveness and limitations. *Front. Water*, 6, 2024.
163. Das, T.K., *et al.*, Ultra-high arsenic adsorption by graphene oxide iron nanohybrid: Removal mechanisms and potential applications. *Chemosphere*, 253, 126702, 2020.
164. De Klerk, R.J., *et al.*, Continuous circuit coprecipitation of arsenic(V) with ferric iron by lime neutralization: Process parameter effects on arsenic removal and precipitate quality. *Hydrometallurgy*, 111–112, 65–72, 2012.

165. Pal, C.A., *et al.*, *Chapter 13 - Carbon dots as adsorbents for removal of toxic chemicals, in Carbon Dots in Analytical Chemistry*, S.K. Kailasa and C.M. Hussain (Eds.), pp. 161–180, Elsevier, Netherlands, 2023.
166. Oehmen, A., *et al.*, Arsenic removal from drinking water through a hybrid ion exchange membrane - Coagulation process. *Sep. Purif. Technol.*, 83, 137–143, 2011.
167. Nicomel, N.R., *et al.*, Technologies for Arsenic Removal from Water: Current Status and Future Perspectives. *Int. J. Environ. Res. Public Health*, 13, 1, ijerph13010062, 2015.
168. Alvarez, C.C., Bravo Gómez, M.E., Hernández Zavala, A., Hexavalent chromium: Regulation and health effects. *J. Trace Elem. Med. Biol.*, 65, 126729, 2021.
169. Shanker, A. and Venkateswarlu, B., Chromium: Environmental Pollution, Health Effects and Mode of Action, in: *Encyclopedia of Environmental Health*, 2019.
170. Alam, N., *et al.*, An overview of chromium removal techniques from tannery effluent. *Appl. Water Sci.*, 10, 205, 2020.
171. Sharma, P., *et al.*, Health hazards of hexavalent chromium (Cr (VI)) and its microbial reduction. *Bioengineered*, 13, 4923–4938, 2022.
172. Wilbur, S., *et al.*, *Agency for Toxic Substances and Disease Registry (ATSDR) Toxicological Profiles*, in: *Toxicological Profile for Chromium*, Agency for Toxic Substances and Disease Registry (US), Atlanta (GA), 2012.
173. Yatera, K., *et al.*, Cancer Risks of Hexavalent Chromium in the Respiratory Tract. *J. uoeh*, 40, 2, 157–172, 2018.
174. Velma, V., Vutukuru, S.S., Tchounwou, P.B., Ecotoxicology of hexavalent chromium in freshwater fish: a critical review. *Rev. Environ. Health*, 24, 2, 129–45, 2009.
175. Bakshi, A. and Panigrahi, A.K., A comprehensive review on chromium induced alterations in fresh water fishes. *Toxicol. Rep.*, 5, 440–447, 2018.
176. Yan, G., *et al.*, Toxicity mechanisms and remediation strategies for chromium exposure in the environment. *Front. Environ. Sci.*, 11, 2023.
177. Zulfiqar, U., *et al.*, Chromium toxicity, speciation, and remediation strategies in soil-plant interface: A critical review. *Front. Plant Sci.*, 13, 1081624, 2022.
178. Saud, S., *et al.*, The impact of chromium ion stress on plant growth, developmental physiology, and molecular regulation. *Front. Plant Sci.*, 13, 994785, 2022.
179. Sharma, A., *et al.*, Chromium Bioaccumulation and Its Impacts on Plants: An Overview. *Plants*, 9, 100, 2020.
180. Zeng, F., *et al.*, Genotypic differences in nutrient uptake and accumulation in rice under chromium stress. *J. Plant Nutr.*, 33, 518–528, 2010.
181. Hasanuzzaman, M., *et al.*, Regulation of Reactive Oxygen Species and Antioxidant Defense in Plants under Salinity. *Int. J. Mol. Sci.*, 22, 17, 2021.

182. Mathur, S., Kalaji, H.M., Jajoo, A., Investigation of deleterious effects of chromium phytotoxicity and photosynthesis in wheat plant. *Photosynthetica*, 54, 185–192, 2016.
183. Oruko Ongon'g, R., *et al.*, The Potential Health Risk Associated with Edible Vegetables Grown on Cr(VI) Polluted Soils. *Int. J. Environ. Res. Public Health*, 17, 2, 2020.
184. Mohan, D. and Pittman, C.U., Activated carbons and low cost adsorbents for remediation of tri- and hexavalent chromium from water. *J. Hazard. Mater.*, 137, 2, 762–811, 2006.
185. Minas, F., Chandravanshi, B.S., Leta, S., Chemical precipitation method for chromium removal and its recovery from tannery wastewater in Ethiopia. *Chem. Int.*, 3, 4, 291–305, 2018.
186. Gheju, M. and Balcu, I., Removal of chromium from Cr(VI) polluted wastewaters by reduction with scrap iron and subsequent precipitation of resulted cations. *J. Hazard. Mater.*, 196, 131–138, 2011.
187. Rengaraj, S., Yeon, K.-H., Moon, S.-H., Removal of chromium from water and wastewater by ion exchange resins. *J. Hazard. Mater.*, 87, 1, 273–287, 2001.
188. Piedra, E., Álvarez, J.R., Luque, S., Hexavalent chromium removal from chromium plating rinsing water with membrane technology. *Desalin. Water Treat.*, 53, 1431–1439, 2015.
189. Hafiane, A., Lemordant, D., Dhahbi, M., Removal of hexavalent chromium by nanofiltration. *Desalination*, 130, 3, 305–312, 2000.
190. Zolfaghari, G. and Kargar, M., Nanofiltration and microfiltration for the removal of chromium, total dissolved solids, and sulfate from water. *MethodsX*, 6, 549–557, 2019.
191. National Center for Biotechnology Information, PubChem Compound Summary for CID 11313, Congo red, 2024.
192. Oladoye, P.O., *et al.*, Toxicity and decontamination strategies of Congo red dye. *Groundwater Sustain. Dev.*, 19, 100844, 2022.
193. Clark, M., *Handbook of textile and industrial dyeing*, Elsevier Woodhead Publishing, Cambridge, United Kingdon, 2011.
194. Yakupova, E.I., *et al.*, Congo Red and amyloids: history and relationship. *Biosci. Rep.*, 39, 1, 2019.
195. Siddiqui, S.I., *et al.*, Investigation of Congo Red Toxicity towards Different Living Organisms: A Review. *Processes*, 11, 3, 2023.
196. Mansour, D., *et al.*, Removal of Congo red dye by electrochemical advanced oxidation process: optimization, degradation pathways, and mineralization. *Sustainable Water Resour. Manage.*, 10, 1, 41, 2024.
197. Hernández-Zamora, M. and Martínez-Jerónimo, F., Congo red dye diversely affects organisms of different trophic levels: a comparative study with microalgae, cladocerans, and zebrafish embryos. *Environ. Sci. Pollut. Res. Int.*, 26, 12, 11743–11755, 2019.

198. Gadekar, M.R. and Ahammed, M.M., Coagulation/flocculation process for dye removal using water treatment residuals: modelling through artificial neural networks. *Desalin. Water Treat.*, 57, 55, 26392–26400, 2016.
199. Tripathi, M., *et al.*, Microbial biosorbent for remediation of dyes and heavy metals pollution: A green strategy for sustainable environment. *Front. Microbiol.*, 14, 1168954, 2023.
200. Ledakowicz, S. and Paździor, K., Recent Achievements in Dyes Removal Focused on Advanced Oxidation Processes Integrated with Biological Methods. *Molecules*, 26, 4, 870, 2021.
201. Adeniyi, A., *et al.*, Trends in Biological Ammonia Production. *BioTech*, 12, 2, 41, 2023.
202. Padappayil, R.P. and Borger, J., Ammonia Toxicity, in: *StatPearls*, StatPearls Publishing, Treasure Island FL, 2024, Copyright © 2024, StatPearls Publishing LLC.
203. Kater, B.J., Dubbeldam, M., Postma, J.F., Ammonium toxicity at high pH in a marine bioassay using Corophium volutator. *Arch. Environ. Contam. Toxicol.*, 51, 3, 347–51, 2006.
204. Soler, P., *et al.*, Improving water quality does not guarantee fish health: Effects of ammonia pollution on the behaviour of wild-caught pre-exposed fish. *PLoS One*, 16, 8, e0243404, 2021.
205. Xiao, C., *et al.*, The alleviation of ammonium toxicity in plants. *J. Integr. Plant Biol.*, 65, 6, 1362–1368, 2023.
206. Esteban, R., *et al.*, Review: Mechanisms of ammonium toxicity and the quest for tolerance. *Plant Science Int. J. Exp. Plant Biol.*, 248, 92–101, 2016.
207. Zhang, Y., *et al.*, Treatment of ammonianitrogen wastewater by the ultrasonic strengthened break point chlorination method. *J. Water Process Eng.*, 45, 1, 2022.
208. Mavhungu, A., *et al.*, Advocating circular economy in wastewater treatment: Struvite formation and drinking water reclamation from real municipal effluents. *J. Environ. Chem. Eng.*, 8, 4, 103957, 2020.
209. Ding, Y. and Sartaj, M., Optimization of ammonia removal by ion-exchange resin using response surface methodology. *Int. J. Environ. Sci. Technol.*, 13, 4, 985–994, 2016.
210. Otero, M., *et al.*, Ammonium Removal in Wastewater Treatments by Adsorbent Geopolymer Material with Granite Wastes: Full-Scale Validation. *Clean Technol.*, 6, 1, 339–364, 2024.
211. Singh, S., *et al.*, Nitrates in the environment: A critical review of their distribution, sensing techniques, ecological effects and remediation. *Chemosphere*, 287, 131996, 2022.
212. Ward, M.H., *et al.*, Drinking Water Nitrate and Human Health: An Updated Review. *Int. J. Environ. Res. Public Health*, 15, 7, 2018.
213. Akinnawo, S.O., Eutrophication: Causes, consequences, physical, chemical and biological techniques for mitigation strategies. *Environ. Challenges*, 12, 100733, 2023.

214. Verma, A., *et al.*, Nitrate contamination in groundwater and associated health risk assessment for Indo-Gangetic Plain, India. *Groundw. Sustain. Dev.*, 23, 100978, 2023.
215. Bibi, S., *et al.*, Environmental Impacts of Nitrogen Use in Agriculture, Nitrate Leaching and Mitigation Strategies, 2016.
216. Schoeman, J.J. and Steyn, A., Nitrate removal with reverse osmosis in a rural area in South Africa. *Desalination*, 155, 1, 15–26, 2003.
217. Wagner, C.A., The basics of phosphate metabolism. *Nephrol.Dial. Transplant*, 39, 2, 190–201, 2024.
218. Rezania, S., *et al.*, Recent advances on the removal of phosphorus in aquatic plant-based systems. *Environ. Technol. Innov.*, 24, 101933, 2021.
219. Rubio-Aliaga, I. and Krapf, R., Phosphate intake, hyperphosphatemia, and kidney function. *Pflugers Arch.*, 474, 8, 935–947, 2022.
220. Zhou, C., *et al.*, Hyperphosphatemia and Cardiovascular Disease. *Front. Cell Dev. Biol.*, 9, 644363, 2021.
221. Carpenter, S.R., Eutrophication of aquatic ecosystems: bistability and soil phosphorus. *Proc. Natl. Acad. Sci. U.S.A.*, 102, 29, 10002–5, 2005.
222. Bashir, I., *et al.*, Concerns and Threats of Contamination on Aquatic Ecosystems. *Biorem. Biotechnol.*, 1–26, 2020.
223. Provin, T. and Pitt, J., Phosphorus–Too Much and Plants May Suffer, 2008.
224. Roncal-Herrero, T., *et al.*, Precipitation of Iron and Aluminum Phosphates Directly from Aqueous Solution as a Function of Temperature from 50 to 200 °C. *Cryst. Growth Des.*, 9, 5197–5205, 2009.
225. Bacelo, H., *et al.*, Performance and prospects of different adsorbents for phosphorus uptake and recovery from water. *Chem. Eng. J.*, 381, 122566, 2020.
226. Wu, X., Li, R., Lin, J., Contrasting effects of MgAl- and MgFe-based layered double hydroxides on phosphorus mobilization and microbial communities in sediment. *Chemosphere*, 346, 140643, 2024.
227. Dorofeev, A., *et al.*, Role of Phosphate-Accumulating Bacteria in Biological Phosphorus Removal from Wastewater. *Appl. Biochem. Microbiol.*, 56, 1–14, 2020.
228. Han, C., *et al.*, Removal of phosphate using calcium and magnesium-modified iron-based adsorbents. *Mater. Chem. Phys.*, 198, 115–124, 2017.
229. Koh, K.Y., Zhang, S., Chen, J.P., Improvement of Ultrafiltration for Treatment of Phosphorus-Containing Water by a Lanthanum-Modified Aminated Polyacrylonitrile Membrane. *ACS Omega*, 5, 13, 7170–7181, 2020.
230. Jiang, S., Li, Y., Ladewig, B.P., A review of reverse osmosis membrane fouling and control strategies. *Sci. Total Environ.*, 595, 567–583, 2017.
231. Leustek, T., Sulfate Metabolism, in: *The Arabidopsis Book*, vol. 1, p. e0017, 2002.
232. Kabil, O., Vitvitsky, V., Banerjee, R., Sulfur as a signaling nutrient through hydrogen sulfide. *Annu. Rev. Nutr.*, 34, 171–205, 2014.

233. Chandini, *et al.*, The Impact of Chemical Fertilizers on our Environment and Ecosystem, in: *Research Trends in Environmental Sciences*, 2019.
234. Bashir, M.T., Ali, S., Bashir, A., Health Effects from Exposure to Sulphates and Chlorides in Drinking Water, 2016.
235. Zak, D.H., *et al.*, Sulphate in freshwater ecosystems: A review of sources, biogeochemical cycles, ecotoxicological effects and bioremediation. *Earth Sci. Rev.*, 212, 103446, 2021.
236. Sharma, I. and Birman, S., Biodiversity Loss, Ecosystem Services, and Their Role, in: *Promoting Sustainable Health, in The Climate-Health-Sustainability Nexus: Understanding the Interconnected Impact on Populations and the Environment*, P. Singh and N. Yadav (Eds.), pp. 163–188, Springer Nature Switzerland, Cham, 2024.
237. Geurts, J.J., *et al.*, Interacting effects of sulphate pollution, sulphide toxicity and eutrophication on vegetation development in fens: a mesocosm experiment. *Environ. Pollut.*, 157, 7, 2072–81, 2009.
238. Fang, P., *et al.*, Removal of High-Concentration Sulfate Ions from the Sodium Alkali FGD Wastewater Using Ettringite Precipitation Method: Factor Assessment, Feasibility, and Prospect. *J. Chem.*, 2018, 1, 1265168, 2018.
239. Öztürk, Y. and Ekmekçi, Z., Removal of sulfate ions from process water by ion exchange resins. *Miner. Eng.*, 159, 106613, 2020.
240. Sawadogo, B., *et al.*, Removal of Sulphate Ions from Borehole Water Using Nanofiltration and Reverse Osmosis. *Water*, 14, 21, 3422, 2022.
241. Pourhakkak, P., *et al.*, Chapter 1 - Fundamentals of adsorption technology, in: *Interface Science and Technology*, M. Ghaedi (Ed.), pp. 1–70, Elsevier, Netherlands, 2021.
242. Tsuchiya, T., Van der Waals Force, in: *Encyclopedia of Geochemistry: A Comprehensive Reference Source on the Chemistry of the Earth*, W.M. White (Ed.), pp. 1473–1474, Springer International Publishing, Cham, 2018.
243. Liu, D., *et al.*, Pore structure characterization and its significance for gas adsorption in coals: A comprehensive review. *Unconv. Resour.*, 2, 139–157, 2022.
244. Bonilla-Petriciolet, A., Mendoza-Castillo, D., Reynel, H., Adsorption Processes for Water Treatment and Purification, 2017.
245. Badran, A.M., *et al.*, Advancements in Adsorption Techniques for Sustainable Water Purification: A Focus on Lead Removal. *Separations*, 10, 11, 565, 2023.
246. Alaqarbeh, M., Adsorption Phenomena: Definition, Mechanisms, and Adsorption Types: Short Review, vol. 13, pp. 43–51, 2021.
247. Tao, J. and Rappe, A., Physical Adsorption: Theory of van der Waals Interactions between Particles and Clean Surfaces. *Phys. Rev. Lett.*, 112, 106101, 2014.
248. Xing, B. and Pignatello, J.J., SORPTION | Organic Chemicals, in: *Encyclopedia of Soils in the Environment*, pp. 537–548, Elsevier, Netherlands, 2005.
249. Petrovic, B., Gorbounov, M., Masoudi Soltani, S., Impact of Surface Functional Groups and Their Introduction Methods on the Mechanisms of

CO2 Adsorption on Porous Carbonaceous Adsorbents. *Carbon Capture Sci. Technol.*, 3, 100045, 2022.
250. Wang, J. and Guo, X., Adsorption kinetic models: Physical meanings, applications, and solving methods. *J. Hazard. Mater.*, 390, 122156, 2020.
251. Al-Harby, N.F., Albahly, E.F., Mohamed, N.A., Kinetics, Isotherm and Thermodynamic Studies for Efficient Adsorption of Congo Red Dye from Aqueous Solution onto Novel Cyanoguanidine-Modified Chitosan Adsorbent. *Polym. (Basel)*, 13, 24, 2021.
252. Waswa, G.A., Adsorption Of Selected Pesticide Residues By Selected Soils From Lake Naivasha Region In Kenya: Adsorption/desorption Properties And Related Thermodynamic Data, 2013.
253. Muttil, N., *et al.*, Production, Types, and Applications of Activated Carbon Derived from Waste Tyres: An Overview. *Appl. Sci.*, 13, 1, 257, 2022.
254. Pérez-Botella, E., Valencia, S., Rey, F., Zeolites in Adsorption Processes: State of the Art and Future Prospects. *Chem. Rev.*, 122, 24, 17647–17695, 2022.
255. Brahmi, D., *et al.*, Preparation of amorphous silica gel from Algerian siliceous by-product of kaolin and its physico chemical properties. *Ceram. Int.*, 40, 10499–10503, 2014.
256. Yusuf, V.F., Malek, N.I., Kailasa, S.K., Review on Metal-Organic Framework Classification, Synthetic Approaches, and Influencing Factors: Applications in Energy, Drug Delivery, and Wastewater Treatment. *ACS Omega*, 7, 49, 44507–44531, 2022.
257. Hao, L., *et al.*, A critical review on arsenic removal from water using iron-based adsorbents. *RSC Adv.*, 8, 69, 39545–39560, 2018.
258. Aragaw, T.A., Bogale, F.M., Aragaw, B.A., Iron-based nanoparticles in wastewater treatment: A review on synthesis methods, applications, and removal mechanisms. *J. Saudi Chem. Soc.*, 25, 8, 101280, 2021.
259. Keshta, B.E., *et al.*, State of the art on the magnetic iron oxide Nanoparticles: Synthesis, Functionalization, and applications in wastewater treatment. *Results Chem.*, 7, 101388, 2024.
260. Bauer, J., *et al.*, Zeolite/aluminum composite adsorbents for application in adsorption refrigeration. *Int. J. Energy Res.*, 33, 13, 1233–1249, 2009.
261. Li, X., *et al.*, Novel Impregnation–Deposition Method to Synthesize a Presulfided MoS2/Al2O3 Catalyst and Its Application in Hydrodesulfurization. *ACS Omega*, 8, 2, 2596–2606, 2023.
262. Gupta, A., *et al.*, A Review of Adsorbents for Heavy Metal Decontamination: Growing Approach to Wastewater Treatment. *Materials*, 14, 16, 4702, 2021.
263. Usman, M.O., *et al.*, A review of adsorption techniques for removal of phosphates from wastewater. *Water Sci. Technol.*, 86, 12, 3113–3132, 2022.
264. He, J., *et al.*, Review of fluoride removal from water environment by adsorption. *J. Environ. Chem. Eng.*, 8, 6, 104516, 2020.
265. Saleh, T.A., Chapter 2 - Adsorption technology and surface science, in: *Interface Science and Technology*, T.A. Saleh (Ed.), pp. 39–64, Elsevier, Netherlands, 2022.

266. Chang, E.E., *et al.*, Adsorption of Selected Pharmaceutical Compounds onto Activated Carbon in Dilute Aqueous Solutions Exemplified by Acetaminophen, Diclofenac, and Sulfamethoxazole. *Sci. World J.*, 2015, 186501, 2015.
267. Sayed, M., Removing Cadmium and lead From Wastewater Using Natural Zeolite Isotherm models, pp. 98–104, 2013.
268. Ebrahimzadeh, S., *et al.*, Removal of organic micropollutants in a drinking water treatment plant by powdered activated carbon followed by rapid sand filtration. *J. Water Process Eng.*, 47, 5, 2022.
269. Arora, R., Adsorption of Heavy Metals–A Review. *Mater. Today Proc.*, 18, 4745–4750, 2019.
270. Ahmed, M., *et al.*, Recent developments in hazardous pollutants removal from wastewater and water reuse within a circular economy. *npj Clean Water*, 5, 1, 12, 2022.
271. Tadda, M.A., *et al.*, A review on activated carbon: process, application and prospects. *J. Adv. Res. Civ. Eng. Prac. Res.*, 2, 7–13, 2016.
272. Sorrels, J.L., *Carbon Adsorbers*, vol. 2018, Air Economics Group. Health and Environmental Impacts Division. Office of Air Quality Planning and Standards. U.S. Environmental Protection Agency. Research Triangle Park, NC, 2771.
273. Rathi, B.S. and Kumar, P.S., Application of adsorption process for effective removal of emerging contaminants from water and wastewater. *Environ. Pollut.*, 280, 116995, 2021.
274. Benettayeb, A., *et al.*, Chitosan Nanoparticles as Potential Nano-Sorbent for Removal of Toxic Environmental Pollutants. *Nanomaterials*, 13, 3, 447, 2023.

11

Recovery of Na_2CO_3 and $Mg(OH)_2$ from Alkali Earth Metal Sulfates

Conny P. Mokgohloa[1,2], Johannes P. Maree[3*], Malose P. Mokhonoana[2] and Mary P. Motaung[4]

[1]Department of Water and Sanitation, University of Limpopo, Sovenga, South Africa
[2]Department of Chemistry, University of Limpopo, Private Bag, Sovenga, South Africa
[3]Institute for Nanotechnology and Water Sustainability (iNanoWS), College of Science, Engineering and Technology, University of South Africa, Private Bag, Florida Science Campus, Johannesburg, South Africa
[4]Department of Chemical, Metallurgical and Materials Engineering, Tshwane University of Technology, Pretoria, South Africa

Abstract

Acid mine drainage is one of mining's most serious threats to water scarcity. It is highly acidic and is a rich source of sulfate and heavy metals. High-sulfate content has a negative impact on humans. The presence of sulfate in acid mine water also contributes to corrosion and gypsum scale formation, as well as corrosion of equipment. However, valuable products such as sodium carbonate (Na_2CO_3) can be recovered from the sulfates, namely, calcium sulfate ($CaSO_4$) and barium sulfate ($BaSO_4$). Na_2CO_3 is used for mine water treatment and glass manufacturing with a current price of R4,500/t (US$345/t). In the reverse osmosis–cooling process, Na_2CO_3 is used in the pre-treatment stage for the formation of pigment and precipitation of other metals. Calcium carbonate ($CaCO_3$) is used to remove acidity and achieve only partial sulfate and metal removal. Magnesium hydroxide ($Mg(OH)_2$) is used to filter out suspended solids and to precipitate dissolved heavy metals. The Mintek Pyrosim software model was used to predict the products formed under the conditions that were studied experimentally. Pyrosim studies showed that $CaSO_4$ can be converted to CaS, and $BaSO_4$ can be converted to BaS through reduction with coal at 1,000°C. This correlates with the muffle furnace studies that showed that $CaSO_4$ and $BaSO_4$ can be converted to calcium sulfide (CaS) and barium sulfide (BaS), respectively. CaS was used to

**Corresponding author*: maree.jannie@gmail.com

Elvis Fosso-Kankeu, Vhahangwele Masindi, Johannes Maree and Bhekie Mamba (eds.) Customized Technologies for Sustainable Management of Industrial Wastewater: A Circular Economy Approach, (467–536) © 2025 Scrivener Publishing LLC

produce $CaCO_3$. The new process developed by Van Vuuren and Maree was used for the indirect processing of sodium sulfate (Na_2SO_4) to sodium sulfide (Na_2S), where Na_2SO_4 was reacted with BaS. This was further used as a reactant to produce sodium carbonate (Na_2CO_3) and $Mg(OH)_2$. The reduction of alkalinity from 136,500 mg/L (as Na_2S) at a CO_2/Na_2S mole ratio of 1 to 113,100 (as Na_2S) at a CO_2/Na_2S mole ratio of 4 confirmed the precipitation of $NaHCO_3$. Na_2CO_3 can be recovered by heating sodium bicarbonate ($NaHCO_3$). The OLI (OLI Systems, Inc.) simulations and beaker studies revealed that $Mg(OH)_2$ formed when magnesium sulfate ($MgSO_4$) was contacted with Na_2S. The Fe^{2+} can be back oxidized to Fe^{3+} through chemical or biological oxidation. H_2S can be converted to S by contacting it with Fe^{3+}.

Keywords: Thermal treatment, gypsum, sodium sulfate, magnesium sulfate, barium sulfate, calcium carbonate, sodium carbonate, magnesium hydroxide

Remark

This chapter is largely based on the Ph.D. thesis of Ms. Conny Mokgohloa at the Faculty of Science and Agriculture in the Department of Chemistry, University of Limpopo. A summarized version of the thesis was also published in the following Proceedings:

Conny P Mokgohloa, Johannes P Maree, M P Motaung, Malose P Mokhonoana (2022) *Processing of Na_2S to Na_2CO_3 and NaHS*, International Institute of Biological Chemical and Environmental Engineering (IICBEE), Johannesburg, South Africa, Nov. 28-29, Editors: S.K. Ntwampe, E Fosso, B. Topcuoğlu, ISBN: 978-989-9121-14-0, 30-40.

11.1 Introduction

Acid mine drainage (AMD) is the most significant environmental pollution problem associated with the mining industry. The high concentration of toxic metals and the high levels of acidity in AMD are harmful to the vegetation, aquatic life, terrestrial wildlife, and human beings. Pyrite (FeS_2) is a major constituent of the strata being mined, and large rock surfaces become exposed to air and water during mining activities [1]. Pyrite is oxidized to soluble iron complexes and sulfuric acid, catalyzed by sulfur-oxidizing bacteria [2], according to the reaction shown in **Equation 11.1.**

$$2FeS_{2(s)} + 7O_{2(aq)} + 2H_2O \rightarrow 2Fe^{2+} + 4SO_{4(aq)}^{2-} + 4H_{(aq)}^{+} \quad (11.1)$$

Because of its detrimental effects on the environment and public health, it is mandatory to treat AMD with the aim to achieve zero waste. AMD can be treated by two methods: active treatment and passive treatment. Active treatment includes the use of chemical agents such as limestone ($CaCO_3$), hydrated lime ($Ca(OH)_2$), caustic soda (NaOH), soda ash (Na_2CO_3), calcium oxide (CaO), anhydrous ammonia (NH_3), magnesium hydroxide ($Mg(OH)_2$), and magnesium oxide (MgO) [3]. Passive treatment can be further classified into two types: conventional and emerging. Examples of conventional passive treatment are constructed wetlands and anaerobic sulfate-reducing bioreactors, whereas an example of an emerging passive treatment is phytoremediation [3].

The reverse osmosis–cooling (ROC) process can be used for the treatment of AMD through neutralization with Na_2CO_3 for the removal of metals, desalination with reverse osmosis (RO), freeze crystallization for the recovery of Na_2SO_4 from the RO brine, and processing of Na_2SO_4 to Na_2S. In the ROC process, Na_2CO_3 is used for selective recovery of metals through a stepwise increase of the pH, resulting in the formation of Na_2SO_4, which can be recovered from the brine after the RO and freeze-crystallization stages. It would be beneficial if the raw material, Na_2CO_3, can be recovered from the product, Na_2SO_4. Because Na_2CO_3 has a high price of 263.3 US$/t, it would be economically beneficial if it can be processed from the product, Na_2SO_4.

Studies by Mashigwana (2022) on the direct conversion of Na_2SO_4 into Na_2CO_3 under thermal conditions found that it was not possible, as Na_2SO_4 melted in the same temperature range where the reduction took place, i.e., 860°C to 920°C [4]. Van Vuuren and Maree proposed that Na_2CO_3 can be produced indirectly from CaS or BaS [5]. Various researchers have studied the reduction of $CaSO_4$ and $BaSO_4$ in depth without experiencing the melting problem of one of the reactants [6–9]. Recently, Mokgohloa *et al.* studied the reduction of $CaSO_4$ and found that Na_2CO_3 can be produced from Na_2SO_4 *via* CaS [10]. In the indirect method, $CaSO_4$ precipitates, due to its relatively low solubility, whereas Na_2S or NaHS remains in solution and is converted into Na_2CO_3 or $NaHCO_3$ by contacting them with CO_2. A limitation of the CaS route is, however, the partial solubility of $CaSO_4{\cdot}2H_2O$ (2,440 mg/L). The Na_2S formed is, therefore, contaminated with Ca^{2+}.

The purpose of this investigation was to:

i. Evaluate BaS as an alternative to CaS for the processing of Na_2S from Na_2SO_4. BaS may offer the following benefits: Na_2S will be pure due to the low solubility of $BaSO_4$ (2.44 mg/L) compared to the 2,640 mg/L in the case of $CaSO_4{\cdot}2H_2O$. More energy is also needed for the conversion of $BaSO_4$ into BaS (226.12 kJ/mol) than for the conversion of $CaSO_4$ into CaS (155.8 kJ/mol).
ii. Summarize the process steps needed for the processing of Na_2SO_4 to its bicarbonates, sulfides, and carbonates as published by Mokgohloa *et al.* [11].
iii. Investigate the process steps needed for the processing of $MgSO_4$ to its bicarbonates, sulfides, carbonates, and hydroxides.

The objectives of this investigation are as follows:

1. Thermal conversion of $BaSO_4$ and $CaSO_4$ to BaS and CaS, respectively.
2. Conversion of Na_2SO_4 to Na_2S using BaS.
3. Processing of Na_2S into $NaHCO_3$ and NaHS.
4. Processing of $NaHCO_3$ into Na_2CO_3.
5. Processing of $MgSO_4$ to $Mg(OH)_2$, Na_2SO_4, and NaHS.

11.2 Literature

This chapter deals with processing of solids, liquids, and gases as well as the reactions that take place from a temperature of 0°C to 1,000°C. This section, therefore, covers the thermodynamic and kinetic principles of various reactions. It further describes the freeze crystallization by the ROC process and the types of freeze crystallization.

11.2.1 Solids, Liquids, and Gases

Solids are one of the three basic states of matter, the others being liquids and gases. Because the solid form of a liquid or gas has relatively ordered atoms, the energy of the atoms decrease as they adopt a three-dimensional structure [12]. Solids have certain properties that distinguish them from liquids and gases. For example, all solids can withstand a force acting normally or parallel to a surface (perpendicular or shear load, respectively). These properties depend on the characteristics of the atoms that form the solid, and the arrangement of the atoms as well as the forces between the atoms. Solids are generally divided into three major classes: crystalline, non-crystalline (amorphous), and semi-crystalline [13].

The liquid state is, sometimes, described as a state between the solid and gaseous states, and, for simple molecules, the distinction is unclear. However, the clear distinction between liquid, gaseous, and solid states apply only to substances whose molecules are composed of a small number of atoms [14]. In a liquid, the attraction between the particles keeps the volume of the liquid constant. Due to the movement of particles, liquids take different forms. The liquid flows and fills the bottom of the container and takes on the shape of the container, but the volume does not change. Liquids have a very limited compressibility because the space between the particles is limited.

A gas is a state of matter that does not have a definite shape or volume. Gases have their own behavior depending on various variables such as temperature, pressure, and volume [15]. While each gas is different, all gases act in a similar manner. The particles that make up a gas can range from individual atoms to complex molecules.

11.2.2 Kinetics

Chemical kinetics is the branch of physical chemistry that deals with the understanding of the speed at which chemical reactions take place. Chemical kinetics has a significant impact, as it encompasses many aspects of cosmology, geology, biology, engineering, and even psychology [16]. The principles of chemical kinetics apply to purely physical processes as well as to chemical reactions.

The kinetic behavior of a general chemical reaction is usually studied to determine the effect of certain external factors, such as reactant concentrations, temperature, and pressure, on the reaction rate [17]. In the case of a reaction in which two substances A and B react with each other, the reaction rate is, sometimes, proportional to their concentrations of A, denoted by [A], and [B]. In that case, the reaction is said to be a second-order reaction; it is first order in both [A] and [B]. In such a case, the reaction rate (v) can be expressed as shown in **Equation 11.2.**

$$v = k[A][B] \tag{11.2}$$

where k is a constant, known as the rate constant for the reaction.

The transformation of substance A into another substance may be the subject of an equation of motion of the form v = k [A], which is a first-order reaction. The kinetics of a reaction simply does not correspond to the coefficients in the balanced chemical equation of the reaction. Therefore, the reaction shown in **Equation 11.3** is not necessarily second order in both directions.

$$A + B \rightleftharpoons Y + Z \tag{11.3}$$

11.2.3 Thermodynamics

Thermodynamics is the science of studying the relationship between heat, work, temperature, and energy. In broad terms, thermodynamics deals with the transfer of energy from one place to another, and transformation from one form to another. The laws of thermodynamics are as follows:

- Zeroth law of thermodynamics: When two systems are each in thermal equilibrium with a third system, the first two systems are in thermal equilibrium with each other. If a = b and b = c, then a = c.
- The first law of thermodynamics or the law of conservation of energy: The change of the internal energy of a system is equal to the sum of the heat added to the system by the environment and the work done to the environment of the system. **Equation 11.4** describes the first law of thermodynamics in terms of internal energy, work, and heat transfer mechanisms.

$$\Delta U = q + W \tag{11.4}$$

where ΔU is the change in internal energy of the system, q is the algebraic sum of heat transfer between system and surroundings, and W is the work interaction of the system with its surroundings.

- The second law of thermodynamics: Heat does not flow spontaneously from a colder area to a warmer area, and heat at a given temperature is not completely converted into work. As a result, the entropy or thermal energy per unit temperature of a closed system increases to its maximum value over time. Therefore, any closed system tends to be in balance with maximum entropy and without energy to do useful work as shown in **Equation 11.5**.

$$\Delta S_{uni} = \Delta S_{sys} + \Delta S_{surr} \geq 0 \tag{11.5}$$

where ΔS_{uni} is the entropy of the universe, ΔS_{sys} is the entropy of the system, and ΔS_{surr} is the entropy of the surrounding.

- The third law of thermodynamics: As the temperature approaches absolute zero, the entropy of a perfect crystal of an element in its most stable form tends to zero. It defines the absolute measure of entropy that statistically defines the degree of randomness or disorder in a system as described in **Equation 11.6.**

$$S = k \log \Omega \tag{11.6}$$

where S = entropy, k = Boltzmann constant, and = number of microstates.

11.2.3.1 Gibbs Free Energy

The Gibbs free energy is a quantity used to measure the maximum amount of work performed in a thermodynamic system when temperature and pressure are kept constant. The Gibbs free energy of a system at any moment in time is defined as the enthalpy of the system minus the product of the temperature times the entropy of the system as shown in **Equation 11.7.**

$$G = H - TS \tag{11.7}$$

Thus, the change in the Gibbs free energy of a system that takes place during a reaction is equal to the change in the enthalpy of the system minus the change in the product of the temperature multiplied by the entropy of the system as shown in **Equation 11.8.**

$$\Delta G = \Delta H - \Delta(TS) \tag{11.8}$$

If the reaction is run at constant temperature, then **Equation 11.8** can be written as shown in **Equation 11.9.**

$$\Delta G = \Delta H - T\Delta S \tag{11.9}$$

The change in free energy of a system that occurs during a reaction can be measured in all cases. When data is collected under standard-state conditions, the result is the reaction energy under standard state ΔG^o as shown in **Equation 11.10.**

$$\Delta G^o = \Delta H^o - T\Delta S^o \tag{11.10}$$

11.2.3.2 Enthalpy and Heat of Reaction

Heat of reaction, or enthalpy of reaction, is an important parameter for the safe and successful scaling of chemical processes. Heat of reaction is the energy released or absorbed when a chemical substance is transformed in a chemical reaction. It describes the change in energy content when reactants are transformed into products. A reaction can be exothermic (heat-releasing) or endothermic (heat-absorbing) [18]. The heat of reaction or reaction enthalpy is typically expressed as molar enthalpy in kJ/mol, or as specific enthalpy in kJ/kg or kJ/L. The enthalpy can be calculated using **Equation 11.11.**

$$H = U + PV \tag{11.11}$$

where U is internal energy, P is pressure and V is volume. In the case of a change in enthalpy (ΔH), **Equation 11.11** changes to **Equation 11.12.**

$$\Delta H = \Delta U + P\Delta V + V\Delta P \tag{11.12}$$

In practice, the pressure is held constant, and **Equation 11.12** changes to **Equation 11.13.**

$$\Delta H = \Delta U + P\Delta V \tag{11.13}$$

However, for a constant pressure system, the change in enthalpy is simply the heat (q) transferred as shown in **Equation 11.14**.

$$\Delta H = q \tag{11.14}$$

11.2.4 Freeze Crystallization by the ROC Process

Freeze crystallization can be used to separate compounds with different solubilities by removing water from the solution through freezing of ice. This method was used in this chapter to separate Na_2SO_4, which has a lower solubility at 0°C from $NaHCO_3$, which has a high solubility. This section provides background information on freeze crystallization.

11.2.4.1 Introduction

The ROC process was developed for the treatment of brines from desalination processes, such as RO [19–21]. In the ROC process, the brine is treated with chemicals such as Na_2CO_3 and/or NaOH in the pre-treatment stage to allow selective precipitation of metals [$CaCO_3$, MnO_2, and $Mg(OH)_2$]. After pre-treatment, the Na-rich and Mg-rich water is passed through a membrane stage to produce drinking water and brine. The brine has a Na_2SO_4 concentration, which is high enough to allow Na_2SO_4 crystallizations upon cooling/freeze desalination.

11.2.4.2 Types of Freeze Crystallization

Freezing crystallization (FC) is used to separate fresh water from a salt solution in the form of ice crystals and then thawed to purify the water. The process was discovered by Thomas Bartolinus (1680), who said that fresh water can be obtained by melting ice from sea water [22]. Its first commercial market was in the 1950s [23]. Karnofsky and Steinhoff and Weigandt *et al.* proposed the first direct freezing process using butane as a coolant for seawater desalination [24, 25]. The freezing method uses the transition of water from liquid to solid. In addition to crystallization, this involves separating ice crystals from brine, washing them to remove salt from the crystal surface, and thawing the ice to obtain the final fresh water [26]. The literature mentions different freezing-crystallization methods through four main methods: direct, indirect, vacuum, and eutectic freezing-crystallization [22]. Recent advances in freezing crystallization have introduced two new methods of ice crystallization that are subject to indirect contact crystallization: suspension crystallization and surface crystallization at low temperature [22, 27–29]. According to the literatures [30, 31], the energy cost of the freezing desalination process is theoretically equal to the cost of the RO membrane unit. However, the investment and operating costs of the freezing technology are high due to the significant reduction in biological contamination using lower temperatures than the installation of RO membranes.

11.2.4.2.1 Direct Contact Freeze Crystallization

Direct contact freezing uses a refrigerant in direct contact with the freezing solution. This method mainly has an operating temperature of −5°C, which is characterized as a low energy process [22]. The main advantages of this method are the efficient design with high production speed, small size, and low driving force. On the other hand, freezing crystallization requires a specific coolant. The refrigerant must have a normal boiling point of −4°C or less, water-resistant, non-toxic, non-flammable, chemically stable, inexpensive, and commercially available [22]. In addition to all these requirements, it is important to mix the refrigerant solution correctly to obtain ice crystals with minimal impurities.

11.2.4.2.2 Indirect Contact Freeze Crystallization

In indirect contact freeze crystallization, the refrigerant is used without direct contact with the crystallization solution by indirect contact with the frozen crystal, and the energy of the refrigerant passes through the wall of the heat exchanger [32]. Generally, indirect freezing can be classified into two main classes: (i) suspension freezing and (ii) freezing on a cold plate [27, 33]. When the suspension crystallizes, many fine ice particles form in two stages from the mother liquor suspension. The crystals first form in the ice core and then smaller crystals are recrystallized by the Ostwald maturation mechanism [33]. However, crystallization in suspension is commonly used in the food concentration industry [32]. This technique has many cons such as cost, complexity, difficult nucleation, and crystal growth control [34, 35]. Alternatively, freezing on a cold surface forms a single crystal layer on a cooled surface.

11.2.4.2.3 Vacuum Freeze Crystallization

Crystallization by vacuum freezing is a specialized freeze-crystallization technique that provides a cooling effect by using a high vacuum to evaporate a little water and then lowering the temperature of the solution to produce ice crystallization [22]. Crystallization by vacuum-freezing is especially used in the food, chemical, and pharmaceutical industries to separate and purify solids from liquids. Ice crystals are created in a solution under low temperature and pressure conditions in this approach. The solution is first frozen to form a solid mass in the process [36]. The frozen solid mass is then placed in a vacuum chamber under reduced temperature and pressure, hence causing the ice to change from solid to vapor without melting. A concentrated solution containing the needed solid is left behind as the sublimation process occurs. This process provides several significant advantages, including the preservation of crystal structure and shape, as well as the production of high-purity solids. However, some of the constraints of this technique include its slow nature and high cost required at its development [31].

11.2.4.2.4 Eutectic Freeze Crystallization

Eutectic freeze crystallization (EFC) dates to the 1950s, when Nelson and Thompson attempted to determine the salt crystallization sequence. However, the use of EFC as a separation technique was introduced by Stepakoff *et al.* in the 1970s. In 1974, EFC techniques was improved by adding coolant directly to the solution [37]. EFC is a new water purification technology that allows the recovery of ice and salty products from a concentrated brine stream. There are several benefits to using EFC as part of overall water treatment process. Crystallization is a physical process of separation, so no chemicals are required [38]. Freeze crystallization is thermodynamically more efficient than evaporative crystallization, as the heat of fusion of water is six times lower than the heat of evaporation of water [39]. The principle of the EFC process is that, when a saturated solution cools slowly below the solidification line, ice crystals begin to form and rise to the surface, and the salt concentrates in the rest of the solution and finally crystallizes at the eutectic temperature [40]. Under the conditions of the eutectic process, ice and salt crystallize simultaneously and can be separated depending on the difference in density [41]. Crystallization of both products occurs in pure form, eliminating the need for a secondary separation process [42]. A significant part of the energy spent to achieve lower operating temperatures can be stored and recovered in the ice product. Lower operating temperatures further reduce the corrosion potential of mechanical equipment, allowing the use of cheaper building materials [43]. Disadvantages of EFC include high capital costs [32] and the formation of an insulating ice scale layer on cooled crystallizer surfaces [44]. Reddy *et al.* (2010) reported that EFC can be utilized to treat a RO retentate stream containing 4 wt.% sodium sulfate and several impurities (F^-, Cl^-, K^+, Li^+, Mg^{2+}, Ca^{2+}, NO_3^-, and NH_3^+), producing pure water and sodium sulfate decahydrate [45]. Theoretically, high performance (100% binary system) and purity can be achieved in EFC, and this method results in the recovery of salt with residual solubility at final operating temperatures over crystallization limited to cooling [39, 46]. In addition, the growth process that produces high-purity crystals can be closely monitored to avoid the formation of solid solutions and impurities [39]. Cheaper materials of construction are required because of minimal corrosion at low temperatures characteristic of EFC processes [37, 47]. For aqueous solutions with very low process temperatures, EFCs can be economically inefficient due to the large surface area required for heat transfer. A major limitation of EFC is the contamination of the surface of the heat exchanger due to the formation of an ice sheet, which significantly reduces the heat transfer coefficient and, consequently, reduces the heat flow, thus reducing the crystallization rate. Layers of ice can also damage the mechanical equipment of the crystallizer. The cost of EFC equipment is still high, but it is expected to decrease over time as better projects develop.

11.3 Materials and Methods

11.3.1 Feedstock, Chemicals, and Reagents

High-purity reagents were used in this work for the preparation of samples. Deionized water was used for all solutions and dilutions. The following chemicals were used: $BaSO_4$, Na_2SO_4, (Protea Chemicals), $CaSO_4 \cdot 2H_2O$ (Foskor), activated carbon (MCL), CO_2 gas (Air Liquide, South Africa), 1 N HCl, 0.1 N I_2, 0.1 N $Na_2S_3O_4$, 1M NaOH, and starch solution.

11.3.2 Equipment

A muffle furnace was used for reduction of $CaSO_4$ to CaS and $BaSO_4$ to BaS. For reaction studies, 500-mL beakers, *BirCraft* agitator, pH and conductivity meter, and CO_2 gas supplied at a pressure of 50 kPa were used.

11.3.3 Experimental and Procedure

11.3.3.1 Thermal Treatment

Thermal studies were carried out by reacting mixtures of $CaSO_4/BaSO_4$ and coal at elevated temperatures (1,000°C to 1,110°C) in the muffle furnace to produce CaS. The weighed, dry $CaSO_4/BaSO_4$ (20 g) and powdered coal (5 g) were mixed thoroughly, placed in ceramic crucibles for all the reduction experimental trials. Solid samples were collected and analyzed for mass loss, sulfide content, and alkalinity. The retention times for all the reduction experiments were 30–60 min.

Mokgohloa *et al.* studied Na_2S formation and Na_2S processing as described below [11]:

11.3.3.2 Na_2S Formation

OLI simulations (OLI ESP software program from OLI Systems, Inc.) and beaker studies were used by Mokgohloa *et al.* to determine which compound will precipitate out and which compounds/elements will stay in solution [11].

- Na_2S was produced by reacting Na_2SO_4 with BaS. BaS was produced from $BaSO_4$. A BaS solution (10.00 g of BaS in 100 mL of H_2O) was mixed with a Na_2SO_4 solution (8.40 g of Na_2SO_4 in 100 mL).
- The formation of Na_2S by contacting a CaS solution with Na_2SO_4 was studied though OLI simulations.

11.3.3.3 Na_2S Processing

Beaker studies were carried out by Mokgohloa *et al.* to convert Na_2S into $NaHCO_3$ and NaHS by reacting it with CO_2 [11]. The following procedure was followed in the experiment by using the laboratory setup shown in Figure 11.1.

- CO_2 was put in a 300-mL measuring cylinder, with the open end on top of a water bath *via* Line 1.
- Na_2S (0.88 g of Na_2S + 4.38 g of H_2O; 200 g of Na_2S/L of H_2O) was put in a 250-mL separation funnel.

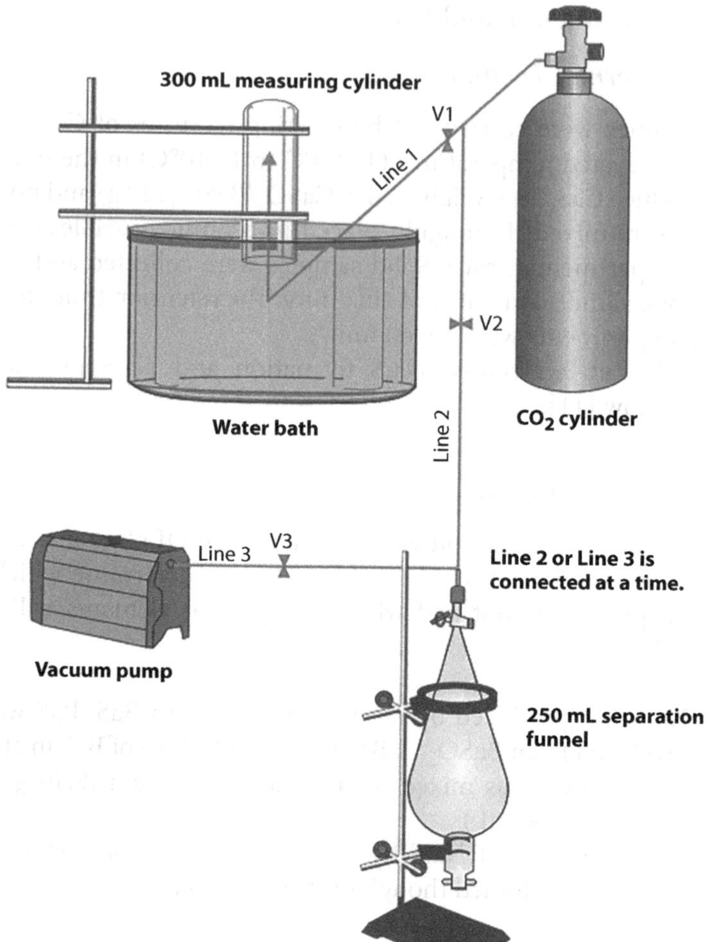

Figure 11.1 Schematic diagram of a laboratory setup used for monitoring the precipitation of aHCO$_3$ when Na$_2$S is reacted with CO_2 [11].

- Air was removed from the separation funnel with a vacuum pump *via* Line 3 and replaced with CO_2 stored in the 300-mL measuring cylinder *via* Line 2.
- The Na_2S and CO_2 in the separation funnel were allowed to react by shaking it for 15 min. CO_2 additions were repeated one, two, and four times to determine the effect of CO_2/Na_2S mole ratio on the completion of the reaction.
- The CO_2 in the measuring cylinder was replaced with air.

- The volume of CO_2 reacted with the Na_2S was determined by allowing the air in the measuring cylinder to flow into the 250-mL separation funnel *via* Line 2, which was partially under vacuum due to the CO_2 reaction with Na_2S. The volume of CO_2 reacted was determined by reading the volume of water that was sucked into the 300-mL measuring cylinder.
- Sulfide and alkalinity analyses were carried out on the filtered liquid samples collected from the separation funnel. Filtered samples were collected by using a 25-mL syringe with a filter tip.

11.3.4 Analysis

Sulfide was determined by standard iodometry [48] as follows: A 50-mg sample of the reduced product was weighed, dissolved, and made up to 100 mL with deionized water. Then, 10 mL (V) of standard iodine solution (C = 0.1 N) was added, followed by 100 mL of deionized water. The resulting mixture was acidified using 2 mL of 1 N HCl solution and subsequently titrated (volume V) with standard (C = 0.1 N) sodium thiosulfate using starch as indicator to a colorless endpoint as shown in **Equation 11.15.**

$$\%NaS = \frac{[(V.C)_{I2} - (V.C)_{S2O3} \times 36]}{Sample\ mass(mg)} \times 100 \tag{11.15}$$

11.3.4.1 Pyrosim Mintek Simulation

The prediction of $CaSO_4/BaSO_4$ was done using the Pyrosim Mintek model. Figure 11.2 shows (a) the thermo page and (b) the Pyrosim page of Pyrosim Mintek model. The Pyrosim Mintek model is a computer simulation program for steady-state simulation of pyrometallurgical and other high-temperature processes. It has a built-in capacity to evaluate multiphase equilibria, with the potential of predicting high-temperature processes more accurately. It also gives proficiency of steady state mass and energy calculations for high-temperature processes [49].

11.3.5 OLI Software Simulations

The OLI System Chemical Analyzer software program was used to predict the behavior of metals dissolved in water when treated with alkalis such as

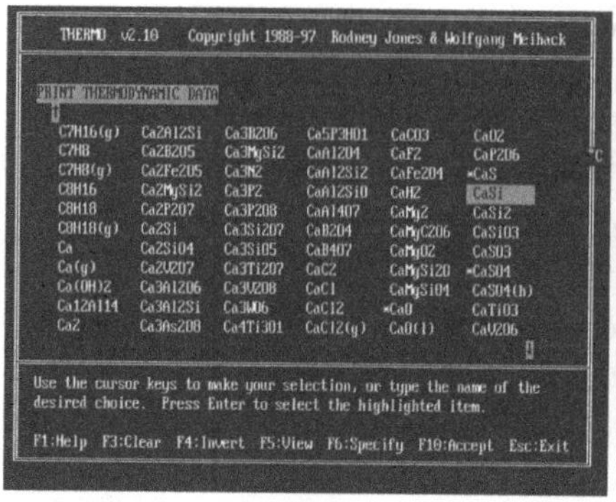

(a)

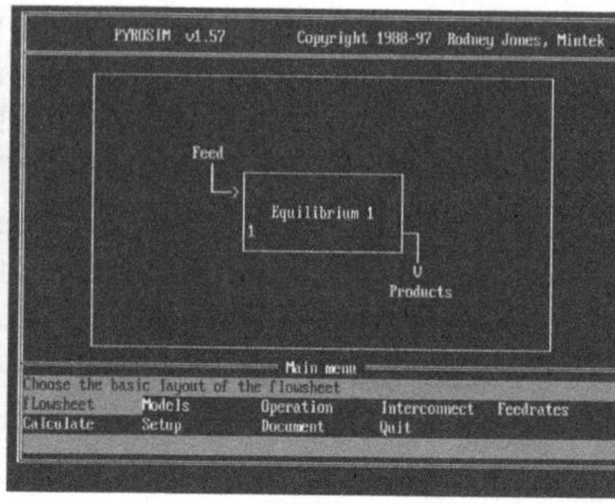

(b)

Figure 11.2 (a) Screen layout during the specification of the chemical species present in the slag phase. (b) Screen layout showing the Pyrosim Mintek operation page.

$MgSO_4$ and Na_2SO_4 [50]. OLI is an aqueous equilibrium chemistry estimator with an interactive and self-instructive interface for clarifying reactions. It has the ability to work with all kinds of common equilibrium reactions; it has a strong solution algorithm, expressive and easily understandable displays of results, and the ability to produce results in multiple formats according to different uses. The OLI System Chemical Analyzer was used to perform single point equilibrium calculations and multiple point survey calculations. The concentration of the various species in solutions or solids was calculated as a function of temperature, pressure, pH, and initial concentrations. The calculations provide liquid- and solid-phase separations for a specialized model.

11.4 Results and Discussion

11.4.1 Thermodynamic Values for $CaSO_4$, $BaSO_4$, and Na_2SO_4

Tables 11.1, 11.2 and 11.3 show the values for heat of formation, standard Gibbs free energy of formation, and standard entropy values for the various reactants and products, respectively.

- The heat of formation, also known as enthalpy of formation, is a measure of the energy released or consumed when one mole of a substance is created from its pure elements under standard conditions. It is given by **Equation 11.16**:

Table 11.1 Standard enthalpy/heat of formation [51].

Compound	Heat of formation		Mol mass
	kJ/mol	kJ/kg	
Na_2SO_4	−1,387.1	−9,768	142
NaHCO3	-932.3	−11,099	84
NaHS	-257.7	−4,602	56
Na_2S	-443.3	−5,683	78
$CaSO_4$	−1,425.2	−10,479	136
$CaSO_4{\cdot}2H_2O$	−2,022.6	−11,759	172
$CaSO_4{\cdot}½H_2O$	−1,575.0	−10,862	145
CaS	−482.4	−68,914	72
$BaSO_4$	−1,473.2	−6,323	233
BaS	−460.0	−2,722	169
CO_2	−393.5	−8,943	44
H_2O	−285.8	−15,879	18
C	0.0	0	12

$$\Delta H^o_{reaction} = \sum H^o_{f(Products)} - \sum H^o_{f(Reactants)} \tag{11.16}$$

- Enthalpy is the sum of the internal energy and the product of the pressure and volume of a thermodynamic system as shown in **Equation 11.17**:

$$H = U + PV \tag{11.17}$$

where U is internal energy, P is pressure, and V is volume.

- Standard Gibbs free energy of formation is the change in Gibbs free energy for the formation of one mole of the compound in its standard state from its constituent elements in their standard states. It is given by **Equation 11.18**:

Table 11.2 Standard Gibbs free energy of formation [51].

Compound	Gibbs free energy kJ/mol	Gibbs free energy kJ/kg	Mol mass g/mol
Na_2SO_4	−1,270.2	−8,945	142
$NaHCO_3$	−848.7	−10,104	84
NaHS	−249.8	−4,461	56
Na_2S	−438.1	−5,617	78
$CaSO_4$	−1,309.1	−9,626	136.0
$CaSO_4{\cdot}2H_2O$	−1,797.5	−10,451	172.0
$CaSO_4{\cdot}½H_2O$	−1,436.8	−9,909	145.0
CaS	−477.4	−6,631	72.0
$BaSO_4$	−1,362.2	−5,846	233.0
BaS	−456.0	−2,698	169.0
CO_2	−394.4	8,963	44.0
H_2O	−237.1	−13,174.4	18.0
C	0.0	0.0	12.0
Steam	−228.6	−12,700.6	18.0

$$\Delta G_f^o = \Delta H_f^o - T\Delta S^o \qquad (11.18)$$

where ΔH_f^o is standard enthalpy of formation, T is standard temperature, ΔS^o is standard entropy, and subscript f indicates that the substance is formed from its elements.

- Standard entropy is the entropy of 1 mole of a substance in its standard state, at 1 atm of pressure and temperature of 298 K. It can be calculated from the difference in standard entropy between the products and the reactants as shown in **Equation 11.19**:

Table 11.3 Standard entropy values [51].

Compound	Standard entropy J/mol·K	Standard entropy J/kg·K	Mol mass g/mol
Na_2SO_4	149.6	1,053.5	142.0
$NaHCO_3$	150.2	1,788.1	84.0
NaHS	121.8	2,175.0	56.0
Na_2S	103.3	1,324.4	78.0
$CaSO_4$	108.4	797.1	136.0
$CaSO_4{\cdot}2H_2O$	194.1	1,128.5	172.0
CaSO4·½H_2O	130.5	900.0	145.0
CaS	56.5	784.7	72.0
$BaSO_4$	132.2	567.4	233.0
BaS	78.2	462.7	169.0
CO_2	213.8	4,858.8	44.0
Steam	188.8	10,490.8	18.0
H_2O	70.0	3,886.1	18.0
C	5.7	478.3	12.0

$$\Delta S^o = \sum S^o_{(\text{Products})} - \sum S^o_{(\text{Reactants})} \tag{11.19}$$

- Specific heat capacity, c, is defined as the quantity of heat absorbed per unit mass of the material when its temperature increases as shown in **Equation 11.20**:

$$c = \frac{Q}{m\Delta T} \tag{11.20}$$

where Q is the heat energy, m is the mass, and ΔT is change in temperature.

Tables 11.4 and 11.5 show the heat of reaction (H_r) (Equation 11.16) and specific heat capacity (C_p) for the various reactions. The following

Table 11.4 Specific heat capacity values for reactants and products.

	Specific heat capacity		Mol mass	
Compound	**kJ/K·mol**	**kJ/K·kg**	**g/mol**	**References**
Na_2SO_4	0.128	0.901	142	Speight, J. G. (2017)
$NaHCO_3$	0.088	1.043	84	
NaHS	-	-	56	
Na_2S	0.083	1.062	78	
$CaSO_4{\cdot}2H_2O$	0.186	1.081	172	Robie, Russell-Robinson, and Hemingway (1989)
$CaSO_4$	0.099	0.728	136	Speight, J. G. (2017)
$CaSO_4{\cdot}½H_2O$	0.442	2.570		
C	0.009	0.710	12	
CaS	0.047	0.658	72	
CO_2	0.037	0.848	44	
H_2O	0.075	4.186	18	
Steam	0.035	1.950	18	
$BaSO_4$	0.102	0.436	233	
BaS	0.049	0.292	169	
Coal		1.260		Engineering tool box

approach was followed to determine how much coal was needed for the processing of $CaSO_4{\cdot}2H_2O$ to CaS and $BaSO_4$ to BaS, respectively:

(i) The enthalpy (sensible heat) for the various reactants ($CaSO_4{\cdot}2H_2O$, $CaSO_4$, C, and H_2O) at the input temperature (30°C) was calculated by using the mass flow rate and the heat of reaction as shown in **Equation 11.21**:

Table 11.5 Reaction energy values for various reactions.

No.	Reaction	Heat of reaction, Hrxn (Equation 11.21)	Heat of reaction, Hrxn	Gibbs free energy of reaction (Equation 11.18)	Entropy of reaction (Equation 11.19)
		kJ/mol	kJ/kg product	kJ/mol	kJ/mol·K
1	$Na2SO_4 + 2C =$ Na2S + 2CO2	156.8	1,104.1	43.3	0.387
2	Na2S + CO2 + H2O --> NaHCO3 + NaHS	−67.4	−864.1	−28.9	0.231
3	$CaSO_4{\cdot}2H_2O \rightarrow CaSO_4 + 2H_2O$	25.7	3,501	14.12	0.054
4	$CaSO_4{\cdot}2H_2O \rightarrow CaSO_4{\cdot}\tfrac{1}{2}H_2O + 1\tfrac{1}{2}H_2O$	18.9	2,734	4.99	0.075
5	$CaSO_4 + 2C \rightarrow CaS + 2CO_2$	155.78	11,216	42.92	0.376
6	$BaSO_4 + 2C \rightarrow BaS + 2CO_2$	226.17	38,223	117.42	0.374

$$H(kW) = \dot{m}\left(\frac{kg}{s}\right) \times Hr\left(\frac{kJ}{kg}\right) \tag{11.21}$$

(ii) The enthalpy for the various products ($CaSO_4$, CaS, BaS, CO_2, and H_2O) at the output temperature (1,000°C) were calculated by using **Equation 11.21**.

(iii) Activation energy

a. The latent heat of vaporization of H_2O was calculated by using the mass flow rate ($\dot{m}$), latent heat of vaporization (Hv), and the temperature change (ΔT) as shown in **Equation 11.22**:

$$H(kW) = \dot{m}\left(\frac{kg}{s}\right) \times Hv\left(\frac{kJ}{K \cdot kg}\right) \times \Delta T\left(Hv = \frac{2260kJ}{kg \cdot K}\right) \quad (11.22)$$

b. The heat of combustion of coal was calculated by using the mass flow rate of coal ($\dot{m}$) and the coal calorific value (H_{CV}) as shown in **Equation 11.23**:

$$H(kW) = \dot{m}\left(\frac{kg}{s}\right) \times H_{CV}\left(\frac{kJ}{kg}\right)\left(H_{CV} = \frac{27MJ}{kg}\right) \quad (11.23)$$

c. The enthalpy for reactions 1 to 5 in Table 11.2, were calculated by using **Equation 11.21**.

(iv) The amounts of coal needed were calculated from the difference in enthalpy-out and enthalpy-in and divided by the calorific value of coal.

11.4.1.1 Standard Enthalpy/Heat of Formation

The standard enthalpy values for the reactants and products during thermal process are given Table 11.5. The standard enthalpy/heat of formation gives the thermodynamic data for reactions that take place. It shows the measure of the energy released or consumed when one mole of a substance is created under standard conditions, i.e., constant pressure of 1 atm and at room temperature of 25°C. As it can be seen in Table 11.1, the standard enthalpies of formations are negative. This is an indication that the formation of a compound from its elements is an exothermic reaction.

11.4.1.2 Standard Gibbs Free Energy of Formation

The standard Gibbs free energy of formation for the thermal reduction of $NaSO_4$, $CaSO_4{\cdot}2H_2O$, and $BaSO_4$ is given in Table 11.2. The standard Gibbs free energy of formation of a compound is the free energy change that occurs when 1 mole of the compound is synthesized from its constituent elements, each in their standard state. It is defined as a measure of a compound's stability relative to its elements. As it can be seen in Table 11.2, the standard Gibbs free energy of formations for all the reactants and products is negative. The negative value indicates that the reaction is spontaneous, and a positive value indicates that the reaction is non-spontaneous.

11.4.1.3 Standard Entropy Values

Entropy is the physical property that is most associated with a state of disorder, randomness, or uncertainty in a reaction. It is the measure of how the energy of a reaction system energy is spread out or dispersed.

As it can be seen in Table 11.3, the standard entropy of the compounds used in this study is positive, indicating that a system is more disordered.

11.4.1.4 Specific Heat Capacity Values for Reactant and Products

Specific heat capacity is defined as the amount of heat it takes to raise the temperature of 1 kg of a substance by 1°C with no change in state.

Table 11.4 shows the specific heat capacities of all the compounds used in this study.

11.4.1.5 Reaction Energy Values for Various Reactions

The heat of reaction can be defined as the amount of energy released or absorbed in the reaction. As it can be seen in Table 11.5, the reactions are endothermic because they require energy from the environment to proceed; hence, a plus sign is put in front of the amount of energy absorbed from the environment into the reactants for the reaction to occur. The only exothermic reaction in Table 11.5 was the one of $NaHCO_3$ and NaHS production *via* Na_2S.

11.4.2 Mass and Energy Balance Models for Reduction of Na_2SO_4, $CaSO_4$, and $BaSO_4$

A material and energy balance model can be used to determine the amount of coal needed for the reduction of metal sulfate compounds. The mass and energy were determined using an easy-to-use Excel-balance template that allows easy modification of the plant model. The reference enthalpies of each species were zero at 25°C. All physical properties and equations used were found in Perry's Handbook of Chemical Engineering (Green and Southard, 2019).

In Table 11.6, it was calculated that 14.9 t/h coal is needed for converting 50 t/h Na_2SO_4 to Na_2S. In Table 11.7, it was calculated that 11.7 t/h coal is needed for converting 31.63 t/h $CaSO_4$ to CaS. In Table 11.8, it was calculated that 9.6 t/h coal is needed for converting 50 t/h $BaSO_4$ to BaS.

Table 11.6 Calculation of coal needed for the processing of 50 ton/h Na_2SO_4 to Na2S.

	Input (Mass) (t/h)	Output (Mass) (t/h)	T1 (°K)	T2 (°K)	Cp (kJ/kg·K)	H_f (kJ/mol)	Hv (kJ/kg)	Enthalpy (MJ/h)	Hrxn (MJ/mol)
ROUTE A	**Step 1. Reduction (Na2SO$_4$ + 2C → Na2S + 2CO$_2$)**								
Total Inputs								97,771	**−1.3871**
Input								285	
Na_2SO_4	50.0		298.0	303.0	0.90	−1,387		225	−1.3871
Coal for reduction	11.27		298.0	303.0	1.06	0.00		59.8	0.0000
Coal for combustion	3.61						27,000.0	97,486	
Total outputs								97,771	**−1.2303**
Outputs								42,567	

(Continued)

Table 11.6 Calculation of coal needed for the processing of 50 ton/h Na_2SO_4 to Na2S. *(Continued)*

	Input (Mass) (t/h)	Output (Mass) (t/h)	T1 (°K)	T2 (°K)	Cp (kJ/kg·K)	H_f (kJ/mol)	Hv (kJ/kg)	Enthalpy (MJ/h)	Hrxn (MJ/mol)
Na_2S		27.46	303.0	1,073.0	1.06	−443.3		22,449	−0.4433
CO_2		30.99	303.0	1,073.0	0.84	−393.5		20,118	−0.7870
Heat of reaction NaS_2		27.46				156.8		**55,204**	0.1568
Total Energy required for step 1	3.6							**97,486**	
Total coal usage	14.9								

(Continued)

Table 11.6 Calculation of coal needed for the processing of 50 ton/h Na_2SO_4 to Na2S. (*Continued*)

	Input (Mass) (t/h)	Output (Mass) (t/h)	T1 (°K)	T2 (°K)	Cp (kJ/kg·K)	H_f (kJ/mol)	Hv (kJ/kg)	Enthalpy (MJ/h)	Hrxn (MJ/mol)
Coal usage (t coal/t wet gypsum)	0.1								
Na_2SO_4 for Route A (MJ/t Na_2SO_4)								1,950	
Coal (MJ/kg)								27.0	
Coal required (kg/t Na_2SO_4)								72.2	

(*Continued*)

Table 11.6 Calculation of coal needed for the processing of 50 ton/h Na_2SO_4 to Na2S. (*Continued*)

	Input (Mass) (t/h)	Output (Mass) (t/h)	T1 (°K)	T2 (°K)	Cp (kJ/kg·K)	H_f (kJ/mol)	Hv (kJ/kg)	Enthalpy (MJ/h)	Hrxn (MJ/mol)
Coal required (kg/h) (equal to cell C9)								3,611	
	Step 2. Carbonation Na2S + CO_2 + H_2O → $NaHCO_3$ + NaHS								
Inputs									−0.0674
Na_2S	27.46		298.0	298.0	1.06	−443.3			−0.4433
H_2O	6.34		298.0	298.0	4.19	−285.8			−0.2858
CO_2	15.49				0.84	−393.5			−0.3935
Outputs									

(*Continued*)

Table 11.6 Calculation of coal needed for the processing of 50 ton/h Na_2SO_4 to Na2S. (*Continued*)

	Input (Mass) (t/h)	Output (Mass) (t/h)	T1 (°K)	T2 (°K)	Cp (kJ/kg·K)	H_f (kJ/mol)	Hv (kJ/kg)	Enthalpy (MJ/h)	Hrxn (MJ/mol)
$NaHCO_3$		29.58	298.0	298.0	1.04	–932.3			–0.9323
NaHS		19.72	298.0	298.0	-	–257.7			–0.2577
Reaction energy								**–23,733**	–0.0674
Total energy for route A								**73,753**	
Note 1. Sensible heat, E = mCpΔT									
Note 2. Carbon content of reductant coal (%)			75						

Table 11.7 Calculation of coal needed for the processing of 50 ton/h $CaSO_4{\cdot}2H_2O$ to CaS.

Parameter	**Input (Mass) (t/h)**	**Output (Mass) (t)**	**T1 (°K)**	**T2 (°K)**	**Cp (kJ/ kg·K)**	**H_f (kJ/ mol)**	**Hv (kJ/ kg)**	**Enthalpy (MJ/h)**	**Hrxn (MJ/ mol)**
Wet gypsum	**50.00**								
Moisture content (%)	20.0								
	Step 1. Drying H_2O (aq) → H_2O (g)								**0.00**
Total inputs								**29,409**	
Input (sensible)								426	
$CaSO_4{\cdot}2H_2O$	**40.00**		298	303.0	1.08	**−2,023**		216	**−2.02**
H_2O	10.00		298	303.0	4.19	−285.8		209	−0.29
Combustion of coal	1.07						27,000	28,984	
Total output								**29,409**	

(Continued)

Table 11.7 Calculation of coal needed for the processing of 50 ton/h $CaSO_4{\cdot}2H_2O$ to CaS. (*Continued*)

Parameter	Input (Mass) (t/h)	Output (Mass) (t)	T1 (°K)	T2 (°K)	Cp (kJ/kg·K)	H_f (kJ/mol)	Hv (kJ/kg)	Enthalpy (MJ/h)	Hrxn (MJ/mol)
Output (sensible)								6,809	
$CaSO_4{\cdot}2H_2O$		40.00	298	378.0	1.08			3,460	**−2.02**
H_2O (free)		10.00	298	378.0	4.19			3,349	**−0.29**
Steam		10.00	378	378.0	1.95	−285.83		0	**−0.24**
H_2O to steam		10.00					2,260	22,600	
	Step 2. Gypsum to Anhydride $CaSO_4{\cdot}2H_2O \rightarrow CaSO_4 + 2H_2O$								**0**
Total inputs								**35,083**	
Input (sensible)								3,460	

(*Continued*)

Table 11.7 Calculation of coal needed for the processing of 50 ton/h $CaSO_4{\cdot}2H_2O$ to CaS. (*Continued*)

Parameter	**Input (Mass) (t/h)**	**Output (Mass) (t)**	**T1 (°K)**	**T2 (°K)**	**Cp (kJ/kg·K)**	**H_f (kJ/mol)**	**Hv (kJ/kg)**	**Enthalpy (MJ/h)**	**Hrxn (MJ/mol)**
$CaSO_4{\cdot}2H_2O$	40.00		298	378.0	1.08	−2,023		3,460	**−2.02**
Coal (for drying)	1.17						27,000	31,623	
Total outputs									
Output (sensible)								**35,083**	**0.00**
$CaSO_4$		31.63	298	673.0	0.73	−1,425		8,634	**−1.43**
H_2O (Free)		8.37	298	373.0	4.19			2,631	
Steam		8.37	373	673.0	1.95			4,898	
Steam		8.37	373	673.0	1.95	−285.8	2,260	18,921	**−0.24**
									0.00

(*Continued*)

Table 11.7 Calculation of coal needed for the processing of 50 ton/h $CaSO_4{\cdot}2H_2O$ to CaS. (*Continued*)

Parameter	Input (Mass) (t/h)	Output (Mass) (t)	T1 (°K)	T2 (°K)	Cp (kJ/kg·K)	H_f (kJ/mol)	Hv (kJ/kg)	Enthalpy (MJ/h)	Hrxn (MJ/mol)
	Step 3. Reduction ($CaSO_4 + 2C \rightarrow CaS + 2CO_2$)								
Total inputs								**63,749**	
Input								8,681	
$CaSO_4$	31.63		298	673.0	0.73	−1,425		8,634	**−1.43**
Coal for reduction	7.44		298	303.0	1.26	0.00		47	
Coal for Step 3	2.04						27,000	55,068	**0.00**
Total outputs								**63,749**	
Outputs (sensible)								27,521	
CaS		16.74	303	1,273.0	0.66	−482.4		10,687	**−0.48**
CO_2		20.47	303	1,273.0	0.85	−393.5		16,834	**−0.79**

(*Continued*)

Table 11.7 Calculation of coal needed for the processing of 50 ton/h $CaSO_4{\cdot}2H_2O$ to CaS. (*Continued*)

Parameter	Input (Mass) (t/h)	Output (Mass) (t)	T1 (°K)	T2 (°K)	Cp (kJ/kg·K)	H_f (kJ/mol)	Hv (kJ/kg)	Enthalpy (MJ/h)	Hrxn (MJ/mol)
Heat of reaction CaS		16.74				155.8		36,228	**0.16**
Total coal usage (t/h)	**11.7**						27,000	**86,900**	
Total energy required	**3.22**							115,675	

Table 11.8 Calculation of coal needed for the processing of 50 ton/h $BaSO_4$ to BaS.

	Input (Mass) (t/h)	Output (Mass) (t/h)	T1 (°K)	T2 (°K)	Cp (kJ/kg·K)	H_f (kJ/mol)	Hv (kJ/kg)	Enthalpy (MJ/h)	Heat of formation Hrxn (MJ/mol)
ROUTE A	**Step 1. Reduction ($BaSO_4 + 2C \rightarrow BaS + 2CO_2$)**								
Total inputs								74,176	**−1.473**
Input								152.2	
$BaSO_4$	50.00		298.0	303.0	0.44	−1,473.2		109.0	−1.473
Coal for reduction	6.87		298.0	303.0	1.26	0.0		43.3	0.000
Coal for combustion	2.74						27,000	74,023	
Total outputs								74,176	**−1.247**
Outputs								25,641	
BaS		36.27	303.0	1,273.0	0.29	−460.0		10,254	−0.460
CO_2		18.88	303.0	1,273.0	0.84	−393.5		15,387	−0.787

(Continued)

Table 11.8 Calculation of coal needed for the processing of 50 ton/h $BaSO_4$ to BaS. (*Continued*)

	Input (Mass) (t/h)	**Output (Mass) (t/h)**	**T1 (°K)**	**T2 (°K)**	**Cp (kJ/kg·K)**	**H_f (kJ/mol)**	**Hv (kJ/kg)**	**Enthalpy (MJ/h)**	**Heat of formation Hrxn (MJ/mol)**
Heat of reaction BaS		36.27				226.17		**48,534**	0.226
Total energy required for step 1								**74,023**	
Total coal usage	9.6								
$BaSO_4$ for Route A (MJ/t $BaSO_4$)								1,480	
Coal (MJ/kg)								27.0	
Coal required (kg/t $BaSO_4$)								54.8	

(*Continued*)

Table 11.8 Calculation of coal needed for the processing of 50 ton/h $BaSO_4$ to BaS. (*Continued*)

	Input (Mass) (t/h)	Output (Mass) (t/h)	T1 (°K)	T2 (°K)	Cp (kJ/kg·K)	H_f (kJ/mol)	Hv (kJ/kg)	Enthalpy (MJ/h)	Heat of formation Hrxn (MJ/mol)
Coal required (kg/h) (equal to cell C9)								2,742	
Standard Heat of Reaction, H_{rxns}($BaSO_4 + 2C = BaS + 2CO_2$) (kJ/mol)								226.2	
	Step 2. Carbonation $BaS + CO_2 + H_2O \rightarrow BaCO_3 + H_2S$)								
Inputs									−0.094
BaS	36.27		298.0	298.0	0.29	−460.0			−0.460
H_2O	3.86		298.0	298.0	4.20	−285.8			−0.286
CO_2	9.44				0.84	−393.5			−0.394
Outputs									
$BaCO_3$		42.35	298.0	298.0	0.44	−1,213.0			−1.213

(*Continued*)

Table 11.8 Calculation of coal needed for the processing of 50 ton/h $BaSO_4$ to BaS. (*Continued*)

	Input (Mass) (t/h)	Output (Mass) (t/h)	T1 (°K)	T2 (°K)	Cp (kJ/ kg·K)	H_f (kJ/ mol)	Hv (kJ/ kg)	Enthalpy (MJ/h)	Heat of formation Hrxn (MJ/mol)
H_2S		7.30	298.0	298.0	1.01	−20.6			−0.021
Reaction energy								**−20,227**	−0.094
Total energy for route A								**53,796**	

Table 11.9 summarizes the previous results and shows that the coal usage per Mmol MSO_4 varies between 42.2 and 50.4 t coal/Mmol compound. It is noted that coal to $BaSO_4$ ratio is less than that for the coal to $CaSO_4$ ratio. This, together with the lower solubility of $BaSO_4$, makes $BaSO_4$ and attractive option for the processing of Na_2SO_4 to Na_2S. In terms of the energy required, it can be concluded that less energy is needed for the conversion of $BaSO_4$ into BaS (1480 MJ/t $BaSO_4$ or 344.9 kJ/mol $BaSO_4$) than for the conversion of $CaSO_4$ into CaS (3,657 MJ/t CaSO4 or 497.4 kJ/mol $CaSO_4$) (Table 11.9). The corresponding values for Na2SO4 amounted to (1,950 MJ/t Na_2SO_4 or 276.9 kJ/mol Na_2SO_4).

11.4.3 Reduction of $CaSO_4$ and $BaSO_4$

Table 11.10 shows the optimum conditions, as determined **experimentally**, for the reduction of $CaSO_4{\cdot}2H_2O$ to CaS. Table 11.11 shows the results for the reduction of $BaSO_4$ to BaS. It was found that the optimum temperature for both reactions was 1,000°C; the optimum reaction time for $BaSO_4$ reduction was 20 min versus 60 min for $CaSO_4{\cdot}2H_2O$; and the higher the carbon-metal sulfate ratio, the better was the conversion. From the results in Table 11.8, at a C/$CaSO_4$ mol/mol ratio of 2 and higher, and, at a temperature of 1,000°C, 100% of the $CaSO_4$ was converted to CaS. For lower C/$CaSO_4$ mol/mol ratios, the CaS yield was linearly related to the ratio. Unreacted $CaSO_4$ remained with the CaS.

Table 11.12 and Table 11.13 show the $CaSO_4$ and $BaSO_4$ reduction using coal as a redundant, simulation result *via* modeling with the Pyrosim Mintek model. From the experimental results obtained through thermal studies in the laboratory (Table 11.10 and Table 11.11), it can be seen that the calculated experimental values corresponded well with the predicted values using the Pyrosim Mintek models (Table 11.12 and Table 11.13).

At lower temperatures (200°C to 600°C), the CaS yield dropped from 92.7% to 0%, whereas the $CaCO_3$ yield increased from 3% to 99%. In a previous study, it was noted that the kinetics of $CaCO_3$ formation at low temperatures were slow and not feasible for implementation [10]. CO was the main gas product from carbon at the higher C/$CaSO_4$ ratios (**Equation 11.24**), whereas CO_2 replaced CO as main product, as the ratio decreased (**Equation 11.25**).

$$CaSO_4 + 4C \rightarrow CaS + 4CO \tag{11.24}$$

Table 11.9 Comparison of coal and energy needed for the reduction of Na_2SO_4, $CaSO_4$, and $BaSO_4$.

	Mol mass	MSO_4	Coal	Coal/MSO_4 ratio	Coal/MSO_4	Energy needed		
Compound	**g/mol**	**t/h**	**t/h**	**t/t**	**t/Mmol**	**MJ/h**	**MJ/t compound**	**kJ/mol compound**
Na_2SO_4	142	50.00	14.88	0.298	42.25	97,486	1,950	276.9
$CaSO_4$	136	31.63	11.73	0.371	50.42	115,675	3,657	497.4
$BaSO_4$	233	50.00	9.61	0.192	44.78	74,023	1,480	344.9

Table 11.10 Effect of various parameters (bold) on the rate of thermal reduction of $CaSO_4{\cdot}2H_2O$.

Parameter	Unit	Exp. no.											
Exp. no.		**6.1**	**6.2**	**6.3**	**6.4**	**7.1**	**7.2**	**7.3**	**7.4**	**8.1**	**8.2**	**8.3**	**8.4**
Chemical		$CaSO_4{\cdot}2H_2O$	$CaSO_4{\cdot}2H_2O$	$CaSO_4{\cdot}2H_2O$	$CaSO_4{\cdot}2H_2O$	$CaSO_4{\cdot}2H_2O$	$CaSO_4{\cdot}2H_2O$	$CaSO_4{\cdot}2H_2O$	$CaSO_4{\cdot}2H_2O$	$CaSO_4{\cdot}2H_2O$	$CaSO_4{\cdot}2H_2O$	$CaSO_4{\cdot}2H_2O$	$CaSO_4{\cdot}2H_2O$
Mol mass		172	172	172	172	172	172	172	172	172	172	172	172
Product		CaS	CaS	CaS	CaS	CaS	CaS	CaS	CaS	CaS	CaS	CaS	CaS
Mol mass		72	72	72	72	72	72	72	72	72	72	72	72
Mass	G	20.00	20.00	20.00	20.00	20.00	20.00	20.00	20.00	20.00	20.00	20.00	20.00
Carbon	G	5.58	5.58	5.58	5.58	5.58	5.58	5.58	5.58	1.40	2.79	4.19	5.58
Ash content	%	8	8	8	8	8	8	8	8	8	8	8	8
Ash	G	0.45	0.45	0.45	0.45	0.45	0.45	0.45	0.45	0.11	0.22	0.33	0.45
Unreacted C	G	4.17	3.46	2.71	3.47	4.76	3.96	2.67	3.51	0.78	1.76	2.44	2.83
C/Chem mole ratio		3.68	3.68	3.68	3.68	3.68	3.68	3.68	3.68	**0.92**	**1.84**	**2.76**	**3.68**
Temperature	°C	**900**	**950**	**1,000**	**1,100**	1,000	1,000	1,000	1,000	1,000	1,000	1,000	1,000
Reaction time	Min	60	60	60	60	**15**	**30**	**60**	**120**	60	60	60	60
Conversion	%	35.16	55.03	72.51	57.02	16.78	48.46	70.30	44.68	21.92	32.18	53.93	71.54

Mass of CaS in product sample used for titration (mg) = (40+32)/2 × 0.1 × (10 – Titr. Vol of $S_2O_3^{2-}$)
Mass of CaS in product after treatment of 20 g of $CaSO_4{\cdot}2H_2O$ (g) = Mass of product in sample/1,000 × 20 g/0.05 g
Conversion (%) = (Mass of CaS in product (g)/72)/(20 g/172) × 100

Table 11.11 Effect of various parameters (bold) on the rate of thermal reduction of $BaSO_4$.

Parameter	Unit	Exp. no.											
Exp. no.		**16.1**	**16.2**	**16.3**	**16.4**	**17.1**	**17.2**	**17.3**	**17.4**	**18.1**	**18.2**	**18.3**	**18.4**
Chemical		$BaSO_4$	$BaSO_4$	$BaSO_4$	$BaSO_4$	$BaSO_4$	$BaSO_4$	$BaSO_4$	$BaSO_4$	$BaSO_4$	$BaSO_4$	$BaSO_4$	$BaSO_4$
Mol mass		233	234	235	236	237	238	239	240	241	242	243	244
Product		BaS	BaS	BaS	BaS	BaS	BaS	BaS	BaS	BaS	BaS	BaS	BaS
Mol mass		169	170	171	172	173	174	175	176	177	178	179	180
Mass	G	20.00	20.00	20.00	20.00	20.00	20.00	20.00	20.00	20.00	20.00	20.00	20.00
Carbon	G	4.11	4.11	4.11	4.11	4.11	4.11	4.11	4.11	1.03	2.06	3.09	4.11
Ash content	%	8	9	10	11	12	13	14	15	16	17	18	19
Ash	G	0.33	0.37	0.41	0.45	0.49	0.53	0.58	0.62	0.16	0.35	0.56	0.78
Unreacted C	G	3.07	2.88	1.99	2.56	3.51	2.92	1.97	2.59	0.33	1.13	1.96	2.25
C/Chem mole ratio		3.67	3.65	3.63	3.60	3.58	3.55	3.52	3.50	**0.87**	**1.72**	**2.56**	**3.39**
Temperature	°C	**900**	**950**	**1,000**	**1,100**	1,000	1,000	1,000	1,000	1,000	1,000	1,000	1,000
Reaction time	Min	20	20	20	20	**5**	**10**	**20**	**40**	20	20	20	20
Conversion	%	34.74	60.08	86.93	59.67	13.34	42.18	88.17	58.16	18.10	28.80	50.74	82.56

Mass of BaS in product sample used for titration (mg) = (137+32)/2 × 0.1 × (10 – Titr. Vol of $S_2O_3^{2-}$)
Mass of BaS in product after treatment of 20 g of $BaSO_4$ (g) = Mass of product in sample/1,000 × 20 g/0.05g
Conversion (%) = (Mass of BaS in product (g)/169)/(20 g/233) × 100

Table 11.12 Thermal reduction of $CaSO_4$ (Pyrosim Mintek simulation).

No	Conversion								Input data				Matte						
	$CaCO_3$	S	CaS	$CaSO_4$	C/ $CaSO_4$	Temp	$CaSO_4$	$CaSO_4$	Coal	Coal	H_2O	Volatiles	$CaCO_3$	CaO	CaS	$CaSO_4$	S	C	Total
	%	%	%	%	%	o C	%	kg/h	kg/h	%	%	%	%	%	%	%	%	%	%
	Effect of C/$CaSO_4$ ratio							142.00											
1	0.26	0.00	100.65	0.00	3.40	1000.00	100.00	100.00	40.00	75.00	5.40	19.60	0.35		98.35	0.00		0.41	99.11
2	0.42	0.00	100.45	0.01	2.13	1000.00	100.00	100.00	25.00	75.00	5.40	19.60	0.57		98.65	0.03		0.00	99.25
3	0.35	0.00	84.35	15.99	1.70	1000.00	100.00	100.00	20.00	75.00	5.40	19.60	0.42		72.88	26.09		0.00	99.39
4	0.25	0.00	63.40	36.95	1.28	1000.00	100.00	100.00	15.00	75.00	5.40	19.60	0.26		47.25	52.02		0.00	99.53
	Effect of temperature																		0.00
5	0.42	0.00	100.45	0.01	2.13	1000.00	100.00	100.00	25.00	75.00	5.40	19.60	0.57		98.65	0.03	0.00	0.00	99.25
6	3.17	0.00	92.72	2.11	2.13	600.00	100.00	100.00	25.00	75.00	5.40	19.60	4.46		94.08	4.05	0.00	0.91	103.50
7	22.26	0.00	77.89	0.00	2.13	400.00	100.00	100.00	25.00	75.00	5.40	19.60	27.57		69.44	0.01	0.00	2.85	99.86
8	99.46	0.00	0.78	0.32	2.13	200.00	100.00	100.00	25.00	75.00	5.40	19.60	92.54		0.52	0.41	0.00	6.49	99.96

(Continued)

Table 11.12 Thermal reduction of $CaSO_4$ (Pyrosim Mintek simulation). (*Continued*)

Matte						Volatiles						Gas						Pyrosim
C	S	CaS	$CaSO_4$	$CaCO_3$	Total	CO	SO_2	H_2S	CO_2	H_2O	S	H_2O	CO	CO_2	H_2S	S	SO_2	MWh/t
kg/h	kg/h	kg/h	kg/h	kg/h	kg/h	%	%	%	%	%	%	%	%	%	%	%	%	
12.00	32.00	78.00	142.00	106.00														
0.42	0.00	53.29	0.00	0.19	53.70	74.97		0.07	16.95	3.40		1.91	65.62	23.31	0.08			1.02
0.00	0.00	53.18	0.01	0.31	53.50	10.92		0.18	81.35	2.94		1.25	7.22	84.53	0.15			1.06
0.00	0.00	44.66	15.99	0.26	60.90	1.97		0.19	90.49	2.73		1.12	1.26	90.84	0.15			0.80
0.00	0.00	33.56	36.95	0.18	70.70	1.50		0.18	91.23	2.45		1.00	0.96	91.27	0.14			0.57
0.00	0.00	53.18	0.01	0.31	53.50	10.92		21.99	81.35	0.18	0.00	1.25	7.22	84.53	0.15	0.00		1.39
0.88	0.00	49.08	2.11	2.33	54.00	6.24		1.49	85.41	2.09		0.14	0.28	86.74	12.85			1.06
2.85	0.00	41.24	0.00	16.37	59.30	2.91		4.11	83.13	0.54	6.74	0.23	0.20	83.94	3.22	4.95		0.87
6.49	0.00	0.41	0.32	73.13	79.00	0.00		5.85	33.72	0.16	53.87	0.08	0.00	38.90	5.24	45.26		0.20

Table 11.13 Thermal reduction of $BaSO_4$ (Pyrosim Mintek simulation).

Yield (% mol/mol $BaSO_4$)						Input data							Matte						
$BaCO_3$	S	BaS	$BaSO_4$	H_2S	C/ BaSO4	Temp	$BaSO_4$	$BaSO_4$	Coal	Coal	H_2O	Volatiles	$BaCO_3$	BaO	BaS	$BaSO_4$	S	C	Total
Effect of C/$BaSO_4$ ratio:							%	kg/h	kg/h	%	%	%	%	%	%	%	%	%	%
0.41	0.00	119.97	0.35	-20.32	5.83	1 000	100.00	100.00	40.00	75.00	5.40	19.60	0.35		87.02	0.00	0.00	11.60	83.00
1.02	0.00	99.09	1.18	-0.27	3.64	1 000	100.00	100.00	25.00	75.00	5.40	19.60	1.18		98.09	0.00	0.00	0.36	99.62
1.59	0.00	98.74	1.84	-0.58	2.91	1 000	100.00	100.00	20.00	75.00	5.40	19.60	1.84		97.78	0.00	0.00	0.05	99.68
1.90	0.00	98.17	2.19	-0.36	2.18	1 000	100.00	100.00	15.00	75.00	5.40	19.60	2.19		97.14	0.40	0.00	0.00	99.72
Effect of Temperature:																			
1.02	0.00	99.35	1.18	-0.53	3.64	1 000	100.00	100.00	25.00	75.00	5.40	19.60	1.18	0.01	98.09	0.00		0.36	99.63
						800	100.00	100.00	25.00	75.00	5.40	19.60							
13.16	0.59	86.97	13.63	-1.19	3.64	600	100.00	100.00	25.00	75.00	5.40	19.60	13.63		77.23	0.00	0.10	8.94	99.90
69.96	0.00	29.86	65.06	5.08	3.64	400	100.00	100.00	25.00	75.00	5.40	19.60	65.06		23.82	0.28	0.00	10.81	99.97
68.09	72.93	0.00	63.23	-36.16	3.64	200	100.00	100.00	25.00	75.00	5.40	19.60	63.23		0.00	24.30	11.00	12.39	110.93

(Continued)

Table 11.13 Thermal reduction of $BaSO_4$ (Pyrosim Mintek simulation). (*Continued*)

Matte					Volatiles								Gas								Pyrosim
S	BaS	$BaSO_4$	$BaCO_3$	Total	CO	SO_2	H_2S	CO_2	H_2O	S	Total	Vol	H_2O	CO	CO_2	H_2S	SO_2	S	Total	Gas	MWh/t product
kg/h	kg/h	kg/h	kg/h	kg/h	%	%	%	%	%	%	%	kg/h	%	%	%	%	%	%	%	kg/h	
0.00	87.02	0.35	0.35	83.00	88.71		0.08	0.89	3.74				3.74	88.71	0.89	0.08					1.16
0.00	71.87	1.18	0.86	73.00	81.62		0.26	10.17	3.30				1.92	73.87	14.46	0.29					1.13
0.00	71.62	1.84	1.35	73.00	57.71		0.51	34.21	2.96	0.00			1.53	46.40	43.22	0.49					0.92
0.00	71.21	2.19	1.60	73.10	16.29		0.80	75.73	2.54	0.00			1.10	10.99	80.24	0.66		0.00			0.69
																			0.00		
0.00	72.06	1.18	0.87	73.20	81.62		0.26	10.17	3.30		95.35		1.92	73.87	14.46	0.29			90.53		1.13
0.08	63.08	13.63	11.13	81.60	18.57		5.42	67.65	0.89		92.52		0.39	12.48	71.42	4.43			88.71		0.53
0.00	21.66	65.06	59.15	90.90	0.51		8.83	53.46	0.11	27.60	90.51		0.05	0.35	56.68	7.25		21.32	85.64		0.45
10.02	0.00	63.23	57.57	101.00	0.00		11.53	32.76	0.28	42.44	87.01		0.13	0.00	35.88	9.78		33.86	79.65		0.11

$$CaSO_4 + 2C \rightarrow CaS + 2CO_2 \tag{11.25}$$

The reduction of $BaSO_4$ was like that of $CaSO_4$, as seen, at a C/$BaSO_4$ mol/mol ratio of 2 and higher, and, at a temperature of 1,000°C, more than 98% of the $BaSO_4$ was converted to BaS. As the C/$BaSO_4$ ratio in the experiments was always greater than 2, the $BaCO_3$ did not exceed 2%. It can also be seen that, at lower temperatures (200°C to 600°C), the BaS yield dropped from 86% to 0% while the $CaCO_3$ and S were formed together with BaS. In addition, CO was the main gas product from carbon at the higher C/$BaSO_4$ ratios (**Equation 11.26**), whereas CO_2 replaced CO as main product, as the ratio decreased (**Equation 11.27**).

$$BaSO_4 + 4C \rightarrow BaS + 4CO \tag{11.26}$$

$$BaSO_4 + 2C \rightarrow BaS + 2CO_2 \tag{11.27}$$

11.4.4 Oxidation of CaS and BaS

The reduction of $CaSO_4$ and $BaSO_4$ requires reducing conditions during the reaction and even in the cooling stage thereafter. If the product is in contact with air/oxygen, then back oxidation may take place (**Equations 11.28** and **11.29**):

$$CaS + O_2 \rightarrow CaSO_4 \tag{11.28}$$

$$BaS + O_2 \rightarrow BaSO_4 \tag{11.29}$$

Figures 11.3 and 11.4 show the effect of oxygen on the products, CaS and BaS, respectively, as determined with TGA studies. In the presence of oxygen, the sulfide is oxidized back to the sulfate compound at temperatures above 600°C, as indicated by the mass increase due to the conversion of sulfides to sulfates. Under reducing condition (N_2 atmosphere), no oxidation was observed. This shows the importance that no oxygen should be in contact with the product, even during the cooling stage.

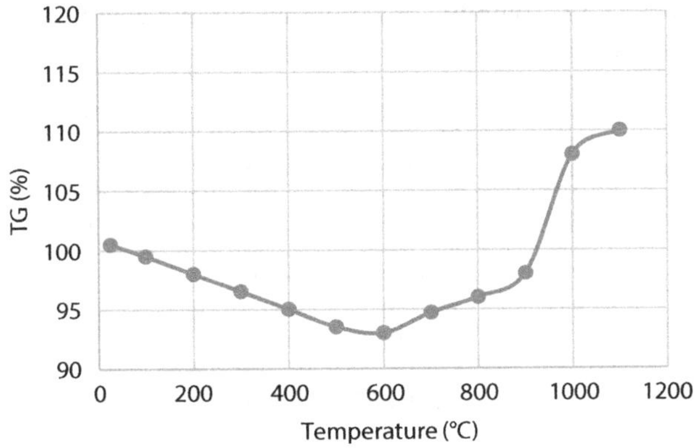

Figure 11.3 Oxidation of CaS when O_2 is present.

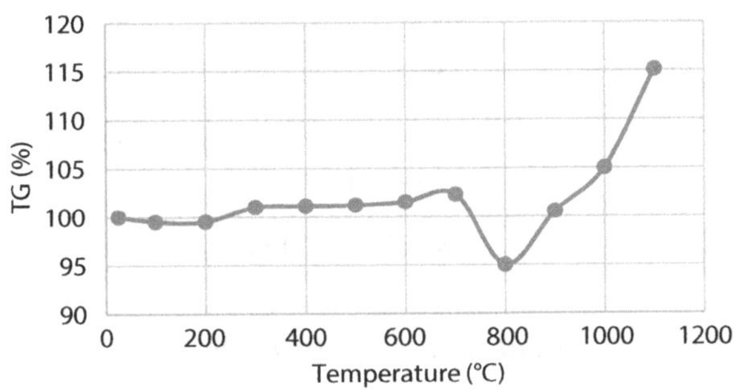

Figure 11.4 Oxidation of BaS when O_2 is present.

11.4.5 Conversion of Na_2SO_4 to Na_2S

Na_2SO_4 melts in the same temperature range where its reduction to Na_2S takes place, namely, 960°C. For this reason, Na_2SO_4 cannot be converted directly to Na_2S, as the case is with $CaSO_4$ and $BaSO_4$. The indirect conversion *via* of Na_2S from Na_2SO_4 *via* CaS or BaS was studied.

11.4.5.1 Solubility of Sulfate, Sulfide, and Bicarbonate Compounds

Alkaline-earth sulfates, except for magnesium sulfate, are characterized by very low solubilities (Table 11.14). The difference in solubility between compounds can be used for separation [52]. The solubility of barium sulfate, for example, is only 0.07% (mole/mole) of that of calcium sulfate (0.0105/15 × 100) (Table 11.14). The solubility of $BaSO_4$ increases the ionic strength of water.

Table 11.14 Solubility of various alkalis and alkali products [11].

	Solubility		Temp.
Compound	**mmol/L**	**g/L**	**°C**
$CaSO_4{\cdot}2H_2O$	15	2.64	25.0
CaS	186	13	25.0
$NaHCO_3$	1,300	109	25.0
$MgSO_4$	2,161.3	260	0.0
$MgSO_4$	2,801.3	337	20.0
$BaSO_4$	0.0105	0.002448	20.0
Na_2S	2,540	198	25.0
Na_2SO_4	2,873	408	25.0
$Ca(HS)_2$	A strong function of H_2S partial pressure		
NaHS	88,200	4,939	25.0

11.4.5.2 Processing of Na_2S via BaS from Na_2SO_4

Mokgohloa *et al.* showed that BaS should rather be used than CaS for the conversion of Na_2SO_4 to Na_2S. The reason is $BaSO_4$ has a low solubility, resulting in a purer Na_2S, than in the case of CaS, where Na_2S will be contaminated with CaS. OLI simulations showed that BaS reacted stoichiometrically with Na_2SO_4 as indicated by the removal of sulfate from 1,000 mmol/L to 0.03 mmol/L (Table 11.15). Table 11.16 shows that this finding was confirmed experimentally as 99.62% of the Na_2SO_4 was converted to Na_2S. In the case of CaS, the conversion of Na_2SO_4 to Na_2S is lower than in the case of BaS. Table 11.17 shows that 1,000 mmol/L Na_2SO_4 is decreased to only 254.8 mmol/L Na_2S, compared to the 0.03 mmol/L in the case of BaS. This difference can be ascribed to the difference in solubilities: 0.0285 mg/L for $BaSO_4$ and 2,640 mg/L for $CaSO_4{\cdot}2H_2O$.

11.4.6 Production of $NaHCO_3$ and NaHS From Na_2S and CO_2

Valuable products, like $NaHCO_3$, NaHS, and Na_2CO_3, can be recovered by reacting Na_2S with CO_2. The CO_2 can be collected from the thermal stage, where coal is used both as the reductant and the source of energy.

Table 11.15 Conversion of Na_2SO_4 to Na_2S *via* BaS (OLI simulation) [11].

BaS (mmol)	Na_2SO_4 (mmol)	$BaSO_4$ (Barite) - Sol (mmol)	$S^{(+6)}$ (aq) (mmol)	$Ba^{(+2)}$ (aq) (mmol)	$S^{(-2)}$ (aq) (mmol)	$Na^{(+1)}$ (aq) (mmol)	pH
0	1,000	0	1,000.0	0	0	2,000	7.4
100	1,000	100	900.0	0	100	2,000	12.5
200	1,000	200	800.0	0	200	2,000	12.7
300	1,000	300	700.0	0	300	2,000	12.8
400	1,000	400	600.0	0	400	2,000	12.9
500	1,000	500	500.0	0	500	2,000	12.9
600	1,000	600	400.0	0	600	2,000	13.0
700	1,000	700	300.0	0	700	2,000	13.0
800	1,000	800	200.0	0	800	2,000	13.0
900	1,000	900	100.0	0	900	2,000	13.1
1,000	1,000	1,000	0.3	0	1,000	2,000	13.1

Temp = 25°C; pH = 12.5 to 13.1

Table 11.16 Conversion of Na_2SO_4 to Na_2S *via* BaS (experimental) [11].

Parameter	Value
BaS (g)	10.00
Purity (%)	95.00
BaS (meq)	112.43
Na_2SO_4 (g)	8.40
H_2O (mL)	200.00
Na_2S (q)	4.37
Na_2S (meq)	112.00
NaS production (%)	99.62

Table 11.17 Conversion of Na_2SO_4 to Na_2S *via* CaS (OLI simulation) [11].

CaS (mmol)	Na_2SO_4 (mmol)	$CaSO_4.2H_2O$ (Gypsum) (mmol)	$Ca(OH)_2$(Portlandite) - (s) (mmol)	$S^{(+6)}$(aq) (mmol)	$Ca^{(+2)}$(aq) (mmol)	$S^{(-2)}$(aq) (mmol)	$Na^{(+1)}$(aq) (mmol)	pH
0	1,000	0.0	0.0	1,000.0	0.0	0	2,000	7.4
100	1,000	81.6	0.0	918.4	18.4	100	2,000	12.5
200	1,000	178.9	0.0	821.1	21.1	200	2,000	12.7
300	1,000	269.1	6.7	730.9	24.2	300	2,000	12.7
400	1,000	346.7	26.5	653.3	26.8	400	2,000	12.7
500	1,000	421.9	48.0	578.1	30.1	500	2,000	12.6
600	1,000	494.5	71.5	505.5	34.0	600	2,000	12.6
700	1,000	563.7	97.3	436.3	39.0	700	2,000	12.6
800	1,000	629.1	125.6	370.9	45.3	800	2,000	12.5
900	1,000	689.9	156.9	310.1	53.2	900	2,000	12.5
1,000	1,000	745.2	191.2	254.8	63.5	1,000	2,000	12.4

Temp = 25°C; pH = 12.5 to 12.4

11.4.6.1 Solubilities of Sodium Compounds

Table 11.18 shows that NaHS has a solubility of 88,203 mmol/L or 4,939 g/L NaHs. Equation 11.30 shows that NaHS decomposes at higher concentrations to produce Na_2S and H_2S. This reaction occurs at an elevated temperature of 60°C-70°C and a pressure of 1-2 atm.

$$2NaHS \rightarrow Na_2S + H_2S \tag{11.30}$$

Table 11.8 shows the solubility of the reactant, Na_2S, and the products: $NaHCO_3$ and NaHS. **Equation 11.31** shows the reaction between Na_2S and CO_2.

$$Na_2S + 2CO_2 + 2H_2O \rightarrow 2NaHCO_3 + H_2S \tag{11.31}$$

11.4.6.2 Separation of $NaHCO_3$ and NaHS

$NaHCO_3$ and NaHS can be separated through its solubility differences as shown in (Table 11.18), namely, 204.3 g/L for Na_2S and 4,939.4 g/kg water for NaHS. Mokgohloa *et al.* showed through OLI simulations in Table 11.24 that, when 10,000 mmol Na_2S (780 g of Na_2S) is reacted with 1 L of H_2O at 25°C, the following soluble compounds will form: 1,377 mmol $Na_2S{\cdot}9H_2O$ and 8,622.9 mmol $Na_2S{\cdot}5H_2O$, with a pH of 12.2, with no free H_2O left [11]. By adding 8,000 mmol CO_2 to the $Na_2S{\cdot}9H_2O$ and $Na_2S{\cdot}5H_2O$, 2,656 mmol $Na_2CO_3{\cdot}NaHCO_3{\cdot}2H_2O$ formed, and, by increasing the CO_2 dosage to 20,000 mmol, 18,417 mmol $NaHCO_3$ and 8,875 mmol H_2S formed (**Equation 11.31**). The solid $NaHCO_3$ can be converted to Na_2CO_3 and H_2S to elemental sulfur.

Due to the difference in solubilities, $NaHCO_3(s)$ will form when a concentration of up to 10,000 mmol/L is reacted with stoichiometric amounts of CO_2 (Table 11.19) [11]. The compound, $Na_2CO_3{\cdot}NaHCO_3{\cdot}2H_2O$, will form at higher stoichiometric dosages. Sulfide will remain in solution up to 10,000 mmol/L when Na_2S and CO_2 are contacted, whereas $H_2S(g)$ will form at higher concentrations.

An experiment was planned to confirm that sulfide (S^-) will remain in solution when contacted with CO_2. Table 11.20 shows how the pH dropped

Table 11.18 Solubility of Na_2S, $NaHCO_3$ and NaHS (OLI simulation) [11].

	Temp.	Solubility		
Compound	**(°C)**	**mmol/L**	**g/L**	**Table no. in [11]**
Na_2S	25	2,620	204.3	18
Na_2S	80	6,101	475.9	20
$NaHCO_3$	25	1,110	86.5	21
$NaHCO_3$	0	640	49.9	22
NaHS	25	88,203	4,939.4	23

Table 11.19 Reaction between 0 and 20,000 mmol Na_2S and between 0 and 20,000 mmol CO_2 at 25°C (OLI simulation) [11].

CO_2 (mmol)	Na_2S (mmol) (Y2)	pH	$NaHCO_3$ (Nahcolite) - (s) (mmol)	$Na_2CO_3NaHCO_3 \cdot 2H_2O$ - (s) (mmol)	H_2S - (vap) (mmol)	CO_2 - (vap) (mmol)	$S^{(-2)}$(aq) (mmol)	$Na^{(+1)}$(aq) (mmol)	$C^{(+4)}$(aq) (mmol)	H_2O (mmol)
0	0	7.0	0	0	0	0.00	0	0	0	55,508
2,000	2,000	8.3	1,358	0	0	0.00	2,000	2,642	642	55,508
4,000	4,000	8.6	3,599	0	29	0.48	3,971	4,401	400	55,508
6,000	6,000	8.8	5,697	0	80	0.39	5,920	6,303	303	55,508
8,000	8,000	9.1	7,722	0	142	0.26	7,858	8,278	278	55,508
10,000	10,000	9.3	9,691	0	225	0.17	9,775	10,309	309	55,508
12,000	12,000	9.3	3,210	4,274	4,446	3.31	7,554	7,968	239	55,508
14,000	14,000	9.5	0	6,903	7,059	2.71	6,941	7,290	191	55,508
16,000	16,000	9.8	0	7,896	8,091	0.92	7,909	8,313	208	55,508
18,000	18,000	10.1	0	8,851	9,143	0.36	8,857	9,446	297	55,508
20,000	20,000	10.5	0	9,706	10,291	0.20	9,709	10,881	587	55,508

Temp = 25°C; Pressure = 1 atm

from 13.1 to 8.2 as CO_2 was added, the Na_2S concentration remained constant around 195,975 mg/L Na_2S, as S^{2-} was converted to HS^-, and alkalinity decreased by 50% as $NaHCO_3$ precipitated (from 198,900 to 93,600), according to **Equation 11.32.**

$$Na_2S + CO_2 + H_2O \rightarrow NaHS + NaHCO_3 \qquad (11.32)$$

11.4.6.3 Na_2CO_3 from $NaHCO_3$

Na_2CO_3 can be produced from $NaHCO_3$ by heating it to 200°C. In the case of dilute $NaHCO_3$ solutions, it can be concentrated by removing water though freeze crystallization or through evaporation. Table 11.21 shows the conversion of $NaHCO_3$ to Na_2CO_3 during evaporation at 100°C. It is

Table 11.20 $NaHCO_3$ precipitation when CO_2 is bubbled through a 200 g/L Na_2S solution (100 mL) [11].

Time	pH	S^{2-}	Alk
min		**mg/L Na_2S**	**mg/L Na_2S**
0	13.1	198,900	198,900
5	12.1	202,800	167,700
10	11.1	191,100	159,900
15	10.0	200,850	140,400
20	9.1	195,000	117,000
21	8.2	187,200	93,600
Average		195,975	

noted that the double salt, $Na_2CO_3{\cdot}NaHCO_3{\cdot}2H_2O$, formed as described in **Equation 11.33.** At a concentration of 500 mmol $NaHCO_3$/20 535 mmol H_2O, it was noted that $NaHCO_3$(aq) start to precipitate as Na_2CO_3/$NaHCO_3{\cdot}2H_2O$, which corresponds to a concentration of 48.7 g/L $NaHCO_3$.

$$3NaHCO_3 + H_2O \rightarrow Na_2CO_3 \cdot NaHCO_3 + CO_2 + 2H_2O \tag{11.33}$$

11.4.7 Production of MgO from $MgSO_4$ and NaHS

The purpose of this section is to identify a way how only $Mg(OH)_2$(s) will be formed from brine streams rich in $MgSO_4$. During freeze crystallization, $MgSO_4$ remains in solution due to its high solubility of 260 g/L (Table 11.14). The use of Na_2S appeared to be attractive as the products, Na_2SO_4, as shown in **Equation 11.34,** is highly soluble (408 g/L) (Table 11.14). The use of BaS is not attractive as $BaSO_4$ would form, which has a low solubility (0.0025 g/L) (Table 11.14).

11.4.7.1 Reaction Between Na_2S and $MgSO_4$

Table 11.22 shows the OLI simulation for the conversion of $MgSO_4$ to MgO with Na_2S. From Table 11.22, it can be seen that $Mg(OH)_2$ formed when $MgSO_4$ was contacted with Na_2S (**Equation 11.34**). When 500 mmol Na_2S

Table 11.21 Conversion of $NaHCO_3$ to Na_2CO_3 [11].

$NaHCO_3$ (mmol] (Y2)	H_2O (mmol)	$NaHCO_3$ (Nahcolite) - (s) (mmol)	$Na_2CO_3 \cdot NaHCO_3 \cdot 2H_2O$ - (s) (mmol)	$Na^{(+1)}$ (aq) (mmol)	H_2O - (vap) (mmol)	CO_2 - (vap (mmol)
500	55,500	0	0	500	29,279	204
500	50,505	0	0	500	28,118	206
500	45,510	0	0	500	26,750	209
500	40,515	0	0	500	25,143	212
500	35,520	0	0	500	23,263	214
500	30,525	0	0	500	21,082	217
500	25,530	0	0	500	18,589	219
500	20,535	0	0	500	15,810	221
500	15,540	0	31.5	406	13,006	211
500	10,545	0	163.4	10	10,321	168

Temp = 100°C; pH = 9.46

was contacted stepwise with $MgSO_4$, it was noted that (i) the pH dropped from 12.9 to 10.1 and lower as $Mg(OH)_2$ formed; (ii) 249 mmol $Mg(OH)_2$ formed from 500 mmol Na_2S, showing that two Na_2S reacted with one $MgSO_4$ to form $Mg(OH)_2$(s), Na_2SO_4(aq) and NaHS(ag); and (iii) the remaining 250 mmol Mg^{2+} was linked to SO_4^{2-}, and (iv) the Na^+ from Na_2S was linked to HS^-:

$$MgSO_4 + 2Na_2S + 2H_2O \rightarrow Mg(OH)_2 + 2NaHS + Na_2SO_4 \quad (11.34)$$

From Table 11.22, OLI simulations further showed that, at concentrations above 2100 mmol/L $MgSO_4$ and Na_2S, solid $Na_2SO_4 \cdot 10H_2O$ precipitated together with $Mg(OH)_2$. In the case of $MgSO_4$ and CaS, solid gypsum precipitated together with $Mg(OH)_2$. The reaction shown in **Equation 11.34**

Table 11.22 Conversion of $MgSO_4$ to MgO with Na_2S (OLI simulation).

$MgSO_4$ (mmol)	Na_2S (mmol)	pH	$MgSO_4$ (mmol)	$Mg(OH)_2$ (Brucite) - Sol (mmol)	$Mg^{(+2)}$ Aq (mmol)	SO_4^{-2} (mmol)	HS^{-1} (mmol)	S^{-2} (mmol)	$S^{(-2)}$ (aq) (mmol)	$S^{(+6)}$ (aq) (mmol)	$Na^{(+1)}$ (aq) (mmol)
0	500	12.9	0	0	0	0	105.8	394.2	500	0	1,000
50	500	12.6	50	50	0	22.8	160.7	339.3	500	50	1,000
100	500	12.4	100	100	0	46.4	233.5	266.5	500	100	1,000
150	500	12.1	150	150	0	70.6	317.4	182.6	500	150	1,000
200	500	11.7	200	200	0	95.3	407.1	92.9	500	200	1,000
250	500	10.1	250	249	1.4	120.5	497	2.8	500	250	1,000
300	500	9.4	300	250	49.8	146.7	498.3	0.5	500	300	1,000
350	500	9.2	350	250	99.5	173.7	498	0.4	500	350	1,000
400	500	9.2	400	251	149.4	201.5	497.8	0.4	500	400	1,000
450	500	9.1	450	251	199.2	230.1	497.6	0.4	500	450	1,000
500	500	9.1	500	251	249.2	259.5	497.4	0.4	500	500	1,000

Temp = 25°C

was confirmed experimentally by reacting Na_2S with $MgSO_4$ at $Na_2S/MgSO_4$ mole ratios of 1, 2, and 4 as shown in Table 11.23. It was also found that Mg^{2+} was removed as $Mg(OH)_2$ at a mole ratio of 1 as indicated by the low alkalinity value of 124,800 and the low Mg^{2+} values where the ratio was and higher. Sulfide remained high at the same concentration of the feed.

$Mg(OH)_2(s)$ can be separated from the NaHS and the Na_2SO_4 through settling or filtration with a belt filter. The separation of Na_2SO_4 and NaHS can be done through freeze crystallization as discussed in the next section.

11.4.7.2 Separation of Na_2SO_4 and NaHS

Na_2SO_4 can be separated from the NaHS through cooling as Na_2SO_4 has a solubility of only 45 g/L at 0°C compared to 400 g/L at room temperature. The solubility of Na_2SO_4 was first studied for a pure Na_2SO_4 solution, with

Table 11.23 Effect of $Na_2S/MgSO_4$ mole ratio on the conversion of $MgSO_4$ to $Mg(OH)_2$.

Parameter	Feed	$Na_2S/MgSO_4$ Mole Ratio			Average
		1	2	4	
Temperature (°C)		25	25	25	25
Na_2S (g/L)	400	400	400	400	400
Na_2S stock sol (g/L)	400	400	400	400	400
$MgSO_4$ stock sol (g/L)	616.9	616.9	616.9	616.9	616.9
Na_2S (g/L)		200	200	200	200
$MgSO_4$ (g/L)		308.46	154.23	77.12	
Na_2S stock sol (mL)	100	100	100	100	100
$MgSO_4$ stock sol (mL)	0	100	50	25	
H_2O (mL)		0.00	50.00	75.00	
Total Vol (mL)		200	200	200	200
Ph	13.13	10.68	10.41	10.28	
Sulfide mg/L Na_2S)	191,100	198,900	195,000	189,150	193,538
Alk (mg/L Na_2S)	202,800	124,800	140,400	167,700	
Mg^{2+} (mg/L Na_2S)		3,861.00	3,705.00	3,549.00	

both OLI simulation (Table 11.24) and experimentally (Table11.24). OLI simulation showed that Na_2SO_4 has a solubility of 44.3 g/L at 0°C. This compared well with the experimentally determined solubility of 49.9 g/L at 0°C as showed in Table 11.25.

The results collected with a pure Na_2SO_4 solution were repeated for a solution containing both Na_2SO_4 and NaHS using OLI simulation.

Table 11.26 shows the effect of temperature when 2,000 mmol NaHS (112.0 g) reacted with 1,000 mmol Na_2SO_4 (142 g) (OLI simulation). It was noted that 928.4 mmol $Na_2SO_4{\cdot}10H_2O$ crystallized from the 1,000 mmol Na_2SO_4 due to its decreased solubility at colder temperatures. NaHS remained in solution as it has a high solubility of 88,203 mmol/L at 25°C.

Table 11.24 Effect of temperature on the solubility of Na_2SO_4 (OLI simulation).

Temperature (°C)	Na_2SO_4 (mmol)	$Na_2SO_4 \cdot 10H_2O$ (Mirabilite) (mmol)	$S^{(+6)}$ (aq) (mmol)	$Na^{(+1)}$ (aq) (mmol)	SO_4 (g/L)	Na (g/L)	Na_2SO_4 (g/L)	H_2O (mmol)
60	2,500	0.0	2,500.0	5,000.0	240.00	115.00	355.0	55,508
54	2,500	0.0	2,500.0	5,000.0	240.00	115.00	355.0	55,508
48	2,500	0.0	2,500.0	5,000.0	240.00	115.00	355.0	55,508
42	2,500	0.0	2,500.0	5,000.0	240.00	115.00	355.0	55,508
36	2,500	0.0	2,500.0	5,000.0	240.00	115.00	355.0	55,508
30	2,500	0.0	2,500.0	5,000.0	240.00	115.00	355.0	55,508
24	2,500	1,012.8	1,487.1	2,974.3	174.63	83.68	258.3	45,380
18	2,500	1,687.0	813.0	1,625.9	112.12	53.72	165.8	38,637
12	2,500	2,027.0	473.0	946.0	71.53	34.27	105.8	35,238
6	2,500	2,211.8	288.2	576.6	45.99	22.05	68.0	33,391
0	2,500	2,318.5	181.5	363.0	29.92	14.34	44.3	32,323

Table 11.25 Effect of temperature on the solubility of Na_2SO_4 (experimentally determined).

Temp.		$Na_2SO_4 \cdot 10H_2O$		
°C	pH	g/L SO_4	g/L Na	g/L Na_2SO_4
25	6.96	131,700	56,300	188,000
20	7.18	103,800	44,600	148,400
15	7.45	58,200	24,800	83,000
10	7.59	51,680	23,040	74,720
5	7.86	41,300	18,600	59,900
0	7.97	32,760	14,210	46,970

Table 11.26 Separation of Na_2SO_4 and NaHS through solubility difference during cooling (OLI simulation).

Temperature (°C)	H_2O (mmol)	Na_2SO_4 (mmol)	$Na_2SO_4{\cdot}10H_2O$ (Mirabilite) (mmol)	NaHS (mmol)	$S^{(-2)}$ (aq) (mmol)	$SO_4^{2-}(+6)$ Aq (mmol)	$Na^{(+1)}$ (aq) (mmol)	$S^{(+6)}$ (aq) (mmol)
25	55,508	1,000	0.0	2,000	2,000	1,000	4,000	1,000
23	55,508	1,000	0.0	2,000	2,000	1,000	4,000	1,000
20	55,508	1,000	86.6	2,000	2,000	913	3,827	913
18	55,508	1,000	338.6	2,000	2,000	661	3,323	661
15	55,508	1,000	519.2	2,000	2,000	481	2,962	481
13	55,508	1,000	649.7	2,000	2,000	350	2,701	350
10	55,508	1,000	744.6	2,000	2,000	255	2,511	255
8	55,508	1,000	813.8	2,000	2,000	186	2,372	186
5	55,508	1,000	864.4	2,000	2,000	136	2,271	136
3	55,508	1,000	901.4	2,000	2,000	99	2,197	99
0	55,508	1,000	928.4	2,000	2,000	72	2,143	72

NaHS can be concentrated through removal of water through freeze crystallization. The freeze-crystallization technology is essentially a freeze desalination process that separates a liquid phase from a dissolved solid phase. A temperature change of approximately 25°C to 0°C is required, which indicates a temperature drop of 25°C or more. The freeze-crystallization principle shows the potential for economic efficiency in terms of low energy demand, low by-product contamination, process flexibility, simplicity, and the possibility of further use of low-temperature energy as a process by-product. The main advantage of freeze crystallization over current wastewater treatment technologies such as distillation and evaporation ponds come from the much lower heat of fusion of ice (333 kJ/kg) compared to the heat of vaporization (2,500 kJ/kg) of water [53].

11.4.7.3 Up-Concentration of NaHS

After removal of $NaHCO_3$ as a solid, NaHS remained in solution [4]. Figure 11.5 shows that, when 1,000 mL of water (55,508 mmol) was reduced to 150 mL (8,117 mmol) by freeze crystallization, the NaHS concentration in solution increased from 188 g/L to 1,180 g/L (21,100 mmol/L). The $NaHCO_3$ concentration would have increased to the same extent if the $NaHCO_3$ was not removed as a solid prior to water removal through freeze crystallization.

11.4.8 Sulfur from H_2S

11.4.8.1 H_2S Oxidation with Fe^{3+}

H_2S gas is produced when BaS, CaS, or Na_2S is contacted with CO_2 (**Equation 11.29**). H_2S can be converted to S by using the Claus process [54]:

$$H_2S + 1.5O_2 \rightarrow SO_2 + H_2O \qquad (11.35)$$

$$2H_2S + SO_2 \rightarrow 3S + 2H_2O \qquad (11.36)$$

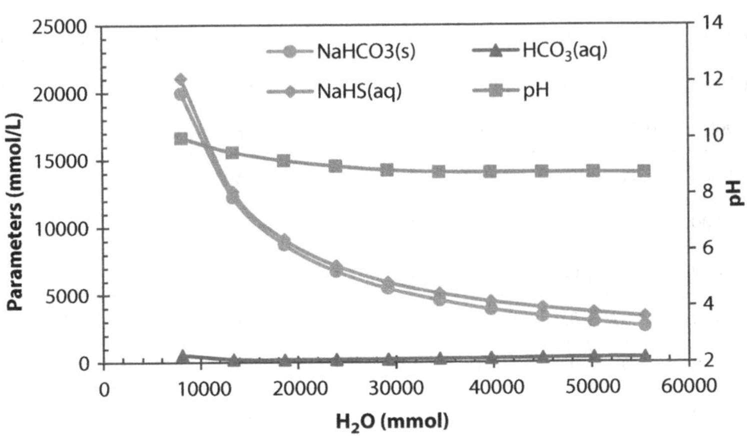

Figure 11.5 Concentration of NaHS by freeze crystallization [4, 10].

$$3H_2S + 1.5O_2 \rightarrow 3S + 3H_2O \tag{11.37}$$

It would be attractive if H_2S can be converted to S by contacting it with a solution. Masukume *et al.* (2013) assumed that S forms when H_2S is contacted with an Fe^{3+} solution (**Equation 11.38**) [6]. OLI software was used to verify if this assumption was correct. If not correct, then the question is which compound was formed?

Table 11.27 showed the oxidation of 600 mmol/L H_2S (20 400 mg/L) with $Fe_2(SO_4)_3$ (Fe^{3+}). The effect of Fe^{3+} concentration was shown by stepwise increasing the dosage of $Fe_2(SO_4)_3$ from 0 to 3,000 mmol/L. It was noted that 300 mmol $Fe_2(SO_4)_3$ (600 mmol Fe^{3+} and 900 mmol SO_4^{2-}) reacted with 600 mmol H_2S to form FeS_2. The 900 mmol S^{+6} (SO_4^{2-}) came from the $Fe_2(SO_4)_3$ that was dosed. FeS_2 formed when the $Fe_2(SO_4)_3$/H_2S mole

Table 11.27 H_2S oxidation with Fe^{3+} for 600 mg/L H_2S (OLI simulation).

H_2S (mmol) (Y2)	$Fe_2(SO_4)_3$ (mmol)	pH	$S^{(-2)}$(aq) (mmol)	S_8 (Sulfur) - (s) (mmol)	FeS_2 (Pyrite) - (s) (mmol)	$S^{(+6)}$(aq) (mmol)	$Fe^{(+2)}$ (aq) (mmol)	$Fe^{(+3)}$ (aq) (mmol)	$FeSO_4.7H_2O$ (Melanterite) (mmol)	H_2O (mmol)
600	0	4.00	99.3	0.0	0.0	0	0	0.0	0.0	55,560
600	300	0.43	0.0	0.0	300.0	900	300	0.0	0.0	55,560
600	600	0.28	0.0	0.0	257.1	1,886	943	0.0	0.0	55,560
600	900	0.27	0.0	62.5	0.0	2,800	1,800	0.0	0.0	55,560
600	1,200	0.04	0.0	50.0	0.0	3,800	2,400	0.0	0.0	55,560
600	1,500	−0.19	0.0	37.5	0.0	4,347	2,547	0.0	453.1	55,560
600	1,800	−0.42	0.0	25.0	0.0	4,796	2,596	0.1	1,004.0	55,560
600	2,100	−0.67	0.0	12.5	0.0	5,171	2,571	0.4	1,628.8	55,560
600	2,400	−0.96	0.0	0.1	0.0	5,536	2,534	2.5	2,263.9	55,560
600	2,700	−1.72	0.0	0.0	0.0	4,507	607	600.0	4,193.4	55,560
600	3,000	−1.77	0.0	0.0	0.0	4,949	149	1,200.0	4,650.5	55,560

ratio was 0.5 (Equation 11.40). Due to the increase in the concentration of $Fe_2(SO_4)_3$ (0 to 3,000 mmol/L), the pH dropped from 4 to negative values.

$$2H_2S + Fe(SO_4)_2 \rightarrow FeS_2 + 2H_2SO_4 \tag{11.38}$$

Upon further dosing of $Fe_2(SO_4)_3$ to 780 mmol/L, the formed FeS_2 was converted to mainly S_8 (**Equations 11.39** and **11.40**) and a smaller portion was converted to SO_4^{2-}.

$$4FeS_2 + 8Fe^{3+} \rightarrow S_8 + 12Fe^{2+} \tag{11.39}$$

$$4FeS_2 + 4Fe_2(SO_4)_2 \rightarrow S_8 + 12FeSO_4 \tag{11.40}$$

With further dosing of $Fe_2(SO_4)_3$, the S_s can be oxidized to H_2SO_4 (**Equations 11.41** and **11.42**). A large excess mole ratio is needed, namely, a Fe^{3+}/S_8 mole ratio of 48.

$$S_8 + 48Fe^{3+} + 32H_2O \rightarrow 8SO_4^{2-} + 48Fe^{2+} + 64H^+ \tag{11.41}$$

$$S_8 + 24Fe_2(SO_4)_3 + 32H_2 \rightarrow 48FeSO_4 + 32H_2SO_4 \tag{11.42}$$

The above experiment was repeated for a lower H_2S concentration of 100 mg/L H_2S. The results are shown in Table 11.28. As it can be seen in Table 11.28, FeS_2 formed when 50 mmol/L $Fe_2(SO_4)_3$ was added to 100 mmol/L H_2S (**Equation 11.42**). Upon further addition of $Fe_2(SO_4)_3$, no S_8 was formed, as in the case of higher concentrations. FeS_2 was oxidized to form $FeSO_4$ and H_2SO_4 (**Equations 11.43** and **11.44**). The lower H_2S concentration can be the reason why no S_8 was formed.

$$FeS_2 + 14Fe^{3+} \rightarrow 2S^{6+} + 15Fe^{2+} \tag{11.43}$$

$$FeS_2 + 7Fe_2(SO_4)_3 + 8H_2O \rightarrow 15FeSO_4 + 8H_2SO_4 \tag{11.44}$$

11.4.8.2 H_2S Oxidation with O_2

In the previous sections, $Fe_2(SO_4)_3$ was used for oxidation. In this section, O_2 has been used for H_2S oxidation. Table 11.29 shows the 100 mol O (50 mmol O_2) was used for oxidation of H_2S to S_8 (**Equation 11.46**).

Table 11.28 H_2S oxidation with Fe^{3+} for 100 mg/L H_2S (OLI simulation).

H_2S (mmol)	$Fe_2(SO_4)_3$ (mmol)	pH	$S^{(-2)}$ (aq) (mmol)	S_8 (Sulfur) - Sol (mmol)	FeS_2 (Pyrite) - Sol (mmol)	$S^{(+6)}$ (aq) (mmol)	$Fe^{(+2)}$ (aq) (mmol)	$Fe^{(+3)}$ (aq) (mmol)	$FeSO_4.7H_2O$ (Melanterite) (mmol)	Enthalpy - Total (cal)
100	0	4	99.3	0	0	0	0	0	0	−3.79E+06
100	50	1.14	0	0	50	150	50	0	0	−3.83E+06
100	100	1.06	0	0	42.86	314	157	0	0	−3.86E+06
100	150	0.98	0	0	35.71	479	264	0	0	−3.89E+06
100	200	0.91	0	0	28.57	643	371	0	0	−3.93E+06
100	250	0.84	0	0	21.43	807	479	0	0	−3.96E+06
100	300	0.77	0	0	14.29	971	586	0	0	−3.99E+06
100	350	0.71	0	0	7.14	1,136	693	0	0	−4.03E+06
100	400	0.65	0	0	0	1,300	800	0	0	−4.06E+06
100	450	0.69	0	0	0	1,450	800	100	0	−4.10E+06
100	500	0.72	0	0	0	1,600	800	200	0	−4.13E+06

With further addition of O_2, S_8 was oxidized to form H_2SO_4 (**Equations 11.45** to **11.47**).

$$8H_2S + 4O_2 \rightarrow S_8 + 8H_2O \tag{11.45}$$

$$S_8 + 12O_2 + 8H_2O \rightarrow 8H_2SO_4 \tag{11.46}$$

$$S + 1.5O_2 + H_2O \rightarrow H_2SO_4 \tag{11.47}$$

11.4.8.3 Fe^{2+}-Oxidation with O_2

In **Section 11.4.5.1**, it was shown that H_2S can be converted to S by contacting it with Fe^{3+}. The products, Fe^{2+}, need to be back oxidized to Fe^{3+}. This could be an alternative to the Claus process for the treatment of

Table 11.29 H_2S oxidation with O_2 (OLI simulation).

H_2S (mmol)	O_2 (mmol)	pH	$S^{(-2)}$ (aq) (mmol)	S_8 (Sulfur) - Sol (mmol)	$S_{(+6)}$ (aq) (mmol)	Enthalpy - Total (cal)	O_2 - (vap) (mmol)
100	0.0	4.00	99.3	0.0	0.0	−3.79E+06	0.0
100	25.0	4.14	50.0	6.2	0.0	−3.80E+06	0.0
100	50.0	3.56	0.4	12.4	0.1	−3.80E+06	0.0
100	75.0	1.69	0.0	10.4	16.7	−3.80E+06	0.0
100	100.0	1.43	0.0	8.3	33.3	−3.80E+06	0.0
100	125.0	1.28	0.0	6.2	50.0	−3.81E+06	0.0
100	150.0	1.18	0.0	4.2	66.7	−3.81E+06	0.0
100	175.0	1.09	0.0	2.1	83.3	−3.81E+06	0.0
100	200.0	1.02	0.0	0.0	100.0	−3.81E+06	0.0
100	225.0	1.02	0.0	0.0	100.0	−3.81E+06	23.8
100	250.0	1.02	0.0	0.0	100.0	−3.81E+06	48.8

H_2S-rich streams. Another application for Fe^{2+} oxidation is when iron-rich acid mine water needs to be neutralized. Acid mine water is normally rich in both Fe^{2+} and Fe^{3+}. The neutralization of Fe^{3+} with $CaCO_3$ is a rapid reaction, whereas the neutralization of Fe^{2+} with $CaCO_3$ is slow. The Fe^{2+} first needs to be oxidized to Fe^{3+}. It would be beneficial if Fe^{2+} can be oxidized before the neutralization stage, so that all Fe is in the Fe^{3+} state. OLI simulations showed that Fe^{2+} can be oxidized to Fe^{3+} (Table 11.30) according to **Equation 11.48.**

$$2Fe^{2+} + \tfrac{1}{2}O_2 + 2H^+ \rightarrow 2Fe^{3+} + H_2O \qquad (11.48)$$

The Fe^{2+} can be back-oxidized to Fe^{3+} through electrolytic, chemical or biological oxidation (**Equations 11.49 and 11.50**). This reaction is important in the pre-treatment of acid mine water before neutralization. Mogashane *et al.* (2023) concluded from their research findings on ferric

Table 11.30 Fe^{2+} oxidation with O_2 (OLI simulation).

$FeSO_4$ (mmol)	O_2 (mmol)	pH	$Fe^{(+2)}$ (aq) (mmol)	$Fe^{(+3)}$ (aq) (mmol)	$S^{(+6)}$ (aq) (mmol)
100	0.0	1.54	100	0	150
100	2.5	1.60	90	10	150
100	5.0	1.66	80	20	150
100	7.5	1.73	70	30	150
100	10.0	1.80	60	40	150
100	12.5	1.89	50	50	150
100	15.0	1.98	40	60	150
100	17.5	2.09	30	70	150
100	20.0	2.22	20	80	150
100	22.5	2.37	10	90	150
100	25.0	2.53	0	100	150

hydroxide recovery from iron-rich acid mine water that that the rate of neutralization with $CaCO_3$ is rapid in the case of mine water rich in Fe^{3+} but slow in the case of Fe^{2+}-rich water [55]. The reasons for this are (i) that Fe^{2+} remains in solution up to pH 8, whereas Fe^{3+} precipitates as $Fe(OH)_3$ at pH 3.5 and (ii) the rate of Fe^{2+} oxidation increases with increase in pH values. At pH 7.2 and higher, the rate of Fe^{2+} oxidation is fast. The difference in behavior is that Fe^{2+} first needs to be oxidized to Fe^{3+}:

$$H_2S + 2Fe^{3+} \rightarrow S + 2Fe^{2+} + 2H^+ \tag{11.49}$$

$$2Fe^{2+} + 0.5O_2 + 2H^+ \rightarrow 2Fe^{3+} + H_2O\left(\text{pressure} + O_2 \text{ reaction}\right) \tag{11.50}$$

11.5 Conclusions

i. $CaSO_4$ and $BaSO_4$ can be thermally reduced to CaS and BaS, respectively, as demonstrated by both muffle furnace and the Pyrosim prediction simulation studies.
ii. BaS is more suitable than CaS to produce Na_2S due to the low solubility of $BaSO_4$.
iii. $NaHCO_3$ and NaHS can be produced by contacting Na_2S and CO_2.
iv. Na_2S can be used to produce $Mg(OH)_2$ by contacting it with $MgSO_4$.
v. MgO can be produced from $Mg(OH)_2$ through heating.
vi. Freeze crystallization can be used to separate (i) NaHS and Na_2SO_4 and (ii) $NaHCO_3$ and NaHS.
vii. The Fe^{2+} can be back oxidized to Fe^{3+} through chemical or biological oxidation.
viii. H_2S can be converted to S by contacting it with Fe^{3+}.

Acknowledgments

The authors thank TIA Seed Fund, F'SAGRI and the Department of Trade and Industry and Competition (DTIC) for financial support, and NIPMO, who sponsored the grant for the innovation award of 2021 NSTF-South32, that was received by ROC Water. Dr Fritz Carlsson provided proof-reading and extensive editorial advice.

References

1. Sasowsky, I., Foos, A., Miller, C., Lithic controls on the removal of iron and remediation of acidic mine drainage. *Water Res.*, 34, 10, 2742–2746, 2000.
2. Sawyer, D., Fleck, R., Lanphere, M., Warren, R., Broxton, D., Hudson, M., Episodic caldera volcanism in the Miocene southwestern Nevada volcanic field: Revised stratigraphic framework, 40Ar/39Ar geochronology, and implications for magmatism and extension. *Geol. Soc. Am. Bull.*, 106, 10, 1304–1318, 1994.
3. RoyChowdhury, A., Sarkar, D., Datta, R., Remediation of acid mine drainage-impacted water. *Curr. Pollut. Rep.*, 3, 1, 131–141, 2015.
4. Mashigwana, L., Maree, J.P., Monyatse, L., Adeniyi, A., Onyango, M., Thermal processing of sodium sulphate to sodium carbonate. *International Institute*

of Biological Chemical and Environmental Engineering (IICBEE), Birchwood Hotel & OR Tambo Conference Centre, Johannesburg, South Africa, Mar. 17–18, 2022.
5. van Vuuren, D.S. and Maree, J.P., *Production of sodium carbonate from sodium sulphate*, 18 November 2020, PCT Patent Provisional Patent Appl. No 2020/07178.
6. Masukume, M., Maree, J.P., Ruto, S., Joubert, H., Processing of Barium Sulphide to Barium Carbonate and Sulphur. *J. Chem. Eng. Process Technol.*, 4, 2013, doi: 10.4172/2157-7048.1000157.
7. Ruto, S., Maree, J.P., Zvinowanda, C.M., Kolesnikov, A., Louw, W., Thermal studies on barium sulphate in a pilot scale, rotary kiln, *9th International Conference on Clean Technologies for the Mining Industry, Sheraton Santiago Hotel & Convention Center, Chile*, 10–12 April, 2011.
8. Ruto, S.J., Maree, J.P., Zvinowanda, C.M., Louw, W.J., Thermal processing of alkali metal sulphates to alkali metal sulphides in a pilot scale, rotary kiln. *WISA Biennial Conference, Cape Town International Conference Centre*, 5 – 9 May, 2012.
9. Ruto, S., Maree, J.P., Zvinowanda, C.M., Kolesnikov, A.V., Louw, W.J., Thermal studies on gypsum in a pilot-scale, rotary kiln, *Water in the South African Minerals Industry Conference, The Southern African Institute of Mining and Metallurgy, Ingwenyama Conference & Sport Resort, White River*, 15 – 17 February, 2011.
10. Mokgohloa, C.P., Maree, J.P., van Vuuren, D.S., Modibane, K.D., Mujuru, M.M.M., Recovery of Na2CO3 and nano CaCO3 from Na2SO4 and CaSO4 wastes, in: *Nano Technology*, pp. 1–2, Wiley Scrivener, New York, 2022.
11. Mokgohloa, C.P., Maree, J.P., Motaung, M.P., Mokhonoana, M.P., Processing of Na2S to Na2CO3 and NaHS, in: *35th JOHANNESBURG International Conference*, Nov 29 - 29, 2022.
12. Ali, M., Atoms in Gaseous and Solid States and their Energy and Force Relationships under Transitional Behaviors. *J. Phys. Sci.*, 2020, Doi.org/10.21203/rs.3.rs-88120/v7.
13. Goodstein, D.L., *States of Matter 1975*, Dover, 2002.
14. Dalarsson, N., Dalarsson, M., Golubovic, L., *Introductory statistical thermodynamics*, Academic Press, San Diego, 2011.
15. San Marchi, C., Somerday, B., Robinson, S., Permeability, solubility and diffusivity of hydrogen isotopes in stainless steels at high gas pressures. *Int. J. Hydrogen Energy*, 32, 1, 100–116, 2007.
16. Zhang, Y., *Geochemical kinetics*, Princeton University Press, Ann Arbor, 2021.
17. Robinson, P., Enzymes: principles and biotechnological applications. *Essays Biochem.*, 59, 1, 2015.
18. Desai, F., Prasad, J., Muthukumar, P., Rahman, M., Thermochemical energy storage system for cooling and process heating applications. *A Review Energy Convers. Manage.*, 229, 113617, 2021.

19. Mtombeni, T. and Maree, J.P., *Treatment of water - ROC process*, 2014, USA Patent 2014/04734Treatment of water - ROC process, ROC Water Technologies, RSA (No 2015/04639), Germany (No 3452414), France (No 3452414), Australia (No 2016405754 - Pending), USA No 16/099,009 - Pending).
20. Mtombeni, T., Maree, J.P., Zvinowanda, C.M., Asante, O., Louw, W., Improved freeze desalination process for recovery of water and salts from brines. *Proceedings of the 2014 WISA Biennial Conference, Mbombela*, 25–29 May, 2014.
21. Mtombeni, T., Maree, J.P., Zikalala, N., F. C. J. S Desalination with a combined membrane filtration/freeze desalination process. *WISA Biennial Conference*, 2016.
22. Kucera, J., *Desalination: Water from water*, Scrivener, Hobokem, USA, 2019.
23. Hendrickson, H. and Moulton, R., *Research and development of processes for desalting water by freezing. R&D Report 10, US Dept. of Commerce*, Office of Saline Water, 1956.
24. Karnofsky, G. and Steinhoff, P., *Saline water conversion by direct freezing with butane. R&D Report 40, US Dept. of Commerce*, Office of Saline Water, 1960.
25. Wiegandt, H., Harriott, P., Leinroth J, P., *Desalting of seawater by freezing. R&D Report 376, US Dept. Of Commerce*, Office of Saline Water, US, 1968.
26. Lu, Z. and Xu, L., Freeze desalination processes. *Incyclopedia Desalin. Water Resour.*, 2, 2010.
27. Randall, D.G., Nathoo, J., Lewis, A.E., A case study for treating a reverse osmosis brine using Eutectic Freeze Crystallization-Approaching a zero waste process. *Desalination*, 266, 256–262, 2011.
28. Williams, P., Ahmad, M., Connolly, B., Radcliffe, D., Technology for freeze concentration in the desalination industry. *Desalination*, 356, 314–327. 2015.
29. Wang, P. and Chung, T., A conceptual demonstration of freeze desalination-membrane distillation (FD-MD) hybrid desalination process utilizing liquefied natural gas (LNG) cold energy. *Water Res.*, 46, 4037–4052, 2012.
30. Rich, A., Youssef, M., Nourimane, B., Denis, M., Souad, A., Christine, B., Naoual, S., Jean-Paul, K., Tijani, B., Ahmed, B., Stephane, V., Freezing desalination of seawater in a static layer crystallizer. *Desalination*, 269, 142–147, 2011.
31. Rahman, M., Ahmed, M., Chen, M., Freezing-Melting Process and Desalination: I. Review of the State-of-the-Art. *Sep. Purif. Rev.*, 32, 2, 59–96, 2006.
32. Randall, D., Nathoo, J., Lewis, A., Seeding for selective salt recovery during eutectic freeze crystallization, *Proceedings of International Mine Water Conference (IMWA)*, no .19, pp. 639–646, Pretoria, South Africa, 2009.
33. Amran, N., Samsuri, S., Safiei, N., Zakaria, Z., Jusoh, M., Parametric Study on the Performance of Progressive Cryoconcentration System. *Chem. Eng. Commun.*, 203, 7, 957–975, 2016.

34. Miyawaki, O., Kato, S., Watabe, K., Yield improvement in progressive freeze-concentration by partial melting of ice. *J. Food Eng.*, 103, 3, 377–382, 2012.
35. Liu, L., Miyawaki, O., Nakamura, K., Progressive freeze concentration of model liquid food. *Food Sci. Technol. Int.*, 3, 348–352, 1997.
36. Roos, A., Verschuur R, J., Schreurs, B., Scholz, R., Jansens, P., Development of a Vacuum Crystallizer for the Freeze Concentration of Industrial Wastewater. *Chem. Eng. Res. Des.*, 81, 8, 881–892, 2003.
37. Stepakoff, G., Siegelmann, D., Johnson, R., Gibson, W., Development of a eutectic freezing process for brine disposal. *Desalination*, 14, 25-38, 1974.
38. Lorain, O., Thiebaud, P., Badorc, E., Aurelle, Y., Potential of freezing in wastewater treatment: soluble pollutant applications. *Water Res.*, 35, 541–547, 2001.
39. Van der Ham, F., Eutectic Freeze Crystallization: Application to process streams and waste water purification. *Chem. Eng. Process.*, 37, 2, 207–213, 1998.
40. Genceli, F., Gartner, R., Witkamp, G., Eutectic freeze crystallization in a 2nd generation cooled disk column crystallizer for the recovery of MgSO4.12H2O. *J. Cryst. Growth*, 275, 1369–1372, 2005.
41. Genceli, F., Scaling-up eutectic freeze crystallization, Delft University, 2008, Ph.D. Thesis.
42. Halde, R., Concentration of impurities by progressive freezing. *Water Res.*, 14, 6, 575–580, 1980.
43. Johnson, W., State-of-the-art of freezing processes, their potential and future. *Desalination*, 19, 1, 349–358, 1976.
44. Pronk, P., Fluidized bed heat exchangers to prevent fouling in ice slurry systems and industrial crystallizers, Delft University of Technology, 2006, Ph.D. Thesis.
45. Reddy, S.T., Lewis, A.E., Witkamp, G.J., Kramer, H.J.M., van Spronsen, J., Recovery of Na2SO4-10H2O from reverse osmosis retentate by euctetic freeze crystallization technology. *Chem. Eng. Res. Des.*, 88, 1153–1157, 2010.
46. Seckler, M., Verdoes, D., Witkamp, G., Application of eutectic freeze to process streams and wastewater purification. *Economy Ecol. Technol. Ministry Econ. Aff.*, 2002, www.Senternovem.nl/mmfiles/EETK97129_tcm24-154587.
47. Huige, N., *Nucleation and growth of ice crystals from water and sugar solutions in continuous stirred tank crystallisers*, Eindhoven, Eindhoven University of Technology, 1972.
48. APHA, *Standard methods for the examination of water and wastewater*, 12th, American Public health Association, Washington, DC, 2012.
49. Abdellatif, M. and Freeman, M., Mintek Thermal Magnesium Process: Status and Prospective. *Adv. Met. Initiative*, Dept of Science and Technology, Johannesburg, 18–19 October, 2008.
50. OLI, 23 11 2015, http://www.olisystems.com/[Online]. [Accessed 2015].

51. Speight, J., Industrial organic chemistry. *Environ. Org. Chem. Eng.*, 87–151, 2017.
52. Stenger, V., Solubilities of various alkali metal and alkaline earth metal compounds in methanol. *J. Chem. Eng. Data*, 41, 5, 1111–1113, 1996.
53. Mtombeni, T., Maree, J., Zvinowanda, C., Asante, J., Oosthuizen, F., Louw, W., Evaluation of the performance of a new freeze desalination technology. *Int. J. Environ. Sci. Technol.*, 10, 3, 545–550, 2013.
54. Elsner, M., Menge, M., Müller, C., Agar, D., The Claus process: teaching an old dog new tricks. *Catal. Today*, 79, 487–494, 2003.
55. Mogashane, T.M., Maree, J.P., Mujuru, M., Mphahlele-Makgwane, M.M., Modibane, K.D., Ferric Hydroxide Recovery from Iron-Rich Acid Mine Water with Calcium Carbonate and a Gypsum Scale Inhibitor. *Minerals*, 13, 2, 167, 2023.
56. Robie, R., Russell-Robinson, S., Hemingway, B., Heat capacities and entropies from 8 to 1000 K of langbeinite (K2Mg2 (SO4) 3), anhydrite (CaSO4) and of gypsum (CaSO4· 2H2O). *Thermochim. Acta*, 139, 67–81, 1989.
57. Engineering-ToolBox, Engineering ToolBox - Resources, Tools and Basic Information for Engineering and Design of Technical Applications, 2001, [Online]. Available: https://www.engineeringtoolbox.com/index.html. [Accessed August 26 2021].
58. Green, D.W. and Southard, M.Z., *Perry's chemical engineers' handbook*, 9, McGraw-Hill Education, New York, 2019.

12

Trends, Prospects, and Challenges of Treatment, Recovering, and Synthesizing Valuable Minerals from Acid Mine Drainage

N. Tshikosi[1], B. Nguegang[2], Vhahangwele Masindi[3]* and M.M. Ramakokovhu[1]

[1]*Department of Chemical, Metallurgy and Materials Engineering, University of Technology, Pretoria, South Africa*
[2]*Department of Chemical Sciences, Faculty of Science, University of Johannesburg, Johannesburg, South Africa*
[3]*Department of Environmental Sciences, College of Agriculture and Environmental Sciences, University of South Africa (UNISA), Florida, South Africa*

Abstract

Acidic and metalliferous mine drainage is a harmful effluent from mining activities and worryingly induces myriads of ecological stresses to receiving ecosystems. This notorious wastewater stream is a common legacy of regions with intensive mining activities, mostly coal and gold mines. Traditionally, active and passive treatment methods were the king modality for mine water management, although these had numerous drawbacks such as high complexity, exorbitant costs, high energy requirements, selectivity, and failure to exhaustively remove contaminants. International research strategies and drivers inspire the concept of recycle-reuse system in wastewater treatment, specifically when considering the waste beneficiation and valorization. In the same context, metalliferous acid mine drainage (AMD) could be seen as viable resource instead of a hazardous and toxic waste. So far, myriads of treatment, recovery, and beneficiation modalities and regimes have also been introduced and meticulously appraised. In this book chapter, the existing body of knowledge and latest development in the composition, environmental impacts, and abatement and treatment methods are discussed and comprehensively reviewed. The emphasis was

**Corresponding author*: emasinv@unisa.ac.za; masindivhahangwele@gmail.com

Elvis Fosso-Kankeu, Vhahangwele Masindi, Johannes Maree and Bhekie Mamba (eds.) Customized Technologies for Sustainable Management of Industrial Wastewater: A Circular Economy Approach, (537–564) © 2025 Scrivener Publishing LLC

sheerly placed on valorization and beneficiation opportunities. This will go a long way in averting ecological impacts of AMD. Future research perspective and outlooks were also elucidated in this book chapter.

Keywords: Acid and metalliferous mine drainage, mining activities, water contamination and pollution, wastewater beneficiation and valorization, circular economy concept, reclamation of valuable minerals, resource recovery

12.1 Introduction

Mining has been a great pillar of the economy of countries with intensive mining industries worldwide, leading to significant environmental effects [1]. Acidic and metalliferous mining effluent, the principal legacy of mining, is a significant cause of water pollution around mining areas [2]. Other sources of acid mine drainage (AMD) include galvanic processing and cleaning of the flue vapors at power stations [3]. Effluents such as AMD are mainly generated from the oxidation of sulfide-bearing minerals and mostly pyrite (FeS_2). Once formed, it is costly and difficult to control because of the dissociation of water-soluble components, thereby leading to the mobility of sulfuric acid, which impacts human health and the environment. AMD has high acidity and low pH, which encourages heavy metals leaching into nearby environments [4–6]. This is owed to the high conductivity and elevated content of iron (Fe), aluminum (Al), manganese (Mn), and sulfates (SO_4^{2-}), as well as low or moderate content of copper (Cu), zinc (Zn), cadmium (Cd), lead (Pb), chromium (Cr), and arsenic (As) [7, 8].

12.2 Formation of Acid Mine Drainage

AMD is formed following the oxidation of sulfide-bearing minerals in the presence of water and oxygen. Those sulfide bearing minerals include: pyrite, marcasite, chalcocite amongst others as reported in Table 12.1 [9, 10]. The presence of oxygen facilitates the oxidation process of pyrite, during which, when in contact with water, acid mine drainage is generated when sufficient electrons are present. Water promotes the mobilization of sulfuric acid when it maneuvers through and over the surfaces of acid-bearing rock. A study by Name and Sheridan [10] reported that different factors, including water, oxygen content, microbial activity, and geological strata, play a significant role in the formation rate of AMD [12]. Overall, the formation process of AMD is shown in Figure 12.1.

AMD generation can be described by reaction equations that explain the oxidation of sulfide-bearing minerals to form toxic chemical species in the presence of water. The dissolved Fe, SO_4^{2-}, and H^+ increase the acidity and total dissolved solids in water, leading to decreased pH. Oxidation of ferrous iron to ferric iron occurs when the environment has sufficient oxygen and bacterial activity as summarized in Equations 12.1–12.6 [1, 12–15]. The oxidation of pyrite in the presence of water is represented by Equation 12.1:

$$FeS_2 + 7/2O_2 + H_2O \rightarrow Fe^{2+} + 2SO_4^{2-} + 2H^+ \quad (12.1)$$

The oxidation of Fe^{2+} to Fe^{3+}, which occurs when there is enough oxygen dissolved in water, is represented by Equation 12.2:

$$Fe^{2+} + 1/4O_2 + H^+ \rightarrow Fe^{3+} + 1/2H_2O \quad (12.2)$$

Ferric iron precipitates to form the $Fe(OH)_3$, where the ferric iron can produce more ferrous iron and acidity by further reacting with pyrite. The hydrolysis of Fe is represented by Equation 12.3:

$$Fe^{3+} + 3H_2O \rightarrow Fe(OH)_3 + 3H^+ \quad (12.3)$$

Reaction equation 12.4 represents the complete process of oxidation of pyrite, where ferric iron is commonly the primary oxidant rather than oxygen. Pyrite oxidation contains many steps, such as oxygen-independent reactions and oxygen-dependent reactions. Oxygen-dependent reactions involve ferrous (Fe^{2+}) re-oxidation to ferric (Fe^{3+}) iron and oxidation of reduced sulfur compounds to produce sulfate:

$$FeS_2 + 15/4O_2 + 7/2H_2O \rightarrow Fe(OH)_3 + 2SO_4^{2-} + 4H^+ \quad (12.4)$$

Reaction equation 12.5 illustrates the complete oxidation reaction of pyrite, where Fe^{3+} does not precipitate during reaction equation 12.3, and Fe^{3+} acts as the reducing agent and promotes a fast reaction [16]:

$$FeS_2 + 14Fe^{3+} + 8H_2O \rightarrow 15Fe^{2+} + 2SO_4^{2-} + 16H^+ \quad (12.5)$$

Sulfide-bearing rocks that have the potential to generate AMD when exposed to water and oxgen in the presence of oxidising bacteria are depicted in Table 12.1.

Table 12.1 Sulfide-bearing rocks that can generate AMD.

Minerals	Chemical formula
Pyrite	FeS_2
Marcasite	FeS_2
Chalcocite	Cu_2S
Covelite	CuS
Pyrrhoite	$Fe_{1-x}S$
Molybdenite	MoS_2
Chalcopyrite	$CuFeS_2$
Millerite	NiS
Galena	PbS
Arsenopyrite	FeAsS
Sphalerite	ZnS

Reaction equation 12.6 represents the complete oxidation of pyrite in the presence of low water content.

$$FeS_2 + 15/4O_2 + 1/2H_2O \rightarrow Fe^{3+} + 2SO_4^{2-} + H^+ \quad (12.6)$$

12.3 Composition of Different Mine Drainages

Mine drainage can be acidic, neutral, or alkaline. However, they are commonly referred to as AMD, although they emanate from different sources, such as coal and metal mines with pH ≤ 4 [16, 17].

12.3.1 Acid Mine Drainage

AMD is characterized by low pH (≤ 4); high concentration of Al, Fe, Mn, and SO_4^{2-}; and low content of Cu, Ni, Zn, As, Mg, Cr, Cd, and Pb. Arsenopyrite (FeAsS) is one source of Fe and S that also produces sulfuric acid [17].

Most trace metallic elements are insoluble and strongly sorbed when the pH of water ranges at 6–9, whereby buffering is attained with bicarbonate equilibria. The lowest restrained pH of AMD is 3.6, where water with that pH is concentrated with dissolved constituents with the density of about 1.43g/cm^3 [12, 18–20].

12.3.2 Neutral Mine Drainage

Neutral mine drainage (NMD) is characterized by pH level ranging from 6 to 9, where oxidation of sulfide-bearing rock results in dissolved heavy metals, and the drainage comprehends with other dissolved constituents such as SO_4^{2-} [18]. NMD contains low concentrations of iron (Fe) and elevated concentrations of zinc (Zn), cadmium (Cd), manganese (Mn), antimony (Sb), arsenic (As), and selenium (Se) [21]. However, the heavy metal content of NMD also depends on the parent rock; hence, some NMDs have elevated concentrations of iron (Fe) and copper (Cu). For example, sulfide-bearing minerals of Cu such as chalcanthite ($CuSO_4.5H_2O$) and brochantite [$Cu_4SO_4(OH)$] are common in many mine drainages. Metal oxides such as hydrous ferric oxides, which adsorb or integrate with other elements such as copper into their structure, are mostly present in NMD [17, 18, 22, 23].

12.3.3 Basic Mine Drainage

Mine drainage that has a pH > 9 is considered to be basic mine drainage. This drainage is characterized by: (i) closed circuits of low penetrability rock with rich feldspars that can interact and interchange protons for the alkalis with elimination of bicarbonate over calcite precipitation; (ii) development of groundwater to a sodium bicarbonate type over ion exchange, reduction of sulfate, and oxidation of organic carbon; (iii) altered marls that contain lime dissolute; and (iv) the pH of water rises to pH value above 9 due to dissolution of villiaumite (NaF) and evaporation. In the early stages of weathering of mine wastes with an acid production capacity greater than the carbonate-based neutralization capacity, neutral drainage can be observed [17, 19, 23].

12.3.4 Saline Drainage

Saline mine drainages are mine water with a moderated concentration of solids or metals; it is characterized by acidic to alkaline pH (pH 3.5–8.35), with moderate sulfate content and low concentration iron and sulfates compared to other types of mine drainages. Most saline mine drainage that occurs in an area near the seacoast with seawater intrusion and mines close to areas in the vicinity of dissolving evaporate deposits is commonly dominated by elevated concentrations sodium chlorides. The elevated salinity of mine water is caused by the reaction between the sodium-calcium and chloride water with the host rock (pyrite) [18, 24].

12.4 Challenges of Acid Mine Drainage

The sulfuric acid produced during AMD formation enhance the leaching of a wide range of ions in the acid mine water [15]. When this toxic effluent percolates into different environmental compartments without being treated, it then causes severe environmental challenges and impacts human health, which vary according to AMD pH and concentration of inorganics contaminants [15, 24].

12.4.1 Effect on Human Health

Contaminants in AMD can adversely affect humans when they are exposed or in contact with them. AMD contains toxic heavy metals that are detrimental to human health. Heavy metals such as As, Cd, Pb, Mn, Zn, Cr, Cu, and mercury have negative impacts on human health as they can cause acute and chronic diseases due to their ability to accumulate in consecutive levels of the biological chain [1, 15, 25]. Table 12.2 summarizes the human health impacts that are caused by different heavy metals found in AMD.

Table 12.2 Human health impacts caused by heavy metals found in AMD.

Heavy metal	Health impact	References
Cadmium	Lung cancer, bone defects, increased blood pressure, kidney damage, and renal dysfunction	[1]
Arsenic	Poisoning, dermatitis, and bronchitis	[15]

(*Continued*)

Table 12.2 Human health impacts caused by heavy metals found in AMD. (*Continued*)

Heavy metal	Health impact	References
Lead	Chronic damage, fatal infant, sensor neural deafness, kidney, liver, and gastrointestinal damage, and mental retardation	[26]
Manganese	Contact and inhalation nervous damage central system	[27]
Mercury	Protoplasm poisoning, damage nervous system, spontaneous abortion, and minor physiological changes	[26]
Zinc	Damage to nervous membrane	[1]
Chromium	Damage to nervous system, fatigue, and irritability	[28, 29]
Copper	Anemia, liver and kidney damage, and stomach and intestinal irritation	[29]

12.4.2 Effect on Aquatic Organisms

As heavy metals in AMD precipitate to form oxides, they block sunlight from reaching aquatic organisms. Also, they disrupt the carbon cycle ($O_2 \leftrightarrow CO_2$) in the aquatic ecosystem, which, in turn, starve aquatic organisms of food. Because aquatic animals and plants obtain food from the surrounding environment, they eventually ingest the metal oxides when heavy metal oxides pollute their territory. Consequently, the entire food chain gets affected because heavy metals, especially ferric hydroxides and other toxic heavy metals, can be accumulated by aquatic organisms directly from polluted water and indirectly *via* bioaccumulation in the food chain [1, 15, 25].

12.4.3 Effect on Plants

Despite being very vital for plant metabolism, heavy metals can also become toxic to plants at elevated concentration. As such, a high concentration of toxic heavy metals can affect plant tissues, leading to irregular plant growth. The plants experience oxidative stress when exposed to heavy metals that leads to the disturbance of cellular mutilation and cellular ionic homeostasis. Nutrient absorption by plants and their growth are highly influenced by the soil pH and a proper balance of macro- and micronutrients in the soil. When soil is contaminated with AMD, most essential nutrients that plants ingest, such as calcium, phosphorus, potassium, and nitrogen, become limited, sometimes not available, due to the low pH of the soil. When toxic heavy metals such as Fe, Mn, Cr, As, and Al leach into the soil at low pH, the probability of bioaccumulation increases [6, 15, 22].

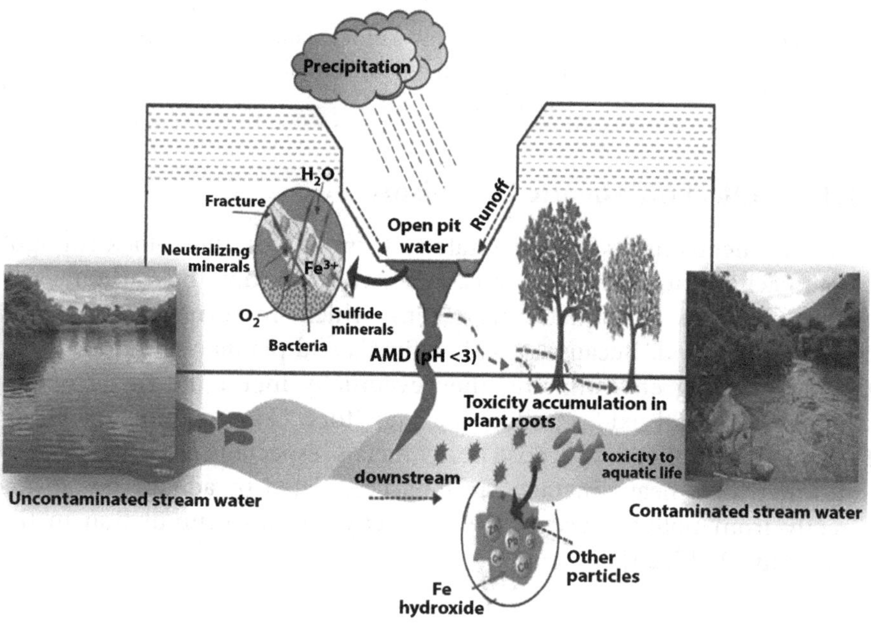

Figure 12.1 AMD generation and pollution pathways [11, 12].

12.5 Treatment Strategies of AMD (Active, Passive, and Integrated)

Mining activities can severely damage the water resources, especially when no action is taken to avoid contamination of the environment or treat the polluted water streams [30]. AMD treatment, control, and remediation strategies remain the focus for many researchers [31]. When AMD is generated, suitable treatment technologies must be implemented to minimize its negative impacts on the different environmental compartments [31]. Remediation of AMD involves consecutive or selective elimination of soluble metals and other toxic environmental contaminants [6]. Numerous treatment technologies exist for the remediation of mine effluents, depending upon the type of drainage and its chemical composition. Treatment technologies that produce water with pH > 7 and low acidity, minimize the level of the heavy metals and sulfates, have compensations such as low cost and ease of operation, and require low maintenance are considered to be effective [32]. AMD that contains an elevated concentration of several dissolved toxic heavy metals is complex and expensive to treat [6]. Most of the existing AMD remediation methods mainly focus on stabilizing, removing, and neutralizing contaminants through various physical, chemical, and biological processes [6]. Membrane technology, ion exchange, adsorption, precipitation, neutralization, biological approach, and electrochemical method are some of the studied remediation techniques for treating AMD.

12.5.1 Passive Technologies

Passive technologies are self-renewable approach that use natural process to treat wastewater [33]. When the contaminated water from the mines gets into contact with alkaline water or carbonate rocks, the pH of mine drainage increases [32]. The advantage of using passive treatment techniques is that it requires relatively small quantities of resources as input [34]. Other advantages of passive technologies include low operational costs, periodic maintenance, and little external energy required. Furthermore, passive treatment technologies can maintain both redox conditions and biological activities by utilizing limestone with or without the aid of organic substances. However, one of the setbacks of passive treatment is that, when the pH of AMD < 2 and the drainage flow rate is above 50 L/s, passive treatment cannot be executed to treat the AMD [35].

12.5.1.1 Wetlands

Wetlands are bio-filters that utilize vegetation for the filtration process and microbial energy as the source of energy. The quality of water is improved through physical, chemical, microbial, and plant-mediated processes. These processes take place in a wetland to reduce the content of inorganic pollutants and neutralize the acidity of the AMD [32]. The mechanisms of the processes that occur in wetlands include oxidation, precipitation, complexation, reduction, sedimentation, filtration, adsorption, chelation, microbial conversion/immobilization, and active plant-intake of heavy metals [31]. There are two comprehensive types of wetlands, namely, aerobic and anaerobic wetlands.

12.5.1.2 Aerobic Wetlands

Aerobic wetland systems are utilized to treat neutral or alkaline water through ion exchange and adsorption [36]. A shallow aerobic wetland system decreases small solid particles that remain in suspension in water and provide substrate and increases the water retention time for reaction to occur between the effluent's alkalinity and acidity. The metal oxidation and precipitation from the AMD within the wetlands generate alkalinity and acidity [31]. Aerobic wetlands contain clay that consists of impermeable sediments with vegetation planted on it. The only difference between aerobic wetlands and other passive treatment technologies is that aerobic wetlands do not neutralize the AMD; therefore, they must obtain net-alkaline water that is often diverted from the passive pretreatment system. Residence time and aeration rate play a major role in the precipitation of heavy metals, such as iron, manganese, arsenic, and chromium, which depend on the redox state of water for their solubility. The precipitates get adsorbed on the surfaces of plants. However, one disadvantage of aerobic wetlands is that they cannot remove various heavy metals from effluents or handle AMD [31].

12.5.1.3 Anaerobic Wetlands

Anaerobic wetlands are retaining ponds consisting of organic matter substrates such as cow or horse dung, sawdust, hay, spent mushroom compost, and limestone aggregate [31]. For the wetland to operate effectively, the root propagation must be inhibited toward the organic substrate because the treatment of AMD is caused by the sub-surface flow in the anaerobic wetland. The root transmission promotes oxygen penetration into the organic substrate, preventing reduction from occurring [36]. Anaerobic wetlands can treat AMD because alkalinity can be generated by reducing sulfate and ferric [36]. Other sources that generate alkalinity in wetlands include sulfate-reducing bacterial (SRB) activity. Organic matter such as carbon sources and sulfates are used as electron acceptors for development. Bicarbonate alkalinity is generated by converting sulfate (SO_4^{2-}) to hydrogen sulfide (H_2S) by bacteria, improving water quality and decreasing acidity. Dissolution of the limestone can also promote alkalinity formation [31]. Reaction equation 12.7 illustrates the generation of H_2S from the reaction of SO_4^{2-} and carbon matter in the presence of water.

$$2H_2O + SO_4^{2-} + 2C_{(\text{organic matter})} \leftrightarrow H_2S + 2HCO_3^- \qquad (12.7)$$

The concentration of heavy metals can be reduced by the precipitation of metal sulfides in the reduced organic layer of the wetland. Some of the metals can be recovered as precipitates of carbonate [31].

12.5.2 Active Technologies

Active or chemical treatment of AMD involves the continuous addition of neutralizing materials/chemicals, such as hydrated lime, magnesite, lime, and limestone, to withstand the process [34]. Chemical treatment is the best method for AMD treatment because it allows to attenuate inorganic pollutants to an acceptable level as required by regulatory body, but at the expense of huge energy input and release of considerable volume of toxic sludge. To increase the rate of chemical oxidation of heavy metals, such as Fe(II), and cause precipitation of metals present in AMD as hydroxides and carbonates, a neutralizing agent such as alkali must be added to raise the pH of the solution [34]. However, like any other technology, the disadvantages of active treatment systems include installing complex plants with precipitators, agitated reactors, clarifiers, and thickeners, leading to an increase in the cost of reagents, maintenance, and operation [32].

12.5.2.1 Chemical Neutralization

Neutralization is water and wastewater treatment technology mostly applied for the treatment and management of AMD and other acidic wastewater. It allows to increase the pH to acceptable level and precipitate inorganic pollutants [31]. Neutralizing agents that are commonly used include lime, ammonia, limestone, sodium carbonate, sodium hydroxide, and magnesium hydroxide [9]. One disadvantage of chemical neutralizing is that the method requires large quantities of chemicals that generate large sludge quantities. As a result, the cost for removing or disposal of sludge increases [37].

12.5.2.2 Membrane Technology

Membranes use selective barriers that prevent the transportation of numerous chemical species between two phases [38]. The membrane process has become a promising process to reduce effluent discharge and maximize water recovery from wastewater [6]. There are four membrane processes, namely, nanofiltration, microfiltration [39], ultrafiltration (UF) [40], and reverse osmosis (RO) [38]. A disadvantage of treating AMD is membrane fouling [6]. However, RO and nanofiltration are considered effective processes because they are applied at real mines [41]. Both processes have gained attention because of their high capacity of salt and metal holding. Integration of the adsorption and membrane processes, for example, UF and RO, has been reported to reduce membrane fouling and risk performances. Filtration is usually done through sand filtration using coal fly-ash and rice husk-ash adsorbent columns as a pre-treatment technique and then further treated by UF and RO [6]. A membrane system has two separate feed streams: the retentate and permeate, and a product stream that can either retentate or permeate. Materials that can generate stable thin films, such as glass, polymers, and ceramics, can be used to synthesize membranes. Mechanisms that membranes utilize for separation are molecular sieving, electrostatic interaction, and adsorption effects. Factors such as equilibrium concentration, temperature difference, electrical potential difference, and pressure difference primarily affect the separation process in a membrane [38]. The membrane processes that are currently in use include electrodialysis, membrane distillation (MD), membrane electrolysis, and anion and cation exchange membranes [42].

The membrane-based methods such as electrodialysis and MD are the most recommended effective technology for AMD treatment to recover water and acid [15, 43]. Electrodialysis is one of the membrane processes that separate the ions across the charged membrane from one solution to another, utilizing the electric field's driving force. Properties of ion such as solution conductivity, mobility, charge, relative concentrations, and applied voltage affect the direction of transport and rate of the ion [15, 22]. Ion exchange membranes are primarily utilized in electrodialysis [15, 22]. MD is the process that has been applied in various applications such as food processing, removal of volatile organic compounds, water, and wastewater treatment, and desalination [15]. The process mentioned above is the most effective membrane technique for acid reclamation. The MD process utilizes the conventional distillation and membrane separation process [42]. It is different from other membrane separation process such as RO, nanofiltration, and microfiltration, which is driven by the absolute pressure difference, or electrodialysis by an electrical potential, or dialysis by the concentration gradient; the MD utilizes the hydrophobic and microporous membrane, which act as the barrier that splits two aqueous solutions at different composition and temperature to get a concentrated solution containing salts and acid. The driving force in this process is vapor pressure difference, which is encouraged by differences between temperature across the hydrophobic membrane [15, 42].

The ion-exchange membrane is considered the semi-permeable membrane that transfers specific dissolved ions while preventing other ions or neutral molecules. The exchange of ions between the surroundings and the substrate is called ion exchange [15]. Ion exchange is the technique where the ions are reversibly exchanged between the solid and liquid phases. The solid phase is the ion exchange resin, which is insoluble in the liquid phase; it does not change the substantial structure during reaction because it carries exchangeable ions. The high cation-exchange capacity materials such as clay and resin are utilized for ion exchange uptake of metals from the solution. The disadvantage of this process is that it relies on several factors such as contact time, the concentration of the solution, temperature, and pH of the supernatants. The efficiency of the ion exchange process is metal concentration–dependent. Some materials have been used to remove metals from aqueous systems and wastewater treatment, such as natural and synthetic clay, zeolites, and synthetic resins [11]. There are two types of ion exchange membranes: anion-exchange membranes and cation-exchange membranes; it depends on the types of an ionic group devoted to the membrane matrix. Ion exchange is the treatment technique that effectively removes trace metals from mine water compared to other treatment techniques. It is necessary to treat a large volume of diluted solutions.

12.5.2.3 Integrated Technologies

Utilization of one technology for AMD treatment can potentially lead to some fraction of the wastewater being untreated [44]. Integrated techniques, therefore, involve the combination of both passive and active technologies [34]. Integration of technologies for AMD treatment involves an active technology, such as upstream neutralization, and passive technology, such as downstream purification of the neutralized water using wetlands.

The removal of heavy metals and sulfates by integrated brucite and barium salts, where heavy metals are removed in the first phase and sulfates in the second phase, has also been studied [19]. Researchers have also reported the neutralization of AMD utilizing slaked lime [$Ca(OH)_2$] or limestone ($CaCO_3$) as the most effective method. The advantages of these methods include simultaneous removal of heavy metals and sulfates, where sulfates precipitate as gypsum ($CaSO_4 \cdot 2H_2O$), whereas heavy metals precipitate as metal hydroxides [45, 46].

12.6 Adsorption

Adsorption is the diffusion of adsorbate of pollutants from an aqueous solution onto the surface of adsorbent. The dominant effect in the adsorption process is the contact area between the adsorbent and adsorbate, and the large surface area leads to large interparticle forces between particles, which results in high adsorption capacity. The interparticle forces control the stability of the colloidal system and can be classified into two categories, namely, attractive or repulsive. There are two ways in which adsorbate can adhere to the adsorbent surface: physisorption and chemisorption [37, 47]. Many factors affect the adsorption process, such as molecular weight, shape, or the polarity of the solutes [20].

12.6.1 Factors that Affect the Adsorption Process

12.6.1.1 Effect of Solution pH

The measure of hydrogen ion concentration is referred to as pH, and it is the most vital parameter that influences the adsorption process. The variation of solution pH changes the surface characteristics (functional groups and charge) of the sorbents and chemistry of sorbate. The adsorption capacity of the adsorbent is pH-dependent, in general, as acidic pH is more suitable for inorganic pollutant removal compared to alkaline pH, and this is due to increase in the electrostatic attraction between the adsorbate and adsorbent. In addition, in a high-pH solution, the removal efficiency of adsorbate by adsorbent decreases because there is a high concentration of OH^- ions in the solution, and they compete with species for sorption sites [48–50].

12.6.1.2 *Effect of Contact Time*

The period at which the adsorbent is in contact with adsorbate ions in the aqueous solution is referred to as contact time. Various studies have revealed that the contact time plays a vital role in the adsorption process. In fact, contact or equilibrium time is a vital factor to implement appropriate technology for pollutant removal using adsorption techniques [50, 51].

12.6.1.3 *Effect of Temperature*

The effect of temperature in adsorption process is vital because it sheds more light on the mechanism involved in the sorption process. To assess the effect of temperature, the thermodynamic parameters, including enthalpy (AH°), entropy (ΔS°), and Gibbs free energy (ΔG°) of adsorption, are used to describe the nature and type of adsorption. There are two mechanisms that control the adsorption process, namely, exothermic and endothermic. When the adsorption capacity of the contaminant increases with an increased temperature, it is endothermic process; whereas if its rates decrease with temperature, then, it is exothermic. The process is spontaneous and depends on the change in entropy and Gibbs free energy of the absorption process. Positive values of entropy change and negative values of free energy change indicate an increase in the randomness of the solution interface and spontaneous, but negative entropy change and positive free energy change reveal that the interface of solution is slow and non-spontaneous sorption process [50, 51].

12.6.1.4 *Effect of Initial Ion Concentration*

Initial ion concentrations play a major role in determining the amount adsorbed onto magnetite during the adsorption process. The removal efficiency (RE) of the adsorbent is highly influenced by the initial ion concentration in the solution. As such, investigating the effect of the initial concentration becomes imperative in order to evaluate the number of ions adsorbed and consequentially the percentage efficiency (%) of ions removed because the increase in ion concentration leads to the decrease in RE as revealed by various studies [50, 51].

12.6.1.5 *Effect of Adsorbent Dose*

To assess the effect of adsorbent dosage, different weight of adsorbent is applied to remove pollutants *via* adsorption process at the optimum

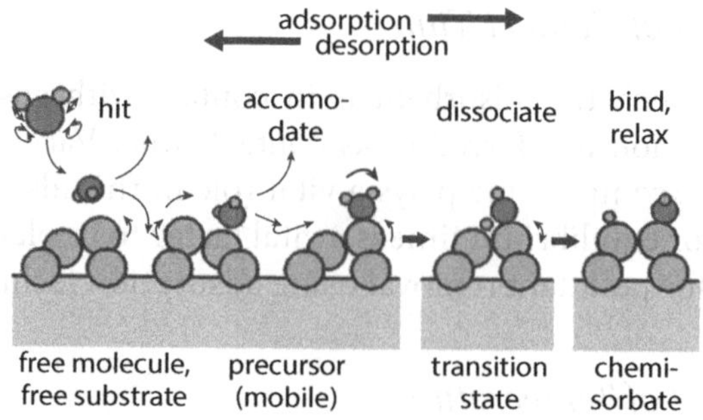

Figure 12.2 Illustration of the adsorption/desorption processes [52].

contact time. Other conditions must be kept constant during the study of the optimum amount of adsorbent dose to be used for adsorption process while varying the sorbent dosage. In general, the removal efficiency (%) of ions is directly proportional to the sorbent dose. As the adsorbent dose increases, the more the adsorption sites available to take up the adsorbate ions increase, and, subsequently, better adsorption takes place [48–50].

12.7 Types of the Adsorption Process

There are two types of adsorption processes available for the removal of pollutants from wastewater. The two processes depend on the type of attraction occurring between the adsorbate and adsorbent. Those processes are physisorption and chemisorption, which are both illustrated in Figure 12.2. Sorbate is removed from the adsorbent surface, allowing regeneration of the solid material, during the desorption process.

12.7.1 Physisorption

The physisorption process is primarily caused by forces that co-exist between the adsorbate molecules and the atoms on the surface of the adsorbent, such as electrostatic force, steric force, and van der Waals force. Low binding energy causes the physiosorbed materials to be highly mobile during the adsorption process [37]. Physisorption is considered as an exothermic process with a low enthalpy ranging from 20 to 40 kJ/mol, and the weak force of attraction causes this during the adsorption process; this process is reversible but not very specific [53].

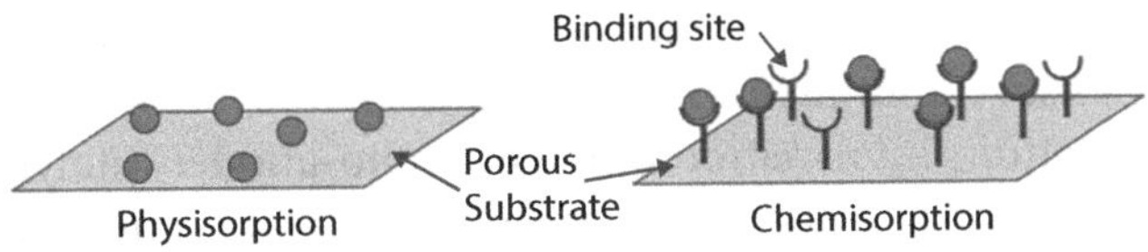

Figure 12.3 Illustration of the physisorption phenomenon and chemisorption phenomenon [55].

12.7.2 Chemisorption

Chemisorption is another adhesion process that involves forming strong bonds between the adsorbate and adsorbent during the adsorption process. These bonds are chemical bonds or chemical forces responsible for attaching the molecules or atoms to the adsorbent surface. Both chemical and physical adsorption processes are exothermic, but chemisorption has higher enthalpy energy ranging between 80 and 240 kJ/mol; this is due to a strong force of attraction. Chemisorption is an irreversible process, but highly specific adsorption occurs with the formation of a monomolecular layer (uni-layer) of the adsorbate on the adsorbent. At high temperatures, the chemisorption operates very well, and chemisorption increases with an increase in temperature. Surface area is another factor that affects the rate of chemisorption, and an increase in the surface area also increases the chemisorption [53, 54]. Illustration of the physiosorption and chemisorption is shown in Figure 12.3.

12.8 Adsorbents

During the adsorption process, many essential components play a major role, and adsorbent is one of the most critical elements. The effectiveness of the adsorption process entirely depends on the adsorbent's performance in both equilibria and kinetics adsorption process, and it even plays a major role in the success or failure of the process. The internal surface area accessible to the contaminant removed from the solution is the primary factor for the successful adsorption process. However, the adsorbent media should also have the capability to transfer adsorbing components rapidly to the adsorption sites. Characteristics of good adsorbent media are high sorption capacity, mechanical properties such as excellent kinetics properties, resistance to attrition, and strength. Cryptocrystalline magnesite can eliminate inorganic pollutants from industrial run-offs and may be a good material for metal elimination and reclaiming due to its low solubility that permits its capacity for metal recovery at varying pH gradients [56, 57]. Magnetite has several industrial applications such as ferrofluid technology, photo-degradation, information storage, photo-anode, catalyst, biomedicine, controlled drug transfer, wastewater treatment, and magnetic creation. Several methods for producing magnetite from Fe-rich salts include thermal decomposition, chemical precipitation, and sonochemical synthesis. Chemical precipitation is the most widely effective technique among these techniques due to its versatility [13, 58].

12.8.1 Magnetite

Magnetite (Fe_3O_4) contains both ferrous iron (Fe^{2+}) and ferric iron (Fe^{3+}), whereas others contain only trivalent iron (Fe^{3+}). The Fe_3O_4 contains a crystal structure with a well-recognized cubic spinel structure, where the Fe(II) and Fe(III) cations are located in interstice with octahedral and tetrahedral coordination. Magnetite nanoparticles have an inverse spinel structure on the tetrahedral sites, and the other half of the trivalent cations is where iron(III) cations are located. In contrast, iron(II) cations are located [59]. Magnetite has been used in many applications or fields due to its advantages such as high durability, good biocompatibility, strong magnetization, low toxicity, and low price [60]. The Fe_3O_4 nanoparticles with a particle size ranging from 10 nm to 15 nm can be used for ultra-sonic adsorption [61, 62].

12.9 Opportunities of Acid Mine Drainage

12.9.1 Recovery of Valuable Minerals from AMD

AMD is considered an environmental pollutant due to the large concentration of heavy metals that are present. However, these contaminants are a great potential source of valuable minerals such as Fe, Zn, Cu, and Ni, which are worth to be recovered. The removal of metals as a pollutant from AMD is mandatory to minimize their negative impact on the environment [54]. The valuable minerals can be recovered from AMD with the aid of chemical, biological, and electrochemical methods [24]. Seo *et al.* [2] explored the recovery of Fe, Al, and Mn from AMD, where AMD was neutralized by lime and a significant volume of sludge was generated. Three neutralizing agents were used, namely, sodium hydroxide, calcium hydroxide, and sodium carbonate. The metals were removed in Al, Fe, and Mn as the pH increased during the neutralization process for selective precipitation. Moreover, the study revealed that each heavy metal could be recovered at a certain pH. At pH 4.5–5.5, the precipitation of Al was observed as NaOH was added. However, at pH 4.5–5.5, Al and Fe showed the exact behavior of being precipitated simultaneously, decreasing their purity. At pH 6.5–7.5, precipitation of Fe with high purity was observed, whereas manganese was precipitated at a pH of 8.5. This may be caused by the oxidizing effect of the oxidizing agent, which oxidized iron(II) to iron(III) [2]. Wei and Viadero [62] investigated the recovery of ferric iron from raw AMD, where selective precipitation was used. The AMD was first oxidized by adding hydrogen peroxide as the oxidizing agent, whereas sodium hydroxide was utilized to raise the pH to 3.5–4.0, where iron was precipitated as ferric hydroxides/oxyhydroxide. The aluminum-rich precipitate was obtained through further neutralization by sodium hydroxide at pH 6.0–7.0, where 97.5% of iron was recovered as $Fe(OH)_3$ [63].

12.9.2 Valorization of Iron Oxides and Adsorbents from Acid Mine Drainage

Literature has revealed that industrial resources can be attained from AMD through selective precipitation. Coagulants, adsorbents, and catalysts can be produced from the sludge obtained from AMD, including magnetite, magnetic particle, and other ferrites [63, 64]. Sorbents recovered from the AMD sludge can be utilized in environmental applications such as removing phosphate and heavy metals from wastewater [66, 67]. Recently, studies have shown that selectively precipitated iron can be utilized to produce pigments such as goethite and hematite, and the colored mortar can be made from the recovered sludge [63]. The recovery of iron oxide can minimize costs for disposal and the volume of iron oxides sent to landfill sites [64]. Literature has reported that iron oxides have been recovered from AMD and reused in different countries such as the United States and United Kingdom [64, 68].

12.9.3 Synthesis of Valuable Minerals from Acid Mine Drainage

As reported in the literature, AMD contains high concentrations of heavy metals worth recovering, including iron. The concentration of minerals present in AMD is high enough to be extracted from the sludge recovered or precipitated during treatment [65, 69]. Some authors have reported that pigments and other industrial products can be synthesized from Fe recovered from mine water [13].

12.9.3.1 Synthesis of Goethite

Silva *et al.* [63] reported that iron(III) could be selectively precipitated at pH 3.6 and 3.8 from AMD by adding alkaline media. The iron is precipitated from AMD as iron hydroxide for the conventional synthesis of goethite. The synthesis of goethite includes dissolving the iron, selective precipitation, and crystallization of goethite. Alkali materials are added to the AMD to accelerate the rate of hydrolysis of soluble iron to selectively precipitate the metal as ferric hydroxide at pH 3.7 ± 0.1. This promotes the separation of iron from other metals present in the AMD. Reaction equations 12.8 and 12.9 illustrate the synthesis of goethite from AMD [19].

Reaction equation 12.8 illustrates the selective precipitation of Fe(III) from AMD:

$$Fe^{3+}_{(aq)} + 3OH^{-}_{(aq)} \rightarrow Fe(OH)_{3(s)} \tag{12.8}$$

Reaction equation 12.9 illustrates how the iron-Hexa-aquo ion is formed when the iron hydroxide is dissolved in water by adding nitric acid:

$$Fe(OH)_{3(s)} + 3HNO_{3(aq)} \rightarrow Fe(NO_3)_{3(aq)+} 3H_2O_{(aq)} \tag{12.9}$$

At the alkaline conditions, the ferric solution (iron-Hexa-aquo-ion) can be precipitated for the synthesis of goethite. At pH 12.0, potassium hydroxide is added to precipitate ferrihydrite $(Fe_5HO_8 \cdot 4H_2O)_{n(s)}$, as shown in reaction equation 12.10. Under these conditions, the iron-Hexa-aquo-ion $(Fe_5HO_8 \cdot 4H_2O)_{n(s)}$ precipitate is dissolved, and crystalline goethite will be formed from released ionic species by nucleation mechanism. After that, the formation of goethite takes place ($\propto$-FeOOH), as shown in reaction equation 12.11 [13, 19, 57].

$$(Fe_5HO_8 \cdot 4H_2O)_{n(s)} \rightarrow [Fe(OH)_4{}^{-}]_{(aq)} \tag{12.10}$$

$$[Fe(OH)_4{}^{-}]_{9aqueous)} \rightarrow \propto -FeOOH_{(s)} \tag{12.11}$$

12.9.3.2 Synthesis of Magnetite

Recently, most researchers have been exploring the synthesis of magnetite nanoparticles from AMD [62], using salts or a combination of iron-salts [13]. Decomposition, ball milling, chemical precipitation, and thermal and sonochemical synthesis are some of the techniques used to synthesize magnetite nanoparticles. However, chemical precipitation is considered the most effective approach for synthesizing magnetite nanoparticles and is reported to influence particle size and morphological properties [62]. Chemical precipitation of metal hydroxides involves several chemical mechanisms such as (1) co-precipitation of ferric and ferrous iron, (2) partial reduction of ferric iron, (3) oxidation of ferrous iron to ferric iron, and (4) precipitation of ferric iron, and (5) unaerated precipitation of ferrous iron, which is followed by co-precipitation of the Fe(III): Fe(II) at a molar ratio of 2:1 to form magnetite nanoparticles [13]. Typical conditions for the synthesis of magnetite are pH > 9 and temperature of 60°C – 100°C in the presence of nitrates that act as proton acceptors to form NO [13]. Akinwekomi *et al.* [13] synthesized the magnetite from iron-rich water (AMD) utilizing sodium carbonate. The recovery of magnetite from the AMD is pH-dependent, and the metals were recovered at pH $\geq$ 10 through selective precipitation. The magnetite nanoparticles were recovered from the AMD at optimum conditions, such as the molar ratio of 2:1 [Fe(III): Fe(II)] and temperature ranging from 25°C – 100°C. Wei and Viadero [62] also conducted the synthesis of magnetite from AMD, where iron was first recovered from AMD through the oxidation-selective precipitation process and used to synthesize magnetite nanoparticles. This study showed the successful synthesis of magnetite through co-precipitation of Fe(III) and Fe(II) at room temperature.

12.9.3.3 Recovery of Water

AMD is rich in heavy metals and sulfates, which makes it unfit for many applications such as industrial and domestic [54]. However, water can be reclaimed from AMD during the removal of heavy metals. Masindi [19] investigated the reclamation of water and valuable minerals utilizing integrated methods such as neutralization of the AMD with lime, soda ash, CO_2 [19].

Ryu *et al.* [41] studied the treatment of AMD by an integrated submerged direct contact MD-zeolite sorption system, where there was the recovery of heavy metals sulfuric acid and treated water. The integration of submerged MD with zeolite for treating AMD enabled the reclamation of 50% pure water within 30 h of the sorption process [41].

12.10 Conclusions and Future Research Directions

12.10.1 Conclusions

AMD is a biorecalcitrant wastewater matrix that causes notable environmental impacts and endemic water scarcity in areas with intense mining activities. Furthermore, high dissolved ions make it a valuable resource especially when considering the concept of resource recovery and beneficiation. In the literature, Fe-(hydro)xide, goethite, hematite, and magnetite can be synthesized from elements recovered from AMD. Drinking water could also be reclaimed along with gypsum as valuable mineral. Rare earth metals could also be recovered. This can open a new stream of revenue and possibly even make AMD treatment systems self-sufficient with revenue from the synthesized minerals and reclamation of clean water that could offset the running costs of the treatment process and ecological footprints. For the recovery of valuable minerals, the most promising recovery technology documented is precipitation, whereas, for clean water reclamation, it is filtration technologies. Filtration and ion exchange technologies have challenges of brine production. Adsorption has quick saturation to multi-component solution due to limited adsorption sites, demanding regeneration requirement, and additional requirements of regenerants. Henceforth, sequential recovery of chemical species emphatically appears to be the most promising treatment regimen especially when viewed under the circular economy concept. However, the premier challenges that relates to the recovery of valuable minerals and the reclamation of clean water comprise high costs and complexity of the systems and processes, along with the generation of heterogeneous sludge and brine, that require proper handling and management. Finally, the purity of recovered minerals due to cross contamination were also highlighted as a major challenge in previous studies.

12.10.2 Future Research Directions

Considering findings from this in-depth and overarching book chapter, the following are numerous avenues of investigation that future research should focus on:

- Highlighting the techno-economic analysis of post recovery process, i.e., the synthesis, in relation to the use of virgin materials
- Explore the feasibility of recovering rare earth metals from AMD and eliminate the over-reliant on dwindling natural resources
- Refinement of existing technologies to minimize the feasibility of cross contamination in precipitation technologies
- Coupling different technologies to minimize ecological footprints of individual systems and acquire synergy of different technologies
- Employment of different technologies including machine learning for the optimization of the treatment process and efficacy.

Acknowledgment

The authors of this manuscript would like to convey their sincere gratitude to the University of South Africa (UNISA) and Magalies Water (MW) for extending their facilities and funding toward the accomplishment of the objectives of this project.

References

1. Akcil, A. and Koldas, S., Acid Mine Drainage (AMD): causes, treatment and case studies. *J. Clean. Prod.*, **14**, 12, 1139–1145, 2006.
2. Seo, E., *et al.*, Recovery of Fe, Al and Mn in acid coal mine drainage by sequential selective precipitation with control of pH. *Catena*, **148**, 11–16, 2017.
3. Johnson, D.B. and Hallberg, K.B., Acid mine drainage remediation options: a review. *Sci. Total Environ.*, **338**, 1, 3–14, 2005.
4. Masindi, V., *et al.*, Efficiency of ball milled South African bentonite clay for remediation of acid mine drainage. *J. Water Process Eng.*, **8**, 227–240, 2015.
5. Masindi, V. and Gitari, W.M., Simultaneous removal of metal species from acidic aqueous solutions using cryptocrystalline magnesite/bentonite clay composite: an experimental and modelling approach. *J. Clean. Prod.*, **112**, 1077–1085, 2016.
6. Kefeni, K.K., Msagati, T.A.M., Mamba, B.B., Acid mine drainage: Prevention, treatment options, and resource recovery: A review. *J. Clean. Prod.*, **151**, 475–493, 2017.

7. Motsi, T., Rowson, N.A., Simmons, M.J.H., Adsorption of heavy metals from acid mine drainage by natural zeolite. *Int. J. Miner. Process.*, **92**, 1–2, 42–48, 2009.
8. Mogashane, T.M., *et al.*, Ferric Hydroxide Recovery from Iron-Rich Acid Mine Water with Calcium Carbonate and a Gypsum Scale Inhibitor. *Minerals*, **13**, 167, 2023.
9. Nguegang, B. and Ambushe, A.A., Sustainable acid mine drainage treatment: A comprehensive review of passive, combined, and emerging technologies. *Environ. Eng. Res.*, 30, 240592, 2025.
10. Name, T. and Sheridan, C.J.M.E., Remediation of acid mine drainage using metallurgical slags. *Miner. Eng.*, **64**, 15–22, 2014.
11. Masindi, V., Remediation of acid mine drainage using magnesite and its bentonite clay composite, University of Venda, 2015.
12. Chen, G., *et al.*, A critical review of prevention, treatment, reuse, and resource recovery from acid mine drainage. *J. Clean. Prod.*, **329**, 129666, 2021.
13. Akinwekomi, V., *et al.*, Synthesis of magnetite from iron-rich mine water using sodium carbonate. *J. Environ. Chem. Eng.*, **5**, 3, 2699–2707, 2017.
14. Yang, Y., *et al.*, A review of treatment technologies for acid mine drainage and sustainability assessment. *J. Water Process Eng.*, **55**, 104213, 2023.
15. Simate, G.S. and Ndlovu, S., Acid mine drainage: Challenges and opportunities. *J. Environ. Chem. Eng.*, **2**, 3, 1785–1803, 2014.
16. Skousen, J.G., Ziemkiewicz, P.F., McDonald, L.M., Acid mine drainage formation, control and treatment: Approaches and strategies. *Extr. Ind. Soc.*, **6**, 1, 241–249, 2019.
17. Majzlan, J., *et al.*, Mineralogy and geochemistry of the copper-dominated neutral mine drainage at the Cu deposit Ľubietová-Podlipa (Slovakia). *Appl. Geochem.*, **92**, 59–70, 2018.
18. Nordstrom, D.K., Blowes, D.W., Ptacek, C.J., Hydrogeochemistry and microbiology of mine drainage: An update. *Appl. Geochem.*, **57**, 3–16, 2015.
19. Masindi, V., Recovery of drinking water and valuable minerals from acid mine drainage using an integration of magnesite, lime, soda ash, CO2 and reverse osmosis treatment processes. *J. Environ. Chem. Eng.*, **5**, 4, 3136–3142, 2017.
20. Aslam, T., *et al.*, Valorization of Acid Mine Drainage into an Iron Catalyst to Initiate the Solar Photo-Fenton Treatment of Municipal Wastewater. *Environments*, **10**, 132, 20238.
21. Tame, C., Hudson-Edwards, K.A., Potter, H.A.B., Weathering of Zinc-(Zn)-bearing Mine Wastes in a Neutral Mine Drainage Setting, Gunnerside Gill, Yorkshire. *Procedia Earth Planet. Sci.*, **17**, 284–287, 2017.
22. Fu, F. and Wang, Q., Removal of heavy metal ions from wastewaters: A review. *J. Environ. Manage.*, **92**, 3, 407–418, 2011.
23. Masindi, V., *et al.*, Challenges and avenues for acid mine drainage treatment, beneficiation, and valorisation in circular economy: A review. *Ecol. Eng.*, **183**, 106740, 2022.

24. Naidu, G., *et al.*, A critical review on remediation, reuse, and resource recovery from acid mine drainage. *Environ. Pollut.*, **247**, 1110–1124, 2019.
25. Wang, X., *et al.*, A novel approach to rapidly purify acid mine drainage through chemically forming schwertmannite followed by lime neutralization. *Water Res.*, **151**, 515–522, 2019.
26. Rahdar, S., *et al.*, Adsorption of arsenic (V) from aqueous solution using modified saxaul ash: isotherm and thermodynamic study. *Appl. Water Sci.*, **9**, 4, 87, 2019.
27. Nguegang, B. and Ambushe, A.A., Insight into the chemical and biochemical mechanisms governing inorganic contaminants removal by selective precipitation and neutralization in acid mine drainage treatment using MgO: A comparative study. *J. Water Process Eng.*, 59, 104924, 2024.
28. Onwudiwe, D.C., *et al.*, Photocatalytic reduction of Cr(VI) using star-shaped Bi2S3 obtained from microwave irradiation of bismuth complex. *J. Environ. Chem. Eng.*, **8**, 4, 103816, 2020.
29. Crane, R.A. and Sapsford, D.J., Towards "Precision Mining" of wastewater: Selective recovery of Cu from acid mine drainage onto diatomite supported nanoscale zerovalent iron particles. *Chemosphere*, **202**, 339–348, 2018.
30. Macías, F., *et al.*, Management strategies and valorization for waste sludge from active treatment of extremely metal-polluted acid mine drainage: A contribution for sustainable mining. *J. Clean. Prod.*, **141**, 1057–1066, 2017.
31. Taylor, J., Pape, S., Murphy, N., A summary of passive and active treatment technologies for acid and metalliferous drainage (AMD), in: *Proceedings of the in Fifth Australian workshop on Acid Mine Drainage*, 2005.
32. Gazea, B., Adam, K., Kontopoulos, A., A review of passive systems for the treatment of acid mine drainage. *Miner. Eng.*, **9**, 1, 23–42, 1996.
33. Nyquist, J. and Greger, M., A field study of constructed wetlands for preventing and treating acid mine drainage. *Ecol. Eng.*, **35**, 5, 630–642, 2009.
34. Masindi, V., *et al.*, Assessing the sustainability of acid mine drainage (AMD) treatment in South Africa. *Sci. Total Environ.*, **635**, 793–802, 2018.
35. Park, I., *et al.*, A review of recent strategies for acid mine drainage prevention and mine tailings recycling. *Chemosphere*, **219**, 588–606, 2019.
36. Fernando, W.A.M., *et al.*, Challenges and opportunities in the removal of sulphate ions in contaminated mine water: A review. *Miner. Eng.*, **117**, 74–90, 2018.
37. Kurniawan, T.A., *et al.*, Physico–chemical treatment techniques for wastewater laden with heavy metals. *Chem. Eng. J.*, **118**, 1, 83–98, 2006.
38. Tawalbeh, M., *et al.*, Membrane separation as a pre-treatment process for oily saline water. *Desalination*, **447**, 182–202, 2018.
39. Abreu, M.M., *et al.*, List of Contributors, in: *Bio-Geotechnologies for Mine Site Rehabilitation*, M.N.V. Prasad, PJC. Favas, S.K. Maiti (Eds.), pp. ix–xiii, Elsevier, Amsterdam, Netherlands, 2018.
40. Zhu, R., *et al.*, Adsorbents based on montmorillonite for contaminant removal from water: A review. *Appl. Clay Sci.*, **123**, 239–258, 2016.

41. Ryu, S., *et al.*, Acid mine drainage treatment by integrated submerged membrane distillation–sorption system. *Chemosphere*, **218**, 955–965, 2019.
42. Nleya, Y., Simate, G.S., Ndlovu, S., Sustainability assessment of the recovery and utilisation of acid from acid mine drainage. *J. Clean. Prod.*, **113**, 17–27, 2016.
43. Buzzi, D.C., *et al.*, Water recovery from acid mine drainage by electrodialysis. *Miner. Eng.*, **40**, 82–89, 2013.
44. Degens, B., *Proposed guidelines for treating acidic drain water in the Avon catchment, Western Australia: adapting acid mine drainage treatment systems for saline acidic drains*, Department of Water, Perth, Australia, 2009.
45. Akinwekomi, V., *et al.*, Integrated acid mine drainage treatment using Mg(OH)2 or Mg(HCO3)2 and Ca(OH)2: Implications for separate removal of metals and sulphate. *Int. J. Miner. Process.*, **155**, 83–90, 2016.
46. Oyewo, O.A., *et al.*, Chapter 6 - Current Methods for the Remediation of Acid Mine Drainage Including Continuous Removal of Metals From Wastewater and Mine Dump, in: *Bio-Geotechnologies for Mine Site Rehabilitation*, M.N.V. Prasad, PJC. Favas, S.K. Maiti (Eds.), pp. 103–114, Elsevier, Amsterdam, Netherlands, 2018.
47. Liao, H. and Wang, Z., Adsorption removal of amaranth by nanoparticles-composed Cu2O microspheres. *J. Alloys Compd.*, 769, 1088–1095, 2018.
48. Hu, J., Lo, I., Chen, G., Removal of Cr (VI) by magnetite. *Water Sci. Technol.*, **50**, 12, 139–146, 2004.
49. Samarghandi, M.R., *et al.*, magnetite nanoparticles synthesis and modified with chitosan 4 O3 Fe biopolymers for removal of hexavalent chromium from aqueous solutions. *J. Chem. Pharm. Res.*, **7**, 4, 933–941, 2015.
50. Ngulube, T., *et al.*, An update on synthetic dyes adsorption onto clay based minerals: A state-of-art review, *J. Environ. Manag.*, 2017.
51. Olatunji, M.A., *et al.*, Influence of adsorption parameters on cesium uptake from aqueous solutions-a brief review. *RSC Adv.*, **5**, 88, 71658–71683, 2015.
52. Ranke, W., Adsorption and desorption, in: *Modern methods in heterogeneous catalysis research*, 2005.
53. Olaniyi, O.T., *Surfactant-modified adsorbents for the removal of toxic anions in water*, Department of Chemical Sciences, Redeemer's University, Ibadan, Nigeria, 2016.
54. Masukume, M., Treatment of high iron and manganese acid mine drainage (AMD) using sea shell and synthetic zeolite, 2016.
55. Kusumastuti, R., *et al.*, Study On The Mechanism of CO2 Adsorption Process on zeolite 5A as a Molecular Sieve In RDE System: An Infrared Investigation. *J. Phys. Conf. Ser.*, 1198, 1–9, 2019, IOP Publishing.
56. Masindi, V. and WMJJCP., Gitari, Removal of arsenic from wastewaters by cryptocrystalline magnesite: complimenting experimental results with modelling. *J. Clean. Prod.*, **113**, 318–324, 2016.

57. Azizi, S., Nguegang, B., *et al.*, Acid mine drainage treatment by means of selective precipitation using magnesium oxide (MgO): An experimental study. *Groundwater Sustain. Develop.*, **25**, 111151, 2024.
58. Akinwekomi, V., *et al.*, Beneficiation of acid mine drainage (AMD): A viable option for the synthesis of goethite, hematite, magnetite, and gypsum–Gearing towards a circular economy concept. *Miner. Eng.*, **148**, 106204, 2020.
59. Morales, F., *et al.*, Structural and magnetic properties of silica-coated magnetite nanoaggregates. *Physica B*, **572**, 214–219, 2019.
60. Nguegang, B., Masindi, V., *et al.*, Acid mine drainage treatment technologies: Challenges and future perspectives. In book: *Hybridized Technologies for the treatment of mining effluents,* Elvis Fosso-Kankeu and Bhekie, B. Mamba (eds.) pp. 245–286, Scrivener Publishing, Berverly, USA.
61. Bhateria, R. and Singh, R., A review on nanotechnological application of magnetic iron oxides for heavy metal removal. *J. Water Process Eng.*, **31**, 100845, 2019.
62. Wei, X. and Viadero, R.C., Synthesis of magnetite nanoparticles with ferric iron recovered from acid mine drainage: Implications for environmental engineering. *Colloids Surf., A*, **294**, 1, 280–286, 2007.
63. Silva, RA., *et al.*, Optimizing the selective precipitation of iron to produce yellow pigment from acid mine drainage. *Miner. Eng.*, **135**, 111–117, 2019.
64. Nguegang, B., Masindi, V., Msagati, T.A.M., Tekere, M., Effective treatment of acid mine drainage using a combination of MgO-nanoparticles and a series of constructed wetlands planted with *Vetiveria zizanioides*: A hybrid and stepwise approach. *J. Environ. Manag.*, **310**, 114751, 2022.
65. Sadasivam, S. and Thomas, H.R., Colour and toxic characteristics of metakaolinite–hematite pigment for integrally coloured concrete, prepared from iron oxide recovered from a water treatment plant of an abandoned coal mine. *J. Solid State Chem.*, **239**, 246–250, 2016.
66. Flores, H., *et al.*, Impact of climate change on Antarctic krill. *Mar. Ecol. Prog. Ser.*, **458**, 1–19, 2012.
67. Nguegang, B., Masindi, V., Msagati, T.A.M., Tekere, M., The Treatment of Acid Mine Drainage Using Vertically Flowing Wetland: Insights into the Fate of Chemical Species. *Minerals*, **11**, 477, 1–24, 2021.
68. Nguegang, B., Masindi, V., Msagati, T.A.M., Tekere M. and Mbue, M., II, Assessing the performance of horizontally flowing subsurface wetland equipped with Vetiveria zizanioides for the treatment of acid mine drainage. *Adv. Environ. Technol.*, **2**, 103–127, 2022b.
69. Nguegang, B. and Ambushe, A.A., The treatment of acid mine drainage (AMD) using a combination of selective precipitation and bio-sorption techniques: A hybrid and stepwise approach for AMD valorization and environmental pollution control. *Environ. Res. Technol.*, **3**, 313–334, 2024.

13

Technical and Economic Feasibility of Pigment and Drinking Water Recovery from Iron-Rich Acid Mine Water

Mokgadi Gladness Rapeta[1], Johannes Philippus Maree[1]* and Tumelo Monty Mogashane[2]

[1]Institute for Nanotechnology and Water Sustainability (iNanoWS), College of Science, Engineering and Technology, University of South Africa, Johannesburg, South Africa
[2]Analytical Chemistry Division, Mintek, Randburg, South Africa

Abstract
The focus of this study was to determine the feasibility of the integrated process using $CaCO_3$ and $Ca(OH)_2$ for removal of Fe^{3+}, Al^{3+}, and Mn^{2+}. Na_2CO_3 was used for the removal of Ca^{2+} as $CaCO_3$ to avoid fouling of the membranes. Reverse osmosis, electro-dialysis, and freeze crystallization were used for desalination to produce salt $Na_2SO_4{\cdot}10H_2O$/L H_2O and minimum brine. $Fe(OH)_3$, recovered at pH 3.5, was processed to pigment, a salable product. The study was carried out by using a 1 m^3/h demonstration plant for neutralization and reverse osmosis and information from the literature, and a spreadsheet-based model was developed to determine the feasibility of close to zero waste treatment. The cost of chemical and electrical power amounts R42.39/m^3 feed water (chemical cost was R27.46/m^3, and power cost was R14.93/m^3). The value of pigment, at a price of R20/kg, was R96.78/m^3 feed water. The cost of brine and sludge disposal, after conventional treatment options, will be R433.00/m^3 feed water at a disposal cost of R2,500/t.

***Keywords*:** Mine wastewater, drinking water, salable products, reverse osmosis cooling, modeling

**Corresponding author*: maree.jannie@gmail.com

Elvis Fosso-Kankeu, Vhahangwele Masindi, Johannes Maree and Bhekie Mamba (eds.) Customized Technologies for Sustainable Management of Industrial Wastewater: A Circular Economy Approach, (565–602)

13.1 Background

Acid mine drainage (AMD) from mining operations in South Africa is a significant environmental hazard and requires neutralization and desalination. Metal-rich sludge produced during neutralization has to be managed as toxic waste as it contains high levels of Fe^{2+}, Fe^{3+}, Al^{3+}, Mn^{2+}, Co^{2+}, and Ni^{2+} [1, 2]. In addition to inadequate or absence of treated sewage, mine water discharge frequently contaminates scarce surface-water and groundwater supplies [3]. The media frequently discuss AMD discharges in relation to the resultant rising salt and heavy metal content of surface water and groundwater. Where water users are exposed to excessive metal concentrations, they pose health problems. The Mpumalanga Province's coal mines generate 250 ML/d of AMD. Desalination of 100 ML/d AMD is already being applied at the eMalahleni, Middelburg, and Optimum Collieries. The 75 ML/d processed at eMalahleni and Optimum is marketed in significant quantities as potable water. The Olifants River unfortunately receives the remainder. Brine is assigned to evaporation ponds [4, 5].

AMD can be treated using passive, active, and desalination techniques [6]. Active treatment techniques can handle huge flow rates and heavily contaminated streams, whereas passive treatment is utilized for small streams with low contamination levels. Active treatment includes the use of chemical agents like lime, limestone, magnesite, caustic soda, soda ash, calcium oxide, anhydrous ammonia, magnesium hydroxide, and magnesium to increase the pH of AMD and precipitate most dissolved transition metal ions [7]. Site specificity, daily AMD burden, and metal concentration are a few factors that could affect how efficiently these alkalis function to treat AMD [8]. The advantage of using active over a passive treatment is that it does not require special structures or excessive space, making AMD treatment cheaper [6, 9, 10].

Mine water has to be treated in accordance with South African law to a level appropriate for drinking water quality, with the additional objective of generating zero waste [11]. The main problems with managing AMD are the physical and chemical stability of acid mine waste and reservoirs [12, 13]. The next issue is the storage of sludge that must first be treated in order to manage and stabilize contaminants before it can be disposed of. However, this process is expensive. If waste is created, then Environmental Impact Assessment studies are necessary before treatment plants can be established. Such drawn-out inquiries and approval of processes can be avoided by halting the generation of any hazardous waste. In order to avoid producing mixed sludge that must be disposed of to landfill or brines consigned to evaporation ponds, operations must be enabled, if possible, to produce salable products. It has been reported, for example, that AMD sludge can be recycled to recover heavy metals, rare earth elements, and pigment from wastewater [2, 13].

The output volume of effluent mine water has been reduced by the identification of cost-effective strategies for treatment [14–17]. Before 1980, mine water had to be neutralized with lime before it could be released into public waterways. The primary candidate technology for treating significantly contaminated mine water is reverse osmosis (RO). Pre-treatment, ultrafiltration (UF), and RO are the three steps that make up Keyplan's High Recovery Precipitating Reverse Osmosis (HiPRO®) Process [4, 18]. Each stage produces potable-quality water. Brine from each stage serves as the feed for the next stage. Gypsum precipitation during lime dosing in the pre-treatment step results in partial sulfate elimination. After the RO step, 98% of the water is recovered by removing soluble sodium sulfate and sodium chloride. Although the HiPRO® technique produces high-quality water, it has several disadvantages. To prevent gypsum scaling of the membranes, the recovery of water in the RO process is restricted to water with a total dissolved solid (TDS) content of less than 3,000 mg/L [19, 20]. Evaporation ponds are used to dispose of brine, which is hazardous for the environment. Additionally, treatment is expensive, and the use of lime/limestone neutralization produces a mixed sludge that is costly to dispose of.

Thus, the treatment of brines, especially those resulting from mining-impacted waters, requires energy-efficient, economical, and ecologically friendly processes. To achieve zero waste using the ROC process, this feasibility study will also incorporate freeze crystallization and thermal treatment of solid sludge in addition to neutralization and desalination *via* membrane treatment. Another advantage of near-zero waste processes is that they lower treatment costs by permitting the recovery of marketable products or the recovery for reuse of process chemicals.

13.1.1 ROC Process Description

Reverse osmosis/cooling (ROC) is a treatment method developed by ROC Water Technologies for mine water, and it is constantly being improved [21]. For the purpose of treating brines from desalination procedures like RO, the pipe freeze crystallization (PFC) process was developed. The ROC and PFC procedures are intended to reduce sludge and brine disposal by increasing water recovery, from approximately 85% to more than 99%, and chemical recovery (e.g., pigment, aluminum hydroxide, calcium carbonate, magnesium hydroxide, and sodium sulfate) (Figure 13.1).

In the pre-treatment stage of the ROC process, acidic or neutral mine water is treated with $CaCO_3$ (and an inhibitor), Na_2CO_3 and/or NaOH, or MgO to allow for the selective precipitation of metals $Al(OH)_3$, $Fe(OH)_3$, $CaCO_3$, $Mg(OH)_2$, and MnO_2 [2, 21–23]. Pre-treated metal-free water is then sent through a membrane step to produce brine and drinking water. The TDS content in the brine is high enough to support Na_2SO_4 crystallization during cooling.

Sodium sulfate–rich brine can be treated *via* cooling for partial removal of Na_2SO_4. Complete removal of both Na_2SO_4 and NaCl can be achieved by freezing the water. The brine has a concentration high enough to allow Na_2SO_4 crystallization upon cooling. Na_2SO_4 has a solubility of 4% at 0°C and 20% at 20°C. The ROC process exploits this peculiarity to treat the brine generated from RO by cooling it to 0°C, thereby precipitating $Na_2SO_4{\cdot}10H_2O$, which, in turn, lowers the osmotic pressure. The lower osmotic pressure enables the use of RO to further treat the brine in a closed loop, resulting in the zero liquid discharge process. It is advantageous to remove the maximum sodium sulfate through cooling as the energy needed for cooling amounts to only 25.3% of that when freezing (Table 13.1).

The positive results were achieved by a Water Research Commission (WRC)–supported study [24], and the objective of this study was to provide a pigment recovery stage, upstream of the neutralization plant, where a portion of the flow is treated for pigment recovery. Laboratory studies showed that $Fe(OH)_3$ can be recovered from mine water rich in iron by

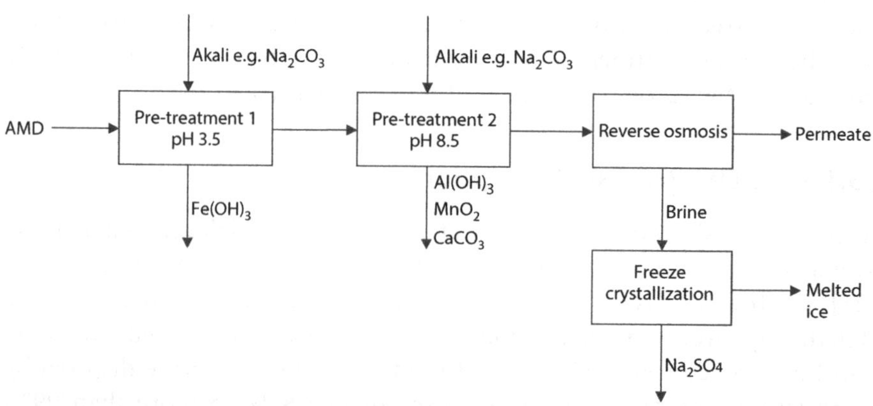

Figure 13.1 Ideal solution for mine water treatment.

Table 13.1 Energy needed for (i) cooling and (ii) cooling plus freezing.

Parameter	Unit	Cooling	Cooling + Freezing
Flow (m)	m^3/h	1.00	1.00
Water temperature	°C	25.00	25.00
Freezing point	°C	−2.00	−2.00
Heat capacity of water (Cp)	$kJ\ kg^{-1}k^{-1}$	4.18	4.18
Latent heat of ice (Hv)	$kJ\ kg^{-1}$	334.00	334.00
Energy needed for cooling (E = m·Cp·dT)	kW	31.35	31.35
Energy needed for freezing (E = m·Hv)	kW		92.78
Total energy needed	kW	31.35	124.13
Total energy needed	kWh/m^3	31.35	124.13
Percentage ratio		25.26	100.00

dosing $CaCO_3$. Gypsum precipitation is prevented for a period by dosing an inhibitor [2].

13.1.2 Biological Iron(II) Oxidation

Ferrous iron, Fe(II), in the mine water must first be oxidized to Fe(III) in order to be neutralized by limestone; otherwise, oxidation and hydrolysis will take place downstream of the neutralization plant, resulting in acidity. Iron-oxidizing bacteria, such *Thiobacillus ferrooxidans*, a bacterium that thrives in acidic environments, catalyze biological oxidation [25, 26]. The effect of several parameters on the biological and chemical rates of iron oxidation was evaluated by Nengovhela and colleagues [27]. In batch trials, they discovered that, while the presence of bacterial support media had no impact on the rate of chemical iron oxidation, it was crucial for biological iron oxidation. The maximum oxidation rate under continuous flow circumstances took place when the initial Fe (II) content ranged between 4.5 and 4.8 g/L. When fertilizers (ammonia and phosphate) were introduced into the reactor, geotextile was used as the bacterial support medium. A hydraulic retention time of 8 h resulted in the best iron oxidation rate. The concentration of suspended $Fe(OH)_3$ and $CaSO_4$, which function as catalysts, determines the rate of chemical iron oxidation. The growth of the bacteria, which was affected by a number of factors including the support media, nutrition, CO_2 and oxygen concentrations, determined the biological iron oxidation rate [27].

13.1.3 Pigment Formation

The $Fe(OH)_3$-sludge precipitated at pH 3.5 can be used to recover pigment. Metal oxide nanoparticles (NPs) are found to be playing a larger role in a variety of applications, such as the production and storage of energy, catalytic reactions, and water treatment [2, 28, 29]. Controlling size and form is of great importance to nanotechnology because it determines how well NPs function, altering properties including reactivity, conductivity, and magnetic behavior. In addition to using environmentally acceptable solvents, supercritical fluid NP synthesis provides a dependable and easy technique to meet the demand to manage size and form for a variety of metal oxide NPs [30, 31]. AMD is known to have greater levels of Fe ions that are important for recovery. Only a few workers have successfully extracted Fe from mine water for the synthesis of pigments and other commercially important compounds [2, 32]. Given the quantity of AMD produced each year, the ability to recover iron compounds from AMD is important, practical, and practicable [11, 33, 34]. It is possible to use iron oxide NPs for industrial scale wastewater treatment because they are affordable, have better stability, and are straightforward to separate [35, 36].

Both laboratory- and field-scale testing have been conducted to examine the efficacy of iron oxide NPs to remove pollutants [35, 36]. Adsorptive/immobilization technologies that use iron oxide NPs as a type of nanosorbent or immobilization carrier to increase removal efficiency and photocatalytic technologies, which use iron oxide nanomaterials (NMs) as photocatalysts to break down or convert contaminants into less toxic forms, make up the most recent uses of iron oxide NPs in wastewater treatment [35]. Iron oxide, including hematite (α-Fe_2O_3), magnetite (Fe_3O_4), and maghemite (γ-Fe_2O_3), has recently undergone a transition to the nanoscale and has generated a great deal of interest due to their unique magnetic and electrical properties for a variety of applications, including the creation of magnetic storage media, inorganic pigments, ferrofluid technology, electronic and optical devices, the development of gas sensors, and wastewater treatment [35–37].

13.1.4 Objectives

The focus of this study was to establish the economic feasibility of the ROC process. The following objectives were set for the study: (i) pilot plant evaluation of the neutralization, desalination, brine treatment, and pigment processing stages; and (ii) economic feasibility of the ROC process for treatment of acid mine water rich in iron.

13.2 Literature

13.2.1 Nanofiltration Technologies

Nanofiltration (NF) and RO, two technologies for membrane liquid separation, have numerous similarities. In contrast to RO, which has a strong rejection of almost all dissolved solutes, NF has a high rejection of multivalent ions like calcium and a poor rejection of monovalent ions like chloride.

The pre-treatment requirements and waste disposal challenges that NF shares with RO are similar. However, NF effectively separates multivalent ions while allowing monovalent ions to pass through the membrane, resulting in a reduced volume of waste brine. A solution containing both monovalent and divalent ions, such as sodium sulfate, behaves like a divalent salt and hinders the passage of sodium ions through the membrane, as the remaining multivalent salt must remain charge neutral. When the primary ionic components of the feed wastewater to be treated are divalent ions like Mn^{2+} and Mg^{2+}, NF is an effective option [38].

13.2.2 Reverse Osmosis

After the removal of the metals during the pre-treatment stage, RO is frequently utilized for desalination. The TDS of the feed and the development of scaling compounds like $CaCO_3$ and $CaSO_4$ affect water recovery in RO. It is possible to form brine with a TDS of 30 g/L by adding inhibitors. Through the ROC process, where Ca^{2+} is replaced with Na^{+}, brine with a TDS of 90 g/L can result. By doing this, 67% less RO brine can be produced. The selective recovery of metals by incremental pH increases is another advantage of the ROC method, which is made possible by the use of Na_2CO_3 in the pre-treatment stage. $Fe(OH)_3$ can be recovered at pH 3.5 [34].

Membrane processes generate two effluent streams: one with a low concentration of dissolved ions and another with a high concentration, utilizing a physical membrane barrier to separate charged ions from clean water [39]. For the neutralization and oxidation of mine water, as well as the removal of metals, RO pre-treatment typically involves adding limestone and/or lime, followed by aeration. This is then succeeded by UF or filtration using sand filters and cartridge filters before the main RO stage. The RO phase is a proven technology for eliminating dissolved solids from brackish or saline feed waters, where pressurized water is forced through a semi-permeable membrane, leaving dissolved ions behind. With a molecular weight cutoff of over 150–250 Daltons, RO effectively retains microorganisms, dissolved salts, carbohydrates, proteins, particles, dyes, and other substances found in the feed water [14].

The potential costs associated with RO are considerable because it can produce water of exceptional quality. However, water recovery is not perfect, making the reduction of TDS expensive. The cost of an RO plant increases with the level of TDS that needs to be reduced. A single-stage RO plant can achieve recovery rates between 50% and 80%, whereas adding more RO stages can boost this rate to as high as 95%, thereby decreasing the volume of waste brine and the associated disposal costs. Depending on the quality of the feed water, multi-stage RO systems can even achieve water recoveries exceeding 99%. High recovery rates are possible when the feed water primarily contains divalent ions that can be precipitated from the brine before being treated in the next RO stage. Incorporating NF membranes in multi-stage RO systems enhances overall water recovery by allowing monovalent ions to pass through the membrane that reduces or eliminates brine formation. For heavily scaling water streams that traditional spiral-wound RO systems cannot handle, tubular RO systems with regular flow direction reversals and sponge ball cleaning can be utilized. Due to their hazardous nature and high concentrations of dissolved salts, sludge and brine waste streams must be disposed of permanently [40].

Three mine water desalination facilities, where RO is the primary component of the process, have been established in South Africa since 2005 to satisfy the needs of AMD management in the Upper Olifants River catchment [39]. These requirements include safeguarding the Upper Olifants River from contamination and providing drinking water to the municipalities of Witbank and Middelburg. The annual growth in demand for drinking water was 3.5% [41]. In September 2007, the initial plant was put into operation. Later, it was updated to treat 30,000 m^3/d [41]. At Optimum Colliery, a second plant with a capacity of 15,000 m^3/d and a peak capacity of 18,750 m^3/d was put into service in August 2011 [40, 42]. The amount of recovered water was 98%. At Middelburg Colliery, South 32 is currently commissioning the third plant. The plant can produce 25,000 m^3/d. The plant's capital cost was R750 million, with a fixed cost of R8.00 per square meter and a variable cost of R4.00 per square meter (pers. commun., M. Koen, 2015).

Pre-treatment, UF, and RO procedures are all included in each stage of the HiPRO® by Keyplan [40, 42]. Each stage produces potable-quality treated water, whereas the brine from each stage serves as the next stage feed. Gypsum precipitation during lime dosing in the pre-treatment phases results in partial sulfate elimination. After the RO step, 98% of the water is recovered by removing soluble sodium sulfate and sodium chloride. When the feed water has a low pH, limestone is utilized in the pre-treatment stage to partially remove H_2SO_4, Fe^{3+}, Al^{3+}, and Fe^{2+}. Manganese, other low-concentration metals, and any remaining Fe^{2+} are removed using lime. At Optimum Colliery, where the feed water contains only trace amounts of iron and manganese, ozone is utilized to remove those elements. In the Stage 1 clarifier, solids are eliminated. To facilitate the agglomeration of solids, an anionic polymer flocculant is dosed. Prior to additional treatment using UF, clarification is followed by sand filtration. Pre-treatment Stages 2 and 3 are distinct from Stage 1 in that the pH is increased to a level that allows for the precipitation of magnesium hydroxide in addition to gypsum. Large agitators in the reactors keep the generated sludge suspended. Hydrocyclones are employed to remove substantially sized sludge particles. While the coarse sludge from the hydrocyclone underflow is later dewatered by a vacuum belt filter, the small particles in the overflow are put into the appropriate Stage 2 or Stage 3 clarifiers.

Before RO treatment, UF is the final step in the solids removal process. In the Optimum plant, each UF skid consists of 38 modules operating in dead-end mode. These skids are regularly backwashed to eliminate trapped particles. The plant's drain sump collects and further processes the backwash water from the UFs and sand filters. In Stage 1, an anti-scalant and sulfuric acid are added upstream of the UFs to prevent gypsum scaling on the RO membranes. In Stages 2 and 3, anti-scalant and sulfuric acid are dosed after the UFs to adjust pH, as the water's pH increases following pre-treatment. The final treatment step, RO, is employed to achieve the desired water recovery and feed water quality. The $CaSO_4$ saturation of the RO supply water in Stage 1 is approximately 90%–95%, with recovery limited by the risk of $CaCO_3$ scaling. The maximum achievable recovery in Stage 1 RO was around 70%, utilizing NF membranes to allow monovalent bicarbonate ions to pass into the permeate [43]. This approach has several benefits: (i) it stabilizes the low salinity permeate, and (ii) it reduces the need for additional lime in Stage 2 precipitation reactors due to lower bicarbonate levels in the reject stream. Given the higher $CaSO_4$ saturation levels in the input water, Stage 2 and Stage 3 RO systems are designed for water yields of 65% and 60%, respectively. The reject from each stage is pumped to the next stage, with a small portion of Stage 3 reject directed to the brine pond, whereas the permeate from each RO skid is collected in a shared water tank. The waste streams consist of gypsum, brine, and mixed sludge. Findings indicate that waste production is minimized, specifically yielding less than 12 L of brine/m^3 feed, less than 6 L of dewatered mixed sludge/m^3 feed, and less than 15 L of dewatered gypsum slurry/m^3 feed. Additionally, potentially useful materials are generated: (i) high-quality drinking water is produced; (ii) 98% of the water is recovered; (iii) minimal waste is generated, specifically less than 12 L of brine/m^3 feed, less than 6 L of dewatered mixed sludge/m^3 feed, and less than 15 L of dewatered gypsum slurry/m^3 feed; (iv) potentially useful by-product gypsum; and (v) variable costs excluding power were R5.00/m^3, whereas fixed costs were R2.50/m^3 [40].

13.2.3 Electro-Dialysis

An electric field is used to selectively transport ions across ion-exchange membranes during the membrane-based separation process known as electro-dialysis. Numerous industries have used it, including the treatment of acid mine water, after extensive research and application. Electro-dialysis can be used to remove dissolved ions from water, especially heavy metals and sulfate, in the context of treating acid mine water [44]. A succession of ion-exchange membranes and electrodes are used in the procedure to generate an electric field that drives the selective transit of ions. Pressure and TDS are two variables to consider when electro-dialysis is used to treat acid mine water. Commonly, pressure is used to increase mass transfer across membranes and increase the efficiency of the operation. Increasing driving forces and permeate flux can result from increasing pressure [44, 45]. The pressure requirements are typically between 1 and 3 bar, which is far lower than the pressures required in RO procedures. In electrodialysis, TDS is a key factor because excessive TDS concentrations can increase electrical resistance and reduce the process' efficiency. The precise TDS level at which electrical resistance becomes important and influences the effectiveness of the process might vary based on several variables, including the particular electro-dialysis system design, membrane properties, and operational circumstances. Consequently, in order to maximize the efficacy of the treatment, it is crucial to monitor and regulate TDS levels during electro-dialysis [46].

13.2.4 Brine Treatment with Freezing Technologies

13.2.4.1 Basics

Freezing water requires only about one-sixth of the energy needed for evaporation, as the heat of fusion for ice is six times lower than that of evaporation. Specifically, around 438.5 kJ of energy is needed to freeze 1 kg of water, compared to 2,570.5 kJ required to evaporate 1 kg of water at 25°C. Consequently, technologies that utilize freezing can be more energy-efficient than those based on evaporation.

Freezing technology gained attention in the late 1950s for its potential in water purification. Despite thorough investigations, it was never widely adopted for seawater desalination. However, in other sectors, particularly the food industry, freeze concentration (FC) has become a standard practice. Navigating Innovation and Research Opportunities (NIRO) (formerly Grenco) has created a commercially viable FC system utilized in food processing and recently in wastewater management. These concentration methods differ from the eutectic freeze crystallization (EFC) process that produces two solid phases simultaneously. Limited literature exists on EFC, likely due to its perceived complexity compared to traditional techniques like evaporative and cooling crystallization. With advancements in wash column technology and the broader application of melt crystallization for ultra-purification, the main drawbacks of EFC, namely, high investment costs and scalability issues, are being addressed, making it a candidate for future commercial use [47].

13.2.4.2 Freeze Desalination

Freeze desalination, a substitute method for salt and water recovery, was created on the basis of the principle of salt rejection when the water passes through a phase change to become solid. The development of a small dimension of ice crystal lattice structure, which rejects salt ions during phase shift rather than incorporating them into the lattice, is what causes the rejection of salt from water. Saline water can be cooled and/or frozen in a number of ways. Precooling of the feed, cooling, freezing in a heat exchanger, and ice separation from brine are a few of the steps in the process. The size of the crystals is another crucial factor in the freezing and separation process; the larger the crystals, the easier it is to separate from brine [48].

13.2.4.3 *Eutectic Freeze Crystallization*

The main challenge is anticipated to be the regulation of the shape and size distribution of the ice crystals that must be supplied to the wash column during EFC, which entails the synthesis of two solids. Either direct cooling or indirect cooling can produce ice crystals from aqueous solutions. Direct cooling involves injecting an inert cooling fluid into the solution, which vaporizes at the required temperature. The heat of evaporation is removed, which results in cooling of the solution. The fact that the cooling fluid is essentially present in every piece of equipment is a significant disadvantage of this way of operation: filters, a wash column, pumps, and a solid/solid separator [16]. The second and more typical principle is indirect cooling, where the cooling liquid runs through pipes or jackets rather than mixing with the solution. The important factor in this case is the heat transfer coefficient as well as the total surface area required for cooling [47].

The temperature at which a solute crystallizes out of solution when it is cooled to that level is known as the eutectic temperature. All dissolved salts in the solution will crystallize at a temperature of −25°C (Figure 13.2). Different dissolved salts have different eutectic temperatures. From a brine combination, relatively pure salts can be recovered using this method. A fluidized bed-type design is employed to increase efficiency [48]. Because some salt is carried over as a result, the water that is collected and turned

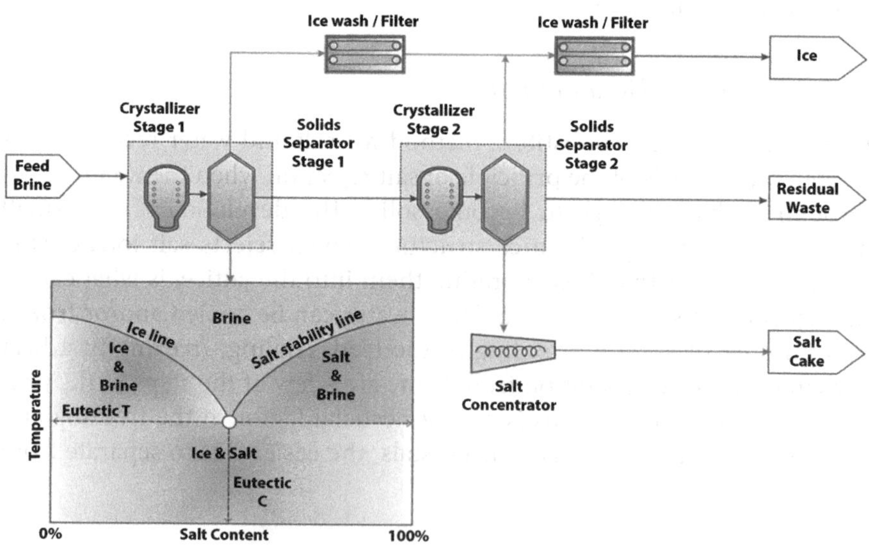

Figure 13.2 Eutectic freeze crystallization [11].

into ice could not be up to drinking water standards. However, it will be of a caliber that allows it to be put back into the feed stream of a normal mine water treatment facility (Figure 13.2) [11].

13.2.4.4 Pipe Freeze Crystallization

In the PFC process, the primary refrigerant is recycled through a heat exchanger in a water bath to cool the secondary refrigerant (30% methanol/water mixture) to −10°C or as required. Brine is pumped through the water bath in a >150-m pipe that is in a coil-shape, immersed in the secondary refrigerant, to the clarifier, and the brine recycling takes place between the water bath and the clarifier [49, 50].

13.3 Materials and Methods

13.3.1 Study Area

The pre-treatment and desalination (RO) steps were run in the Khwezela Colliery near eMalahleni (Figure 13.3).

13.3.2 Feedstock

The Top Dam water was rich in Fe^{3+} as a result of considerable exposure to oxygen over an extended period of time in a shallow pond, whereas the fresh leachate from the coal dump that is collected in the trench was rich in Fe^{2+} due to oxidation of pyrites (FeS) under limited exposure to oxygen,

Figure 13.3 Site where project was carried out at Khwezela Colliery [Photos: Google Maps (left), J Maree (right)].

which was studied by various researchers [1, 24, 51, 52]. The Top Dam's 4,500 mg/L Fe^{3+} concentration was high for the following reasons:

- The iron is in the Fe^{3+} state, which eliminates the need for Fe^{2+} oxidation.

Table 13.2 Chemical composition of water in the trench, Top dam, and Kopseer dam.

Parameter	**Trench**	**Kopseer**	**Top dam**
pH	3.7	2.4	2.3
Acidity (mg/L $CaCO_3$)	18,460.0	5,100.0	14,981.0
H^+ (mg/L H)	0.8	13.3	40.0
Na^+ (mg/L)	52.7	21.5	50.0
K^+ (mg/L)	31.1	0.0	30.0
Mg^{2+} (mg/L)	722.0	333.0	300.0
Ca^{2+} (mg/L)	486.0	287.0	500.0
Mn^{2+} (mg/L)	118.0	73.3	200.0
Fe^{2+} (mg/L)	9,270.0	335.1	400.0
Fe^{3+} (mg/L)	0.0	893.6	4,500.0
t-Fe (mg/L)	7,580.0	1,228.7	5,000.0
Al^{3+} (mg/L)	328.0	258.0	300.0
Si^{4+} (mg/L)	139.0	68.1	60.0
Sr^{2+} (mg/L)	25.6	0.0	0.0
Ba^{2+} (mg/L)	0.0	0.0	0.0
SO_4^{2-} (mg/L)	23,073.1	7,268.5	19,095.2
Cl^- (mg/L)	200.0	200.0	200.0
TDS (mg/L)	34,246.3	9,551.4	25,475.2
Cations (meq/L)	480.7	157.1	397.8
Anions (meq/L)	480.7	157.1	397.8
Sample date	12-Mar-21	12-Mar-21	12-Mar-21

- The Fe^{3+} has a high concentration of 4,500 mg/L due to evaporation and will provide a higher yield of pigment compared to other waters.

Table 13.2 shows the chemical compositions of the Trench water, Top Dam water, and the combined water in the Kop seer Dam. The Top Dam had a high concentration (4,500 mg/L Fe^{3+}) that made it the preferred water to treat. The Fe^{3+} had a high concentration due to evaporation and would provide a higher yield of pigment compared to other waters.

13.3.3 Equipment

Letjiane evaluated the neutralization and desalination stages on-site using a 1 m^3/h demonstration plant [24]. The brine from the RO stage was treated at the freeze crystallization plant of ROC Water Technologies at Olifantsfontein. Ferric hydroxide sludge from the neutralization stage was converted to pigment through thermal studies at TUT.

Figure 13.4 and Table 13.3 show the Wader pilot plant that was operated to demonstrate that drinking water and pigment can be recovered from Toe Seep water at the Kopseer Dam. The continuous operation of the pre-treatment stage was supported by laboratory studies. The brine and $Fe(OH)_3$ sludge were transported to A-Thermal Retort Technologies, Olifantsfontein. Through freeze crystallization, the brine was further concentrated. Thermal processing was used to convert $Fe(OH)_3$ into pigment.

13.3.4 Analytical

Samples were taken at various stages, filtered using (Whatman No. 1), and analyzed for levels of Fe(II), Fe(III), pH, Ca, and alkalinity according to [53]. Metals were quantified using inductively coupled plasma-atomic emission spectroscopy (ICP-OES) with the iCAP 7000 Series from ANATECH, South Africa. To determine Fe(II) concentrations, 5 mL of the filtered sample was mixed with 0.1 N H_2SO_4 and 5 mL of Zimmerman-Reinhardt (ZR) reagent in an Erlenmeyer flask and then titrated with 0.05 N $KMnO_4$ until a pale pink color appeared. Calcium served as the indicator for overall total hardness because magnesium was absent. For this analysis, a 100-mL Erlenmeyer flask was filled with 5 mL of the filtered sample, 45 mL of deionized water, 5 mL of diluted NH_3, and two drops of the Eriochrome Black T indicator [53] and then titrated with 0.01 M EDTA until reaching a blue endpoint. Acidity was measured by titrating a 5-mL

Figure 13.4 Photograph of the Wader mine water treatment demonstration plant (courtesy J. P. Maree).

Table 13.3 RO process configurations that were tested.

Configuration	Feed	Target
1	20 g/L Na_2SO_4	Determine highest brine concentration
2	Pre-treated water	Determine membrane fouling

sample to pH 8.3 using 0.1/1 N NaOH. The ICP-OES was used to analyze Fe, Al, Na, Ca, Ni, Mg, K, and Mn levels. Calibration of pH/EC meters was performed before and during each experimental set using calibration buffers.

13.3.5 OLI Software

In this work, the OLI ESP software program was used to predict the behavior of metals dissolved in water when treated with alkalis such as NaOH and MgO [54]. The OLI is an aqueous equilibrium estimator that features a self-explanatory interface for interpreting reactions, the ability to work with all common equilibrium reaction types, easily comprehensible results display, and the potential to produce results in various formats to suit various uses. The Stream Analyzer of the OLI was used to perform single point equilibrium calculations, multiple point survey calculations, and a thorough trend analysis for features including temperature, pH, pressure, composition effects, and separate capabilities. The computations gave a specialized model of liquid- and solid-phase separations.

By running a simulated AMD sample with assumed parameters for temperature, pressure, and pH, the reactions were simulated using the OLI Analyzer 9.0 System. MgO, $CaCO_3$, $MgCO_3$, Na_2CO_3, and $Mg(OH)_2$ were among the base titrants used. When MgO is used to neutralize to pH 3.2, followed by Na_2CO_3 to elevate the pH to 8.6, the effect of temperature on the solubility of $MgSO_4$ and Na_2SO_4 was found to represent this scenario. An analytical summary of the results from the simulation would emerge after the input data had been employed in a computation in the OLI Systems Chem Analyzer. The final concentration of the various factors as well as the actual reactions that will occur under particular circumstances could be predicted using this.

13.4 Results and Discussion

13.4.1 Neutralization

The mining industry currently utilizes $CaCO_3$ for neutralizing AMD. Limestone is the most cost-effective option (approximately R680/t), followed by lime [$Ca(OH)_2$] at around R3,000/t. Selective metal recovery using calcium alkalis is feasible only if the co-precipitation of gypsum can be prevented by adding inhibitors [2]. Replacing Na_2CO_3 with $CaCO_3$ or $Ca(OH)_2$ for selective recovery could reduce alkali costs due to their lower prices: Na_2CO_3 costs about ZAR5,000/t, whereas $CaCO_3$ and $Ca(OH)_2$ are priced at ZAR750/t and ZAR2,500/t, respectively. To avoid simultaneous precipitation of metal hydroxides and gypsum during the metal removal process, an anti-scalant is used to keep gypsum in solution. Furthermore, research was conducted to determine whether neutralizing actual mine water with $CaCO_3$ could limit gypsum crystallization. Several studies indicated that increasing the pH of the reacting solution from 4 to 7 enhanced the effectiveness of low molecular weight poly(acrylic acid) in inhibiting gypsum formation; however, it negatively affected the performance of higher molecular weight poly(acrylic acid) by forming a "net-structure" in the solution [2, 55].

13.4.2 Desalination

Two feed waters were evaluated in the RO section of the Wader plant. In the case of **Feed Water 1**, tap water was supplemented with 20 g/L Na_2SO_4 (Tables 13.4–13.6). The aim was to determine the maximum brine concentration that can be produced and the TDS of the permeate under those condition. The TDS of the permeate was less than 150 mg/L in all the runs showing that drinking water quality can be produced from Toe Seep water.

Table 13.4 Chemical composition when 20 g/L Na_2SO_4 was treated with RO (Water recovery set point = 40%).

Parameter	Unit	Feed	Permeate	Brine
Flow	L/h	1,500.0	79.4	1,420.6
Water recovery	%		5.3	
Salt rejection	%			100.0
Ca	mg/L	38.2	0.0	68.0
K	mg/L	12.6	6.7	13.9
Na	mg/L	8,080.0	7.3	8,500.0
Si	mg/L	0.0	0.0	0.0
SO_4	mg/L	16,969.8	23.4	17,919.4
TDS	mg/L	25,100.6	37.4	26,501.3

Water recovery reading = 40%

Table 13.5 Chemical composition when 20 g/L Na_2SO_4 was treated with RO (Water recovery set point = 55%).

Parameter	Unit	Feed	Permeate	Brine
Flow	L/h	1,500.0	426.9	1,073.1
Water recovery	%		28.5	
Salt rejection	%			99.9
Ca	mg/L	29.4	0.0	80.0
K	mg/L	12.6	7.3	10.5
Na	mg/L	6,290.0	3.2	8,750.0
Si	mg/L	0.0	1.3	0.0
SO4	mg/L	13,213.0	24.5	18,465.8
TDS	mg/L	19,545.0	36.3	27,306.3

Water recovery reading = 5%

Table 13.6 Chemical composition when 20 g/L Na_2SO_4 was treated with RO (Water recovery set point = 71%).

Parameter	Unit	Feed	Permeate	Brine
Flow	L/h	1,500.0	852.6	647.4
Water recovery	%		56.8	
Salt rejection	%			99.8
Ca	mg/L	33.8	0.0	112.0
K	mg/L	13.0	5.5	0.0
Na	mg/L	6,490.0	7.0	15,000.0
Si	mg/L	0.9	4.1	0.0
SO_4	mg/L	13,647.6	49.4	31,573.1
TDS	mg/L	20,185.3	66.0	46,685.1

In the case of **Feed Water 2**, treated water from the pre-treatment stage was treated in the RO stage (Table 13.7). The aim was to determine whether any fouling of membranes took place. The TDS of the brine in Table 13.7 amounted to 40,884 mg/L. With further optimization studies, the aim was to increase this concentration to 80,000 mg/L. With the ROC process, the aim was to achieve maximum salt removal with RO so that less salt needed to be removed with freeze desalination. The energy usage of freeze crystallization (100 kWh/t) is higher than that of RO (4.5 kWh/t). Freeze crystallization should, therefore, only be used after the maximum salt has been removed with RO. Good results were also achieved when treated Kopseer dam water was used for treatment in the RO plant.

13.4.3 Brine Treatment

Brine has a negative effect on the environment owing to its high salinity. Zero waste generation is the goal of mine water treatment. Freeze crystallization was examined by Ramothole *et al.* (2020) for the treatment of sodium sulfate–rich RO brine [48]. The aim was to recover potable water and $Na_2SO_4 \cdot 10H_2O$. The brine was cooled to freezing point using a chiller. Ice alone was produced while Na_2SO_4 was below its saturation level, about 45 g/L (as Na_2SO_4). Further ice removal caused $Na_2SO_4 \cdot 10H_2O$ to crystallize. As water recovery increased from zero to 80%, it was noted that: (i) the

Table 13.7 Chemical composition when neutralized Kopseer dam water was treated with RO.

Parameter	Unit	Feed	Permeate	Brine
Flow Feed	m3/h	40.0	28.0	12.0
Water recovery			70.0	
Salt rejection				99.0
pH		7.7		
TDS	mg/L	12,389.1	177.0	40,884.1
H^+	mg/L	0.0		
Na+	mg/L	3,297.6	47.1	10,882.1
K+	mg/L	19.7	0.3	65.0
Mg^{2+}	mg/L	308.7	4.4	1,018.7
Ca^{2+}	mg/L	195.0	2.8	643.5
Fe^{3+}	mg/L	0.0	0.0	0.0
Al^{3+}	mg/L	0.0	0.0	0.1
Fe^{2+}	mg/L	1.0	0.0	3.3
Mn^{2+}	mg/L	0.3	0.0	1.0
Cu^{2+}	mg/L	0.1	0.0	0.3
Si^{4+}	mg/L	5.0	0.1	16.5
SO_4^{2-}	mg/L	8,361.7	119.5	27,593.6
Cl^-	mg/L	200.0	2.9	660.0
HCO3–	mg/L	0.0	0.0	0.0
OH–	mg/L	0.0	0.0	0.0
Cations	meq/L	179.8	2.6	593.3
Anions	meq/L	179.8	2.6	593.5

freezing point decreased from −2°C at the start of ice formation, to −4°C when $Na_2SO_4 \cdot 10H_2O$ started to crystallize; (ii) ice purity was determined by measuring TDS, which increased from 2,000 mg/L to 3,000 mg/L; and (iii) energy utilizations were 10.45 kWh/m³ during cooling down to freezing point and 91.7 kWh/m³ when ice and $Na_2SO_4 \cdot 10H_2O$ were formed.

Table 13.8 shows the results when brine from the RO plant was treated by freeze crystallization. The feed water had a TDS of 12,540 mg/L. The TDS values of the ice samples were <3,900 mg/L and lowered over time. The sample collected after 2 h had a TDS of 3,868 mg/L and after 6 h a TDS of 2,989 mg/L. The lower TDS in the ice over time was ascribed to the formation of larger ice crystals. If this TDS concentration of the melted ice is too high for reuse or as drinking water, then the water can be recycled to the RO plant. The brine concentration increased with time from 13,430 mg/L to 14,070 mg/L. The brine concentration increased slightly but it also contained some precipitate. *OLI Software* showed that the precipitate that formed was $Na_2SO_4 \cdot 10H_2O$. As water recovery increased from 7.2% in Sample 1 to 57.3%, in Sample 3, the saturation level of Na_2SO_4 was exceeded, resulting in the crystallization of $Na_2SO_4 \cdot 10H_2O$ (Table 13.9). It was noted that Na_2SO_4 remained in solution up to its solubility level of 44,310 mg/L Na_2SO_4.

Table 13.8 Water quality of feed, brine, and melted ice samples with increased water recovery (actual results).

			Brine			Ice		
Parameter	**Unit**	**Feed**	**Sample 1**	**Sample 2**	**Sample 3**	**Sample 1**	**Sample 2**	**Sample 3**
Water recovery	%		0.0	20.0	80.0	0.0	20.0	80.0
Solution								
pH		9.3	8.9	8.8	8.8	9.0	9.0	9.1
TDS	mg/L	12,540	13,430	13,550	14,070	3,868	3,480	2,989
EC	mS/cm	22.1	23.2	23.6	26.0	7.8	6.8	6.0
Volume	L	200.0	185.5	141.6	85.3			
Ice	kg		14.5	58.4	114.7			
Ice recovery	%		7.2	29.2	57.3			

Table 13.9 OLI predicted values of various parameters during water removal though ice crystallization.

H_2O (mg)	H_2O (mg)	Temperature (°C)	pH	Specific Electrical Conductivity - Aqueous (µmho/cm)	SO_4^{2-} Aq (mg)	Na(+1) Aq (mg)	Na_2SO_4 Aq (mg/L)	$Na_2SO_4{\cdot}10H_2O$ (Mirabilite) (mg)	Na_2SO_4 (mg)
1,000,000	1,000,000	0	8	12,592	20,317	9,711	30,028	0	30,000
920,000	920,000	0	8	13,389	20,317	9,711	32,640	0	30,000
840,000	840,000	0	8	14,311	20,317	9,711	35,748	0	30,000
760,000	760,000	0	8	15,391	20,317	9,711	39,511	0	30,000
680,000	680,000	0	8	16,676	20,317	9,711	44,159	0	30,000
600,000	595,378	0	8	16,717	17,849	8,532	44,310	8,266	30,000
520,000	510,620	0	8	16,717	15,308	7,317	44,310	16,776	30,000
440,000	425,861	0	8	16,717	12,767	6,102	44,310	25,287	30,000
360,000	341,102	0	8	16,717	10,226	4,888	44,310	33,798	30,000
280,000	256,343	0	8	16,717	7,685	3,673	44,310	42,309	30,000
200,000	171,585	0	8	16,717	5,144	2,459	44,310	50,820	30,000

13.4.4 Thermal Processing of Na_2SO_4 to Na_2CO_3 to Further Improve the Salability

The product obtained from freeze crystallization, $Na_2SO_4{\cdot}10H_2O$, can be transformed into higher-value products such as Na_2SO_4 or Na_2CO_3 (Mokgohloa, *et al.*, 2022). Na_2CO_3 is utilized as a raw material in the soap and glass industries and for water neutralization. The cost of importing Na_2CO_3 from Botswana is R5,000/t, and this expense, along with the high disposal costs of waste, drives the need to process Na_2SO_4 into Na_2CO_3. However, converting Na_2SO_4 to Na_2CO_3 is challenging because Na_2SO_4 melts within the same temperature range where the reduction occurs.

Additionally, the high reactivity of Na_2S causes it to stick to the crucible's surface. An innovative solution was proposed by Van Vuuren (Provisional patent, 2020), wherein Na_2SO_4 is reacted with CaS or $Ca(HS)_2$ to facilitate the processing of Na_2CO_3 through Na_2S (Reactions 13.1 and 13.2). He also suggested that nano-$CaCO_3$ could be produced similarly by reacting $Ca(HS)_2$ with Na_2CO_3 (Reaction 13.3) (D.S. van Vuuren, pers. commun., 2020).

$$Ca(HS)_2 + Na_2SO_4 \rightarrow CaSO_4 + 2NaHS \quad (13.1)$$

$$2NaHS + CO_2 \rightarrow Na_2CO_3 + H_2S \quad (13.2)$$

$$Ca(HS)_2 + Na_2CO_3 \rightarrow CaCO_3 + 2NaHS \quad (13.3)$$

13.4.5 Feasibility

Any investment decision must begin with an economic feasibility study. The attractiveness of a project hinges on its anticipated economic returns; the greater the potential for future profits, the more enticing the project becomes [56]. In this study, the total cost was estimated based on changes in water quality throughout various process stages, along with the costs of chemicals and electricity. Simulations were performed to analyze operation and maintenance costs, water quality, and energy requirements for each treatment method evaluated. Additionally, the costs of sludge disposal and power consumption were included in the operating and maintenance expenses.

The feasibility of the ROC is enhanced by the recovery of pigment from $Fe(OH)_3$ at pH 3.5, which can be sold as a product. $Fe(OH)_3$ is recovered by using $CaCO_3$ as an alkali to raise the pH to 3.5, along with an inhibitor to keep gypsum in solution during the separation of $Fe(OH)_3$ from water. $CaCO_3$, $Ca(OH)_2$, or Na_2CO_3 was employed to remove remaining metals at higher pH levels. The pigment is widely utilized in the paint, tile, and cement industries and has a market value ranging from R6/kg to R32/kg.

Figure 13.5 and Table 13.10 show the following:

i. 8,307 mg/L $CaCO_3$ (as 100%) is used for the recovery of 5,739.5 mg/L $Fe(OH)_3$ at pH 3.5.
ii. 1,667 mg/L $CaCO_3$ (as 100%) is used for recovery of 866.7 mg/L $Al(OH)_3$ at pH 5.

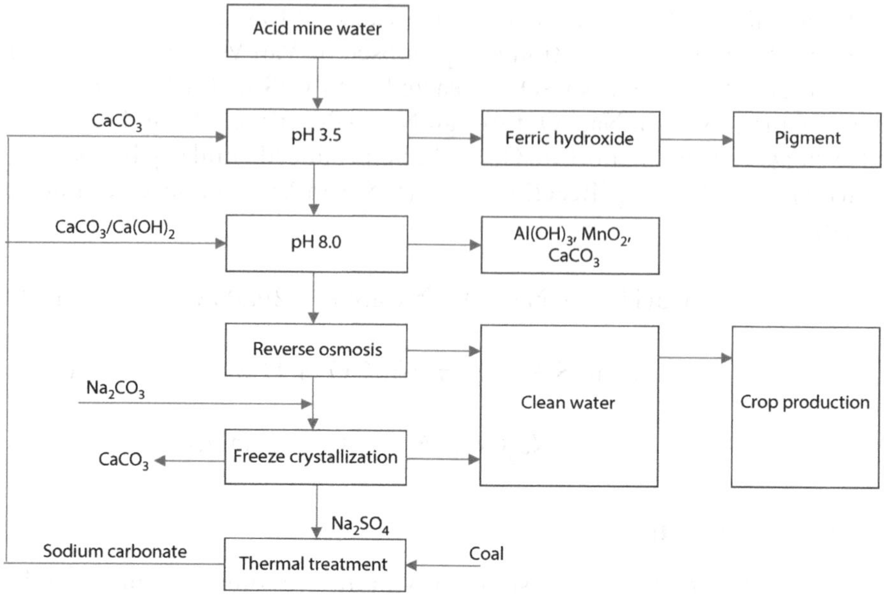

Figure 13.5 Process configuration when acid mine water is treated with $CaCO_3$ for removal of Fe^{3+} and Al^{3+}; $Ca(OH)_2$ for removal of Fe^{2+}, Mn^{2+}, and Ca^{2+} during pre-treatment.

iii. 34 mg/L $Ca(OH)_2$ (as 100%) is used for removal of 645.5 mg/L $Fe(OH)_2$ and $Mn(OH)_2$ at pH 8.5.
iv. Gypsum crystallization is allowed to occur (16,636 mg/L) for partial removal of Ca^{2+} and SO_4^{2-} by dosing Na_2CO_3.
v. Na_2CO_3 is used for removal of Ca^{2+} through formation of 245.9 mg/L $CaCO_3$.
vi. RO is used for desalination to produce permeate with a TDS of 121 mg/L and brine with a TDS 74,470 mg/L. The water recovery is 95% and the salt rejection 97%.
vii. ED (electro-dialysis) is used for up concentration of RO brine to 150 g/L. The water recovery is 53.3% and the salt rejection is 94%.
viii. Freeze crystallization is used recovery of 0.22 m^3/h water with a TDS 183,546 mg/L and 885,456 mg $Na_2SO_4{\cdot}10H_2O$/L H_2O (s). Due to large salt and ice crystals, easy dewatering is possible.

Table 13.10 shows the cost of the chemicals (R27.46/m^3 feed) and electricity (R14.79/m^3 feed) and the value of pigment (R85.78/m^3 feed) and

Table 13.10 Chemical cost and water quality when acid mine water is treated with $CaCO_3$ for removal of Fe^{3+} and Al^{3+}, $Ca(OH)_2$ for the removal of Fe^{2+} and Mn^{2+}, and Na_2CO_3 for the removal of Ca^{2+} associated with SO_4^{2-}.

Compound	Unit	Composition												Feasibility		
														Price	Cost	Value
		Feed	$Fe(OH)_3$	$Al(OH)_3$	Other metals	$CaSO_4$ Crystallization	$CaCO_3$/RO feed	RO perm	RO brine/ED feed	ED perm	ED brine	F Des Ice	F Des brine	R/t	R/m³ feed	R/m³ feed
Flow Feed	m³/h	40.0	40.0	39.2	39.1	39.0	36.9	35.0	**1.84**	0.98	**0.86**	0.65	0.22			
$CaCO_3$ (calc)	mg/L		8,307.3	1,666.7										750	7.48	
$Ca(OH)_2$ (calc)	mg/L				534.4									2,500	1.34	
Na_2CO_3	mg/L						2,288.1							5,000	11.44	
Inhibitor	mg/L		120.0											60,000	7.20	
Water recovery	%							95.0		53.3		75.0				
Salt rejection	%							97.0		94.0		90.0				
Product water																
pH		2.8	3.5	4.5	8.0	8.0	10.0									
TDS	mg/L	16,239	16,557	17,088	16,977	3,822	3,839	121.2	74,470	**8,378**	150,000	21,277	183,546			

(Continued)

Table 13.10 Chemical cost and water quality when acid mine water is treated with $CaCO_3$ for removal of Fe^{3+} and Al^{3+}, $Ca(OH)_2$ for the removal of Fe^{2+} and Mn^{2+}, and Na_2CO_3 for the removal of Ca^{2+} associated with SO_4^{2-}. (*Continued*)

Compound	Unit	Composition												Feasibility		
														Price	Cost	Value
		Feed	$Fe(OH)_3$	$Al(OH)_3$	Other metals	$CaSO_4$ Crystallization	$CaCO_3$/RO feed	RO perm	RO brine/ED feed	ED perm	ED brine	F Des Ice	F Des brine	R/t	R/m³ feed	R/m³ feed
H^+	mg/L	5	0	0	0	0	0	0.0	0.0	0.0	0.0	0.0	0.0			
Na+	mg/L	100	100	100	100	100	980	30.9	19,009	2,139	38,288	5,431	20,000			
Mg^{2+}	mg/L	200	200	200	200	200	200	6.3	3,880	437	7,815	1,109	29,930.6			
Fe^{3+}	mg/L	3,000	0	0	0	0	0	0.0	0.0	0.0	0.0	0.0	0.0			
Al^{3+}	mg/L	300.0	300.0	0.0	0.0	0.0	0.0	0.0	0.0	0.0	0.0	0.0	0.0			
Fe^{2+}	mg/L	200.0	200.0	200.0	0.0	0.0	0.0	0.0	0.0	0.0	0.0	0.0	0.0			
Mn^{2+}	mg/L	200.0	200.0	200.0	0.0	0.0	0.0	0.0	0.0	0.0	0.0	0.0	0.0			
Ca^{2+}	mg/L	300	3,623	4,454	4,742	873	10	0.3	194	22	391	55	1,496.5			
SO_4^{2-}	mg/L	11,784	11,784	11,784	11,784	2,499	2,499	78.9	48,477	5,454	97,644	13,850	109,671.1			
Cl^-	mg/L	150	150	150	150	150	150	4.7	2,910	327	5,861	831	22,448.0			

(*Continued*)

Table 13.10 Chemical cost and water quality when acid mine water is treated with $CaCO_3$ for removal of Fe^{3+} and Al^{3+}, $Ca(OH)_2$ for the removal of Fe^{2+} and Mn^{2+}, and Na_2CO_3 for the removal of Ca^{2+} associated with SO_4^{2-}. *(Continued)*

Compound	Unit	Composition												Feasibility		
														Price	Cost	Value
		Feed	$Fe(OH)_3$	$Al(OH)_3$	Other metals	$CaSO_4$ Crystallization	$CaCO_3$/RO feed	RO perm	RO brine/ED feed	ED perm	ED brine	F Des Ice	F Des brine	R/t	R/m^3 feed	R/m^3 feed
HCO_3^-	mg/L			500.0	500.0	500.0	200.0	6.3	3,880	437	7,815	1,109	29,930.6			
OH^-	mg/L						0.0	0.0	0.0	0.0	0.0		0.0			
Cations	meq/L	249.7	249.7	257.9	257.9	64.5	59.6	1.9	1,156	130	2,327	330	3,408			
Anions	meq/L	249.7	249.7	257.9	257.9	64.5	59.6	1.9	1,156	130	2,327	330	3,408			
SO_4^{2-} (OLI)						1,503.0										
Sludge																
$Fe(OH)_3$	mg/L		5,739.5		0.0											
$Al(OH)_3$			0.0	866.7	0.0											
Fe(OH)2			0.0	0.0	321.8											
Mn(OH)2			0.0	0.0	323.8											

(Continued)

Table 13.10 Chemical cost and water quality when acid mine water is treated with $CaCO_3$ for removal of Fe^{3+} and Al^{3+}, $Ca(OH)_2$ for the removal of Fe^{2+} and Mn^{2+}, and Na_2CO_3 for the removal of Ca^{2+} associated with SO_4^{2-}. (*Continued*)

Compound	Unit	Composition												Feasibility		
														Price	Cost	Value
		Feed	$Fe(OH)_3$	$Al(OH)_3$	Other metals	$CaSO_4$ Crystallization	$CaCO_3$/RO feed	RO perm	RO brine/ED feed	ED perm	ED brine	F Des Ice	F Des brine	R/t	R/m³ feed	R/m³ feed
$CaSO_4·2H_2O$	mg/L					16,636.4								0.0		0.00
$CaCO_3$	mg/L						245.9									
$Na_2SO_4·10H_2O$	mg/L												886,456			
Solids	mg/L		5,739.5	866.7	645.5	16,636.4	245.9									
Solids	%		30.0	30.0	30.0	30.0	30.0									
Solids	m3/h		0.8	0.1	0.1	2.2	0.0									
Solids																
Pigment (Fe_2O_3)	mg/L		4,289.2											20,000		85.78
Water	mg/L							875,720		24,581		16,132		12.0		11.00
Energy usage	kWh/m³		0.6		0.6		0.6	4.5		7.0		100.0				

(*Continued*)

Table 13.10 Chemical cost and water quality when acid mine water is treated with $CaCO_3$ for removal of Fe^{3+} and Al^{3+}, $Ca(OH)_2$ for the removal of Fe^{2+} and Mn^{2+}, and Na_2CO_3 for the removal of Ca^{2+} associated with SO_4^{2-}. (*Continued*)

Compound	Unit	Composition												Feasibility		
														Price	Cost	Value
		Feed	$Fe(OH)_3$	$Al(OH)_3$	Other metals	$CaSO_4$ Crystallization	$CaCO_3$/RO feed	RO perm	RO brine/ED feed	ED perm	ED brine	F Des Ice	F Des brine	R/t	R/m³ feed	R/m³ feed
Energy usage	kWh/m³ Feed		0.6		0.6		0.6	3.9		0.2		1.6		2.0	14.93	
Chemical cost	R/m³ Feed														27.46	
Total Cost/Value															42.39	96.78
Bruto income	R/m³ Feed														54.39	

SO_4 crystallized calculated using OLI

Table 13.11 Waste disposal cost in the case of conventional treatment.

Parameter	Unit	Sludge	Brine
Feed flow	m^3/h	40.0	40.0
Solids in sludge	mg/L	24,322	
30% sludge	mg/L	81,073	
Waste disposal cost	R/t	2,500	2,500
Brine concentration	mg/L		35,406
Brine volume	m^3/h		3.7
Liquid waste	t/m^3 Feed		0.09
Waste disposal cost	R/m^3	202.68	230.31
Total waste disposal cost	R/m^3	432.99	

water (R11.00/m^3 feed). The bruto income amounted to R54.39. This compared favorably when waste streams need to be disposed of at toxic waste disposal sites. Zero-waste disposal is required when mine water desalination plants are constructed. Transporting the waste streams to a toxic waste disposal site is necessary if they cannot be processed to products that can be sold. Costs for transport and disposal average R1,000/t and R1,500/t, respectively. This serves as justification for the necessity to handle waste streams. Table 13.11 shows the waste disposal cost if the sludge and brine from the leachate used in Table 13.10 in the case of conventional treatment. The figures were based on the disposal of solids (30% dry base) and brine with a TDS of 37,237 mg/L. The cost of solid disposal will amount to R202.68/m^3 feed and that of brine to R230.31/m^3 feed, a total cost of R432.99/m^3 feed.

13.5 Intensive Farming

Understanding the significance of employing unconventional water resources in agriculture is strongly motivated by concern over the sustainability of future water use. Due to its acceptable quality, treated wastewater has been used as an alternative to conventional water sources for irrigation in arid locations with intense competition among water users [57].

Intensive farming can utilize treated water effectively. This method, characterized by the heavy use of pesticides and chemical fertilizers, aims to maximize yields from available land through agricultural intensification and mechanization land [58]. As a result of this approach, millions of animals, including cattle, pigs, and chickens, are housed indoors in what are referred to as factory farms. However, intensive farming practices pose significant threats to the global environment, contributing to the loss of ecosystem services and climate change, which have led to the emergence of new parasites and the resurgence of previously controlled ones. While intensive farming generates more affordable food per acre and per animal, thereby helping to feed a growing population and potentially preventing the conversion of additional land into agricultural use, it also has detrimental effects. These practices eliminate beneficial plants and insects, degrade and deplete soil quality, cause polluted runoff and clogged waterways, increase flood risks, reduce biodiversity, destroy natural habitats, and contribute significantly to the accumulation of genetically modified organisms in the environment according to the World Wide Fund for Nature. Despite these challenges, certain aspects of intensive farming have mitigated climate change by preventing further land destruction and enhancing yields on already cleared but underperforming land. While intensive farming has both benefits and drawbacks, its advantages considerably outweigh the disadvantages. The world is transitioning from an era of food abundance to one of potential shortages. To meet the needs of a rapidly growing population sustainably, a food production strategy should be implemented that minimizes negative impacts on the environment, animal welfare, and human health. Currently, human food production occupies 40% of the planet's land area, up from just 7% in 1700. By 2050, global food demand is projected to rise by 70% [59].

13.6 Conclusions

The following conclusions were made from results of this study:

i. $CaCO_3$ can be used for removal of $Fe(OH)_3$ at pH 3.5. Simultaneous gypsum crystallization can be prevented by dosing of in inhibitor.
ii. $CaCO_3$ can be used for the removal of $Al(OH)_3$ at pH 4.5, followed by $Ca(OH)_2$ dosing for removal of $Mn(OH)_2$.
iii. Partial desalination can be done through gypsum crystallization after $CaCO_3$ and $Ca(OH)_2$ treatment.

iv. Ca^{2+} removal, as $CaCO_3$, can be achieved by dosing Na_2CO_3, to avoid membrane fouling due to gypsum or $CaCO_3$ crystallization.

v. RO can be used for desalination to produce permeate with a TDS of 121 mg/L and brine with a TDS of 74,470 mg/L.

vi. RO brine can be treated with electro-dialysis to produce permeate with a TDS of 8,278 mg/L and a brine with a TDS of 150,000 mg/L.

vii. Freeze crystallization can be used to concentrate the ED brine to produce melted ice with a TDS of 21,277mg/L, brine with a TDS of 183,546 mg/L and 886,456 mg $Na_2SO_4 \cdot 10H_2O$/L.

viii. Pigment can be recovered from $Fe(OH)_3$. Goethite (yellow) is formed at 800° and hematite at 200°.

ix. The cost of chemicals and electrical power amounts to R42.39/m^3 feed water (chemicals, R27.46/m^3; and power, R14.93/m^3).

x. The value of pigment, at a price of ZAR20/kg, amounts to R96.78/m^3 feed water.

xi. The cost of brine and sludge disposal, after conventional treatment options, will amount to R432.99/m^3 feed water, at a disposal cost of R2,500/t.

Acknowledgments

The authors express their gratitude to: the *South African Water Research Commission* for financial support through its Wader Programme; the *Khwezela Colliery* who provided the site for the study as well as financial support for accommodation, safety, and supply of water and electricity. We are grateful to *NIPMO*, who sponsored the grant for the innovation award of 2021 *NSTF-South32*, under the category SMME, for acknowledgment of the ROC technology. The *Innovation Hub* are thanked for supporting business development; and also *F'SAGri* for financial support for mine water treatment. Dr. Fritz Carlsson is thanked for proof-reading and editorial and language input.

References

1. Maree, J., Mujuru, M., Bologo, V., Daniels, N., Mpholoane, D., Neutralisation treatment of AMD at affordable cost. *Water SA*, 39, 2, 245–50, 2013.
2. Mogashane, T., Maree, J., Mujuru, M., Mphahlele-Makgwane, M., Modibane, K., Ferric Hydroxide Recovery from Iron-Rich Acid MineWater with Calcium Carbonate and a Gypsum Scale Inhibitor. *Minerals*, 13, 2, 1–23, 2023.
3. Muller, M., Water: Decide now on Lesotho 2, acid mine drainage projects. Creamer Media's Eng. News, Pretoria, 3 December 2015.
4. Gunther, P., Mey, W., Selection of mine water treatment technologies for the eMalahleni (Witbank) water reclamation project, in: *WISA Biennial Conference*, WISA (Ed.), South Africa, Sun City, 2008.
5. Tshwete, L., Gunther, P., Mey, W., van Niekerk, A., Emalahleni (Witbank) mine water reclamation project, in: *WISA biennial Conference*, WISA (Ed.), Durban, South Africa, 2006.
6. Nguegang, B., Masindi, V., Msagati, T.A.M., Tekere, M., The Treatment of Acid Mine Drainage Using Vertically Flowing Wetland: Insights into the Fate of Chemical Species. *Minerals*, 11, 5, 477, 2021.
7. Masindi, V., Gitari, M., Tutu, H., DeBeer, M., Synthesis of cryptocrystalline magnesite–bentonite clay composite and its application for neutralization and attenuation of inorganic contaminants in acidic and metalliferous mine drainage. *J. Water Process Eng.*, 15, 2–17, 2017.
8. Amanda, N. and Moersidik, S.S., Characterization of sludge generated from acid mine drainage treatment plants. *J. Phys.*, 1, 012113, 2019.
9. Akinwekomi, V., Maree, J.P., Wolkersdorfer, C., Using Calcium Carbonate/ Hydroxide and Barium Carbonate to Remove Sulphate from Mine Water. *Mine Water Environ.*, 36, 264–72, 2017.
10. Sulaiman, A., Othman, A., Ibrahim, I., The use of magnesium oxide in acid mine drainage treatment, in: *The 3rd International Conference on Green Chemical Engineering and Technology (3rd GCET): Materials Science*, GCET (Ed.), Melaka, Malaysia, pp. 21566–73, 2017.
11. Mogashane, T.M., Maree, J.P., Mujuru, M., Mphahlele-Makgwane, M.M., Technologies that can be Used for the Treatment of Wastewater and Brine for the Recovery of Drinking Water and Saleable Products, in: *Recovery of Byproducts from Acid Mine Drainage Treatment*, E. Fosso-Kankeu, J. Burgess, C. Wolkersdorfer (Eds.), pp. 97–151, Wiley, New York, 2020.
12. Kefeni, K.K., Msagati, A.T., Nkambule, T.T., Mamba, B.B., Synthesis and application of hematite nanoparticles for acid mine drainage treatment. *J. Environ. Chem. Eng.*, 6, 2, 1865–74, 2018.
13. Munyengabe, A., Zvinowanda, C., Zvimba, J.N., Ramontja, J., Characterization and reusability suggestions of the sludge generated from a synthetic acid mine drainage treatment using sodium ferrate (VI). *Heliyon*, 6, 10, 1–6, 2020.

14. Aveng, W., eMalahleni Water Reclamation Plant, 2014, Available from: http://www.avengwater.co.za/projects/emalahleni-water-reclamation-plant.
15. Zvinowanda, C.M., Zikalala, N.N., Mtombeni, T., Maree, J.P., Louw, W.J., Oosthuizen, F., Process optimization of freeze desalination of brine using HybridICETM pilot plant. *Desalin. Water Treat.*, 10, 1–10, 2013.
16. Reddy, S.T., Lewis, A.E., Witkamp, G.J., Kramer, H.J.M., S, J., Recovery of Na2SO4•10H2O from a reverse osmosis retentate by eutectic freeze crystallisation technology. *Chem. Eng. Res. Des.*, 88, 1153–7, 2010.
17. Hasan, M., Rotich, N., John, M., Louhi-Kultanen, M., Salt recovery from wastewater by air-cooled eutectic freeze crystallization. *Chem. Eng. J.*, 326, 192–200, 2017.
18. Karakatsanis, E. and Cogho, V.E., Drinking Water from Mine Water Using the HiPrO® Process – Optimum coal Mine Water reclamation Plant, in: *International Mine Water Association Congress*, IMWA (Ed.), Sydney, Australia, 2010.
19. Adeniyi, A., Mbaya, R.K.K., Onyango, M.S., Popoola, A.P.I., Sethoga, M.E., Hlongwane, V.T., *et al.*, Water Recovery and Rejection of Organochloride Pesticides During Reverse Osmosis Process. *J. Comput. Theor. Nanosci.*, 14, 5104–9, 2017.
20. Khawaji, A.D., Kutubkhanah, I.K., Wie, J.M., Advances in seawater desalination technologies. *Desalination*, 221, 1, 47–69, 2008.
21. Maree, J.P. and Mtombeni, T., Treatment of brine for the recovery of drinking water, calcium carbonate and sodium sulphate, in: *WISA Biennial Conference*, WISA (Ed.), Cape Town, South Africa, 2018.
22. Akinwekomi, V., Maree, J.P., Strydom, A., Zvinowanda, C., Recovery of magnetite from iron-rich mine water, in: *WISA Biennial Conference*, WISA (Ed.), Durban, 2016.
23. Magagane, N., Masindi, V., Ramakokovhu, M.M., Shongwe, M.B., Muedi, K.L., Facile thermal activation of non-reactive cryptocrystalline magnesite and its application on the treatment of acid mine drainage. *J. Environ. Manage.*, 236, 499–509, 2019.
24. Maree, J.P., Letjiane, S.L., Ramothole, R., Oyewo, A.O., Mogashane, T.M., Sebothoma, P.S., *et al.*, *Treatment for the recovery of drinking water and valuable products from mine water.* Project C2019/2020-00033, Wader, Water Research Commission, Pretoria, 2021.
25. Nemati, M. and Webb, C., Combination biological chemical oxidation of ferrous sulphate using immobilised Thiobacillus ferrooxidans. *J. Chem. Technol. Biotechnol.*, 74, 562–70, 1999.
26. Nemati, M. and Webb, C., Effect of ferrous iron concentration on the catalytic activity of mobilized cells of Thiobacillus ferrooxidans. *Appl. Microbiol. Biotechnol.*, 46, 250–5, 1996.
27. Nengovhela, N.R., Strydom, C.A., Maree, J.P., Greben, H.A., Chemical and Biological Oxidation of Iron in Acid Mine Water. *Mine Water Environ.*, 23, 76–80, 2004.

28. Cornell, R.M. and Schwertmann, U., *The iron oxides: structure, properties, reactions, occurrences and uses*, Wiley, Weinheim, 2000.
29. Legodi, M. and Dewaal, D., The preparation of magnetite, goethite, hematite and maghemite of pigment quality from mill scale iron waste. *Dyes Pigm.*, 74, 1, 161–8, 2007.
30. Meng, J.Y., G, Yan, L., Wang, X., Synthesis and characterization of magnetic nanometer pigment Fe3O4. *Dyes Pigm.*, 66, 109–13, 2004.
31. Campos, E., Pinto, D., de Oliveira, J., Mattos, D., Dutra, R., Synthesis, Characterization and Applications of Iron Oxide Nanoparticles – a Short Review. *J. Aerosp. Technol. Manag.*, 7, 3, 267–76, 2015, https://doi.org/10.5028/jatm.v7i3.471.
32. Seo, E.Y., Cheong, Y.W., Yim, G.J., Min, K.W., Geroni, J.N., Recovery of Fe, Al and Mn in acid coal mine drainage by sequential selective precipitation with control of pH. *CATENA (Part 1)*, 148, 11–6, 2017.
33. Akinwekomi, V., Maree, J., Zvinowanda, C., Masindi, V., Synthesis of magnetite from iron-rich mine water using sodium carbonate. *J. Environ. Chem. Eng.*, 5, 2699–707, 2017.
34. Mogashane, T.M., Maree, J.P., Letjiane, L., Masindi, V., Modibane, K.D., Mujuru, M., *et al.*, Recovery of drinking water and nano-sized Fe2O3 pigment from iron rich acid mine water, in: *Application of Nanotechnology in Mining Processes: Beneficiation and Sustainability*, E. Fosso-Kankeu, M. Mkandawire, B. Mamba (Eds.), pp. 237–88, Wiley, New York, 2022.
35. Xu, P., Zeng, G.M., Huang, D.L., Feng, C.L., Hu, S., Zhao, M.H., *et al.*, Use of iron oxide nanomaterials in wastewater treatment: A review. *Sci. Total Environ.*, 424, 1–10, 2012.
36. Mariani, F., Borth, K., Müller, M., Dalpasquale, M., Anaissi, F., Sustainable innovative method to synthesize different shades of iron oxide pigments. *Dyes Pigm.*, 137, 403–9, 2017.
37. Mirzaei, A., Janghorban, K., Hashemi, B., Hosseini, S., Bonyani, M., Leonardi, S., *et al.*, Synthesis and characterization of mesoporous Fe2O3 nanoparticles and investigation of electrical properties of fabricated thick films. *Process. Appl. Ceram.*, 10, 4, 209–17, 2016.
38. Dama-Fakir, P., Sithole, Z., van Niekerk, A.M., Dateling, J., Maree, J.P., Rukuni, T., *et al.*, *Mine water treatment technology selection tool: Users' Guide*, WRC Report No. TT 711/17, Water Research Commission, Pretoria, 2017.
39. Gunther, P. and Mey, W. (Eds.), Selection of mine water treatment technologies for the eMalahleni (Witbank) water reclamation project. *WISA Biennial Conference*, South Africa, Sun City, 2008, 18th-22nd May.
40. Karakatsanis, E. and Cogho, V.E. (Eds.), *Drinking Water from Mine Water Using the HiPrO® Process – Optimum coal Mine Water reclamation Plant*, IMWA, Sydney, Nova Scotia, Canada, 2010, http://www.imwa.info/docs/imwa_2010/IMWA2010_Karakatsanis_430.pdf.

41. Hutton, B., Kahan, I., Naidu, T., Gunther, P., Operating and maintenance experience at the eMalahleni water reclamation plant, in: *International Mine Water Association Congress*, IMWA (Ed.), Pretoria, South Africa, 2009.
42. Günther, P., Naidu, T., Mey, W., *Emalahleni mine water reclamation project - Key learnings*, WISA, South Africa, 2008.
43. Gunther, P. and Naidu, T., Mine water reclamation - towards zero disposal, in: *WISA Biennial Conference*, WISA (Ed.), South Africa, Sun City, 2008.
44. Zhang, Y., Ghyselbrecht, K., Meesschaert, B., Pinoy, L., Van der Bruggen, B., Electrodialysis on RO concentrate to improve water recovery in wastewater reclamation. *J. Membr. Sci.*, 378, 101–10, 2011.
45. Juby, G. and Pulles, W., *Evaluation of electrodialysis reversal for desalination of brackish mine water*, WRC Report 179/1/90. Water Research Commission, South Africa, Pretoria, 1990.
46. Alvarado, L. and Chen, A., Electrodeionization: Principles, strategies and applications. *Electrochim. Acta*, 132, 583–97, 2014.
47. Van der Ham, F., Witkamp, G.J., de Graauw, J., van Rosmalen, G.M., Eutectic freeze crystallization: Application to process streams and wastewater purification. *Chem. Eng. Process.*, 37, 207–13, 1998.
48. Ramothole, P., Maree, J.P., Onyango, M.S., Adeniyi, A., Performance Evaluation of Freeze Crystallization for removal of Water and Sodium Sulfate from Mine Wastewater, in: *18th Johannesburg Int'l Conference on Science, Engineering, Technology & Waste Management (SETWM-20)*, SETWM (Ed.), Johannesburg, South Africa, 2020.
49. Maree, J.P., inventorPipe Freeze Crystallization. South Africa, Australia, Canada, USA. patent Provisional Appl. No: 2018/04084. RSA No: 2019/03795, Australia No: 201204324 (pending), Canada No: 3,092,764 (pending), USA No: 17,027,873, 2018.
50. More, K.S., Maree, J.P., Mahlangu, M., Indirect Freeze Crystallization—An Emerging Technology for Valuable Resource Recovery from Wastewater. *Minerals*, 14, 4, 427, 2024, https://doi.org/10.3390/min14040427.
51. Letjiane, S.L., Maree, J.P., Onyango, M.S., Adeniyi, A., Pre-treatment of coal dump leachate with MgO for pigment recovery, in: *IMWA 2019 Conference - Mine Water: Technological and Ecological Challenge*, E. Khayrulina, C. Wolkersdorfer, S. Polaykova, A. Bogush (Eds.), Perm, Russia, International Mine Water Association, pp. 101–6, 2019.
52. Letjiane, S.L., *Selective recovery of metals compounds from acid mine water with Na2CO3 and MgO [Master of Science]*, Tshwane University of Technology, Pretoria, 2020.
53. APHA, *Standard methods for the examination of water and wastewater (American Public Health Association)*, American Water Works Association, Water Environment Federation, Washington, 2012.
54. OLI, *OLI Software*, OLI Systems Inc., New Jersey, USA, 2022. http://www.olisystems.com/. [Online] [Accessed 2015].

55. Rabizadeh, T., Stawski, T.M., Morgan, D.J., Peacock, C.L., Benning, L.G., The effects of inorganic additives on the nucleation and growth kinetics of calcium sulfate dihydrate crystals. *Cryst. Growth Des.*, 17, 582–9, 2017.
56. Goffi, A.S., Trojan, F., de Lima, J.D., Lizot, M., Thesari, S., Economic feasibility for selecting wastewater treatment systems. *Water Sci. Technol.*, 78, 12, 2518–31, 2019.
57. Zolfaghary, P., Zakerinia, M., Kazemi, H., A model for the use of urban treated wastewater in agriculture using multiple criteria decision making (MCDM) and geographic information system (GIS). *Agric. Water Manage.*, 243, 106490, 1–11, 2021.
58. Abedi-Koupai, J., Mostafazadeh-Fard, B., Afyuni, M., Bagheri, M.R., Effect of treated wastewater on soil chemical and physical. *Plant Soil Environ. Soil Environ.*, 52, 8, 335–44, 2006.
59. Andrews, D.M., Robb, T., Elliott, H., Watson, J.E., Impact of long-term wastewater irrigation on the physicochemical properties of humid region soils: "the living Filter" site case study. *Agric. Water. Manag.*, 178, 239–47, 2016.

Index

Printed in the USA/Agawam, MA
September 16, 2026

910661.164